Springer Proceedings in Mathematics & Statistics

Volume 465

This book series features volumes composed of selected contributions from workshops and conferences in all areas of current research in mathematics and statistics, including data science, operations research and optimization. In addition to an overall evaluation of the interest, scientific quality, and timeliness of each proposal at the hands of the publisher, individual contributions are all refereed to the high quality standards of leading journals in the field. Thus, this series provides the research community with well-edited, authoritative reports on developments in the most exciting areas of mathematical and statistical research today.

Eric Carlen · Patrícia Gonçalves ·
Ana Jacinta Soares
Editors

From Particle Systems to Partial Differential Equations

PSPDE X, Braga, Portugal, June 2022

Springer

Editors
Eric Carlen
Department of Mathematics
Rutgers University
Piscataway, NJ, USA

Patrícia Gonçalves
Instituto Superior Técnico
Universidade de Lisboa
Lisboa, Portugal

Ana Jacinta Soares
Centre of Mathematics
University of Minho
Braga, Portugal

ISSN 2194-1009 ISSN 2194-1017 (electronic)
Springer Proceedings in Mathematics & Statistics
ISBN 978-3-031-65194-6 ISBN 978-3-031-65195-3 (eBook)
https://doi.org/10.1007/978-3-031-65195-3

Mathematics Subject Classification: 35Q20, 35Q70, 60F17, 82C40, 35Q35, 60G60, 82C22, 35R60, 60K35

This Springer imprint is published by the registered company Springer Nature Switzerland AG
The registered company address is: Gewerbestrasse 11, 6330 Cham, Switzerland

Preface

This volume gathered the proceedings of the 10th edition of the international conference on Particle Systems and Partial Differential Equations (PSPDE) that was held in July 2022 at the Centre of Mathematics of the University of Minho, in Braga, Portugal. This was a special edition completely dedicated to the memory of our esteemed colleague, Maria Conceição Carvalho.

Maria Conceição passed away just a couple of days before the 9th edition, which was held at the Centre of Mathematics of the University of Minho back in July 2021. Maria Conceição was not only a notable Portuguese mathematician and an extremely respected colleague in the mathematics community, but also an important player in the organization of these meetings, as she has been part of the organizing committee of many of the previous editions. She was as strongly enthusiastic about these meetings and she had a preference for having them in the North of Portugal, in Braga, inspiring our decision to organize the following edition at the University of Minho, a place that she enjoyed very much visiting.

The 10th edition of the conference was completely dedicated to her memory, by inviting colleagues who have interacted with her scientifically through joint research collaborations, or through joint writing of articles or even researchers who have considered the work of Maria Conceição as milestones in their fields.

These meetings have always had the flavor of bringing together distinguished active researchers working in the areas of probability and partial differential equations, toward the goal of presenting and discussing their latest scientific results in both areas.

This volume includes 17 contributed papers authored by conference participants mostly on topics in the fields of partial differential equations and kinetic theory.

We believe that this volume will be of great interest to probabilists, analysts and also to other mathematicians with general interest in mathematical physics, stochastic processes and differential equations, as well as those physicists whose works intersect with statistical mechanics, statistical physics and kinetic theory.

We would like to take this opportunity to extend our thanks to all the speakers, and the participants, for contributing to the success of these meetings. They are now being put on the map of world conferences in probability and kinetic theory and typically count on the participation of leading experts in those fields.

Lastly, we wish to gratefully acknowledge the financial support provided by the following sponsors:

Agence Nationale de la Recherche, France.
ERC StG 715734.
Centre for Mathematical Analysis, Geometry, and Dynamical Systems of the University of Lisbon.
Centre of Mathematics, Fundamental Applications and Operations Research of the University of Lisbon.
Centre of Mathematics of the University of Minho.

We hope that you enjoy reading this book as much as we did preparing it!

Piscataway, USA — Eric Carlen
Lisbon, Portugal — Patrícia Gonçalves
Braga, Portugal — Ana Jacinta Soares

Contents

Necessary and Sufficient Conditions for Strong Stability of Explicit Runge–Kutta Methods

Franz Achleitner, Anton Arnold, and Ansgar Jüngel

Abstract Strong stability is a property of time integration schemes for ODEs that preserve temporal monotonicity of solutions in arbitrary (inner product) norms. It is proved that explicit Runge–Kutta schemes of order $p \in 4\mathbb{N}$ with $s = p$ stages for linear autonomous ODE systems are not strongly stable, closing an open stability question from [Z. Sun and C.-W. Shu, SIAM J. Numer. Anal. 57 (2019), 1158–1182]. Furthermore, for explicit Runge–Kutta methods of order $p \in \mathbb{N}$ and $s > p$ stages, we prove several sufficient as well as necessary conditions for a less restrictive notion of strong stability. These conditions involve both the stability function and the hypocoercivity index of the ODE system matrix. This index is a structural property combining the Hermitian and skew-Hermitian part of the system matrix.

Keywords Strong stability · Linear ordinary differential equations · Hypocoercivity (index) · Stability function · MSC 65L06 · 65L20

1 Introduction

Explicit Runge–Kutta methods are standard tools in the numerical solution of ordinary differential equations and semi-discrete approximations of partial differential equations. Of particular interest are strongly stable schemes for which the norm of its numerical solution is nonincreasing in time, guaranteeing that the numerical error in each time step is not amplified during time integration. Surprisingly, the characterization of strongly stable explicit schemes is not complete. For instance, while it is well-known [27] that explicit Runge–Kutta methods for linear systems of order

F. Achleitner · A. Arnold (✉) · A. Jüngel
TU Wien, Institute of Analysis and Scientific Computing, Wiedner Hauptstr. 8-10, 1040 Wien, Austria
e-mail: anton.arnold@tuwien.ac.at

F. Achleitner
e-mail: franz.achleitner@tuwien.ac.at

A. Jüngel
e-mail: ansgar.juengel@tuwien.ac.at

E. Carlen et al. (eds.), *From Particle Systems to Partial Differential Equations*,
Springer Proceedings in Mathematics & Statistics 465,
https://doi.org/10.1007/978-3-031-65195-3_1

p are *not* strongly stable if $p \in 4\mathbb{N}_0 + 1$ or $p \in 4\mathbb{N}_0 + 2$ and that they are strongly stable for $p \in 4\mathbb{N}_0 + 3$, the case $p \in 4\mathbb{N}$ is still open.

In this work, we consider a less restrictive notion of strong stability of explicit Runge–Kutta schemes, see Definition 1 of *strong stability with respect to (w.r.t.) Lyapunov stable matrices* and Remark 1 below. We show that the proof of strong stability w.r.t. Lyapunov stable matrices can be reduced to testing strong stability w.r.t. the set of asymptotically stable, semi-dissipative matrices (see below for the definitions) and w.r.t. the set of purely imaginary scalars. We deduce that explicit Runge–Kutta methods for linear systems with order $p \in 4\mathbb{N}$ and $p = s$ stages are *not* strongly stable w.r.t. Lyapunov stable matrices, hence, also not strongly stable in the sense of [27]. In this way, we fill the gap left open in [27, Theorem 4.2]. Furthermore, we prove that strong stability w.r.t. asymptotically stable, semi-dissipative matrices holds if the so-called hypocoercivity index of the system matrix is small enough.

We consider linear time-invariant systems of ordinary differential equations,

$$\frac{\mathrm{d}u}{\mathrm{d}t} = \mathbf{L}u, \quad t > 0, \quad u(0) = u^0 \in \mathbb{C}^n, \tag{1}$$

with Lyapunov stable matrices $\mathbf{L} \in \mathbb{C}^{n\times n}$; see [6, Definition 15.9.1], [10, p.172]. In this paper, we use several notions of stability for matrices $\mathbf{L} \in \mathbb{C}^{n\times n}$:

- $\mathbf{L}$ is *Lyapunov stable* if all eigenvalues have nonpositive real part and those eigenvalues with vanishing real part are non-defective (i.e., the algebraic and geometric multiplicities coincide) [6, Definition 15.9.1].
- $\mathbf{L}$ is *asymptotically stable* if all eigenvalues have negative real part [6, Definition 15.9.1].
- $\mathbf{L}$ is *dissipative* (resp. *semi-dissipative*) if its Hermitian part $(\mathbf{L} + \mathbf{L}^*)/2$ is negative definite (resp. negative semi-definite) [6, Definition 4.1.1].

Dissipative matrices are asymptotically stable, and semi-dissipative matrices are Lyapunov stable. Lyapunov stability characterizes all system matrices $\mathbf{L} \in \mathbb{C}^{n\times n}$ such that $u \equiv 0$ is a Lyapunov stable solution to (1).

We use the following notation: The conjugate transpose of a matrix $\mathbf{L} \in \mathbb{C}^{n\times n}$ is denoted by $\mathbf{L}^*$. Matrices $\mathbf{L} \in \mathbb{C}^{n\times n}$ have a unique decomposition $\mathbf{L} = \mathbf{L}_H + \mathbf{L}_S$ into its Hermitian part $\mathbf{L}_H := (\mathbf{L} + \mathbf{L}^*)/2$ and skew-Hermitian part $\mathbf{L}_S := (\mathbf{L} - \mathbf{L}^*)/2$. We write $\mathbb{H}_n$ for the set of all Hermitian matrices in $\mathbb{C}^{n\times n}$. Positive definiteness (resp. semi-definiteness) of Hermitian matrices $\mathbf{P} \in \mathbb{H}_n$ is denoted by $\mathbf{P} > 0$ (resp. $\mathbf{P} \geq 0$), and $\mathbb{H}_n^> := \{\mathbf{H} \in \mathbb{H}_n : \mathbf{H} > 0\}$ is the set of positive definite Hermitian matrices in $\mathbb{C}^{n\times n}$. Moreover, $\langle \cdot, \cdot \rangle$ is the standard inner product in $\mathbb{C}^n$ with norm $\|\cdot\| = \sqrt{\langle \cdot, \cdot \rangle}$, and $\|\cdot\|_2$ denotes the spectral norm in $\mathbb{C}^{n\times n}$.

It is known that a matrix $\mathbf{L} \in \mathbb{C}^{n\times n}$ is Lyapunov stable if and only if there exists a positive definite Hermitian matrix $\mathbf{P} \in \mathbb{H}_n^>$ such that

$$\mathbf{L}^*\mathbf{P} + \mathbf{P}\mathbf{L} \leq 0; \tag{2}$$

see, e.g., [25, Corollary to Theorem 5], [10, Theorem 3.18]. Under this condition, the solution $u(t)$ of (1) is nonincreasing in the norm $\|\cdot\|_{\mathbf{P}} := \sqrt{\langle\cdot, \mathbf{P}\cdot\rangle}$:

$$\frac{\mathrm{d}}{\mathrm{d}t}\|u(t)\|_{\mathbf{P}}^2 = \frac{\mathrm{d}}{\mathrm{d}t}\langle u, \mathbf{P}u\rangle = \langle \mathbf{L}u, \mathbf{P}u\rangle + \langle u, \mathbf{PL}u\rangle = \langle u, (\mathbf{L}^*\mathbf{P} + \mathbf{PL})u\rangle \le 0. \quad (3)$$

Often, the aim of numerical schemes is to reproduce this property on the discrete level, i.e. to show that a certain norm of an approximation u^k of $u(k\tau)$, where $\tau > 0$ is the uniform time step, is nonincreasing in k. For this, we consider the explicit Runge–Kutta methods

$$u^k = u^{k-1} + \tau\sum_{i=1}^{s} b_i K_i^k, \quad K_i^k = \mathbf{L}\Big(u^{k-1} + \tau\sum_{j=1}^{i-1} a_{ij}K_j^k\Big), \quad i = 1, \dots, s, \quad (4)$$

where $b_i \in \mathbb{R}$ are the weights, $a_{ij} \in \mathbb{R}$ are the coefficients of the Runge–Kutta matrix, and $s \in \mathbb{N}$ is the number of stages. Stability of explicit Runge–Kutta schemes can be expected only under a restriction on the time step. In this article, we give necessary and sufficient conditions such that explicit Runge–Kutta methods preserve the monotonicity property (3) in the following sense.

Definition 1 (*Strong stability with respect to Lyapunov stable matrices*)

(a) The Runge–Kutta scheme (4) is *strongly stable with respect to Lyapunov stable matrices* if for all matrix dimensions $n \in \mathbb{N}$, for all Lyapunov stable matrices $\mathbf{L} \in \mathbb{C}^{n\times n}$, and for all $\mathbf{P} \in \mathbb{H}_n^>$ such that (2) holds, the numerical solution to (1) satisfies $\|u^1\|_{\mathbf{P}} \le \|u^0\|_{\mathbf{P}}$ for all initial data $u^0 \in \mathbb{C}^n$ and sufficiently small time steps.

(b) For practical reasons, we introduce the short-hand notation that a Runge–Kutta scheme (4) is *strongly stable w.r.t.* $\mathscr{L}$ if it is strongly stable w.r.t. the set $\mathscr{L}$ of Lyapunov stable matrices of any matrix dimension $n \in \mathbb{N}$,

$$\mathscr{L} := \bigcup_{n\in\mathbb{N}}\{\mathbf{L} \in \mathbb{C}^{n\times n} : \exists\mathbf{P} \in \mathbb{H}_n^> \text{ such that (1.2) holds}\}.$$

(c) Similarly, the Runge–Kutta scheme (4) is called *strongly stable w.r.t. a subset* $\mathscr{L}_0$ of $\mathscr{L}$ if the above property holds for all $\mathbf{L} \in \mathscr{L}_0$.

Remark 1 (a) The study of (a stricter notion of) strong stability of Runge–Kutta schemes is motivated by the semidiscretization of time-dependent partial differential equations. An explicit Runge–Kutta scheme is *strongly stable* (in the sense of [29, Sect. 4], [27]) if there exists $\alpha > 0$ such that $\|R(\tau\mathbf{L})\|_{\mathbf{P}} \le 1$ for all $\mathbf{L} \in \mathscr{L}_\alpha$ with

$$\mathscr{L}_\alpha := \bigcup_{n\in\mathbb{N}}\{\mathbf{L} \in \mathbb{C}^{n\times n} : \exists\mathbf{P} \in \mathbb{H}_n^> \text{ and } \tau > 0 \text{ such that (1.2) and } \|\tau\mathbf{L}\|_{\mathbf{P}} \le \alpha \text{ hold}\}.$$

Thus, for a strongly stable Runge–Kutta scheme and matrix $\mathbf{L} \in \mathscr{L}_\alpha$, the maximal time step τ_0 depends on $\|\mathbf{L}\|_\mathbf{P}$, the matrix norm induced by the vector norm $\|\cdot\|_\mathbf{P}$.

In contrast, we study strong stability w.r.t. $\mathscr{L}$ of explicit Runge–Kutta schemes via a Taylor expansion of $\|R(\tau\mathbf{L})\|_2$ for $0 \le \tau \le \tau_0$ using that $\|\mathbf{L}\|_\mathbf{P}$ (instead of $\|\tau\mathbf{L}\|_\mathbf{P}$) is bounded. Due to our approach, the Taylor expansion is valid up to a maximal time step τ_0 which depends on $\mathbf{L}$ and not only on $\|\mathbf{L}\|_\mathbf{P}$.

(b) Strong stability of explicit Runge–Kutta schemes w.r.t. $\mathscr{L}$ (as studied here) is "less restrictive" than strong stability (w.r.t. $\mathscr{L}_\alpha$ for some $\alpha > 0$) as studied in [29, Sect. 4], [27] since $\mathscr{L} \subseteq \mathscr{L}_\alpha$ for any $\alpha > 0$: In particular, for explicit Runge–Kutta schemes, necessary conditions for strong stability w.r.t. $\mathscr{L}$ are also necessary to ensure strong stability (w.r.t. $\mathscr{L}_\alpha$ for some $\alpha > 0$). Conversely, sufficient conditions for strong stability (w.r.t. $\mathscr{L}_\alpha$ for some $\alpha > 0$) are also sufficient to ensure strong stability w.r.t. $\mathscr{L}$.

(c) Originally, strong stability w.r.t. (uniformly) coercive system matrices $\mathbf{L}$ has been studied in [17], [29, Sect. 3].

(d) For explicit Runge–Kutta methods, strong stability w.r.t. $\mathscr{L}$ is equivalent to strong stability w.r.t. *semi-dissipative* matrices $\mathbf{L} \in \mathscr{L}_{SD}$ with

$$\mathscr{L}_{SD} := \bigcup_{n\in\mathbb{N}} \{\mathbf{L} \in \mathbb{C}^{n\times n} : \mathbf{L} + \mathbf{L}^* \le 0\} \subset \mathscr{L};$$

see the first step in the proof of Theorem 1 below.

The goal of this paper is twofold: On the one hand, we give novel necessary conditions for strong stability of explicit Runge–Kutta schemes. On the other hand, we derive sufficient conditions for strong stability in a weaker sense: In contrast to Definition 1, we then require the monotonicity property only in *some* weighted norm $\|\cdot\|_\mathbf{P}$ with $\mathbf{P} \in \mathbb{H}_n^>$.

Definition 2 (*Strong stability in weak form*)

(a) The Runge–Kutta scheme (4) is *strongly stable in weak form* if for all $n \in \mathbb{N}$ and for all Lyapunov stable matrices $\mathbf{L} \in \mathbb{C}^{n\times n}$, there exists a matrix $\mathbf{P} \in \mathbb{H}_n^>$ satisfying (2), such that the numerical solution (4) to (1) satisfies $\|u^1\|_\mathbf{P} \le \|u^0\|_\mathbf{P}$ for all initial data $u^0 \in \mathbb{C}^n$ and sufficiently small time steps.

(b) Similarly, the Runge–Kutta scheme (4) is called *strongly stable in weak form w.r.t. a subset $\mathscr{L}_0$ of $\mathscr{L}$* if the above property holds for all $\mathbf{L} \in \mathscr{L}_0$.

This notion of *strong stability in weak sense* is inspired by an example in [17, Sect. 3.5]; see Example 1.

Most of the practical interest in strongly stable Runge–Kutta methods has been for explicit schemes, e.g. in the integration of hyperbolic conservation laws [23, 24]. For implicit or implicit-explicit strongly stable Runge–Kutta methods, we refer to, e.g., [7–9, 13, 28]. Furthermore, nonlinear problems are considered in, e.g., [21, 22].

Note that a Runge–Kutta method of linear order p (i.e., the order for linear systems) may possess a lower order when applied to nonlinear systems.

Strong stability of the (implicit) midpoint rule can be seen as a corollary to Remark 1(d) and [5, Lemma 47(ii)].

Let us summarize the main results of this paper:

- An explicit Runge-Kutta scheme is strongly stable with respect to Lyapunov stable matrices if and only if it is strongly stable w.r.t. the set $\mathcal{L}_{AS}$ of asymptotically stable, semi-dissipative matrices and w.r.t. the set $\mathcal{L}_{IM}$ of purely imaginary scalars, see Theorem 1.
- Explicit Runge–Kutta schemes of order $p \in 4\mathbb{N}$ and $s = p$ stages are *not* strongly stable (even in the stricter sense of [29, Sect. 4], [27]). While these Runge–Kutta schemes are strongly stable w.r.t. $\mathcal{L}_{IM}$, they fail to be strongly stable w.r.t. $\mathcal{L}_{AS}$: see Corollary 1. For this analysis, we derive novel necessary conditions for strong stability with respect to Lyapunov stable matrices of explicit Runge–Kutta schemes using the precise short-time asymptotics for the spectral norm of the matrix exponential as derived in [2, Theorem 2.7].
- Concerning strong stability w.r.t. $\mathcal{L}_{AS}$, we show a remarkable connection between the order $p \in \mathbb{N}$ of the Runge–Kutta scheme and the hypocoercivity (HC) index of asymptotically stable, semi-dissipative matrices; see Theorem 2. Roughly speaking, the HC-index of a matrix $\mathbf{L}$ describes the structural complexity of the intertwining of the Hermitian part $\mathbf{L}_H$ and skew-Hermitian part $\mathbf{L}_S$; see Definition 3.
- Finally, we show that each explicit Runge–Kutta method of order $p \in \mathbb{N}$ is strongly stable in weak form if and only if it is locally stable on the imaginary axis as defined in [16, Definition 2.1]; see Theorem 4.

2 Main Results

Before stating our main theorems, we recall that the Runge–Kutta scheme (4) for ODE systems (1) can be written as $u^k = R(\tau\mathbf{L})u^{k-1}$, where $R(z)$ is the stability function [11, Definition 2.1]. For linear time-invariant systems, $s = p$ stages are sufficient to obtain a scheme of order p [11, Sect. IV.2]. Generally, if an explicit Runge–Kutta method is of order p, its stability function can be written as

$$R(z) = \sum_{j=0}^{p} \frac{z^j}{j!} + \sum_{j=p+1}^{s} c_j \frac{z^j}{j!}, \quad z \in \mathbb{C}, \ c_{p+1} \neq 1, \tag{5}$$

with some constants $c_{p+1}, \ldots, c_s \in \mathbb{R}$ [11, Theorem IV.2.2] and $c_j = 0$ for $j \geq s+1$. For an explicit Runge–Kutta scheme with $s = p$, this expression reduces to the truncated exponential $R(z) = \sum_{j=0}^{p} z^j/j!$; see, e.g., [11, (2.12)].

Our first main result shows that, to establish strong stability w.r.t. $\mathscr{L}$ for an explicit Runge–Kutta scheme, it is (necessary and) sufficient to prove strong stability w.r.t. two distinct subsets of Lyapunov stable matrices.

Theorem 1 *Consider an explicit Runge–Kutta method with stability function (4). Then the Runge–Kutta scheme is strongly stable with respect to the set $\mathscr{L}$ of Lyapunov stable matrices if and only if it is strongly stable w.r.t.*

(a) the set $\mathscr{L}_{AS}$ of asymptotically stable, semi-dissipative matrices, and
(b) the set $\mathscr{L}_{IM}$ of purely imaginary scalars $\mathbf{L} \in i\mathbb{R} \subset \mathbb{C}$.

This means that the Runge–Kutta scheme is strongly stable w.r.t. $\mathscr{L}$ if and only if for all linear systems (1) with $\mathbf{L}$ either in $\mathscr{L}_{AS}$ or $\mathscr{L}_{IM}$, for sufficiently small time steps, the numerical solution to (1) satisfies $\|u^1\| \leq \|u^0\|$ for all initial data $u^0 \in \mathbb{C}^n$.

The (strong) stability of Runge–Kutta schemes for scalar differential equations is well studied; see, e.g., [16]. In particular, strong stability w.r.t. $\mathscr{L}_{IM}$ is equivalent to *local stability on the imaginary axis* (i.e., there exists $Z > 0$ such that $|R(z)| \leq 1$ for all $z \in i\mathbb{R}$ with $|z| \leq Z$). For example, the Runge–Kutta method with $p = s = 4$ is locally stable on the imaginary axis.

Therefore, we shall focus here on strong stability w.r.t. asymptotically stable, semi-dissipative matrices $\mathbf{L} \in \mathscr{L}_{AS}$. We shall show that a Runge–Kutta method with $p = s = 4$ fails to be strongly stable for certain asymptotically stable, semi-dissipative matrices $\mathbf{L} \in \mathscr{L}_{AS}$. We also discuss the (counter-)examples given in [17, Sect. 3.5] and [26, Proposition 1.1].

Our second main result is concerned with strong stability w.r.t. asymptotically stable, semi-dissipative matrices $\mathscr{L}_{AS}$. We show a remarkable connection between the order $p \in \mathbb{N}$ of the Runge–Kutta scheme and the hypocoercivity index (HC-index) of asymptotically stable, semi-dissipative matrices $\mathbf{L} \in \mathscr{L}_{AS}$, which is defined as follows; see [2, Definition 3.1], [4, Definition 3].

Definition 3 Let $\mathbf{L} \in \mathbb{C}^{n\times n}$ be semi-dissipative. The *hypocoercivity index (HC-index) $m_{HC} = m_{HC}(\mathbf{L})$ of the (semi-dissipative) matrix* $\mathbf{L}$ is defined as the smallest integer $m \in \mathbb{N}_0$ (if it exists) such that

$$T_m := \sum_{j=0}^{m} \mathbf{L}_S^j \mathbf{L}_H (\mathbf{L}_S^*)^j < 0. \tag{6}$$

Remark 2 Originally, the HC-index has been defined for accretive matrices $\mathbf{C} \in \mathbb{C}^{n\times n}$ (i.e., matrices with positive semi-definite Hermitian part $\mathbf{C}_H \geq 0$) in [1]. For practical reasons, we present the hypocoercivity theory here for semi-dissipative matrices (using the fact that for a semi-dissipative matrix $\mathbf{L} \in \mathbb{C}^{n\times n}$, the matrix $\mathbf{C} := -\mathbf{L}$ is accretive).

By definition, a semi-dissipative matrix $\mathbf{L}$ is dissipative if and only if $m_{HC}(\mathbf{L}) = 0$. We recall from [1, Lemma 2.4] that every matrix $\mathbf{L} \in \mathscr{L}_{AS}$ has a finite HC-index. For the next theorem, we introduce the following subsets of $\mathscr{L}_{AS}$:

$$\mathscr{L}_{AS}^m := \{\mathbf{L} \in \mathscr{L}_{AS} \ : \ m_{HC}(\mathbf{L}) \le m\}, \quad m \in \mathbb{N}_0.$$

Theorem 2 *Consider explicit Runge–Kutta schemes of order $p \in \mathbb{N}$ for (1) with stability function (5). Then the following results for strong stability w.r.t. asymptotically stable, semi-dissipative matrices $\mathbf{L} \in \mathscr{L}_{AS}$ hold:*

(a) All explicit Runge–Kutta schemes of order $p \in \mathbb{N}$ are strongly stable w.r.t. $\mathscr{L}_{AS}^m$, if $m \in \mathbb{N}_0$ satisfies $2m + 1 \le p$.

(b) An explicit Runge–Kutta scheme of order $p \in \mathbb{N}$ is not *strongly stable w.r.t. $\mathscr{L}_{AS}$ if the coefficient c_{p+1} in (5) satisfies*

$$(-1)^{(p+1)/2}(1 - c_{p+1}) < 0 \quad \text{for } p \text{ odd;} \tag{7a}$$

$$\text{or} \quad 1 + (-1)^{p/2}(c_{p+1} - 1)\binom{p}{p/2} < 0 \quad \text{for } p \text{ even.} \tag{7b}$$

(c) In case (7) holds and the matrix $\mathbf{L} \in \mathscr{L}_{AS}$ satisfies $2m_{HC}(\mathbf{L}) + 1 > p$, the considered Runge–Kutta scheme is not *strongly stable w.r.t. the singleton $\{\mathbf{L}\}$. This means that it is strongly stable w.r.t. $\mathscr{L}_{AS}^m$ but* not *strongly stable w.r.t. $\mathscr{L}_{AS} \setminus \mathscr{L}_{AS}^m$ for any $m \in \mathbb{N}$ with $2m + 1 \le p$.*

The conditions (7) on the coefficients of the stability functions directly yield the following result (which is also included in Table 1 and illustrated in the examples in Sect. 4).

Corollary 1 *Explicit Runge–Kutta schemes of order $p \in 4\mathbb{N}$ or $p \in 4\mathbb{N}_0 + 1$ with $s = p$ stages (and hence with $c_{p+1} = 0$) are* not *strongly stable (w.r.t. $\mathscr{L}_\alpha$ for some $\alpha > 0$).*

Next, we shall combine two sufficient criteria for explicit Runge–Kutta schemes to *fail* strong stability w.r.t. $\mathscr{L}$: the criterion to fail strong stability w.r.t. $\mathscr{L}_{AS}$ (given in Theorem 2(b)), and w.r.t. $\mathscr{L}_{IM}$ (given in Theorem 5—and taken from the literature: [16, Theorem 3.1]).

Theorem 3 *Explicit Runge–Kutta schemes of order $p \in \mathbb{N}$ for (1) are* not *strongly stable w.r.t. $\mathscr{L}$ if*

Table 1 Shows whether or not explicit Runge–Kutta methods of order $p \in \mathbb{N}$ with $s = p$ stages are strongly stable w.r.t. asymptotically stable, semi-dissipative matrices $\mathscr{L}_{AS}$, purely imaginary scalars $\mathscr{L}_{IM}$, and Lyapunov stable matrices $\mathscr{L}$, respectively; new results are in bold face. Theorem 3 shows that explicit Runge–Kutta schemes for linear systems of order $p \in 4\mathbb{N}$ and $s = p$ stages are ***not*** strongly stable. This closes the open case in [27, Sect. 4.2]

$s = p$	$\mathscr{L}_{AS}$	$\mathscr{L}_{IM}$	$\mathscr{L}$
$4\mathbb{N}_0 + 1$	No	No	No
$4\mathbb{N}_0 + 2$	?	No	No
$4\mathbb{N}_0 + 3$	Yes	Yes	Yes
$4\mathbb{N}$	**No**	Yes	**No**

$$(-1)^{(p+1)/2}(1 - c_{p+1}) < 0, \quad \textit{for } p \textit{ odd;} \tag{8a}$$

$$\left.\begin{array}{l} (-1)^{p/2}\big(c_{p+2} - (p+2)c_{p+1} + (p+1)\big) < 0 \\ \textit{or} \quad 1 + (-1)^{p/2}(c_{p+1} - 1)\binom{p}{p/2} < 0 \end{array}\right\} \textit{ for } p \textit{ even.} \tag{8b}$$

Remark 3 The negation of Theorem 3 represents necessary conditions for strong stability w.r.t. $\mathscr{L}$. As such it complements the sufficient condition for strongly stable, explicit Runge–Kutta schemes of even order $p \in 2\mathbb{N}$ given in [27, Theorem 4.5]. More precisely, the negation of the first condition of (8b) coincides with the first condition in [27, Theorem 4.5], except of the (non-)strict relation sign. The negation of the second condition from (8b) and its analog from [27, Theorem 4.5] both give a bound on c_{p+1}, but they leave a gap between the sufficient and necessary conditions. We notice that condition (8a) is sufficient and necessary in [27]. In Table 1, we summarize the results for explicit Runge–Kutta schemes of order $p \in \mathbb{N}$ and $s = p$ stages.

Our third main result is concerned with sufficient conditions for strong stability in weak form.

Theorem 4 *An explicit Runge–Kutta scheme of order $p \in \mathbb{N}$ is strongly stable in weak form if and only if it is strongly stable w.r.t. $\mathscr{L}_{IM}$ (i.e., if and only if it is locally stable on the imaginary axis).*

Due to Theorem 1, an explicit Runge–Kutta method is strongly stable w.r.t. $\mathscr{L}$ if and only if it is strongly stable w.r.t. to the sets $\mathscr{L}_{AS}$ and $\mathscr{L}_{IM}$. For asymptotically stable, semi-dissipative matrices $\mathbf{L} \in \mathscr{L}_{AS}$, using a suitable, modified inner product norm ensures monotonicity/strong stability.

This paper is organized as follows. In Sect. 3 we collect examples which have been used to test strong stability of explicit Runge–Kutta schemes. We prove Theorem 1 in Sect. 4, while Sect. 5 is concerned with the proof of Theorem 2. Finally, we present the proof of Theorem 4 in Sect. 6.

3 Examples

The following two examples were used in the literature to investigate the strong stability of the explicit Runge–Kutta method with $s = p = 4$. Due to the new result in Corollary 1, it is clear that this scheme is *not* strongly stable. However, in view of Theorem 5, since $c_{p+1} = c_{p+2} = 0$ and hence $\delta_{p+1} > 0$, the method is strongly stable w.r.t. $\mathscr{L}_{IM}$. We deduce from Theorem 2(a) that the method is strongly stable w.r.t. $\mathscr{L}_{AS}^{m}$ with $m = 1$. Since (7b) holds, the considered Runge–Kutta scheme is *not* strongly stable w.r.t. $\{\mathbf{L}\}$ if the semi-dissipative matrix $\mathbf{L} \in \mathscr{L}_{AS}$ satisfies $2m_{HC}(\mathbf{L}) + 1 > p$, i.e., if $m_{HC}(\mathbf{L}) > 3/2$; see Theorem 2(b).

Example 1 Levy and Tadmor [17, Sect. 3.5] use the example

$$\mathbf{L} = -5 \begin{pmatrix} 1&2&2&2&2\\ 0&1&2&2&2\\ 0&0&1&2&2\\ 0&0&0&1&2\\ 0&0&0&0&1 \end{pmatrix} \quad \text{with } \mathbf{L}_H = -5 \begin{pmatrix} 1&1&1&1&1\\ 1&1&1&1&1\\ 1&1&1&1&1\\ 1&1&1&1&1\\ 1&1&1&1&1 \end{pmatrix}$$

to test numerically the strong stability of the explicit Runge–Kutta method with $s = p = 4$. The matrix $\mathbf{L}$ is semi-dissipative, but $\|R(\tau\mathbf{L})\|_2 > 1$ for some $\tau > 0$. The staircase form of $(\mathbf{J}, \mathbf{R}) = (\mathbf{L}_S, \mathbf{L}_H)$ is

$$\mathbf{VLV}^* = -5 \begin{bmatrix} 5 & -2\sqrt{2} & 0 & 0 & 0\\ 2\sqrt{2} & 0 & \frac{\sqrt{35}}{5} & 0 & 0\\ 0 & -\frac{\sqrt{35}}{5} & 0 & -\frac{4\sqrt{35}}{35} & 0\\ 0 & 0 & \frac{4\sqrt{35}}{35} & 0 & \frac{\sqrt{7}}{7}\\ 0 & 0 & 0 & -\frac{\sqrt{7}}{7} & 0 \end{bmatrix}, \quad \mathbf{V} = \begin{bmatrix} \frac{\sqrt{5}}{5} & \frac{\sqrt{5}}{5} & \frac{\sqrt{5}}{5} & \frac{\sqrt{5}}{5} & \frac{\sqrt{5}}{5}\\ \frac{\sqrt{10}}{5} & \frac{\sqrt{10}}{10} & 0 & -\frac{\sqrt{10}}{10} & -\frac{\sqrt{10}}{5}\\ -\frac{\sqrt{14}}{7} & \frac{\sqrt{14}}{14} & \frac{\sqrt{14}}{7} & \frac{\sqrt{14}}{14} & -\frac{\sqrt{14}}{7}\\ -\frac{\sqrt{10}}{10} & \frac{\sqrt{10}}{5} & 0 & -\frac{\sqrt{10}}{5} & \frac{\sqrt{10}}{10}\\ \frac{\sqrt{70}}{70} & -\frac{2\sqrt{70}}{35} & \frac{3\sqrt{70}}{35} & -\frac{2\sqrt{70}}{35} & \frac{\sqrt{70}}{70} \end{bmatrix}.$$

In particular, $r = 6$, $n_1 = \cdots = n_5 = 1$, $n_6 = 0$. Hence, the matrix $\mathbf{L}$ has the HC-index $m_{HC} = r - 2 = 4$ (see Remark 4).

Example 2 To show that the explicit Runge–Kutta method with $s = p = 4$ is not strongly stable, Sun and Shu [26, Proposition 1.1] use the counter-example

$$\mathbf{L} = -\begin{pmatrix} 1&2&2\\ 0&1&2\\ 0&0&1 \end{pmatrix} \quad \text{with } \mathbf{L}_H = -\begin{pmatrix} 1&1&1\\ 1&1&1\\ 1&1&1 \end{pmatrix}.$$

The matrix $\mathbf{L}$ is semi-dissipative, but $\|R(\tau\mathbf{L})\|_2 > 1$ for $\tau > 0$ sufficiently small. The staircase form of $(\mathbf{J}, \mathbf{R}) = (\mathbf{L}_S, \mathbf{L}_H)$ is

$$\mathbf{VLV}^* = -\begin{bmatrix} 3 & \frac{2\sqrt{3}\sqrt{2}}{3} & 0\\ -\frac{2\sqrt{3}\sqrt{2}}{3} & 0 & -\frac{\sqrt{2}\sqrt{6}}{6}\\ 0 & \frac{\sqrt{2}\sqrt{6}}{6} & 0 \end{bmatrix}, \quad \mathbf{V} = \begin{bmatrix} \frac{\sqrt{3}}{3} & \frac{\sqrt{3}}{3} & \frac{\sqrt{3}}{3}\\ -\frac{\sqrt{2}}{2} & 0 & \frac{\sqrt{2}}{2}\\ -\frac{\sqrt{6}}{6} & \frac{\sqrt{6}}{3} & -\frac{\sqrt{6}}{6} \end{bmatrix}.$$

Here, we have $r = 4$, $n_1 = n_2 = n_3 = 1$, $n_4 = 0$, and $\mathbf{L}$ has the HC-index $m_{HC} = r - 2 = 2$. Due to Theorem 2(b), matrix $\mathbf{L}$ has the minimal HC-index $m_{HC}(\mathbf{L})$ to detect the failure of strong stability, such that $m_{HC}(\mathbf{L}) > 3/2$.

Following our results in Theorem 2(c), semi-dissipative matrices with a specific HC-index are useful to test strong stability w.r.t. $\mathscr{L}_{AS}$ of explicit Runge–Kutta methods. We consider such a family of matrices in the following example.

Example 3 Consider the family of matrices $\mathbf{L} \in \mathbb{R}^{N\times N}$, $N \in \mathbb{N}$ in staircase form

$$\mathbf{L} = \begin{pmatrix} 0 & -1 & & & & \\ 1 & \ddots & \ddots & & & \\ & \ddots & \ddots & -1 & & \\ & & 1 & 0 & -1 & \\ & & & & 1 & -1 \end{pmatrix} \in \mathbb{R}^{N\times N}. \tag{9}$$

Using the Kalman rank condition [1, Proposition 1], one easily verifies that $\mathbf{L}$ has (maximal) HC-index $N-1$. Moreover, the matrix $\mathbf{L}$ is asymptotically stable, i.e., it has only eigenvalues with negative real part. This family of matrices (9) has been used to give a direct proof that explicit Runge–Kutta methods with $s = p \in 4\mathbb{N}$ and stability function $R(z)$ are not strongly stable w.r.t. $\mathscr{L}$: Following [3, Lemma 4], for $p \in 4\mathbb{N}$, consider the semi-dissipative matrix $\mathbf{L}$ with $N = 1 + p/2$. Then, $\mathbf{M}(\tau) := \mathbf{I} - R(\tau\mathbf{L})^* R(\tau\mathbf{L})$ satisfies

$$\det \mathbf{M}(\tau) = c\tau^{N^2} + \mathscr{O}(\tau^{N^2+1}) \quad \text{as } \tau \to 0\,, \tag{10}$$

with $R(z)$ defined in (5) and

$$c = 2^N \left(\prod_{i=1}^{N} \frac{1}{(p/2 - i + 1)!} \right)^2 \frac{\prod_{1\le i<j\le N} (i-j)^2}{\prod_{i,j=1}^{N} (i + (j-1))} \left(1 - \binom{p}{p/2} \right) < 0.$$

Thus, the matrix function $\mathbf{M}(\tau)$ has a negative determinant on $(0, \varepsilon)$ for sufficiently small $\varepsilon > 0$. Hence, $R(\tau\mathbf{L})^* R(\tau\mathbf{L})$ has at least one eigenvalue larger than one, and $\|R(\tau\mathbf{L})\|_2 > 1$ follows. Therefore, explicit Runge–Kutta methods with $s = p \in 4\mathbb{N}$ and stability function $R(z)$ are not strongly stable.

4 Proof of Theorem 1

The proof is based on a reduction of the matrix $\mathbf{L}$ to block diagonal form. First, we recall some preparatory results and then detail the reduction strategy.

Preparations. It is well-known that a similarity transformation of a Lyapunov stable matrix yields a semi-dissipative transformed matrix.

Lemma 1 *Let $\mathbf{L} \in \mathbb{C}^{n\times n}$ be Lyapunov stable. Then there exists $\mathbf{P} \in \mathbb{H}_n^>$ such that $\widehat{\mathbf{L}} := \mathbf{P}^{1/2}\mathbf{L}\mathbf{P}^{-1/2}$ is semi-dissipative and $\widehat{u} := \mathbf{P}^{1/2}u$ transforms (1) into the semi-dissipative ODE system $\mathrm{d}\widehat{u}/\mathrm{d}t = \widehat{\mathbf{L}}\widehat{u}$. Here, $\mathbf{P}^{-1/2} := (\mathbf{P}^{-1})^{1/2}$.*

Proof Since $\mathbf{L}$ is Lyapunov stable, there exists $\mathbf{P} \in \mathbb{H}_n^>$ such that $\mathbf{L}^*\mathbf{P} + \mathbf{P}\mathbf{L} \le 0$. A congruence transformation with the Hermitian matrix $\mathbf{P}^{-1/2}$ yields

$$0 \ge \mathbf{P}^{-1/2}(\mathbf{L}^*\mathbf{P} + \mathbf{P}\mathbf{L})\mathbf{P}^{-1/2} = \mathbf{P}^{-1/2}\mathbf{L}^*\mathbf{P}^{1/2} + \mathbf{P}^{1/2}\mathbf{L}\mathbf{P}^{-1/2} = 2(\mathbf{P}^{1/2}\mathbf{L}\mathbf{P}^{-1/2})_H,$$

proving that $\widehat{\mathbf{L}}$ is semi-dissipative. Finally, we multiply (1) from the left by $\mathbf{P}^{1/2}$ to find that $\mathrm{d}\widehat{u}/\mathrm{d}t = \mathbf{P}^{1/2}\mathbf{L}u = (\mathbf{P}^{1/2}\mathbf{L}\mathbf{P}^{-1/2})\widehat{u} = \widehat{\mathbf{L}}\widehat{u}$. □

We can characterize semi-dissipative matrices that are *not* asymptotically stable; see [18, Lemma 3.1], [1, Lemma 2.4 with Prop. 1 (B2), (B4)].

Proposition 1 *Let $\mathbf{L} \in \mathbb{C}^{n\times n}$ be semi-dissipative. Then $\mathbf{L}$ has an eigenvalue on the imaginary axis (and is hence* not *asymptotically stable) if and only if $\mathbf{L}_H v = 0$ for some eigenvector v of $\mathbf{L}_S$.*

Note that, due to the assumption on $\mathbf{L}$, purely imaginary eigenvalues of semi-dissipative matrices are necessarily semi-simple, i.e., their algebraic and geometric multiplicities coincide; see [19, 20].

Unitary congruence transformations $\mathbf{L} \mapsto \mathbf{U}\mathbf{L}\mathbf{U}^*$ with unitary matrices $\mathbf{U}$, preserve (asymptotic) stability, semi-dissipativity, and the HC-index (cf. (6)). For semi-dissipative matrices $\mathbf{L} \in \mathbb{C}^{n\times n}$, the pair $(\mathbf{J}, \mathbf{R}) := (\mathbf{L}_S, \mathbf{L}_H)$ can be transformed into a (so-called) *staircase form* by using unitary congruence transformations [5, Lemma 57].

Lemma 2 (Staircase form for $(\mathbf{J}, \mathbf{R})$) *Let $\mathbf{J} \in \mathbb{C}^{n\times n}$ be a skew-Hermitian matrix and $\mathbf{R} \in \mathbb{C}^{n\times n}$ be a nonzero Hermitian matrix. Then there exists a unitary matrix $\mathbf{V} \in \mathbb{C}^{n\times n}$ such that $\mathbf{VJV}^*$ and $\mathbf{VRV}^*$ are block tridiagonal matrices of the form*

$$
\mathbf{VJV}^* = \left[\begin{array}{ccccccc|c}
\mathbf{J}_{1,1} & -\mathbf{J}_{2,1}^* & & \cdots & & & 0 & 0 \\
\mathbf{J}_{2,1} & \mathbf{J}_{2,2} & -\mathbf{J}_{3,2}^* & & & & & \\
 & \ddots & \ddots & \ddots & & & \vdots & \vdots \\
 & & \mathbf{J}_{k,k-1} & \mathbf{J}_{k,k} & -\mathbf{J}_{k+1,k}^* & & & \vdots \\
\vdots & & & \ddots & \ddots & \ddots & & \\
 & & & \mathbf{J}_{r-2,r-3} & \mathbf{J}_{r-2,r-2} & -\mathbf{J}_{r-1,r-2}^* & & \\
0 & & \cdots & & \mathbf{J}_{r-1,r-2} & \mathbf{J}_{r-1,r-1} & & 0 \\
\hline
0 & & \cdots & & & 0 & & \mathbf{J}_{r,r}
\end{array}\right]
\begin{array}{c} n_1 \\ n_2 \\ \vdots \\ n_k \\ \vdots \\ n_{r-2} \\ n_{r-1} \\ n_r \end{array}
\qquad (11a)
$$
$$
\begin{array}{cccc} n_1 & n_{r-2} & n_{r-1} & n_r, \end{array}
$$

$$
\mathbf{VRV}^* = \left[\begin{array}{cccc|c}
\mathbf{R}_1 & 0 & \cdots & 0 & 0 \\
0 & 0 & & \vdots & \vdots \\
\vdots & & \ddots & \vdots & \vdots \\
0 & \cdots & \cdots & 0 & 0 \\
\hline
0 & \cdots & \cdots & 0 & 0
\end{array}\right]
\begin{array}{c} n_1 \\ n_2 \\ \vdots \\ n_{r-1} \\ n_r \end{array}
\qquad (11b)
$$
$$
\begin{array}{ccccc} n_1 & n_2 & \cdots & n_{r-1} & n_r, \end{array}
$$

where $n_1 \geq n_2 \geq \cdots \geq n_{r-1} > 0$, $n_r \geq 0$, and $\mathbf{R}_1 \in \mathbb{C}^{n_1,n_1}$ is nonsingular.

If $\mathbf{R}$ is nonsingular, then $r = 2$ and $n_2 = 0$. For example, $\mathbf{V} = \mathbf{I}$, $\mathbf{J}_{1,1} = \mathbf{J}$, and $\mathbf{R}_1 = \mathbf{R}$ is an admissible choice.

If $\mathbf{R}$ is singular, then $r \geq 3$ and the matrices $\mathbf{J}_{i,i-1}$, $i = 2, \ldots, r-1$, in the subdiagonal have full row rank and are of the form

$$\mathbf{J}_{i,i-1} = \begin{bmatrix} \Sigma_{i,i-1} & 0 \end{bmatrix}, \quad i = 2, \ldots, r-1,$$

with nonsingular matrices $\Sigma_{i,i-1} \in \mathbb{C}^{n_i \times n_i}$, and moreover $\Sigma_{r-1,r-2}$ is a real-valued diagonal matrix.

For semi-dissipative matrices $\mathbf{L} \in \mathbb{C}^{n \times n}$, the staircase form of $(\mathbf{J}, \mathbf{R}) = (\mathbf{L}_S, \mathbf{L}_H)$ can be used to derive a unitary congruence transformation to a block diagonal form.

Lemma 3 *Let $\mathbf{L} \in \mathbb{C}^{n \times n}$ be semi-dissipative. Then there exists a unitary matrix $\mathbf{V} \in \mathbb{C}^{n \times n}$ such that $\mathbf{VLV}^*$ is a block diagonal matrix of the form*

$$\mathbf{V}\,\mathbf{L}\,\mathbf{V}^* = \left[\begin{array}{c|c} \mathbf{L}_1 & 0 \\ \hline 0 & \mathbf{L}_2 \end{array}\right], \tag{12}$$

where $\mathbf{L}_1 \in \mathbb{C}^{\tilde{n}_1 \times \tilde{n}_1}$, $\mathbf{L}_2 \in \mathbb{C}^{\tilde{n}_2 \times \tilde{n}_2}$ with $\tilde{n}_1, \tilde{n}_2 \in \mathbb{N}_0$ such that $\tilde{n}_2 = n - \tilde{n}_1$. Moreover, the semi-dissipative matrix $\mathbf{L}_1 \in \mathbb{C}^{\tilde{n}_1 \times \tilde{n}_1}$ is asymptotically stable, and the semi-dissipative matrix $\mathbf{L}_2 \in \mathbb{C}^{\tilde{n}_2 \times \tilde{n}_2}$ is skew-Hermitian.

In the block-diagonal form (12), not both matrices $\mathbf{L}_1, \mathbf{L}_2$ have to be present. Indeed, if $\mathbf{L}$ is skew-Hermitian then $\tilde{n}_1 = 0$, $\tilde{n}_2 = n$ and $\mathbf{L}_2 = \mathbf{L}$, and if $\mathbf{L}$ is asymptotically stable then $\tilde{n}_1 = n$, $\tilde{n}_2 = 0$ and $\mathbf{L}_1 = \mathbf{L}$.

Proof (*of Lemma* 3) If the Hermitian part $\mathbf{L}_H$ of $\mathbf{L}$ is the null matrix then we set $\mathbf{V} = \mathbf{I}$, $\mathbf{L}_2 = \mathbf{L} = \mathbf{L}_S$, $\tilde{n}_1 = 0$, and $\tilde{n}_2 = n$.

If the Hermitian part $\mathbf{L}_H$ is not the null matrix, we deduce from Lemma 2 for the pair $(\mathbf{J}, \mathbf{R}) := (\mathbf{L}_S, \mathbf{L}_H)$ the existence of a unitary matrix $\mathbf{V} \in \mathbb{C}^{n \times n}$ such that $\mathbf{VLV}^* = \mathbf{VJV}^* + \mathbf{VRV}^*$ given in (11) holds with $\mathbf{R}_1 < 0$ (although the matrix $\mathbf{J}_{r,r}$ may not be present). If the matrix $\mathbf{J}_{r,r}$ is present, then $\mathbf{L}_2 = \mathbf{J}_{r,r}$, $\tilde{n}_2 = n_r$, $\tilde{n}_1 = n - n_r > 0$, otherwise $\tilde{n}_2 = 0$ and $\tilde{n}_1 = n$. If the matrix $\mathbf{L}_2$ is present, it is skew-Hermitian, since it is a (non-leading) principal minor of the skew-Hermitian matrix $\mathbf{VL}_S\mathbf{V}^*$.

Because of $\mathbf{L}_H \neq 0$, the matrix $\mathbf{L}_1 \neq 0$ is the leading principal minor of $\mathbf{VLV}^*$ given as

$$\mathbf{L}_1 := \begin{bmatrix} \mathbf{J}_{1,1} + \mathbf{R}_1 & -\mathbf{J}_{2,1}^* & & \cdots & & & 0 \\ \mathbf{J}_{2,1} & \mathbf{J}_{2,2} & -\mathbf{J}_{3,2}^* & & & & \\ & \ddots & \ddots & \ddots & & & \vdots \\ & & \mathbf{J}_{k,k-1} & \mathbf{J}_{k,k} & -\mathbf{J}_{k+1,k}^* & & \\ \vdots & & & \ddots & \ddots & \ddots & \\ & & & & \mathbf{J}_{r-2,r-3} & \mathbf{J}_{r-2,r-2} & -\mathbf{J}_{r-1,r-2}^* \\ 0 & & \cdots & & & \mathbf{J}_{r-1,r-2} & \mathbf{J}_{r-1,r-1} \end{bmatrix} \begin{matrix} n_1 \\ n_2 \\ \vdots \\ n_k \\ \vdots \\ n_{r-2} \\ n_{r-1} \end{matrix},$$

$$\begin{matrix} n_1 & \qquad\qquad\qquad\qquad & n_{r-2} & n_{r-1} \end{matrix}$$

and $\mathbf{L}_1 \in \mathbb{C}^{\tilde{n}_1 \times \tilde{n}_1}$ is semi-dissipative.

Finally, we apply Proposition 1 in $\mathbb{C}^{\tilde{n}_1}$ to deduce that $\mathbf{L}_1$ is asymptotically stable. If this were not true, there would exist a vector $\tilde{v}_1 = (0, v_2, ..., v_{r-1})^T$ in the kernel of $(\mathbf{L}_1)_H = \mathrm{diag}(\mathbf{R}_1, 0, ..., 0)$ (see (11b)) that is an eigenvector of $(\mathbf{L}_1)_S$. The latter has the block tridiagonal shape in (11a). Since $\mathbf{J}^*_{i,i-1}$, $i = 2, ..., r-1$, has full column rank, we obtain iteratively $\tilde{v}_1 = 0$, which cannot be an eigenvector of $(\mathbf{L}_1)_S$. □

Remark 4 A semi-dissipative matrix $\mathbf{L} = \mathbf{L}_S + \mathbf{L}_H$ is asymptotically stable if and only if $n_r = 0$ holds in the staircase form (11b) of $(\mathbf{J}, \mathbf{R}) = (\mathbf{L}_S, \mathbf{L}_H)$; see the final step in the proof of Lemma 3. If this is the case, the HC-index satisfies $m_{HC}(\mathbf{L}) = r - 2$; see [4, Lemma 4].

Reduction steps. We prove the structural result in Theorem 1 by performing several reduction steps.

Step 1. Reduction to semi-dissipative matrices. First, we show that strong stability of explicit Runge–Kutta schemes w.r.t. Lyapunov stable matrices $\mathbf{L} \in \mathscr{L}$ is equivalent to strong stability w.r.t. semi-dissipative matrices $\widehat{\mathbf{L}} \in \mathscr{L}_{SD}$ (as already stated in Remark 1). To this end, we consider a Lyapunov stable matrix $\mathbf{L} \in \mathbb{C}^{n\times n}$ and $\mathbf{P} \in \mathbb{H}_n^>$ satisfying (2). By Lemma 1, $\widehat{\mathbf{L}} := \mathbf{P}^{1/2}\mathbf{L}\mathbf{P}^{-1/2}$ is semi-dissipative. We define $\widehat{u}^0 := \mathbf{P}^{1/2}u^0$, $u^1 := R(\tau\mathbf{L})u^0$, and $\widehat{u}^1 := R(\tau\widehat{\mathbf{L}})\widehat{u}^0$. Then $\|u^0\|_{\mathbf{P}} = \|\mathbf{P}^{1/2}u^0\| = \|\widehat{u}^0\|$ and

$$\begin{aligned} \|u^1\|_{\mathbf{P}} &= \|R(\tau\mathbf{L})u^0\|_{\mathbf{P}} = \|\mathbf{P}^{1/2}R(\tau\mathbf{L})u^0\| \\ &= \|R(\tau\mathbf{P}^{1/2}\mathbf{L}\mathbf{P}^{-1/2})\mathbf{P}^{1/2}u^0\| = \|R(\tau\widehat{\mathbf{L}})\widehat{u}^0\| = \|\widehat{u}^1\|. \end{aligned} \tag{13}$$

Hence, $\|u^1\|_{\mathbf{P}} \le \|u^0\|_{\mathbf{P}}$ if and only if $\|\widehat{u}^1\| \le \|\widehat{u}^0\|$.

Step 2. Reduction to semi-dissipative matrices in block diagonal form. Considering the unitary congruence transformation from Lemma 3, we note that $\|\mathbf{V}u^j\| = \|u^j\|$, with the unitary matrix $\mathbf{V}$ from (12). Therefore, an explicit Runge–Kutta scheme is strongly stable if and only if it is strongly stable w.r.t. semi-dissipative matrices $\mathbf{L}$ of the form (12).

Step 3. Reduction to two distinct subsets of semi-dissipative matrices. Due to the block diagonal structure of $\mathbf{L}$ in (12), an explicit Runge–Kutta scheme is strongly stable if and only if it is strongly stable w.r.t. two distinct subsets of semi-dissipative matrices, namely

(a) asymptotically stable, semi-dissipative matrices $\mathbf{L} \in \mathbb{C}^{n\times n}$, and
(b) skew-Hermitian matrices $\mathbf{L} \in \mathbb{C}^{n\times n}$ (hence satisfying $\mathbf{L} + \mathbf{L}^* = 0$).

Step 4. Reduction of skew-Hermitian matrices to (purely imaginary) scalar problems. Skew-Hermitian matrices $\mathbf{L} \in \mathbb{C}^{n\times n}$ are unitarily congruent to a diagonal matrix with purely imaginary eigenvalues, i.e., there exists a unitary matrix $\mathbf{U} \in \mathbb{C}^{n\times n}$ such that $\mathbf{U}\mathbf{L}\mathbf{U}^* = \Lambda$ with $\Lambda = \mathrm{diag}(\lambda_1^{\mathbf{L}}, \dots, \lambda_n^{\mathbf{L}})$, where $\lambda_j^{\mathbf{L}} \in \mathrm{i}\mathbb{R}$. Therefore, in case b, the analysis can be reduced to scalar ODEs of the form

$$\frac{\mathrm{d}u}{\mathrm{d}t} = \lambda u, \quad t \ge 0; \quad \lambda \in \mathrm{i}\mathbb{R}.$$

This finishes the proof of Theorem 1. □

We already mentioned that the (strong) stability of Runge–Kutta schemes for scalar ODEs is well studied; see, e.g., [16]. By definition, strong stability w.r.t. $\mathscr{L}_{IM}$ is equivalent to *local stability on the imaginary axis* as defined in [16, Definition 2.1] (and recalled in Sect. 2). Thus, we deduce from [16, Theorem 3.1] the following statement.

Theorem 5 *The explicit Runge–Kutta method (4) of order $p \in \mathbb{N}$ is strongly stable w.r.t. $\mathscr{L}_{IM}$ if*

$$\gamma_{p+1} := (-1)^{(p+1)/2}(1 - c_{p+1}) > 0 \quad \textit{for } p \textit{ odd},$$
$$\delta_{p+1} := (-1)^{p/2}\big(c_{p+2} - (p+2)c_{p+1} + (p+1)\big) > 0 \quad \textit{for } p \textit{ even}.$$

If $\gamma_{p+1} < 0$ or $\delta_{p+1} < 0$, the Runge–Kutta method is not strongly stable.

Notice that the case $\delta_{p+1} = 0$ is left open in [16, Theorem 3.1]. Complementary, $\gamma_{p+1} = 0$ implies $c_{p+1} = 1$, this would contradict that the Runge–Kutta method has order p, see (5).

5 Proof of Theorem 2

The proof is based on the short-time decay of the spectral norm of the matrix exponential $t \mapsto e^{t\mathbf{L}} \in \mathbb{C}^{n\times n}$, characterized by the HC-index of $\mathbf{L}$. We recall from [2, Theorem 2.7(a)]:

Proposition 2 *Let the matrix $\mathbf{L} \in \mathbb{C}^{n\times n}$ be semi-dissipative. Then $\mathbf{L}$ is asymptotically stable (with HC-index $m_{HC} \in \mathbb{N}_0$) if and only if*

$$\|e^{t\mathbf{L}}\|_2 = 1 - ct^a + \mathscr{O}(t^{a+1}) \quad \textit{for } t \in [0, \epsilon), \tag{14}$$

for some $a, c, \epsilon > 0$. In this case, necessarily $a = 2m_{HC} + 1$.

The sharp multiplicative factor c in (14) has been determined explicitly in [2, Theorem 2.7(b)]. Definition 3 immediately yields the following observation.

Remark 5 Let $\mathbf{L}$ be semi-dissipative with HC-index $m_{HC} \in \mathbb{N}_0$. In view of (6), there exists a normalized vector u_0 such that

$$u_0 \in \ker\left(T_{m_{HC}-1}\right) = \ker\left(\sum_{j=0}^{m_{HC}-1} \mathbf{L}_S^j \mathbf{L}_H (\mathbf{L}_S^*)^j\right), \tag{15}$$

but none that would satisfy instead also $(-\mathbf{L}_H)^{1/2}(\mathbf{L}_S)^{m_{HC}} u_0 = 0$. Hence, there exists a normalized vector u_0 such that

$$\|(-\mathbf{L}_H)^{1/2}(\mathbf{L}_S)^j u_0\| = 0 \text{ for } 0 \le j \le m_{HC}-1, \quad (-\mathbf{L}_H)^{1/2}(\mathbf{L}_S)^{m_{HC}} u_0 \neq 0. \tag{16}$$

Preparation. We claim that $t \mapsto \|e^{t\mathbf{L}}\|_2$ and $\tau \mapsto \|R(\tau\mathbf{L})\|_2$ are real analytic on a sufficiently small time interval. In particular, the Taylor expansions of order $p+1$ of $\|e^{t\mathbf{L}}\|_2$ and $\|R(\tau\mathbf{L})\|_2$ about zero exist. Indeed, according to [15, Lemma 1], for sufficiently small time $t_0 > 0$, there exists a real analytic function $\Phi : [0, t_0] \to \mathbb{R}$ such that $\|e^{t\mathbf{L}}\|_2 = \Phi(t)$ for all $t \in [0, t_0]$. This statement can also be derived from [14, Chap. 2, Sect. 6] or [12, Theorem 4.3.17].

Similarly, we prove that $\|R(\tau\mathbf{L})\|_2$ is (at least) locally real analytic for sufficiently small $\tau \ge 0$. The stability function $R(\tau\mathbf{L})$ is an analytic matrix function w.r.t. τ, hence $G(\tau) := R(\tau\mathbf{L}^*)R(\tau\mathbf{L})$ is again an analytic matrix function. Due to [12, Theorem 4.3.17], there exists a neighborhood of $\tau_0 = 0$ and locally analytic functions $\lambda_j(\tau) = \sigma_j(G(\tau))$, where σ_j denotes the singular values of (the self-adjoint) matrix $G(\tau)$, and $\lambda_j(0) = 1$ for $j = 1, ..., n$. Hence,

$$\|R(\tau\mathbf{L})\|_2 = \max_{j=1,\dots,n} \left\{\sqrt{\lambda_j(\tau)}\right\}.$$

If two real analytic functions are the same on some converging sequence then they are identical. Since $\lambda_j(0) = 1$, we use null-sequences to deduce that any two real analytic functions λ_k, λ_ℓ on $[0, \infty)$ are either identical or there exists $\tau_*^{k\ell} > 0$ such that they do not intersect on a finite interval $[0, \tau_*^{k\ell})$. Since there are only finitely many λ_j, there exists $\tau_* := \min\{\tau_*^{k\ell}\} > 0$ such that $\|R(\tau\mathbf{L})\|_2$ is real analytic on $[0, \tau_*)$, see also the discussion in [14, Chap. 2, Sect. 6.4].

Proof of Theorem 2, statement (a). Following the proof of [2, Theorem 2.7], we consider the spectral norm $\|e^{t\mathbf{L}}\|_2^2 = \|Q(t)\|_2 = \lambda_{\max}(Q(t))$ for small $t > 0$, where

$$Q(t) := e^{t\mathbf{L}^*}e^{t\mathbf{L}} = \sum_{j=0}^{\infty} \frac{t^j}{j!} U_j, \quad U_j = \sum_{k=0}^{j} \binom{j}{k} (\mathbf{L}^*)^k \mathbf{L}^{j-k},$$

satisfying $\|U_j\|_2 \le (2\|\mathbf{L}\|_2)^j$ for $j \in \mathbb{N}_0$.

First, we write the stability function $R(z)$ in (5) in the form

$$R(z) = \sum_{j=0}^{\infty} c_j \frac{z^j}{j!}, \quad z \in \mathbb{C}, \ c_{p+1} \neq 1, \tag{17}$$

with constant coefficients $c_j = 1$ for $0 \le j \le p$; $c_{p+1}, \dots, c_s \in \mathbb{R}$, and $c_j = 0$ for $j \ge s+1$, see [11, Theorem IV.2.2]. Then, we compute

$$
\begin{aligned}
G(\tau) = R(\tau\mathbf{L}^*)R(\tau\mathbf{L}) &= \sum_{j=0}^{2s}\frac{1}{j!}\sum_{k=0}^{j}\binom{j}{k}c_k c_{j-k}(\tau\mathbf{L}^*)^k(\tau\mathbf{L})^{j-k} \\
&= \sum_{j=0}^{2s}\frac{\tau^j}{j!}\sum_{k=0}^{j}\binom{j}{k}c_k c_{j-k}(\mathbf{L}^*)^k\mathbf{L}^{j-k} \\
&= \sum_{j=0}^{p}\frac{\tau^j}{j!}U_j + \sum_{j=p+1}^{2s}\frac{\tau^j}{j!}\sum_{k=0}^{j}\binom{j}{k}c_k c_{j-k}(\mathbf{L}^*)^k\mathbf{L}^{j-k} \\
&= \sum_{j=0}^{p}\frac{\tau^j}{j!}U_j + \frac{\tau^{p+1}}{(p+1)!}\Bigg(\sum_{k=1}^{p}\binom{p+1}{k}(\mathbf{L}^*)^k\mathbf{L}^{p+1-k} \\
&\quad + c_{p+1}\mathbf{L}^{p+1} + c_{p+1}(\mathbf{L}^*)^{p+1}\Bigg) + \mathscr{O}(\tau^{p+2}).
\end{aligned}
$$

We subtract

$$
\begin{aligned}
Q(\tau) = \sum_{j=0}^{p}\frac{\tau^j}{j!}U_j &+ \frac{\tau^{p+1}}{(p+1)!}\Bigg(\sum_{k=1}^{p}\binom{p+1}{k}(\mathbf{L}^*)^k\mathbf{L}^{p+1-k} \\
&+ \mathbf{L}^{p+1} + (\mathbf{L}^*)^{p+1}\Bigg) + \mathscr{O}(\tau^{p+2})
\end{aligned}
$$

from $G(\tau)$ and set $\widetilde{c}_k = c_k - 1$ to find that

$$
G(\tau) - Q(\tau) = \frac{\tau^{p+1}}{(p+1)!}\widetilde{c}_{p+1}\left(\mathbf{L}^{p+1} + (\mathbf{L}^*)^{p+1}\right) + \mathscr{O}(\tau^{p+2}). \tag{18}
$$

Consequently, the Taylor expansions for $G(\tau)$ and $Q(\tau)$ differ starting from the τ^{p+1}-term. Now, consider an explicit Runge–Kutta scheme of order $p \in \mathbb{N}$ with stability function $R(z)$, and matrices $\mathbf{L} \in \mathscr{L}_{AS}$ with HC-index m_{HC} satisfying $2m_{HC} + 1 \leq p$. Then, taking into account Proposition 2, there exist $c, \epsilon > 0$ such that

$$
\begin{aligned}
\|R(\tau\mathbf{L})\|_2 = \|G(\tau)\|_2^{1/2} &= \|Q(\tau)\|_2^{1/2} + \mathscr{O}(\tau^{p+1}) \\
&= \|e^{\tau\mathbf{L}}\|_2 + \mathscr{O}(\tau^{p+1}) = 1 - c\tau^{2m_{HC}+1} + \mathscr{O}(\tau^{2m_{HC}+2})
\end{aligned}
$$

for all $\tau \in [0, \epsilon)$. We conclude that $\|R(\tau\mathbf{L})\|_2 \leq 1$ for sufficiently small τ, hence proving that all explicit Runge–Kutta schemes of order $p \in \mathbb{N}$ are strongly stable w.r.t. matrices $\mathbf{L} \in \mathscr{L}_{AS}$ with HC-index m_{HC} satisfying $2m_{HC} + 1 \leq p$.

Proof of Theorem 2**, statement (b) for even order** p. Let $p \in 2\mathbb{N}$ and let $\mathbf{L} \in \mathscr{L}_{AS}$ with HC-index $m_{HC} = p/2$. The proof uses the following result, which follows from [2, Lemma A.4], when using the simple observation from [2, Lemma 2.4] that $\ker(T_m) = \ker[\sum_{j=0}^{m}(\mathbf{L}^*)^j\mathbf{L}_H\mathbf{L}^j]$.

Lemma 4 *Let* $\mathbf{L} \in \mathbb{C}^{n\times n}$ *be semi-dissipative with HC-index* $m_{HC} \in \mathbb{N}$. *Then, for each* $u_0 \in \ker(T_{m_{HC}-1})$, *there exists a polynomial vector function* $u_\tau \in \mathbb{C}^n$, $\tau \in [0, 1]$, *of the form*

$$u_\tau = u_0 + \sum_{\ell=1}^{m_{HC}} b_\ell \tau^\ell (-\mathbf{L})^\ell u_0 \tag{19}$$

for suitable numbers $b_\ell \in \mathbb{R}$, *such that for* $\tau \in [0, 1]$,

$$g(u_\tau; \tau) := u_\tau^* \Bigg(\sum_{j=1}^{\infty} \frac{\tau^j}{j!} U_j \Bigg) u_\tau = -2c_1(u_0)\tau^a + \mathscr{O}(\tau^{a+1}), \tag{20}$$

where $a = 2m_{HC} + 1$ *and* $c_1(u_0) := \|(-\mathbf{L}_H)^{1/2}\mathbf{L}^{m_{HC}} u_0\|^2/((2m_{HC} + 1)!\binom{2m_{HC}}{m_{HC}})$.

The vector function u_τ of this lemma is used as an approximation of an eigenfunction of $G(\tau) = R(\tau\mathbf{L})^* R(\tau\mathbf{L})$, whose associated eigenvalue is larger than one for sufficiently small $\tau > 0$. Since $g(u_\tau; \tau) = u_\tau^* Q(\tau)u_\tau - \|u_\tau\|^2$, we can write

$$u_\tau^* G(\tau)u_\tau = \|u_\tau\|^2 + g(u_\tau; \tau) + u_\tau^*\big(G(\tau) - Q(\tau)\big)u_\tau, \tag{21}$$

where here and in the following, the function g is considered on $\mathbb{C}^n \times [0, \infty)$. As $\mathbf{L}$ is semi-dissipative, Remark 5 yields the existence of some $u_0 \in \ker(T_{m_{HC}-1}) \setminus \{0\}$ satisfying (16). This shows that

$$\begin{aligned} u_0^*\big(\mathbf{L}^{p+1} + (\mathbf{L}^*)^{p+1}\big)u_0 &= u_0^*\big((\mathbf{L}_H + \mathbf{L}_S)^{p+1} + (\mathbf{L}_H - \mathbf{L}_S)^{p+1}\big)u_0 \\ &= 2(-1)^{m_{HC}}\, u_0^*(\mathbf{L}_S^*)^{m_{HC}}\mathbf{L}_H(\mathbf{L}_S)^{m_{HC}} u_0 \\ &= -2(-1)^{m_{HC}} \|(-\mathbf{L}_H)^{1/2}\mathbf{L}_S^{m_{HC}} u_0\|^2 \neq 0, \end{aligned} \tag{22}$$

where we used from (16) that $\mathbf{L}_H(\mathbf{L}_S)^j u_0 = 0$ and $u_0^*(\mathbf{L}_S^*)^j \mathbf{L}_H = 0$ for $0 \le j \le m_{HC} - 1$. Therefore, we infer from (18) and (19) that

$$\begin{aligned} u_\tau^*\big(G(\tau) - Q(\tau)\big)u_\tau = -2(-1)^{m_{HC}} &\frac{\tau^{2m_{HC}+1}}{(2m_{HC}+1)!}\widetilde{c}_{p+1} \|(-\mathbf{L}_H)^{1/2}\mathbf{L}_S^{m_{HC}} u_0\|^2 \\ &+ \mathscr{O}(\tau^{2m_{HC}+2}). \end{aligned}$$

Inserting this expression and (20) into (21) yields

$$\begin{aligned} u_\tau^* G(\tau)u_\tau &= \|u_\tau\|^2 + g(u_\tau; \tau) + u_\tau^*\big(G(\tau) - Q(\tau)\big)u_\tau \\ &= \|u_\tau\|^2 - \frac{2\tau^{2m_{HC}+1}}{(2m_{HC}+1)!}\bigg(\binom{2m_{HC}}{m_{HC}}^{-1} + (-1)^{m_{HC}}\widetilde{c}_{p+1}\bigg) \|(-\mathbf{L}_H)^{1/2}\mathbf{L}_S^{m_{HC}} u_0\|^2 \\ &\quad + \mathscr{O}(\tau^{2m_{HC}+2}). \end{aligned} \tag{23}$$

By condition (7b), the expression in the round brackets is negative,

$$\binom{2m_{HC}}{m_{HC}}^{-1} + (-1)^{m_{HC}}\widetilde{c}_{p+1} = \binom{p}{p/2}^{-1} + (-1)^{p/2}(c_{p+1}-1) < 0,$$

implying that $u_\tau^* G(\tau)u_\tau > \|u_\tau\|^2$ for all $\tau \in (0, \tau_*)$ for some $\tau_* > 0$. We conclude that $\|R(\tau)\|_2 > 1$ on that interval. Thus, the Runge–Kutta scheme with even order p is not strongly stable under condition (7b).

Proof of Theorem 2, **statement (b) for odd order** p. Let $p \in 2\mathbb{N}_0 + 1$ and let $\mathbf{L} \in \mathscr{L}_{AS}$ with HC-index $m_{HC} = (p+1)/2$. As before, Remark 5 yields the existence of some $u_0 \in \ker(T_{m_{HC}-1}) \setminus \{0\}$ satisfying (16) and hence, proceeding similarly as in (22),

$$\begin{aligned} u_0^*\big(\mathbf{L}^{p+1} + (\mathbf{L}^*)^{p+1}\big)u_0 &= u_0^*\big((\mathbf{L}_H + \mathbf{L}_S)^{2m_{HC}} + (\mathbf{L}_H - \mathbf{L}_S)^{2m_{HC}}\big)u_0 \\ &= 2u_0^*\mathbf{L}_S^{2m_{HC}}u_0 = 2(-1)^{m_{HC}}\|(\mathbf{L}_S)^{m_{HC}}u_0\|^2 \neq 0. \end{aligned}$$

Then, with the vector function u_τ as in Lemma 4 and taking into account (18) and $2m_{HC} = p+1$,

$$\begin{aligned} u_\tau^* G(\tau)u_\tau &= \|u_\tau\|^2 + g(u_\tau;\tau) + u_\tau^*\big(G(\tau) - Q(\tau)\big)u_\tau \\ &= \|u_\tau\|^2 - 2c_1(u_0)\tau^{2m_{HC}+1} + \mathscr{O}(\tau^{2m_{HC}+2}) \\ &\quad + \frac{\tau^{p+1}}{(p+1)!}\widetilde{c}_{p+1}u_0^*\big(\mathbf{L}^{p+1} + (\mathbf{L}^*)^{p+1}\big)u_0 + \mathscr{O}(\tau^{p+2}) \\ &= \|u_\tau\|^2 + \frac{\tau^{p+1}}{(p+1)!}\widetilde{c}_{p+1}u_0^*\big(\mathbf{L}^{p+1} + (\mathbf{L}^*)^{p+1}\big)u_0 + \mathscr{O}(\tau^{p+2}) \\ &= \|u_\tau\|^2 + \frac{2\tau^{p+1}}{(p+1)!}(-1)^{(p+1)/2}\widetilde{c}_{p+1}\|\mathbf{L}_S^{(p+1)/2}u_0\|^2 + \mathscr{O}(\tau^{p+2}). \qquad (24) \end{aligned}$$

Since $\|\mathbf{L}_S^{(p+1)/2}u_0\|^2 \neq 0$, condition (7a) implies that

$$(-1)^{(p+1)/2}\widetilde{c}_{p+1} = (-1)^{(p+1)/2}(c_{p+1}-1) > 0$$

and consequently, $u_\tau^* G(\tau)u_\tau > \|u_\tau\|_2^2$ for all $\tau \in (0, \tau_*)$ for some $\tau_* > 0$. Thus, the Runge–Kutta scheme with odd order p is not strongly stable under condition (7a).

Proof of Theorem 2, **statement** (c). Consider an explicit Runge–Kutta scheme of order $p \in \mathbb{N}$ which is not strongly stable and let (7) hold. If a matrix $\mathbf{L} \in \mathscr{L}_{AS}$ with HC-index $m_{HC} = m_{HC}(\mathbf{L})$ satisfies $2m_{HC} + 1 > p$, then we can repeat the construction in the proof of statement (b): Consider $m \in \mathbb{N}$ such that $2m+1 > p$ but $2m - 1 \le p$. Due to $m_{HC} \ge m$, there exists a vector $u_0 \in \mathbb{C}^n$ such that

$$\|(-\mathbf{L}_H)^{1/2}\mathbf{L}_S^j u_0\| = 0 \text{ for } 0 \le j \le m-1, \quad \|(-\mathbf{L}_H)^{1/2}\mathbf{L}_S^m u_0\| \neq 0,$$

see (16), but with m replacing m_{HC}. Using a vector function u_τ for $\tau \in [0, 1]$ as constructed in (19), we can use the above Eqs. (23), (24), again with m replacing m_{HC}, to prove statement (c) in Theorem 2. □

6 Proof of Theorem 4

As pointed out in [17, Sect. 3.5], an explicit Runge–Kutta method of order $p = 4$ with $s = 4$ stages can be made strongly stable w.r.t. $\mathscr{L}_{AS}$ when restricting to some adapted **P**-norm. Motivated by this observation, we prove Theorem 4.

First, we observe that for strong stability in weak form, the analogous statement as in Theorem 1 holds, i.e., an explicit Runge–Kutta method is strongly stable in weak form if and only if it is strongly stable in weak form w.r.t. $\mathscr{L}_{AS}$ and $\mathscr{L}_{IM}$.

Second, let an asymptotically stable, semi-dissipative matrix $\mathbf{L} \in \mathscr{L}_{AS}$ be given. A matrix $\mathbf{L} \in \mathbb{C}^{n\times n}$ is asymptotically stable if and only if there exists a matrix $\mathbf{P} \in \mathbb{H}_n^>$ such that

$$\mathbf{L}^*\mathbf{P} + \mathbf{P}\mathbf{L} < 0, \tag{25}$$

see e.g. [6, Corollary 15.10.1], [10, Theorem 3.17]. Following the proof of Lemma 1, we find that $\widehat{\mathbf{L}} := \mathbf{P}^{1/2}\mathbf{L}\mathbf{P}^{-1/2}$ is dissipative. Hence, $\widehat{\mathbf{L}}$ has the HC-index $m_{HC}(\widehat{\mathbf{L}}) = 0$. Consider the solution $(u^k)_{k\in\mathbb{N}_0}$ in (4) in the weighted norm $\|\cdot\|_{\mathbf{P}}$ with $\mathbf{P} \in \mathbb{H}_n^>$ satisfying (25). By Step 1 of the proof of Theorem 1, strong stability in weak form for the Runge–Kutta scheme, i.e. $\|u^1\|_{\mathbf{P}} \le \|u^0\|_{\mathbf{P}}$, is equivalent to

$$\|\widehat{u}^1\| \le \|\widehat{u}^0\|, \quad \text{where } \widehat{u}^1 = R(\tau\widehat{\mathbf{L}})\widehat{u}^0. \tag{26}$$

The inequality in (26) indeed holds due to Theorem 2(a), since $\widehat{\mathbf{L}}$ has the HC-index $m_{HC} = 0$. Thus, $\|u^1\|_{\mathbf{P}} \le \|u^0\|_{\mathbf{P}}$ holds without (further) constraints on $\mathbf{L}$.

Finally, consider strong stability w.r.t. $\mathscr{L}_{IM}$, the set of purely imaginary scalars $\lambda \in i\mathbb{R}$. Introducing a "weighted norm" $\|\cdot\|_{\mathbf{P}}$ for some positive constant $\mathbf{P} \in \mathbb{R}^+$ does not change the stability criterion $\|u^1\|_{\mathbf{P}} \le \|u^0\|_{\mathbf{P}}$ on the considered explicit Runge–Kutta scheme. This finishes the proof. □

Example 4 (*Continuation of Example* 2) While the explicit Runge-Kutta scheme with $s = p = 4$ stages is *not* strongly stable (due to Corollary 1), it is strongly stable in weak form (due to Theorem 4, since it is strongly stable w.r.t. $\mathscr{L}_{IM}$; see Table 1). Notice that the authors of [26] study this problem in the Euclidean norm, whereas we use an adapted **P**-norm to guarantee the decay estimate $\|u^1\|_{\mathbf{P}} \le \|u^0\|_{\mathbf{P}}$.

In fact, the Jordan normal form $-\mathbf{L} = (\mathbf{W}^*)^{-1}\mathbf{J}\mathbf{W}^*$ of $-\mathbf{L}$ holds with

$$\mathbf{J} = \begin{pmatrix} 1 & 1 & 0 \\ 0 & 1 & 1 \\ 0 & 0 & 1 \end{pmatrix} \quad \text{and} \quad \mathbf{W} = \begin{pmatrix} 1 & 0 & 0 \\ 0 & 2 & 0 \\ 0 & 2 & 4 \end{pmatrix}.$$

Moreover, a suitable positive definite Hermitian matrix $\mathbf{P} = \mathbf{P}(\mathbf{L})$, such that (25) holds, is given as

$$\mathbf{P} = \mathbf{W}\mathbf{W}^* = \begin{pmatrix} 1 & 0 & 0 \\ 0 & 4 & 4 \\ 0 & 4 & 20 \end{pmatrix}, \quad \text{and} \quad \mathbf{P}^{1/2} = \sqrt{\tfrac{2}{5}} \begin{pmatrix} \sqrt{\tfrac{5}{2}} & 0 & 0 \\ 0 & 3 & 1 \\ 0 & 1 & 7 \end{pmatrix}.$$

The transformed matrix

$$\widehat{\mathbf{L}} = \mathbf{P}^{1/2}\mathbf{L}\mathbf{P}^{-1/2} = -\tfrac{1}{10} \begin{pmatrix} 10 & \sqrt{90} & \sqrt{10} \\ 0 & 7 & 9 \\ 0 & -1 & 13 \end{pmatrix}$$

is dissipative, since its Hermitian part $\big(\widehat{\mathbf{L}}\big)_H$ has the negative eigenvalues -1 and $-1 \pm \sqrt{2}/2$. Hence, $\widehat{\mathbf{L}}$ has HC-index $m_{HC}(\widehat{\mathbf{L}}) = 0$. Due to Theorem 2, Eq. (13), and [5, Corollary 26], we compute

$$\|R(\tau\mathbf{L})\|_{\mathbf{P}} = \|R(\tau\widehat{\mathbf{L}})\|_2 = 1 + (-1 + \sqrt{2}/2)\tau + \mathcal{O}(\tau^2)$$

for sufficiently small $\tau > 0$. Thus, the explicit Runge-Kutta scheme with $s = p = 4$ stages is strongly stable w.r.t. $\{\mathbf{L}\}$ and this specific $\mathbf{P}$-norm.

Acknowledgements The authors were supported by the Austrian Science Fund (FWF) via the FWF-funded SFB # F65. The third author acknowledges support from the FWF, # P33010. This work received funding from the European Research Council (ERC) under the European Union's Horizon 2020 research and innovation programme, ERC Advanced Grant NEUROMORPH, no. 101018153.

References

1. Achleitner, F., Arnold, A., Carlen, E.: On multi-dimensional hypocoercive BGK models. Kinet. Relat. Models **11**, 953–1009 (2018)
2. Achleitner, F., Arnold, A., Carlen, E.A.: The hypocoercivity index for the short time behavior of linear time-invariant ODE systems. J. Differ. Equ. **371**, 83–115 (2023)
3. F. Achleitner, A. Arnold, and A. Jüngel. Hypocoercivity for Linear ODEs and Strong Stability for Runge–Kutta Methods. *arXiv preprint* arXiv:2310.19758, (2023)
4. Achleitner, F., Arnold, A., Mehrmann, V.: Hypocoercivity and controllability in linear semi-dissipative Hamiltonian ordinary differential equations and differential-algebraic equations. Z. Angew. Math. Mech. **103**, e202100171, 30 pages (2023)
5. Achleitner, F., Arnold, A., Mehrmann, V.: Hypocoercivity and hypocontractivity concepts for linear dynamical systems. Electronic J. Linear Algebra **39**, 33–61 (2023), https://doi.org/10.13001/ela.2023.7531
6. Bernstein, D.S.: Scalar, Vector, and Matrix Mathematics. Theory, Facts, And Formulas. Revised and expanded edition. Princeton University Press, Princeton, NJ (2018)
7. Conda, S., Gottlieb, S., Grant, Z.J., Shadid, J.N.: Implicit and implicit-explicit strong stability preserving Runge-Kutta methods with high linear order. J. Sci. Comput. **73**, 667–690 (2017)

8. Ferracina, L., Spijker, M.N.: Stepsize restrictions for the total-variation-diminishing property in general Runge-Kutta methods. SIAM J. Numer. Anal. **42**, 1073–1093 (2004)
9. Gottlieb, S., Shu, C.-W., Tadmor, E.: Strong-stability-preserving high-order time discretization methods. SIAM Rev. **43**, 89–112 (2001)
10. Haddad, W.M., Chellaboina, V.: Nonlinear Dynamical Systems and Control. A Lyapunov-Based Approach, Princeton University Press, Princeton, NJ (2008)
11. Hairer, E., Wanner, G.: Solving Ordinary Differential Equations II. Stiff and Differential-Algebraic Problems, 2nd edn. Springer, Berlin (1996)
12. Hinrichsen, D., Pritchard, A.J.: Mathematical Systems Theory. I. Texts in Applied Mathematics, vol. 48. Springer, Berlin (2005)
13. Higueras, I.: On strong stability preserving time discretization methods. J. Sci. Comput. **21**, 193–223 (2004)
14. Kato, T.: Perturbation Theory For Linear Operators. Classics in Mathematics. Springer, Berlin (1995). Reprint of the 1980 edition
15. Kohaupt, L.: Differential calculus for some p-norms of the fundamental matrix with applications. J. Comput. Appl. Math. **135**, 1–21 (2001)
16. Kreiss, H.-O., Scherer, G.: Method of lines for hyperbolic differential equations. SIAM J. Numer. Anal. **29**, 640–646 (1992)
17. Levy, D., Tadmor, E.: From semidiscrete to fully discrete: Stability of Runge-Kutta schemes by the energy method. SIAM Rev. **40**, 40–73 (1998)
18. Mehl, C., Mehrmann, V., Sharma, P.: Stability radii for linear Hamiltonian systems with dissipation under structure-preserving perturbations. SIAM J. Matrix Anal. Appl. **37**, 1625–1654 (2016)
19. Mehl, C., Mehrmann, V., Wojtylak, M.: Linear algebra properties of dissipative Hamiltonian descriptor systems. SIAM J. Matrix Anal. Appl. **39**, 1489–1519 (2018)
20. Mehl, C., Mehrmann, V., Wojtylak, M.: Distance problems for dissipative hamiltonian systems and related matrix polynomials. Linear Algebra Appl. **623**, 335–366 (2021)
21. Ranocha, H.: On strong stability of explicit Runge-Kutta methods for nonlinear semibounded operators. IMA J. Numer. Anal. **41**, 654–682 (2021)
22. Ranocha, H., Ketcheson, D.I.: Energy stability of explicit Runge-Kutta methods for nonautonomous or nonlinear problems. SIAM J. Numer. Anal. **58**, 3382–3405 (2020)
23. Shu, C.-W.: Total-variation-diminishing time discretizations. SIAM J. Sci. Stat. Comput. **9**, 1073–1084 (1988)
24. Shu, C.-W., Osher, S.: Efficient implementation of essentially nonoscillatory shock-capturing schemes. J. Comput. Phys. **77**, 439–471 (1988)
25. Ström, T.: On logarithmic norms. SIAM J. Numer. Anal. **12**, 741–753 (1975)
26. Sun, Z., Shu, C.-W.: Stability of the fourth order Runge-Kutta method for time-dependent partial differential equations. Ann. Math. Sci. Appl. **2**, 255–284 (2017)
27. Sun, Z., Shu, C.-W.: Strong stability of explicit Runge-Kutta time discretizations. SIAM J. Numer. Anal. **57**, 1158–1182 (2019)
28. Sun, Z., Wei, Y., Wu, K.: On energy laws and stability of Runge-Kutta methods for linear seminegative problems. SIAM J. Numer. Anal. **60**, 2448–2481 (2022)
29. Tadmor, E.: From semidiscrete to fully discrete: Stability of Runge–Kutta schemes by the energy method. II. In: Estep, D., Tavener, S. (eds.) Collected Lectures on the Preservation of Stability under Discretization. Proceedings of Applied Mathematics, vol. 109, pp. 25–49. SIAM, Philadelphia (2002)

Radial Laplacian on Rotation Groups

Pierre Degond

Abstract The Laplacian on the rotation group is invariant by conjugation. Hence, it maps class functions to class functions. A maximal torus consists of block diagonal matrices whose blocks are planar rotations. Class functions are determined by their values on this maximal torus. Hence, the Laplacian induces a second order operator on the maximal torus called the radial Laplacian. In this paper, we derive the expression of the radial Laplacian. Then, we use it to find the eigenvalues of the Laplacian, using that characters are class functions whose expressions are given by the Weyl character formula. Although this material is familiar to Lie-group experts, we gather it here in a synthetic and accessible way which may be useful to non experts who need to work with these concepts.

Keywords Representations · Characters · Eigenfunctions · Class functions · Well-posedness · Spectral theory

1 Introduction

Rotation groups are of fundamental physical importance because they encode symmetries of the underpinning space. Recently, they have emerged as important players in the description of large systems of collectively moving self-propelled particles. Such systems are frequently encountered in life science, such as swarming bacteria, migrating cells or schooling fish. While motion of physical particles is determined by their momenta which span a flat space, collectively moving particles offer a greater variety of situations. It often happens that particle dynamics is described in terms of elements of a Lie group or of a homogeneous space (i.e. a quotient of a Lie group by one of its subgroups).

In fond memory of São Carvalho, estimated colleague and friend

---

P. Degond (✉)
Institut de Mathématiques de Toulouse; UMR5219, Université de Toulouse; CNRS, UPS, 31062 Toulouse Cedex 9, France
e-mail: pierre.degond@math.univ-toulouse.fr

E. Carlen et al. (eds.), *From Particle Systems to Partial Differential Equations*,
Springer Proceedings in Mathematics & Statistics 465,
https://doi.org/10.1007/978-3-031-65195-3_2

One of the earliest examples of such models is the Vicsek model [31], where the particles are moving with constant speed. Hence, their dynamics is determined by the direction of their velocity which belongs to $(n-1)$-dimensional sphere $\mathbb{S}^{n-1}$ which can be realized as the homogeneous space $\mathbb{S}^{n-1} \cong \mathrm{SO}_n\mathbb{R}/\mathrm{SO}_{n-1}\mathbb{R}$. More recently, a model in which particle dynamics is determined by an element of the rotation group $\mathrm{SO}_n\mathbb{R}$ has been proposed [10]. Several variants of this model have been explored [6, 7, 11, 12]. One may find other collective dynamics models involving the rotation group in [17], or involving other groups or homogeneous spaces, such as $\mathbb{S}^1 \times \mathbb{R}$ in [13], $\mathbb{S}^n \times \mathbb{S}^1$ in [8] or the unitary group SU_n in [21, 23]. More generally, the interplay between collective dynamics and geometry is an active subject, see e.g. [2, 18].

Here, we focus on the rotation group in relation to the model exposed in [10]. In this paper, a particle system for rigid bodies interacting through body attitude alignment was proposed and a kinetic model was derived when the number of particles is large. Finally, a hydrodynamic model was deduced from the kinetic model assuming that the alignment interaction is strong. At the kinetic level, the interaction operator is of Fokker-Planck type and thus, involves the Laplacian on $\mathrm{SO}_n\mathbb{R}$. The derivation of the hydrodynamic model made in [10] strongly used the peculiar structure of $\mathrm{SO}_3\mathbb{R}$ (Rodrigues formula or the quaternion representation like in [12]) and thus, was restricted to the case of dimension $n = 3$. A further step towards an arbitrary dimension n was made in [6], but in this case, the interaction operator was replaced by a relaxation operator of BGK type, thus avoiding to deal with the Laplacian on $\mathrm{SO}_n\mathbb{R}$. The goal of this paper is to collect the appropriate mathematical framework in view of extending the result of [10] to arbitrary dimensions. Here we focus on the rotation group. We refer to [9] for its use in the model of [10].

Although the material presented here is well-known to Lie-group experts, we were not able to find it anywhere exposed in an accessible fashion to non experts. In particular, we could not find anywhere the expression of the radial Laplacian as exposed in Sect. 3 below. The present work owes a lot to the presentation made in [16] but [16] restricts itself to the unitary group SU_n so we had to reprocess all the details for the rotation groups. By applying the radial Laplacian to the characters of the irreducible representations (which are known thanks to the Weyl character formula), one determines all the eigenvalues of the Laplacian, as explained in Sect. 4. Reference [19] gives a comprehensive introduction to Lie group and Lie algebra representations. Another point of view, focused on compact Lie group, can be found in [30]. Finally, the role of the radial Laplacian in the spectral theory of the Laplacian is treated in [22, 25], however in an abstract way which makes it hardly tractable in practice. Of course, Lie group theory is a very mature subject and there are numerous books and treatises on it. The above is just a small selection. Let us also mention the classical books [5, 33]. Other useful references are (without being exhaustive) [1, 3, 4, 15, 24, 27–29].

The organization of this paper is as follows. Section 2 introduces the basic notations and facts about the rotations groups, the Laplacian operator acting on them and the radial Laplacian. Section 3 is devoted to the computation of the radial Laplacian while Sect. 4 develops the spectral theory of the Laplacian and radial Laplacian.

Finally, a conclusion is drawn in Sect. 5. Two appendices provide proofs of auxilliary results.

2 Laplacian on Rotation Groups

The following presentation is succinct. We refer to e.g. [16] for detail. We denote by $\mathrm{SO}_n\mathbb{R}$, the group of rotations of $\mathbb{R}^n$, i.e. the subset of the space $M_n\mathbb{R}$ of $n \times n$ real matrices defined by

$$\mathrm{SO}_n\mathbb{R} = \{A \in M_n\mathbb{R} \mid AA^T = A^T A = \mathrm{I}, \quad \det A = 1\},$$

where A^T denotes the transpose of A, I is the identity matrix of $\mathbb{R}^n$ and $\det A$ is the determinant of A. The space $M_n\mathbb{R} \approx \mathbb{R}^{2n}$ is a vector space which can be endowed with a Euclidean structure by means of the following inner product

$$A \cdot B = \frac{1}{2}\mathrm{Tr}\{A^T B\}, \quad \forall A,\ B \in M_n\mathbb{R}, \tag{1}$$

where $\mathrm{Tr}A$ denotes the trace of the matrix A. We note that (1) is the usual Frobenius inner product multiplied by the factor $\frac{1}{2}$. This turns out to be the convenient definition of a matrix inner product when dealing with rotations groups. The set $\mathrm{SO}_n\mathbb{R}$ is at the same time a group for matrix multiplication and an embedded manifold into $M_n\mathbb{R}$ where matrix multiplication is C^∞. It is therefore a Lie group. Its tangent space at I is the Lie algebra $\mathfrak{so}_n\mathbb{R}$ of antisymmetric matrices endowed with the Lie bracket $[P, Q] = PQ - QP$, for all $P,\ Q \in \mathfrak{so}_n\mathbb{R}$. Its tangent space at any $A \in \mathrm{SO}_n\mathbb{R}$ is given by $T_A\mathrm{SO}_n\mathbb{R} = A\,\mathfrak{so}_n\mathbb{R} = \mathfrak{so}_n\mathbb{R}\,A$. From now on, we will abbreviate $T_A\mathrm{SO}_n\mathbb{R}$ into simply T_A. As a Lie group, $\mathrm{SO}_n\mathbb{R}$, can be endowed with a unique (up to a multiplicative factor) Haar measure, i.e. a measure which is invariant by left multiplication. Being compact, $\mathrm{SO}_n\mathbb{R}$ is a modular group. Thus, the Haar measure is also right invariant, as well as invariant by group inversion and hence, by transposition. Also, the Haar measure can be normalized, and the resulting normalized Haar measure is unique.

As an embedded manifold in $M_n\mathbb{R}$, $\mathrm{SO}_n\mathbb{R}$ is also endowed with a Riemannian inner-product and metric (for Riemannian geometry, we refer to [14, 20]). For two elements AP, and AQ in T_A, with P and Q in $\mathfrak{so}_n\mathbb{R}$, the Riemannian inner-product is given by $(AP, AQ) = AP \cdot AQ = P \cdot Q$. On oriented manifolds, the Riemannian metric gives rise to a Riemannian volume form ω and thus to a positive measure written $d\omega$ by abuse of notation. As any Lie group is orientable [32, Sect. 4.3, Example (a)], $\mathrm{SO}_n\mathbb{R}$ is thus endowed with a Riemannian volume form ω. On general Lie groups, the Riemannian measure $d\omega$ and the Haar measure may not be multiple of each other. However, they are indeed multiple of each other on $\mathrm{SO}_n\mathbb{R}$ as stated in the following Lemma, whose proof is shown in Appendix 1:

Lemma 1 *Let $d\omega$ and $d\mu$ be the Riemannian measure and the Haar measure on $SO_n\mathbb{R}$ respectively. Then, there exists a constant $C_n > 0$ such that $d\omega = C_n d\mu$.*

On a Riemannian manifold, we can define the gradient and divergence operators. The gradient of a smooth map $\mathrm{SO}_n\mathbb{R} \to \mathbb{R}$, $A \mapsto f(A)$ at A is an element $\nabla f(A)$ of T_A defined by

$$\nabla f(A) \cdot X = df_A(X), \quad \forall X \in T_A, \tag{2}$$

where df_A is the differential of f at A. We recall that df_A is a linear map $T_A \to \mathbb{R}$ and $df_A(X)$ is the image of the tangent vector X by this map. A smooth tangent vector field ϕ is a smooth map $\mathrm{SO}_n\mathbb{R} \to M_n\mathbb{R}$, $A \mapsto \phi(A)$ such that $\phi(A) \in T_A$, for all $A \in \mathrm{SO}_n\mathbb{R}$. For instance, the gradient of a smooth map $\mathrm{SO}_n\mathbb{R} \to \mathbb{R}$ is a smooth vector field. The divergence $\nabla \cdot \phi$ of the tangent vector field ϕ is defined by duality as follows:

$$\int_{\mathrm{SO}_n\mathbb{R}} \nabla \cdot \phi \, \psi \, d\omega = - \int_{\mathrm{SO}_n\mathbb{R}} \phi \cdot \nabla \psi \, d\omega,$$

where ψ is an arbitrary smooth map from $\mathrm{SO}_n\mathbb{R}$ to $\mathbb{R}$. Here, we note the use of the Riemannian measure in the integrals. However, because of Lemma 1, the Riemannian measure $d\omega$ can be replaced by the Haar measure $d\mu$.

The Riemannian Laplacian Δ_M is defined as the composition of the gradient and the divergence, i.e.

$$\Delta_M f = \nabla \cdot (\nabla f).$$

If ϕ is a tangent vector field, and f is a real function on $\mathrm{SO}_n\mathbb{R}$, it is customary in differential geometry [32, Sect. 1.22, formula (6)] to define the map $\phi(f)$, $\mathrm{SO}_n\mathbb{R} \to \mathbb{R}$, by

$$\phi(f)(A) =: df_A\big(\phi(A)\big) = \nabla f(A) \cdot \phi(A). \tag{3}$$

In other words, $\phi(f)(A)$ is the directional derivative of f in the direction of ϕ at A. For i, j in $\{1, \ldots, \mathrm{n}\}$, let $E_{ij} = e_i \otimes e_j \in M_n\mathbb{R}$ where e_i is the i-th canonical basis vector of $\mathbb{R}^n$ and $\otimes$ is the tensor product. In other words, $(E_{ij})_{k\ell} = \delta_{ij}\delta_{k\ell}$, where δ is the Kronecker delta. Then, $\big(\sqrt{2}E_{ij}\big)_{(i,j)\in\{1,\ldots,n\}^2}$ is an orthonormal basis of $M_n\mathbb{R}$ for the inner product (1). Denote by P_{T_A} the orthogonal projection of $M_n\mathbb{R}$ on T_A. We can easily check that

$$P_{T_A} M = A \frac{A^T M - M^T A}{2}, \quad \forall M \in M_n\mathbb{R}, \quad \forall A \in \mathrm{SO}_n\mathbb{R}. \tag{4}$$

Then, by [26, Theorem 3.1.4], we have

$$\Delta_M f(A) = 2 \sum_{(i,j)\in\{1,\ldots,n\}^2} (P_{T_A} E_{ij})^2(f)(A), \quad \forall A \in \mathrm{SO}_n\mathbb{R}, \tag{5}$$

where the square means that the operator $P_{T_A} E_{ij}$ (see (3) for its definition) is acted twice on f. The right-hand side of (5) can be shown to be independent of the choice of the orthonormal basis of $M_n\mathbb{R}$. As is clear from above, this definition of the Laplacian does not depend on the Lie-group structure of $\mathrm{SO}_n\mathbb{R}$.

On Lie groups, there is an alternate concept of the Laplacian (see [16, Sect. 8.2]). We first remind that the matrix exponential maps $\mathfrak{so}_n\mathbb{R}$ into $\mathrm{SO}_n\mathbb{R}$ (and actually onto, but this will not be used). For any $X \in \mathfrak{so}_n\mathbb{R}$, $\rho(X)$ denotes the linear map: $C^\infty(\mathrm{SO}_n\mathbb{R}) \to C^\infty(\mathrm{SO}_n\mathbb{R})$, $f \mapsto \rho(X)f$ such that

$$\big(\rho(X)f\big)(A) = \frac{d}{dt}\big(f(Ae^{tX})\big)|_{t=0}.$$

It is readily seen that

$$(\rho(X)f)(A) = df_A(AX) = \nabla f(A) \cdot AX = AX(f). \tag{6}$$

In particular, ρ is linear with respect to X. The map ρ: $\mathfrak{so}_n\mathbb{R} \to \mathscr{L}\big(C^\infty(\mathrm{SO}_n\mathbb{R})\big)$, $X \mapsto \rho(X)$ is actually a Lie algebra representation, i.e. on top of linearity, it satisfies

$$\rho([X, Y])f = [\rho(X), \rho(Y)]f, \quad \forall X, Y \in \mathfrak{so}_n\mathbb{R}, \quad \forall f \in C^\infty(\mathrm{SO}_n\mathbb{R}),$$

where $\mathscr{L}\big(C^\infty(\mathrm{SO}_n\mathbb{R})\big)$ is the space of linear maps from $C^\infty(\mathrm{SO}_n\mathbb{R})$ into itself. Let us now introduce $F_{ij} = E_{ij} - E_{ji} = e_i \wedge e_j$, where the wedge product of two vectors a and b is their anti-symmetrized tensor product: $a \wedge b = a \otimes b - b \otimes a$. Then, $(F_{ij})_{1 \le i < j \le n}$ is an orthonormal basis of $\mathfrak{so}_n\mathbb{R}$ for the inner product (1). The Lie group Laplacian Δ_G is defined for a function $f \in C^\infty(\mathrm{SO}_n\mathbb{R})$ as follows:

$$(\Delta_G f)(A) = \sum_{1 \le i < j \le n} \frac{d^2}{dt^2} f(Ae^{tF_{ij}})\Big|_{t=0} = \sum_{1 \le i < j \le n} \big(\rho(F_{ij})^2 f\big)(A). \tag{7}$$

It can be shown that the orthonormal basis $(F_{ij})_{1 \le i < j \le n}$ can be replaced by any other orthonormal basis of $\mathfrak{so}_n\mathbb{R}$ without changing the operator Δ_G.

Denote by "Ad" and "ad" the adjoint representations of $\mathrm{SO}_n\mathbb{R}$ and $\mathfrak{so}_n\mathbb{R}$, respectively. Ad maps $\mathrm{SO}_n\mathbb{R}$ into the group $\mathrm{Aut}(\mathfrak{so}_n\mathbb{R})$ of automorphisms of $\mathfrak{so}_n\mathbb{R}$ while ad maps linearly $\mathfrak{so}_n\mathbb{R}$ into the space $\mathscr{L}(\mathfrak{so}_n\mathbb{R})$ of endomorphisms of $\mathfrak{so}_n\mathbb{R}$. They are defined for $A \in \mathrm{SO}_n\mathbb{R}$ and $X, Y \in \mathfrak{so}_n\mathbb{R}$ by

$$\mathrm{Ad}(A)(Y) = AYA^{-1}, \quad \mathrm{ad}(X)(Y) = [X, Y] = XY - YX.$$

Ad is a Lie-group representation of $\mathrm{SO}_n\mathbb{R}$ while ad is a Lie-algebra representation of $\mathfrak{so}_n\mathbb{R}$, i.e. they satisfy

$$\mathrm{Ad}(A)\mathrm{Ad}(B) = \mathrm{Ad}(AB), \quad [\mathrm{ad}(X), \mathrm{ad}(Y)] = \mathrm{ad}([X, Y]),$$

for all $A, B \in \mathrm{SO}_n\mathbb{R}$ and all $X, Y \in \mathfrak{so}_n\mathbb{R}$. The latter formula is a re-expression of the Jacobi identity of the Lie bracket using its skew-commutativity. Now, for $A \in \mathrm{SO}_n\mathbb{R}$, we have $A^{-1} = A^T$, and thus the inner product on $\mathfrak{so}_n\mathbb{R}$ satisfies

$$\mathrm{Ad}(A)(X) \cdot \mathrm{Ad}(A)(Y) = X \cdot Y, \quad \mathrm{ad}(Z)(X) \cdot Y + X \cdot \mathrm{ad}(Z)(Y) = 0. \tag{8}$$

An inner product satisfying (8) is said to be invariant. Furthermore, denote by L and R the left and right regular representations of $\mathrm{SO}_n\mathbb{R}$ on $C^\infty(\mathrm{SO}_n\mathbb{R})$ defined for any $A, B \in \mathrm{SO}_n\mathbb{R}$ and any $f \in C^\infty(\mathrm{SO}_n\mathbb{R})$ by $(L(B)f)(A) = f(B^{-1}A)$ and $(R(B)f)(A) = f(AB)$. We have $L(B)(\Delta_G f) = \Delta_G(L(B)f)$, which means that the Lie-group Laplacian is left-invariant. Because the inner product on $\mathfrak{so}_n\mathbb{R}$ is invariant (see (8)), we have $R(B)(\Delta_G f) = \Delta_G(R(B)f)$, which means that the Lie-group Laplacian is also right-invariant.

It turns out that the two concepts of Laplacians coincide on $\mathrm{SO}_n\mathbb{R}$ as stated in the following lemma, proved in Appendix 2.

Lemma 2 *We have* $\Delta_M = \Delta_G$.

As a consequence, we can drop the subscript M or G of Δ.

Two elements A and B of $\mathrm{SO}_n\mathbb{R}$ are said to be conjugate if and only if there exists $g \in \mathrm{SO}_n\mathbb{R}$ such that $B = gAg^{-1}$. Conjugation is an equivalence relation and we write $A \sim B$ whenever A and B are conjugate. A class function is a function on $\mathrm{SO}_n\mathbb{R}$ which is invariant by conjugation, i.e. which satifies:

$$f(gAg^{-1}) = f(A), \quad \forall A,\ g \in \mathrm{SO}_n\mathbb{R}.$$

A class function thus only depends on the conjugation classes. On $\mathrm{SO}_n\mathbb{R}$ conjugation classes are well-known thanks to classical results on matrix reduction. Let $\theta \in \mathbb{R}$. We define the planar rotation matrix R_θ by

$$R_\theta = \begin{pmatrix} \cos\theta & -\sin\theta \\ \sin\theta & \cos\theta \end{pmatrix}.$$

For $(\theta_1, \ldots, \theta_p) \in \mathbb{R}^p$, $A_{\theta_1,\ldots,\theta_p}$ denotes the following matrices defined by blocks:

– in the case $n = 2p$, $p \geq 2$,

$$A_{\theta_1,\ldots,\theta_p} = \begin{pmatrix} R_{\theta_1} & & & 0 \\ & R_{\theta_2} & & \\ & & \ddots & \\ 0 & & & R_{\theta_p} \end{pmatrix} \in \mathrm{SO}_{2p}\mathbb{R}, \tag{9}$$

– in the case $n = 2p + 1$, $p \geq 1$,

$$A_{\theta_1,\ldots,\theta_p} = \begin{pmatrix} R_{\theta_1} & & & \mathbf{0} & 0 \\ & R_{\theta_2} & & & \vdots \\ & & \ddots & & \vdots \\ \mathbf{0} & & & R_{\theta_p} & 0 \\ 0 & \ldots & \ldots & 0 & 1 \end{pmatrix} \in \mathrm{SO}_{2p+1}\mathbb{R}. \tag{10}$$

We write $\Theta = (\theta_1, \theta_2, \ldots, \theta_p) \in \mathbb{R}^p$. We define the subset $\mathbb{T}$ of $\mathrm{SO}_{2p}\mathbb{R}$ or $\mathrm{SO}_{2p+1}\mathbb{R}$ by

$$\mathbb{T} = \{A_\Theta \mid \Theta \in \mathbb{R}^p\}.$$

$\mathbb{T}$ is an abelian subgroup of $\mathrm{SO}_{2p}\mathbb{R}$ or $\mathrm{SO}_{2p+1}\mathbb{R}$. Consider the map $\psi \colon \Theta \to A_\Theta$. Since we have $A_\Theta A_{\Theta'} = A_{\Theta+\Theta'}$, $A_0 = \mathrm{I}$, ψ is a surjective group morphism $\mathbb{R}^p \to \mathbb{T}$ whose kernel is $(2\pi\mathbb{Z})^p$. Thus, it induces a group isomorphism $\tilde{\psi}$, $\mathcal{T} =: (\mathbb{R}/(2\pi\mathbb{Z}))^p \to \mathbb{T}$. We note that $\mathcal{T} \cong (\mathbb{S}^1)^p$ is a p-dimensional torus. It can be shown that $\mathbb{T}$ is maximal among abelian subgoups of $\mathrm{SO}_n\mathbb{R}$. For that reason, it is called a maximal torus.

Then, by matrix reduction theory, any element of $\mathrm{SO}_{2p}\mathbb{R}$ or $\mathrm{SO}_{2p+1}\mathbb{R}$ is conjugate to an element of $\mathbb{T}$. Let f be a C^∞ class function on $\mathrm{SO}_{2p}\mathbb{R}$ or $\mathrm{SO}_{2p+1}\mathbb{R}$. Defining $\varphi_f(\Theta) = f(A_\Theta)$, we have

$$f(A) = \varphi_f(\Theta), \quad \forall A \text{ such that } A \sim A_\Theta, \tag{11}$$

and $\varphi_f \in C^\infty(\mathcal{T})$. Thus, the knowledge of a class function only requires the knowledge of a function of $\Theta \in \mathcal{T}$. This involves a dramatic reduction of complexity.

However, elements of $\mathbb{T}$ which are conjugate to a given A are not unique. The list of such elements differs in the even and odd dimensional cases:

– In the case of $\mathrm{SO}_{2p}\mathbb{R}$, $(\theta_1, \ldots, \theta_p)$ can undergo any permutations of $\{1, \ldots, p\}$. Permutation of θ_i and θ_j involves conjugation by the permutation matrix which exchanges e_{2i-1} and e_{2j-1} on the one hand and e_{2i} and e_{2j} on the other hand. Since the corresponding permutation is the product of two transpositions, it has signature $+1$ and so, the associated permutation matrix is an element of $\mathrm{SO}_{2p}\mathbb{R}$. But one can also exchange e_{2i-1} and e_{2i} on the one hand, and e_{2j-1} and e_{2j} on the other hand. Again the corresponding permutation matrix is an element of $\mathrm{SO}_{2p}\mathbb{R}$ and conjugating A_Θ by it amounts to making the change $(\ldots, \theta_i, \ldots, \theta_j, \ldots) \to (\ldots, -\theta_i, \ldots, -\theta_j, \ldots)$. Note that we couldn't just conjugate with the permutation matrix associated with the single transposition which exchanges e_{2i-1} and e_{2i} because the associated matrix is of determinant -1 and does not belong to $\mathrm{SO}_{2p}\mathbb{R}$. Thus, only changes of an even number of signs in the θ_i's is allowed. We denote by $\mathcal{G}$ the group of bijections $\mathcal{T} \to \mathcal{T}$ generated by the transformations $(\ldots, \theta_i, \ldots, \theta_j, \ldots) \to (\ldots, \theta_j, \ldots, \theta_i, \ldots)$ and

$(\ldots, \theta_i, \ldots, \theta_j, \ldots) \to (\ldots, -\theta_i, \ldots, -\theta_j, \ldots)$ where (i, j) ranges over all pairs of integers in $\{1, \ldots, p\}$ such that $i \neq j$.

- In the case of $\mathrm{SO}_{2p+1}\mathbb{R}$, $(\theta_1, \ldots, \theta_p)$ can again undergo any permutations of $\{1, \ldots, p\}$ for the same reason as in the even-dimensional case. But one can also conjugate by a matrix which exchanges e_{2i-1} and e_{2i} and bears a -1 in the last diagonal element (which was not possible in the odd-dimensional case). This matrix has determinant unity and is thus an element of $\mathrm{SO}_{2p+1}\mathbb{R}$. Conjugating A_Θ by it amounts to making the change $(\ldots, \theta_i, \ldots) \to (\ldots, -\theta_i, \ldots)$. Thus, changes of an arbitrary number of signs of the θ_i's are allowed in this case. In this case, $\mathscr{G}$ is generated by the transformations $(\ldots, \theta_i, \ldots, \theta_j, \ldots) \to (\ldots, \theta_j, \ldots, \theta_i, \ldots)$ where (i, j) ranges over all pairs of integers in $\{1, \ldots, p\}$ such that $i \neq j$ and by $(\ldots, \theta_i, \ldots) \to (\ldots, -\theta_i, \ldots)$, where i ranges over all integers in $\{1, \ldots, p\}$.

In both the even and odd dimensional cases, all conjugations mapping two elements of $\mathbb{T}$ are generated by the conjugations described above. To see that we introduce the set of all the elements of $\mathrm{SO}_n\mathbb{R}$ which conjugate an element of $\mathbb{T}$ into an element of $\mathbb{T}$. This set is a group called the normalizer of $\mathbb{T}$ and is defined by

$$N(\mathbb{T}) = \big\{g \in \mathrm{SO}_n\mathbb{R} \mid g\mathbb{T}g^{-1} = \mathbb{T}\big\}.$$

Let us introduce the group homomorphism Φ: $N(\mathbb{T}) \to \mathrm{Aut}(\mathbb{T})$, $g \mapsto \Phi_g$ (where $\mathrm{Aut}(\mathbb{T})$ denotes the group of group automorphisms of $\mathbb{T}$), such that $\Phi_g(h) = ghg^{-1}$, $\forall h \in \mathbb{T}$. This homomorphism has kernel equal to $\mathbb{T}$ (that $\mathbb{T} \subset \ker \Phi$ is obvious. The converse is due to the fact that $\mathbb{T}$ is a maximal abelian subgroup of $\mathrm{SO}_n\mathbb{R}$). Thus, Φ induces a group isomorphism $\tilde{\Phi}$: $N(\mathbb{T})/\mathbb{T} \to \mathrm{im}(\Phi) \subset \mathrm{Aut}(\mathbb{T})$. Now, define the isomorphism Ψ: $\mathrm{Aut}(\mathbb{T}) \to \mathrm{Aut}(\mathscr{T})$, $\varphi \mapsto \tilde{\psi}^{-1} \circ \varphi \circ \tilde{\psi}$, $\forall \varphi \in \mathrm{Aut}(\mathbb{T})$, where $\tilde{\psi}$: $\mathscr{T} \to \mathbb{T}$ was defined above. Then, $\Psi \circ \tilde{\Phi}$ is an isomorphism between $N(\mathbb{T})/\mathbb{T}$ and the group of transformations of $\mathscr{T}$ that are induced by conjugations. It turns out that the group $N(\mathbb{T})/\mathbb{T}$ is well-known to be isomorphic to the Weyl group $\mathfrak{W}$ [19, Claim 26.15]. $\mathfrak{W}$ is a finite subsgroup of the orthogonal group of the space $\mathbb{R}^p$. If $\mathbb{R}^p$ is endowed with the canonical basis $(e_i)_{i=1}^p$, then $\mathfrak{W}$ is generated by the orthogonal symmetries in the hyperplanes $\{e_i \pm e_j\}^\perp$, with $i, j \in \{1, \ldots, p\}$, $i \neq j$ in the case $n = 2p$ and by the orthogonal symmetries in the hyperplanes $\{e_i \pm e_j\}^\perp$ and $\{e_i\}^\perp$ in the case $n = 2p + 1$ [19, p. 271]. But it is readily checked that the group $\mathscr{G}$ of transformations of $\mathscr{T}$ defined above is also generated by the same transformations. Hence, $\mathscr{G} = \mathfrak{W}$ showing the claim.

It results that if $f \in C^\infty$ is a class function, the function $\varphi_f \in C^\infty(\mathscr{T})$ associated with it through (11) must be invariant by the action of $\mathfrak{W}$. Hence, φ_f is a function defined on the quotient $\tilde{\mathscr{T}} =: \mathscr{T}/\mathfrak{W}$.

Now, because Δ is left and right invariant, it is invariant by conjugation (indeed, conjugation by g is the composition $L(g^{-1})R(g^{-1})$). It results that if f is a class function, Δf is a class function. For any $\varphi \in C^\infty(\tilde{\mathscr{T}})$, there exists a unique function $f \in C^\infty(\mathrm{SO}_n\mathbb{R})$ such that $\varphi = \varphi_f$. Indeed, f is given by $f(A) = \varphi(\Theta)$, for any Θ such that $A \sim A_\Theta$. Then, we can define a linear operator L: $C^\infty(\tilde{\mathscr{T}}) \to C^\infty(\tilde{\mathscr{T}})$ by

$$L\varphi_f = \varphi_{\Delta f}, \quad \forall f \in C^\infty(\mathbb{T}).$$

The operator L is called the radial Laplacian. The goal of this paper is to find an explicit expression of L and to state some of its properties.

3 Radial Laplacian

3.1 Expression

The expression of the radial Laplacian is given by the following:

Theorem 1 *The radial Laplacian on the rotation groups is given by the following expressions:*
(i) Case of $\mathrm{SO}_{2p}\mathbb{R}$ *for* $p \geq 2$*:*

$$L = \sum_{j=1}^{p} \frac{\partial^2}{\partial\theta_j^2} + \sum_{1\leq j<k\leq p} \frac{2}{\cos\theta_k - \cos\theta_j} \Big(\big(\sin\theta_j \frac{\partial}{\partial\theta_j} - \sin\theta_k \frac{\partial}{\partial\theta_k}\big)\Big); \tag{1}$$

(ii) Case of $\mathrm{SO}_{2p+1}\mathbb{R}$ *for* $p \geq 1$*:*

$$L = \sum_{j=1}^{p} \Big(\frac{\partial^2}{\partial\theta_j^2} + \frac{\sin\theta_j}{1-\cos\theta_j}\frac{\partial}{\partial\theta_j}\Big) + \sum_{1\leq j<k\leq p} \frac{2}{\cos\theta_k - \cos\theta_j} \Big(\big(\sin\theta_j \frac{\partial}{\partial\theta_j} - \sin\theta_k \frac{\partial}{\partial\theta_k}\big)\Big). \tag{2}$$

We note that in the case $p = 1$*, the second sum simply disappears in* (2)*. We leave aside the case* $\mathrm{SO}_2\mathbb{R}$ *which is trivial, the group being abelian.*

To prove this theorem, we will first review the cases of $\mathrm{SO}_3\mathbb{R}$ and $\mathrm{SO}_4\mathbb{R}$ and then, generalize the analysis to the general cases $\mathrm{SO}_{2p}\mathbb{R}$ and $\mathrm{SO}_{2p+1}\mathbb{R}$. We will adopt the method presented in [16] (cf. Proof of Proposition 8.3.3 and Sect. 10.2). This method relies on the following lemma:

Lemma 3 *If* f *is a class function on* $SO_n\mathbb{R}$*, for all* $X \in \mathfrak{so}_n\mathbb{R}$ *and all* $A \in SO_n\mathbb{R}$*, we have*

$$\rho\big(Ad(A^{-1})X - X\big)^2 f(A) - \rho\big([Ad(A^{-1})X, X]\big) f(A) = 0. \tag{3}$$

Proof If f is a class function, we have $f(A) = f(e^{tX}Ae^{-tX})$, for all $A \in \mathrm{SO}_n\mathbb{R}$ and all $X \in \mathfrak{so}_n\mathbb{R}$. This formula can be recast into $f(A) = f(Ae^{t\mathrm{Ad}(A^{-1})X}e^{-tX})$. Taking the second derivative of this formula with respect to t at $t = 0$ and using that

$$\frac{d^2}{dt^2} f(Ae^{tX}e^{tY})\big|_{t=0} = \rho(X+Y)^2 f(A) + \rho\big([X,Y]\big) f(A), \quad \forall X,\, Y \in \mathfrak{so}_n\mathbb{R}, \quad (4)$$

we get (3). Eq. (4) is proved in [16, Sect. 8.2 (d)]. ■

We will use the following formula, whose proof is straightforwad:

$$[F_{ij}, F_{k\ell}] = \big(\delta_{jk}F_{i\ell} + \delta_{i\ell}F_{jk} - \delta_{ik}F_{j\ell} - \delta_{j\ell}F_{ik}\big). \quad (5)$$

We will also use the notation $\{A, B\}$ for the anti-commutator of two operators A, B, i.e.

$$\{A, B\} = AB + BA.$$

3.2 Proof of Theorem 1: Case of $SO_3\mathbb{R}$

This is the case $SO_{2p+1}\mathbb{R}$ with $p = 1$. Thus, the matrix (10) has a single block R_θ. So, the vector Θ has a single coordinate, θ and we use the notation A_θ for A_Θ. Let f be a C^∞ class function on $SO_3\mathbb{R}$ and $\varphi_f = \varphi_f(\theta)$ the corresponding function of θ.

Define

$$\begin{aligned} G^+ &= \frac{1}{\sqrt{2}}(F_{13} + F_{23}) = \frac{1}{\sqrt{2}}(e_1 \otimes e_3 + e_2 \otimes e_3 - e_3 \otimes e_1 - e_3 \otimes e_2), \\ G^- &= \frac{1}{\sqrt{2}}(F_{13} - F_{23}) = \frac{1}{\sqrt{2}}(e_1 \otimes e_3 - e_2 \otimes e_3 - e_3 \otimes e_1 + e_3 \otimes e_2). \end{aligned}$$

Then, (F_{12}, G^+, G^-) is still an orthonormal basis of $\mathfrak{so}_3\mathbb{R}$. We choose this basis to express the Laplacian of a C^∞ function f on $SO_3\mathbb{R}$:

$$\Delta f(A) = \big(\rho(F_{12})^2 f\big)(A) + \big(\rho(G^+)^2 f\big)(A) + \big(\rho(G^-)^2 f\big)(A). \quad (6)$$

We now use (3) with $A = A_\theta$ and $X = G^+$ and $X = G^-$ successively. We have

$$\mathrm{Ad}(A_{-\theta})G^+ = \cos\theta\, G^+ + \sin\theta\, G^-, \quad \mathrm{Ad}(A_{-\theta})G^- = -\sin\theta\, G^+ + \cos\theta\, G^-.$$

Then, using the linearity of the map $X \mapsto (\rho(X)f)(A)$ for any $A \in SO_2\mathbb{R}$, we get:

$$\begin{aligned} \rho\big(\mathrm{Ad}(A_{-\theta})G^+ - G^+\big)^2 f(A_\theta) &= \Big(\big((\cos\theta - 1)^2\, \rho(G^+)^2 + \sin^2\theta\, \rho(G^-)^2 \\ &\qquad + (\cos\theta - 1)\, \sin\theta\, \{\rho(G^+), \rho(G^-)\}\big) f\Big)(A_\theta), \\ \rho\big(\mathrm{Ad}(A_{-\theta})G^- - G^-\big)^2 f(A_\theta) &= \Big(\big((\cos\theta - 1)^2\, \rho(G^-)^2 + \sin^2\theta\, \rho(G^+)^2 \\ &\qquad - (\cos\theta - 1)\, \sin\theta\, \{\rho(G^+), \rho(G^-)\}\big) f\Big)(A_\theta). \end{aligned}$$

Furthermore, using (5), we get

$$[\mathrm{Ad}(A_{-\theta})G^+, G^+] = [\mathrm{Ad}(A_{-\theta})G^-, G^-] = \sin\theta\,[G^-, G^+]$$
$$= \sin\theta\,[F_{13}, F_{23}] = -\sin\theta\, F_{12}.$$

Thus, adding the equations (3) corresponding to G^+ and G^- together leads to

$$2(1-\cos\theta)\Big(\big(\rho(G^+)^2 + \rho(G^-)^2\big)f\Big)(A_\theta) = -2\,\sin\theta\,\big(\rho(F_{12})f\big)(A_\theta). \quad (7)$$

Thus, with (6), we get

$$(\Delta f)(A_\theta) = \Big(\big(\rho(F_{12})^2 - \frac{\sin\theta}{1-\cos\theta}\,\rho(F_{12})\big)f\Big)(A_\theta). \quad (8)$$

Now, we remark that $e^{tF_{12}} = A_{-t}$. Thus, by the definition of ρ:

$$\big(\rho(F_{12})f\big)(A_\theta) = \frac{d}{dt}f(A_\theta e^{tF_{12}})\big|_{t=0} = \frac{d}{dt}f(A_{\theta-t})\big|_{t=0} \quad (9)$$
$$= \frac{d}{dt}\varphi_f(\theta - t)\big|_{t=0} = -\frac{d}{d\theta}\varphi_f(\theta). \quad (10)$$

Inserting this expresion into (8), we get the expression of the radial Laplacian on $\mathrm{SO}_3\mathbb{R}$:

$$L = \big(\frac{d^2}{d\theta^2} + \frac{\sin\theta}{1-\cos\theta}\,\frac{d}{d\theta}\big),$$

which corresponds to (2) in the case $p = 1$. ∎

3.3 Proof of Theorem 1: Case of $SO_4\mathbb{R}$

This is the case $\mathrm{SO}_{2p}\mathbb{R}$ with $p = 2$. Thus, the matrix (9) has two blocks R_{θ_1} and R_{θ_2}. So, $\Theta = (\theta_1, \theta_2)$. Let f be a C^∞ class function on $\mathrm{SO}_4\mathbb{R}$ and $\varphi_f = \varphi_f(\theta_1, \theta_2)$ the corresponding function of (θ_1, θ_2).

We now define

$$H^+ = \frac{1}{\sqrt{2}}(F_{13} + F_{24}), \quad H^- = \frac{1}{\sqrt{2}}(F_{13} - F_{24}),$$
$$K^+ = \frac{1}{\sqrt{2}}(F_{14} + F_{23}), \quad K^- = \frac{1}{\sqrt{2}}(F_{14} - F_{23})$$

Then, $(F_{12}, F_{34}, H^+, H^-, K^+, K^-)$ is still an orthonormal basis of $\mathfrak{so}_4\mathbb{R}$. We choose this basis to express the Laplacian of a C^∞ function f on $\mathrm{SO}_4\mathbb{R}$:

$$\Delta f(A) = \Big(\big(\rho(F_{12})^2 + \rho(F_{34})^2 + \rho(H^+)^2 + \rho(H^-)^2 + \rho(K^+)^2 + \rho(K^-)^2 \big) f \Big)(A). \quad (11)$$

We now use (3) with $A = A_{\theta_1,\theta_2}$ and $X = H^+,\ H^-,\ K^+,\ K^-$ successively. We have

$$\begin{aligned}
\mathrm{Ad}(A_{-\theta_1,-\theta_2})H^+ &= (c_1c_2 + s_1s_2)H^+ - (c_1s_2 - s_1c_2)K^-, \\
\mathrm{Ad}(A_{-\theta_1,-\theta_2})H^- &= (c_1c_2 - s_1s_2)H^- - (c_1s_2 + s_1c_2)K^+, \\
\mathrm{Ad}(A_{-\theta_1,-\theta_2})K^+ &= (c_1s_2 + s_1c_2)H^- + (c_1c_2 - s_1s_2)K^+, \\
\mathrm{Ad}(A_{-\theta_1,-\theta_2})K^- &= (c_1s_2 - s_1c_2)H^+ + (c_1c_2 + s_1s_2)K^-,
\end{aligned}$$

with $c_i = \cos\theta_i$, $s_i = \sin\theta_i$ for $i = 1,\ 2$. Then, we get:

$$\begin{aligned}
\rho\big(\mathrm{Ad}(A_{-\theta_1,-\theta_2})H^+ - H^+\big)^2 f(A_{\theta_1,\theta_2}) &= \Big(\big((c_1c_2 + s_1s_2 - 1)^2 \, \rho(H^+)^2 \\
&\quad + (c_1s_2 - s_1c_2)^2 \, \rho(K^-)^2 \\
&\quad - (c_1c_2 + s_1s_2 - 1)\,(c_1s_2 - s_1c_2) \, \{\rho(H^+), \rho(K^-)\} \big) f \Big)(A_{\theta_1,\theta_2}), \\
\rho\big(\mathrm{Ad}(A_{-\theta_1,-\theta_2})H^- - H^-\big)^2 f(A_{\theta_1,\theta_2}) &= \Big(\big((c_1c_2 - s_1s_2 - 1)^2 \, \rho(H^-)^2 \\
&\quad + (c_1s_2 + s_1c_2)^2 \, \rho(K^+)^2 \\
&\quad - (c_1c_2 - s_1s_2 - 1)\,(c_1s_2 + s_1c_2) \, \{\rho(H^-), \rho(K^+)\} \big) f \Big)(A_{\theta_1,\theta_2}), \\
\rho\big(\mathrm{Ad}(A_{-\theta_1,-\theta_2})K^+ - K^+\big)^2 f(A_{\theta_1,\theta_2}) &= \Big(\big((c_1s_2 + s_1c_2)^2 \, \rho(H^-)^2 \\
&\quad + (c_1c_2 - s_1s_2 - 1)^2 \, \rho(K^+)^2 \\
&\quad + (c_1s_2 + s_1c_2)\,(c_1c_2 - s_1s_2 - 1) \, \{\rho(H^-), \rho(K^+)\} \big) f \Big)(A_{\theta_1,\theta_2}), \\
\rho\big(\mathrm{Ad}(A_{-\theta_1,-\theta_2})K^- - K^-\big)^2 f(A_{\theta_1,\theta_2}) &= \Big(\big((c_1s_2 - s_1c_2)^2 \, \rho(H^+)^2 \\
&\quad + (c_1c_2 + s_1s_2 - 1)^2 \, \rho(K^-)^2 \\
&\quad + (c_1s_2 - s_1c_2)\,(c_1c_2 + s_1s_2 - 1) \, \{\rho(H^+), \rho(K^-)\} \big) f \Big)(A_{\theta_1,\theta_2}).
\end{aligned}$$

Furthermore, using (5), we get

$$\begin{aligned}
[\mathrm{Ad}(A_{-\theta_1,-\theta_2})H^+, H^+] &= [\mathrm{Ad}(A_{-\theta_1,-\theta_2})K^-, K^-] \\
&= -(c_1s_2 - s_1c_2)[K^-, H^+] = (c_1s_2 - s_1c_2)(F_{12} - F_{34}), \\
[\mathrm{Ad}(A_{-\theta_1,-\theta_2})H^-, H^-] &= [\mathrm{Ad}(A_{-\theta_1,-\theta_2})K^+, K^+] \\
&= -(c_1s_2 + s_1c_2)[K^+, H^-] = -(c_1s_2 + s_1c_2)(F_{12} + F_{34}),
\end{aligned}$$

Thus, adding the equations (3) corresponding to H^+ and K^- (resp. H^- and K^+) together leads to

$$\Big(\big(\rho(H^+)^2+\rho(K^-)^2\big)f\Big)(A_{\theta_1,\theta_2}) = \frac{\sin(\theta_2-\theta_1)}{1-\cos(\theta_2-\theta_1)}\Big(\big(\rho(F_{12})-\rho(F_{34})\big)f\Big)(A_{\theta_1,\theta_2}),$$

$$\Big(\big(\rho(H^-)^2+\rho(K^+)^2\big)f\Big)(A_{\theta_1,\theta_2}) = -\frac{\sin(\theta_1+\theta_2)}{1-\cos(\theta_1+\theta_2)}\Big(\big(\rho(F_{12})+\rho(F_{34})\big)f\Big)(A_{\theta_1,\theta_2}),$$

Thus, with (11), we get

$$(\Delta f)(A_{\theta_1,\theta_2}) = \Big(\Big(\rho(F_{12})^2+\rho(F_{34})^2+\frac{\sin(\theta_2-\theta_1)}{1-\cos(\theta_2-\theta_1)}\big(\rho(F_{12})-\rho(F_{34})\big) - \frac{\sin(\theta_1+\theta_2)}{1-\cos(\theta_1+\theta_2)}\big(\rho(F_{12})+\rho(F_{34})\big)\Big)f\Big)(A_{\theta_1,\theta_2}). \quad (12)$$

Now, similarly to (10), we have

$$\big(\rho(F_{12})f\big)(A_{\theta_1,\theta_2}) = -\frac{\partial\varphi_f}{\partial\theta_1}(\theta_1,\theta_2), \quad \big(\rho(F_{34})f\big)(A_{\theta_1,\theta_2}) = -\frac{\partial\varphi_f}{\partial\theta_2}(\theta_1,\theta_2).$$

Inserting it into (12) leads to

$$L = \frac{\partial^2}{\partial\theta_1^2}+\frac{\partial^2}{\partial\theta_2^2}+\frac{\sin(\theta_1-\theta_2)}{1-\cos(\theta_1-\theta_2)}\Big(\frac{\partial}{\partial\theta_1}-\frac{\partial}{\partial\theta_2}\Big) + \frac{\sin(\theta_1+\theta_2)}{1-\cos(\theta_1+\theta_2)}\Big(\frac{\partial}{\partial\theta_1}+\frac{\partial}{\partial\theta_2}\Big).$$

This can be recast into

$$L = \frac{\partial^2}{\partial\theta_1^2}+\frac{\partial^2}{\partial\theta_2^2}+\frac{2}{\cos\theta_2-\cos\theta_1}\Big(\sin\theta_1\frac{\partial}{\partial\theta_1}-\sin\theta_2\frac{\partial}{\partial\theta_2}\Big),$$

which corresponds to (1) in the case $p=1$. ■

3.4 Proof of Theorem 1: Case of $SO_{2p}\mathbb{R}$, $p \geq 1$

For $1 \leq j < k \leq p$, we define

$$H_{jk}^{+} = \frac{1}{\sqrt{2}}(F_{2j-1\,2k-1} + F_{2j\,2k}), \quad H_{jk}^{-} = \frac{1}{\sqrt{2}}(F_{2j-1\,2k-1} - F_{2j\,2k}), \tag{13}$$

$$K_{jk}^{+} = \frac{1}{\sqrt{2}}(F_{2j-1\,2k} + F_{2j\,2k-1}), \quad K_{jk}^{-} = \frac{1}{\sqrt{2}}(F_{2j-1\,2k} + F_{2j\,2k-1}). \tag{14}$$

Then, $\big((F_{2j-1\,2j})_{j=1,\ldots,p}, (H_{jk}^{+}, H_{jk}^{-}, K_{jk}^{+}, K_{jk}^{-})_{1\leq j<k\leq p}\big)$ is still an orthonormal basis of $\mathfrak{so}_{2p}\mathbb{R}$. We choose this basis to express the Laplacian of a C^∞ function f on $\mathrm{SO}_{2p}\mathbb{R}$:

$$\Delta f(A) = \Big(\Big(\sum_{j=1}^{p} \rho(F_{j\,j+1})^2 + \sum_{1\leq j<k\leq p} \big(\rho(H_{jk}^{+})^2 + \rho(H_{jk}^{-})^2 + \rho(K_{jk}^{+})^2 + \rho(K_{jk}^{-})^2\big)\Big) f\Big)(A). \tag{15}$$

We note that the computations of $\mathrm{Ad}(A_{-\Theta})H_{jk}^{\pm}$ and $\mathrm{Ad}(A_{-\Theta})K_{jk}^{\pm}$ only involve the four 2×2 matrix subblocks corresponding to line and column indices belonging to $\{2j-1, 2j\} \times \{2j-1, 2j\}$, $\{2j-1, 2j\} \times \{2k-1, 2k\}$, $\{2k-1, 2k\} \times \{2j-1, 2j\}$ and $\{2k-1, 2k\} \times \{2k-1, 2k\}$. Thus, the computation reduces to computations on 4×4 matrices that are identical to what was done in the case of $\mathrm{SO}_4\mathbb{R}$. Thus, we can directly write that for any $1 \leq j < k \leq p$, we have

$$\Big(\big(\rho(H_{jk}^{+})^2 + \rho(K_{jk}^{-})^2 + \rho(H_{jk}^{-})^2 + \rho(K_{jk}^{+})^2\big) f\Big)(A_\Theta)$$
$$= -\frac{2}{\cos\theta_k - \cos\theta_j} \Big(\big(\sin\theta_j\, \rho(F_{2j-1\,2j}) - \sin\theta_k\, \rho(F_{2k-1\,2k})\big) f\Big)(A_\Theta). \tag{16}$$

Now, similarly to (10), for $j \in \{1, \ldots, p\}$, we have

$$\big(\rho(F_{2j-1\,2j}) f\big)(A_\Theta) = -\frac{\partial \varphi_f}{\partial \theta_j}(\Theta).$$

Inserting these into (15), we get (1). ∎

3.5 Proof of Theorem 1: Case of $SO_{2p+1}\mathbb{R}$, $p \geq 1$

For $1 \leq j < k \leq p$, we define $H_{jk}^{\pm}$ and $K_{jk}^{\pm}$ in a similar way as in (13), (14). For $j \in \{1, \ldots, p\}$, we define

$$G_j^+ = \frac{1}{\sqrt{2}}(F_{2j-1\,2p+1} + F_{2j\,2p+1}), \quad G_j^- = \frac{1}{\sqrt{2}}(F_{2j-1\,2p+1} - F_{2j\,2p+1}).$$

Then, $\big((F_{2j-1\,2j}, G_j^+, G_j^-)_{j=1,\dots,p}, (H_{jk}^+, H_{jk}^-, K_{jk}^+, K_{jk}^-)_{1\le j<k\le p}\big)$ is still an orthonormal basis of $\mathfrak{so}_{2p+1}\mathbb{R}$. We choose this basis to express the Laplacian of a C^∞ function f on $\mathrm{SO}_{2p+1}\mathbb{R}$:

$$\Delta f(A) = \Big(\Big(\sum_{j=1}^{p}\big(\rho(F_{j\,j+1})^2 + \rho(G_j^+)^2 + \rho(G_j^-)^2\big) + \sum_{1\le j<k\le p}\big(\rho(H_{jk}^+)^2 + \rho(H_{jk}^-)^2 + \rho(K_{jk}^+)^2 + \rho(K_{jk}^-)^2\big)\Big)f\Big)(A). \tag{17}$$

The computations involving $H_{jk}^\pm$ and $K_{jk}^\pm$ are identical to those of the preceding section. Those involving $G_j^\pm$ are identical to those done in Section 3.2 for $\mathrm{SO}_3\mathbb{R}$. Thus, we can collect (16) for $1 \le j < k \le p$ and (7) with $G^\pm$ replaced by $G_j^\pm$, θ by θ_j and F_{12} by $F_{2j-1\,2j}$, for all $j \in \{1, \dots, p\}$, and get (2). ■

4 Some Properties of the Radial Laplacian

4.1 Alternate Expressions

As a preliminary to this section, we state the Weyl integration formula [30, Theorems IX.9.4 & IX.9.5]. We introduce the following functions:

$$\Pi_{2p}(\Theta) = \prod_{1\le j<k\le p}\big(\cos\theta_j - \cos\theta_k\big), \quad \text{for } p \ge 2, \tag{1}$$

$$\Pi_{2p+1}(\Theta) = \prod_{1\le j<k\le p}\big(\cos\theta_j - \cos\theta_k\big)\prod_{j=1}^{p}\sin\frac{\theta_j}{2}, \quad \text{for } p \ge 1. \tag{2}$$

We define

$$u_{2p} = \frac{2^{(p-1)^2}}{p!}\Pi_{2p}^2, \quad \text{for } p \ge 2; \qquad u_{2p+1} = \frac{2^{p^2}}{p!}\Pi_{2p+1}^2, \quad \text{for } p \ge 1. \tag{3}$$

Then, we have

Proposition 1 (Weyl integration formula)
Let $n \in \mathbb{N}$, $n \ge 3$. Let $p \in \mathbb{N}$ such that $n = 2p$ or $n = 2p + 1$. For any integrable class function f on $\mathrm{SO}_n\mathbb{R}$, we have

$$\int_{\mathrm{SO}_n\mathbb{R}} f(A)\,dA = \frac{1}{(2\pi)^p}\int_{[0,2\pi]^p} f(A_\Theta)\,u_n(\Theta)\,d\Theta, \tag{4}$$

with $d\Theta = d\theta_1 \dots d\theta_p$.

A vector field $\mathscr{X}\colon \mathscr{T} \to \mathbb{R}^p$ is said to be invariant by the Weyl group $\mathfrak{W}$ if and only if we have $X(W\Theta) = WX(\theta)$, $\forall \Theta \in \mathscr{T}$, $\forall W \in \mathfrak{W}$. For a smooth function φ: $\tilde{\mathscr{T}} \to \mathbb{R}^p$, $\nabla_\Theta \varphi$ denotes the vector field $\mathscr{T} \to \mathbb{R}^p$ of coordinates $(\frac{\partial \varphi}{\partial \theta_i})_{i=1}^p$. We note that this vector field is invariant by $\mathfrak{W}$. Similarly, for a smooth $\mathfrak{W}$-invariant vector field $\mathscr{X}$ of coordinates $(\mathscr{X}_i)_{i=1}^p$, the divergence $\nabla_\Theta \cdot \mathscr{X}$ is the scalar $\sum_{i=1}^p \frac{\partial \mathscr{X}_i}{\partial \theta_i}$ which is a function $\tilde{\mathscr{T}} \to \mathbb{R}$. The Laplacian Δ_Θ is naturally defined by $\Delta_\Theta = \nabla_\Theta \cdot \nabla_\Theta \varphi$. Finally, we introduce the constants

$$\gamma_{2p} = \sum_{j=1}^p (p-j)^2, \quad \gamma_{2p+1} = \sum_{j=1}^p (p-j+\frac{1}{2})^2. \tag{5}$$

We note that γ_{2p} (for $p \ge 2$) and γ_{2p+1} (for $p \ge 1$) are strictly positive constants. We have the following

Lemma 4 *(i) Case $n = 2p$, $p \ge 2$: we have*

$$\nabla_\Theta u_{2p}(\Theta) = \Big(\big(\sum_{k \ne j} \frac{2}{\cos\theta_k - \cos\theta_j}\big) \sin\theta_j \, u_{2p}(\Theta) \Big)_{j=1}^p. \tag{6}$$

(ii) Case $n = 2p+1$, $p \ge 1$: we have

$$\nabla_\Theta u_{2p+1}(\Theta) = \Big(\big(\sum_{k \ne j} \frac{2}{\cos\theta_k - \cos\theta_j} + \frac{1}{1-\cos\theta_j}\big) \sin\theta_j \, u_{2p+1}(\Theta) \Big)_{j=1}^p. \tag{7}$$

(iii) In both the cases $n = 2p$ ($p \ge 2$) and $n = 2p+1$ ($p \ge 1$), we have

$$\Delta_\Theta \Pi_n = -\gamma_n \Pi_n. \tag{8}$$

Proof (i) We compute

$$\frac{\partial}{\partial\theta_j} \log u_{2p} = 2\frac{\partial}{\partial\theta_j}\log \Pi_{2p} = 2\frac{\partial}{\partial\theta_j}\Big(\sum_{1\le k<\ell\le p} \log(\cos\theta_k - \cos\theta_\ell)\Big)$$
$$= -\sum_{j<\ell\le p} \frac{2\sin\theta_j}{\cos\theta_j - \cos\theta_\ell} + \sum_{1\le k<j} \frac{2\sin\theta_j}{\cos\theta_k - \cos\theta_j} = \sum_{k\ne j} \frac{2\sin\theta_j}{\cos\theta_k - \cos\theta_j},$$

which leads to (6).

(ii) We remark that, up to an unimportant multiplicative constant, we have

$$u_{2p+1}(\Theta) = \prod_{1\le k<\ell\le p} \left(\cos\theta_k - \cos\theta_\ell\right)^2 \prod_{k=1}^{p}(1-\cos\theta_k),$$

and we proceed like in the previous point.

(iii) case $n = 2p$: this proof is adapted from that of [16, Lemma 12.5.2]. We have, up to an unimportant multiplicative constant,

$$\Pi_{2p}(\Theta) = \prod_{k<\ell}(e^{i\theta_k} + e^{-i\theta_k} - e^{i\theta_\ell} - e^{-i\theta_\ell}). \tag{9}$$

We note that the right-hand side of (9) is the Vandermonde determinant of the unknowns $(e^{i\theta_k} + e^{-i\theta_k})_{k=1}^{p}$. We recall that if $(X_1, \dots, X_p)$ are elements of $\mathbb{C}$, the Vandermonde determinant is

$$V(X_1, \dots, X_n) = \det\big((X_k^{p-\ell})_{k\ell}\big) = \prod_{k<\ell}(X_k - X_\ell).$$

Thus, we have

$$\begin{aligned}\Pi_{2p}(\Theta) &= V\left(e^{i\theta_1} + e^{-i\theta_1}, \dots, e^{i\theta_p} + e^{-i\theta_p}\right)\\ &= \det\Big(\big((e^{i\theta_k} + e^{-i\theta_k})^{p-\ell}\big)_{k\ell}\Big).\end{aligned} \tag{10}$$

By developing the powers involved in all the terms of the determinant and using elementary manipulations on the rows and columns (see [19, Eq. (24.39)]), one realizes that

$$\begin{aligned}\Pi_{2p}(\Theta) &= \det\big((e^{i(p-\ell)\theta_k} + e^{-i(p-\ell)\theta_k})_{k\ell}\big)\\ &= \sum_{\sigma\in\mathfrak{S}_p}(-1)^\sigma \prod_{k=1}^{p}\left(e^{i(p-\sigma(k))\theta_k} + e^{-i(p-\sigma(k))\theta_k}\right),\end{aligned}$$

where $\mathfrak{S}_p$ is the symmetric group of p elements and for $\sigma \in \mathfrak{S}_p$, $(-1)^\sigma$ stands for its signature. Now, taking two derivatives with respect to θ_j, we get

$$\frac{\partial^2\Pi_{2p}}{\partial\theta_j^2}(\Theta) = -\sum_{\sigma\in\mathfrak{S}_p}(-1)^\sigma(p-\sigma(j))^2\prod_{k=1}^{p}\left(e^{i(p-\sigma(k))\theta_k} + e^{-i(p-\sigma(k))\theta_k}\right).$$

Summing over j and realizing that

$$\sum_{j=1}^{p}(p-\sigma(j))^2 = \sum_{j=1}^{p}(p-j)^2 = \gamma_{2p},$$

is independent of σ, we finally get (8) for $n = 2p$.

(iii) case $n = 2p+1$: the proof is similar. We note that, up to an unimportant multiplicative constant, we have

$$\begin{aligned}\Pi_{2p+1} &= \prod_{k<\ell}(e^{i\theta_k} + e^{-i\theta_k} - e^{i\theta_\ell} - e^{-i\theta_\ell}) \prod_{k=1}^{p}(e^{i\theta_k/2} - e^{-i\theta_k/2}) \\ &= \det\big((e^{i(p-\ell+\frac{1}{2})\theta_k} + e^{-i(p-\ell+\frac{1}{2})\theta_k})_{k\ell}\big),\end{aligned} \tag{11}$$

where the second inequality again requires elementary manipulations of the determinant (see [19, Exercise 24.27]). The remainder of the proof is similar. ■

We now provide alternate expressions of the radial Laplacian in the following

Proposition 2 *(i) In the case $n = 2p$, for any $\varphi \in C^\infty(\tilde{\mathscr{T}})$, we have*

$$L\varphi = \sum_{j=1}^{p}\Big[\frac{\partial^2\varphi}{\partial\theta_j^2} + \Big(\sum_{k\neq j}\frac{2}{\cos\theta_k - \cos\theta_j}\Big)\sin\theta_j\frac{\partial\varphi}{\partial\theta_j}\Big]. \tag{12}$$

(ii) In the case $n = 2p+1$, for any $\varphi \in C^\infty(\tilde{\mathscr{T}})$, we have

$$L\varphi = \sum_{j=1}^{p}\Big[\frac{\partial^2\varphi}{\partial\theta_j^2} + \Big(\sum_{k\neq j}\frac{2}{\cos\theta_k - \cos\theta_j} + \frac{1}{1-\cos\theta_j}\Big)\sin\theta_j\frac{\partial\varphi}{\partial\theta_j}\Big]. \tag{13}$$

(iii) In both cases $n = 2p$ (for $p \geq 2$) and $n = 2p+1$ (for $p \geq 1$), we have

$$L\varphi = \frac{1}{u_n}\nabla_\Theta\cdot\big(u_n\nabla_\Theta\varphi\big), \tag{14}$$

$$= \frac{1}{\Pi_n}\Big(\Delta_\Theta + \gamma_n\Big)(\Pi_n\varphi), \tag{15}$$

Proof (i) and (ii): formulas (12) and (13) are obtained from (1) and (2) respectively, by exchanging i and j inside the second terms of the sums over $i < j$.

(iii) Inserting (6) and (7) into (12) and (13) respectively, we can write

$$L\varphi = \Delta_\Theta\varphi + \frac{1}{u_n}\nabla_\Theta u_n\cdot\nabla_\Theta\varphi = \frac{1}{u_n}\nabla_\Theta\cdot(u_n\nabla_\Theta\varphi),$$

which is (14). Equivalently, we can write

$$
\begin{aligned}
L\varphi &= \frac{1}{\Pi_n^2}\nabla_\Theta \cdot (\Pi_n^2 \nabla_\Theta \varphi) = \frac{1}{\Pi_n}\Big(\nabla_\Theta \Pi_n \cdot \nabla_\Theta \varphi + \nabla_\Theta \cdot (\Pi_n \cdot \nabla_\Theta \varphi)\Big) \\
&= \frac{1}{\Pi_n}\Big(\nabla_\Theta \Pi_n \cdot \nabla_\Theta \varphi + \nabla_\Theta \cdot \big(\nabla_\Theta(\Pi_n \varphi) - \varphi \nabla_\Theta \Pi_n\big)\Big) \\
&= \frac{1}{\Pi_n}\Big(\Delta_\Theta(\Pi_n \varphi) - \varphi \Delta_\Theta(\Pi_n)\Big),
\end{aligned}
$$

which, with (8), leads to (15). ■

4.2 Spectral Theory and Inversion of the Radial Laplacian

In this and the following sections, we denote by G either of the groups $\mathrm{SO}_{2p}\mathbb{R}$ or $\mathrm{SO}_{2p+1}\mathbb{R}$. We state a few facts concerning the spectral decomposition of the Laplacian $-\Delta$ and the radial Laplacian $-L$.

We first start with the Laplacian $-\Delta$. We know [20] that the spectrum of $-\Delta$ on a Riemannian manifold consists of an increasing sequence of discrete eigenvalues $\lambda_1 < \lambda_2 < \ldots \lambda_n \ldots$, tending to $+\infty$, that the associated eigenspaces are finite dimensional and that their direct sum is dense in $L^2(G)$. Furthermore, $\lambda_1 = 0$ and its eigenspace is one-dimensional and consists of the constant functions. However, in the case of the Laplacian on G, we can be more specific about the eigenspaces. We first need to introduce a few concepts from representation theory.

Here, we refer to [16, Chap. 6]. A representation (V, π) of G is a pair with V a vector space over $\mathbb{C}$ and π a group morphism $G \to \mathrm{Aut}(V)$ where $\mathrm{Aut}(V)$ is the group of linear automorphisms of V. If W is a subspace of V such that $\pi(g)W = W, \forall g \in G$, then $(W, \tilde{\pi})$ where $\tilde{\pi}(g) = \pi(g)|_W$ is a representation called a subrepresentation of V. An irreducible representation is a representation which has no proper subrepresentation. When G is compact, any irreducible representation is finite dimensional and any finite-dimensional representation can be decomposed in the direct sum of irreducible representations. Two representations (V, π) and (V', π') are equivalent if and only if there exists a linear isomorphism $A\colon V \to V'$ which intertwines the two representations, i.e. $A\pi(g) = \pi'(g)A, \forall g \in G$. We denote by $\hat{G}$ the set of equivalence classes of irreducible representations of G. For any $\lambda \in \hat{G}$, let us choose a representative (V_λ, π_λ) in λ. We denote by $\mathscr{M}_\lambda$ the space of functions $f\colon G \to \mathbb{C}$ such that there exists $A \in \mathrm{Aut}(V_\lambda)$ and $f(g) = \mathrm{Tr}\{A\pi_\lambda(g)\}$. Then, the Peter-Weyl theorem asserts that

$$
L^2(G) = \overline{\bigoplus_{\lambda \in \hat{G}} \mathscr{M}_\lambda}, \tag{16}
$$

where $L^2(G)$ is the space of square integrable functions on G. Furthermore [16, Proposition 8.2.1], all functions $f \in \mathscr{M}_\lambda$ are eigenfunctions of the Laplace operator associated with the same eigenvalue. So, the decomposition (16) is the eigenspace decomposition of the Laplace operator on G.

Thanks to this, we can easily solve the following problem: given $h \in L^2(G)$, find f such that

$$-\Delta f = h, \tag{17}$$

Obviously, h must be chosen to satisfy the compatibility equation

$$\int_G h(x)\, dx = 0, \tag{18}$$

where we now denote the normalized Haar measure on G by dx. A usual method is to use a variational formulation. Supposing $f \in H^2(G)$ (where $H^k(G)$ is defined recursively by $H^k(G) = \{f \in H^{k-1}(G) \mid \nabla f \in H^{k-1}(G)\}$, and $H^0(G) = L^2(G)$) and taking a test function $v \in H^1(G)$, multiplying (17) by v and using Stokes formula, we get

$$\int_G \nabla f \cdot \nabla v\, dx = \int_G h\, v\, dx, \quad \forall v \in H^1(G). \tag{19}$$

Thanks to (18), Eq. (19) is equivalent to the same formulation but with v taken in $H^1_0(G)$ where $H^1_0(G) = \{v \in H^1(G) \mid \int_G v(x)\, dx = 0\}$. Now, since clearly, a solution f to (17) is defined up to constants, it is reasonable to search for a unique f in the space $H^1_0(G)$. Indeed, we have the following

Theorem 2 *Given $h \in L^2(G)$ satisfying the compatibility condition* (18)*, the variational formulation*

$$f \in H^1_0(G), \tag{20}$$

$$\int_G \nabla f \cdot \nabla v\, dx = \int_G h\, v\, dx, \quad \forall v \in H^1_0(G), \tag{21}$$

has a unique solution, referred to as the variational solution of the problem $-\Delta f = h$. Furthermore, the H^1 norm of f is controlled by the L^2 norm of h.

Proof By the Lax-Milgram theorem, it is enough to show a Poincare inequality. But because there is a spectral gap, i.e. $\lambda_2 > 0$, we have

$$\int_G |\nabla f|^2\, dx \geq \lambda_2 \int_G |f|^2\, dx, \quad \forall f \in H^1_0(G).$$

This shows the result. ■

Remark 1 Thanks to the Peter-Weyl theorem and a Plancherel theorem [16, Theorem 6.4.2], we could express the Fourier coefficients of f in terms of the Fourier coefficients of h and of the eigenvalues of Δ, where the Fourier coefficient $\hat{f}(\lambda)$ is defined as an endomorphism of V_λ given by $\hat{f}(\lambda) = \int_G f(x)\, \pi_\lambda(x^{-1})\, dx$.

Now, we focus on class functions. The character of a representation (V, π) is the map $G \to \mathbb{C}$ defined by $\chi_\pi(g) = \mathrm{Tr}\{\pi(g)\}$, $\forall g \in G$. It is a class function. Now, let $\lambda \in \hat{G}$ and denote by χ_λ the character of any representative (V_λ, π_λ) of λ. Then, the subspace of $\mathscr{M}_\lambda$ consisting of class functions is one-dimensional and spanned by χ_λ. The system $\{\chi_\lambda\}_{\lambda\in\hat{G}}$ is a Hilbert basis of $L^2(G)_{\text{class}}$, the subspace of $L^2(G)$ consisting of class functions [16, Prop. 6.5.3] (the fact that $\{\chi_\lambda\}_{\lambda\in\hat{G}}$ is an orthonormal system is known as Schur's orthogonality relations). Now, since χ_λ is an eigenfunction of Δ and is a class function, $\varphi_\lambda =: \varphi_{\chi_\lambda}$ is an eigenfunction of the radial Laplacian L (we recall that for a class function f, φ_f is defined by (11)).

We introduce $L^2(\tilde{\mathscr{T}})$, the set of functions $\varphi\colon \tilde{\mathscr{T}} \to \mathbb{R}$, such that $\int_{\mathscr{T}} |\varphi(\Theta)|^2\, u_n(\Theta)\, d\Theta < \infty$ and $H^k(\tilde{\mathscr{T}})$, defined recursively by $\varphi \in H^k(\tilde{\mathscr{T}}) \Longleftrightarrow (\varphi \in H^{k-1}(\tilde{\mathscr{T}})$ and $\nabla_\Theta \varphi \in H^{k-1}(\mathscr{T}))$, with $H^0(\tilde{\mathscr{T}}) = L^2(\tilde{\mathscr{T}})$. With these notations, $(\varphi_\lambda)_{\lambda\in\hat{G}}$ is a Hilbert basis of $L^2(\tilde{\mathscr{T}})$ consisting of eigenfunctions of L. It follows that L has the same eigenvalues as Δ. Consequently, analogously to the case of the Laplacian, we can define a variational solution of the problem $-L\varphi = \eta$. Let us define $H^1_0(\tilde{\mathscr{T}}) = \{v \in H^1(\tilde{\mathscr{T}}) \mid \int_{\mathscr{T}} v(\Theta)\, u_n(\Theta)\, d\Theta = 0\}$. Then, given $\eta \in L^2(\tilde{\mathscr{T}})$ satisfying the compatibility condition

$$\int_{\mathscr{T}} \eta(\Theta)\, u_n(\Theta)\, d\Theta = 0, \tag{22}$$

we consider the variational formulation

$$\varphi \in H^1_0(\mathscr{T}), \tag{23}$$

$$\int_{\mathscr{T}} \nabla_\Theta \varphi \cdot \nabla_\Theta v\, u_n\, d\Theta = \int_{\mathscr{T}} \eta\, v\, u_n\, d\Theta, \quad \forall v \in H^1_0(\tilde{\mathscr{T}}). \tag{24}$$

Because $-L$ has the same eigenvalues as $-\Delta$, its smallest eigenvalue is 0 with a one-dimensional eigenspace spanned by the constants and the next eigenvalue is strictly positive. Thus, $-L$ has a spectral gap, which leads to the fact that the bilinear form at the left-hand side of (24) is coercive. Thus, we deduce the following

Theorem 3 *The variational formulation* (23), (24) *has a unique solution. Furthermore, the* H^1 *norm of* φ *is controlled by the* L^2 *norm of* η.

The unique solution of (23), (24) is referred to as the variational solution of the problem $-L\varphi = \eta$.

Remark 2 (i) The singularities when $\cos\theta_j = \cos\theta_k$ (or when $\cos\theta_j = 1$ in the case $n = 2p + 1$) which appear in the expression of the radial Laplacian in strong form (1), (2) disappear in the weak form (24). This is due to the fact that the radial Laplacian can be put in divergence form (14). This allows us to multiply both sides of the equation $-L\varphi = \eta$ by u_n, which removes the singularity.
(ii) The coercivity of the bilinear form at the left-hand side of (24) could be directly verified thanks to a classical compactness argument. The proof is omitted.

(iii) In (24), the integrals appear to be over the domain $\mathscr{T}$. In fact, since all the involved functions are $\mathfrak{W}$-invariant, all integrals could be reduced to a fundamental domain of $\mathscr{T}$ under the action of $\mathfrak{W}$ up to a multiplicative factor equal to the cardinal of $\mathfrak{W}$. Such a fundamental domain can be identified (up to a zero-measure set) to the closed Weyl chambers whose definitions are given at (25) and (27) (with θ_i substituted for λ_i).

Similarly to the case of the inversion of $-\Delta$, we can also recover the Fourier coefficients of φ (in the Hilbert basis $\{\varphi_\lambda\}_{\lambda\in\hat{G}}$) from those of η and the eigenvalues of the Laplace operator. The next section is devoted to the computation of the eigenvalues of Δ and L.

4.3 Characters and Eigenvalues of the Laplacian

As seen in the previous section, characters are at the same time eigenfunctions of the Laplacian and class functions. Expressions of the characters on the maximal torus $\mathbb{T}$ are available thanks to the Weyl character formula. Applying the radial Laplacian to them, we will deduce expressions of the eigenvalues of the Laplacian. All this is expressed in the following theorem:

Theorem 4 *(i) Case $n = 2p$ for $p \geq 2$. The irreducible representations of $SO_{2p}\mathbb{R}$ are indexed by a p-tuple of integers $\lambda = (\lambda_1, \dots, \lambda_p) \in \mathbb{Z}^p$ such that*

$$\lambda_1 \geq \lambda_2 \geq \dots \geq \lambda_{p-1} \geq |\lambda_p| \geq 0. \tag{25}$$

Let χ_λ be its character and define $\varphi_\lambda =: \varphi_{\chi_\lambda}$ with φ_f defined by (11). Then φ_λ is given by the Weyl character formula [19, Eq. (24.40)]:

$$\varphi_\lambda(\Theta) = \frac{det\big((e^{i\theta_k(\lambda_\ell+\delta_\ell)} + e^{-i\theta_k(\lambda_\ell+\delta_\ell)})_{k\ell}\big)}{\Pi_{2p}(\Theta)} + \frac{det\big((e^{i\theta_k(\lambda_\ell+\delta_\ell)} - e^{-i\theta_k(\lambda_\ell+\delta_\ell)})_{k\ell}\big)}{\Pi_{2p}(\Theta)}, \tag{26}$$

with $\delta_\ell = p - \ell$, $\forall \ell \in \{1, \dots, p\}$. We denote by δ the p-tuple $\delta = (\delta_\ell)_{\ell=1}^p$.
(ii) Case $n = 2p + 1$ for $p \geq 1$. The irreducible representations of $SO_{2p+1}\mathbb{R}$ are indexed by a p-tuple of integers $\lambda = (\lambda_1, \dots, \lambda_p) \in \mathbb{Z}^p$ such that

$$\lambda_1 \geq \lambda_2 \geq \dots \geq \lambda_{p-1} \geq \lambda_p \geq 0. \tag{27}$$

Then, the Weyl character formula [19, Eq. (24.28)] is written in this case:

$$\varphi_\lambda(\Theta) = \frac{det\big((e^{i\theta_k(\lambda_\ell+\delta_\ell)} - e^{-i\theta_k(\lambda_\ell+\delta_\ell)})_{k\ell}\big)}{\Pi_{2p+1}(\Theta)}, \tag{28}$$

with $\delta_\ell = p - \ell + \frac{1}{2}$, $\forall \ell \in \{1, \dots, p\}$. We denote by δ the p-tuple $\delta = (\delta_\ell)_{\ell=1}^p$.

(iii) In both the cases $n = 2p$ for $p \geq 2$ and $n = 2p + 1$ for $p \geq 1$, φ_λ and χ_λ are eigenfunctions of $-L$ and $-\Delta$ respectively, associated with the eigenvalue κ_λ given by

$$\kappa_\lambda = \|\lambda + \delta\|^2 - \|\delta\|^2, \tag{29}$$

where $\|\cdot\|$ is the Euclidean norm in $\mathbb{R}^p$. Note that $\|\delta\|^2 = \gamma_{2p}$ or γ_{2p+1} according to the cases considered (see Eq. (5)).

Proof (i) & (ii) lie at the heart of Lie group representation theory and we refer the reader to e.g. [19]. Here, we just make sure that Eqs. (26) and (28) make sense, i.e. that they define functions φ_λ in $\tilde{\mathcal{T}}$, in other words that φ_λ are 2π periodic in each variable θ_j and invariant by the Weyl group.

2π-periodicity follows immediately from the assumption that $\lambda \in \mathbb{Z}^p$. Indeed, in the case $n = 2p$, changing one of the θ_j into $\theta_j + 2\pi$ leaves both the numerator and the denominator of (26) invariant. In the case $n = 2p + 1$, both change sign so that their quotient (28) remains invariant. In fact, it can be shown that $\lambda \in \mathbb{Z}^p$ is also a necessary condition for φ_λ to be 2π-periodic.

Invariance by the Weyl group will be shown by showing the following stronger property, which is an interesting observation in its own right:

- Case $n = 2p + 1$: there exists a symmetric polynomial $P(X_1, \ldots, X_p)$ such that

$$\varphi_\lambda(\Theta) = P(e^{i\theta_1} + e^{-i\theta_1}, \ldots, e^{i\theta_p} + e^{-i\theta_p}), \tag{30}$$

- Case $n = 2p$: there exist two symmetric polynomials $P(X_1, \ldots, X_p)$ and $Q(X_1, \ldots, X_p)$ such that

$$\varphi_\lambda(\Theta) = P(e^{i\theta_1} + e^{-i\theta_1}, \ldots, e^{i\theta_p} + e^{-i\theta_p}) + \prod_{k=1}^{n}(e^{i\theta_k} - e^{-i\theta_k}) Q(e^{i\theta_1} + e^{-i\theta_1}, \ldots, e^{i\theta_p} + e^{-i\theta_p}). \tag{31}$$

It is clear that (31) and (30) imply invariance by the Weyl group (see discussion after (11) for the description of the Weyl group for even and odd orthogonal groups).

In the case $n = 2p + 1$, we first focus on the numerator of (28). We have

$$\begin{aligned} e^{i\theta_k(\lambda_\ell+\delta_\ell)} - e^{-i\theta_k(\lambda_\ell+\delta_\ell)} &= e^{i\theta_k(\lambda_\ell+p-\ell+\frac{1}{2})} - e^{-i\theta_k(\lambda_\ell+p-\ell+\frac{1}{2})} \\ &= (e^{i\frac{\theta_k}{2}} - e^{-i\frac{\theta_k}{2}})\Big(1 + \sum_{m=1}^{\lambda_\ell+p-\ell} (e^{im\theta_k} + e^{-im\theta_k})\Big). \end{aligned}$$

It is easy to show by induction that there exists a polynomial p_m of degree m such that

$$e^{im\theta_k} + e^{-im\theta_k} = p_m(e^{i\theta_k} + e^{-i\theta_k}).$$

Thus, there exists a polynomial q_m of degree m such that

$$e^{i\theta_k(\lambda_\ell+\delta_\ell)} - e^{-i\theta_k(\lambda_\ell+\delta_\ell)} = (e^{i\frac{\theta_k}{2}} - e^{-i\frac{\theta_k}{2}})\, q_{\lambda_\ell+p-\ell}(e^{i\theta_k} + e^{-i\theta_k}).$$

It follows that

$$\begin{aligned} &\det\Big((e^{i\theta_k(\lambda_\ell+\delta_\ell)} - e^{-i\theta_k(\lambda_\ell+\delta_\ell)})_{k\ell}\Big) \\ &\quad = \Big(\prod_{k=1}^{p}(e^{i\frac{\theta_k}{2}} - e^{-i\frac{\theta_k}{2}})\Big) \det\Big(\big(q_{\lambda_\ell+p-\ell}(e^{i\theta_k} + e^{-i\theta_k})\big)_{k\ell}\Big) \qquad (32) \\ &\quad = \Big(\prod_{k=1}^{p}(e^{i\frac{\theta_k}{2}} - e^{-i\frac{\theta_k}{2}})\Big) R(e^{i\theta_1} + e^{-i\theta_1}, \ldots, e^{i\theta_p} + e^{-i\theta_p}), \end{aligned}$$

where $R(X_1, \ldots, X_p)$ is an alternating polynomial (i.e. it is changed in its opposite when X_j and X_k are exchanged, for any pair $j \neq k$). Thanks to (11), we get

$$\varphi_\lambda(\Theta) = \frac{R(e^{i\theta_1} + e^{-i\theta_1}, \ldots, e^{i\theta_p} + e^{-i\theta_p})}{\prod_{k<\ell}(e^{i\theta_k} + e^{-i\theta_k} - e^{i\theta_\ell} - e^{-i\theta_\ell})}. \qquad (33)$$

From (32), it is clear that any factor $(X_j - X_k)$ for $j < k$ divides R. Since all these factors are mutually prime polynomials, the denominator divides the numerator. Hence, the right-hand side of (33) is a polynomial. Since both the numerator and denominator of (33) are alternating, their quotient is symmetric, which ends the proof in the case $n = 2p + 1$.

The case $n = 2p$ is similar. The first term at the right-hand side of (26) gives rise to the first term in (31). Indeed, the numerator is clearly an alternating polynomial of $(e^{i\theta_k} + e^{-i\theta_k})_{k=1}^{p}$ and the denominator too thanks to (10). Furthermore, each factor of the denominator divides the numerator and are mutually prime, so the quotient is a symmetric polynomial. The second term at the right-hand side of (26) gives rise to the second term in (31) thanks to similar computations as those made in the case $n = 2p + 1$.

(iii) The general methodology is outlined in [22, Sect. 5.7] but the proof is quite abstract. On the other hand, proving (29) is quite easy, using (15). Let us consider the case $n = 2p$ for instance. With (26), we have

$$\begin{aligned} \Pi_{2p}\varphi_\lambda(\Theta) = &\sum_{\sigma\in\mathfrak{S}_p}(-1)^\sigma \prod_{j=1}^{p}\Big(e^{i\ell_{\sigma(j)}\theta_j} + e^{-i\ell_{\sigma(j)}\theta_j}\Big) \\ &+ \sum_{\sigma\in\mathfrak{S}_p}(-1)^\sigma \prod_{j=1}^{p}\Big(e^{i\ell_{\sigma(j)}\theta_j} - e^{-i\ell_{\sigma(j)}\theta_j}\Big). \end{aligned}$$

Now, like in the proof of Lemma 4 (iii), we have

$$-\Delta_{\Theta}(\Pi_{2p}\varphi_{\lambda}) = \Big(\sum_{k=1}^{p} \ell_k^2\Big)\Pi_{2p}\varphi_{\lambda}.$$

Then, with (15) and (8), we get (29). The case $n = 2p + 1$ is similar. ■

We end this section with some terminology. The p-tuple λ lies on the lattice $\mathbb{Z}^p$, which is called the weight lattice. Any $\lambda \in \mathbb{Z}^p$ is called a weight. The weights satisfying the restrictions (25) or (27) are called dominant weights, and the regions of $\mathbb{R}^p$ defined by the inequalities (25) or (27) are called the closed Weyl chambers. All irreducible representations of $\mathrm{SO}_n\mathbb{R}$ induce irreducible representations of the Lie algebra $\mathfrak{so}_n\mathbb{R}$. The converse is not true. The weight lattice of $\mathfrak{so}_n\mathbb{R}$ is equal to $\mathbb{Z}^p \cup (\frac{1}{2} + \mathbb{Z})^p$. The Weyl chambers of $\mathfrak{so}_n\mathbb{R}$ and $\mathrm{SO}_n\mathbb{R}$ on the other hand are the same. All representations of $\mathfrak{so}_n\mathbb{R}$ lift to representations of the spin group $\mathrm{Spin}_n\mathbb{R}$ which is a simply connected doubly-sheeted covering of $\mathrm{SO}_n\mathbb{R}$ and is such that $\mathrm{SO}_n\mathbb{R} = \mathrm{Spin}_n\mathbb{R}/\{\pm 1\}$. Half-integer weights correspond to representations of $\mathrm{Spin}_n\mathbb{R}$ that are not trivial on $\{\pm 1\}$ and thus, cannot induce representations of $\mathrm{SO}_n\mathbb{R}$ (but some authors call these representations doubled-valued representations of $\mathrm{SO}_n\mathbb{R}$).

5 Conclusion

In this work, we have reviewed elements of the spectral theory of the Laplacian on rotation groups. One key element is to find the expression of the radial Laplacian, which is the restriction of the Laplacian on the maximal torus. Adequate expressions of this radial Laplacian, together with the use of the Weyl character formula lead to explicit expressions of the eigenvalues of the Laplacian. Although the material presented in these notes is familiar to Lie group experts, it is not easily found in the literature in a concise and accessible way. We hope these notes will prove useful to those working with rotation groups without being experts of Lie group theory.

Acknowledgements PD holds a visiting professor association with the Department of Mathematics, Imperial College London, UK.

Appendix 1: Proof of Lemma 1

By uniqueness of the Haar measure, to show that there is a positive constant C such that $d\omega = C d\mu$, it is enough to show that the Riemannian volume form is left invariant. First, we define an orientation of $\mathfrak{so}_n\mathbb{R}$ associated with the lexicographic ordering of the orthonormal basis $(F_{ij})_{1\leq i<j\leq n}$. For $A \in \mathrm{SO}_n\mathbb{R}$, the tangent space

T_A is oriented by the same lexicographic ordering of the basis $(AF_{ij})_{i<j}$, and this gives rise to a global orientation of $\mathrm{SO}_n\mathbb{R}$. We note that $(AF_{ij})_{i<j}$ is an orthonormal basis of T_A precisely because A is an orthogonal matrix. Then, we construct the Riemannian volume form of $\mathrm{SO}_n\mathbb{R}$ associated with this orientation, following the construction of [32, Sect. 4.10]. We denote by $(\omega_{ij}(A))_{i<j}$ the dual basis of $(AF_{ij})_{i<j}$. The inner-product on T_A generates a pairing between T_A and its dual T_A^*. This pairing is such that the image $\omega_X \in T_A^*$ of an element $X \in T_A$ is such that $\langle \omega_X, Y\rangle = X \cdot Y$ for all $Y \in T_A$, where the bracket indicates duality between T_A^* and T_A. In turn, this pairing induces an inner product on T_A^*, such that $\omega_X \cdot \omega_Y = X \cdot Y$. Thus, since $(AF_{ij})_{i<j}$ is an orthonormal basis of T_A, $(\omega_{ij}(A))_{i<j}$ is an orthonormal basis of T_A^*, so $\omega(A) = \bigwedge_{i<j} \omega_{ij}(A)$ is the volume form where $\bigwedge_{i<j}$ stands for the exterior product of the forms ω_{ij} in lexicographic order.

Now, we show that this volume form is left invariant. Let $\sigma \in \mathrm{SO}_n\mathbb{R}$ and ℓ_σ the morphism of $\mathrm{SO}_n\mathbb{R}$ such that $\ell_\sigma(A) = \sigma A$. The form ω is left invariant if and only if $\delta\ell_\sigma(\omega) = \omega$ for all $\sigma \in \mathrm{SO}_n\mathbb{R}$, where $\delta\ell_\sigma(\omega)$ is the pullback of ω by ℓ_σ. Let $(AX_{ij})_{i<j}$, with $X_{ij} \in \mathfrak{so}_n\mathbb{R}$ for all $i < j$ be elements of T_A. Then, $\delta\ell_\sigma(\omega)$ is defined by

$$\begin{aligned}&\delta\ell_\sigma(\omega)(A)\big(AX_{12}, AX_{13}, \ldots, AX_{n-1\,n}\big)\\&\qquad = \omega(\sigma A)\big(d\ell_\sigma(A)(AX_{12}), d\ell_\sigma(A)(AX_{13}), \ldots, d\ell_\sigma(A)(AX_{n-1\,n})\big),\end{aligned}$$

where $d\ell_\sigma(A)$ is the differential of ℓ_σ at A. It is readily checked that $d\ell_\sigma(A)(AX) = \sigma AX$, for any $X \in \mathfrak{so}_n\mathbb{R}$. Thus,

$$\begin{aligned}\delta\ell_\sigma(\omega)(A)\big(AX_{12}, AX_{13}, \ldots, AX_{n-1\,n}\big) &= \omega(\sigma A)(\sigma AX_{12}, \sigma AX_{13}, \ldots, \sigma AX_{n-1\,n})\\ &= \sum_{\tau \in \mathfrak{S}_N} \varepsilon(\tau) \prod_{i<j} \omega_{ij}(\sigma A)\big(\sigma AX_{\tau(i,j)}\big),\end{aligned}$$

where $N = \frac{n(n-1)}{2}$ is the dimension of T_A, $\mathfrak{S}_N$ is the permutation group of N elements, $\varepsilon(\tau)$ is the signature of the permutation τ and where the elements of $\{1, \ldots, N\}$ are labeled by the pairs (i, j) with $1 \le i < j \le n$. Hence, $\tau(i, j)$ denotes the image of the element (i, j) of $\{1, \ldots, N\}$ by the permutation τ.

Now, if we show that

$$\omega_{ij}(\sigma A)\big(\sigma AX\big) = \omega_{ij}(A)\big(AX\big), \ \forall A, \sigma \in \mathrm{SO}_n\mathbb{R}, \ \forall X \in \mathfrak{so}_n\mathbb{R}, \ \forall (i, j) \,|\, 1 \le i < j \le n, \quad (1)$$

then, we obviously deduce that $\delta\ell_\sigma(\omega) = \omega$, which concludes the proof. To prove (1), it is enough to prove it for all basis vectors of a basis of $\mathfrak{so}_n\mathbb{R}$. We take the basis $(F_{k\ell})_{k<\ell}$. We have $\omega_{ij}(\sigma A)\big(\sigma AF_{k\ell}\big) = \delta_{ij}\delta_{k\ell}$ because $(\omega_{ij}(\sigma A))_{i<j}$ is the dual basis of $(\sigma AF_{k\ell})_{k<\ell}$. But this being true for all σ, it is true for $\sigma = \mathrm{I}$ which gives $\omega_{ij}(A)\big(AF_{k\ell}\big) = \delta_{ij}\delta_{k\ell}$. Hence, $\omega_{ij}(\sigma A)\big(\sigma AF_{k\ell}\big) = \omega_{ij}(A)\big(AF_{k\ell}\big)$, which shows (1). ■

Appendix 2: Proof of Lemma 2

From (4) and (6), we have, for any C^∞ function f on $\mathrm{So}_n\mathbb{R}$ and any $A \in \mathrm{So}_n\mathbb{R}$:

$$(P_{T_A} E_{ij})(f)(A) = \Big(\rho\big(\frac{A^T E_{ij} - E_{ij}^T A}{2}\big) f\Big)(A).$$

A straightforward computation shows that

$$A^T E_{ij} - E_{ij}^T A = \sum_{k=1}^{n} A_{ik} F_{jk},$$

where A_{ik} denote the entries of the matrix A. Since $\rho(X)$ is linear with respect to X, we have

$$\rho(A^T E_{ij} - E_{ij}^T A) = \sum_{k=1}^{n} A_{ik} \rho\big(F_{jk}\big),$$

and

$$\big(\rho(A^T E_{ij} - E_{ij}^T A)\big)^2 = \sum_{k,\ell} A_{ik} A_{im}\, \rho\big(F_{jk}\big)\, \rho\big(F_{jm}\big).$$

Now, with (5), we get

$$\Delta_M f = \frac{1}{2} \sum_{i,j,k,\ell} A_{ik} A_{im}\, \rho\big(F_{jk}\big)\, \rho\big(F_{jm}\big) = \frac{1}{2} \sum_{j,k} \rho\big(F_{jk}\big)^2 = \sum_{1 \le j < k \le n} \rho\big(F_{jk}\big)^2 = \Delta_G f,$$

where, in the second equality, we have used that A is orthogonal, in the third equality, that $F_{ji} = -F_{ij}$ and in the fourth one, Eq. (7). This ends the proof. ■

References

1. Adams, J.F.: Lectures on Lie Groups. University of Chicago Press (1982)
2. Ahn, H., Ha, S.-Y., Shim, W.: Emergent dynamics of a thermodynamic Cucker-Smale ensemble on complete Riemannian manifolds. Kinet. Relat. Models **14**(2), 323 (2021)
3. Bröcker, T., Tom Dieck, T.: Representations of Compact Lie Groups. vol. 98. Springer Science & Business Media (2013)
4. Bump, D.: Lie Groups, vol. 8. Springer (2004)
5. Chevalley, C.: Theory of Lie Groups. Courier Dover Publications (2018)
6. Degond, P., Diez, A., Frouvelle, A.: Body-attitude coordination in arbitrary dimension, not yet published (2021). arXiv:2111.05614
7. Degond, P., Diez, A., Frouvelle, A., Merino-Aceituno, S.: Phase transitions and macroscopic limits in a BGK model of body-attitude coordination. J. Nonlinear Sci. **30**, 2671–2736 (2020)
8. Degond, P., Diez, A., Walczak, A.: Topological states and continuum model for swarmalators without force reciprocity. Anal. Appl. **20**, 1215–1270 (2022)

9. Degond, P., Frouvelle, A.: Macroscopic limit of a Fokker-Planck model of swarming rigid bodies. European J. Appl. Math., appeared online (2024). https://doi.org/10.1017/S0956792524000111
10. Degond, P., Frouvelle, A., Merino-Aceituno, S.: A new flocking model through body attitude coordination. Math. Models Methods Appl. Sci. **27**(06), 1005–1049 (2017)
11. Degond, P., Frouvelle, A., Merino-Aceituno, S., Trescases, A.: Alignment of self-propelled rigid bodies: from particle systems to macroscopic equations. In: International Workshop on Stochastic Dynamics out of Equilibrium, pp. 28–66. Springer (2017)
12. Degond, P., Frouvelle, A., Merino-Aceituno, S., Trescases, A.: Quaternions in collective dynamics. Multiscale Model. Simul. **16**(1), 28–77 (2018)
13. Degond, P., Motsch, S.: A macroscopic model for a system of swarming agents using curvature control. J. Stat. Phys. **143**, 685–714 (2011)
14. Do Carmo, M.:.Riemannian Geometry. Springer (1992)
15. Duistermaat, J.J., Kolk, J.A.: Lie Groups. Springer Science & Business Media (2012)
16. Faraut, J.: Analysis on Lie Groups, An Introduction. Cambridge University Press (2008)
17. Fetecau, R.C., Ha, S.-Y., Park, H.: Emergent behaviors of rotation matrix flocks. SIAM J. Appl. Dyn. Syst. **21**(2), 1382–1425 (2022)
18. Fetecau, R.C., Zhang, B.: Self-organization on Riemannian manifolds. J. Geom. Mech. **11**(3), 397–426 (2019)
19. Fulton, W., Harris, J.: Representation Theory: A First Course. Springer (2013)
20. Gallot, S., Hulin, D., Lafontaine, J.: Riemannian Geometry. Springer (1990)
21. Golse, F., Ha, S.-Y.: A mean-field limit of the Lohe matrix model and emergent dynamics. Arch. Ration. Mech. Anal. **234**(3), 1445–1491 (2019)
22. Gurarie, D.: Symmetries and Laplacians: Introduction to Harmonic Analysis, Group Representations and Applications. Courier Corporation (2007)
23. Ha, S.-Y., Ko, D., Ryoo, S.W.: Emergent dynamics of a generalized Lohe model on some class of Lie groups. J. Stat. Phys. **168**(1), 171–207 (2017)
24. Hall, B.C.: Lie Groups, Lie Algebras, and Representations. Springer (2013)
25. Helgason, S.: Groups and Geometric Analysis: Integral Geometry, Invariant Differential Operators, and Spherical Functions, vol. 83. American Mathematical Society (2022)
26. Hsu, E.P.: Stochastic Analysis on Manifolds, vol. 38. American Mathematical Society (2002)
27. Humphreys, J.E.: Introduction to Lie Algebras And Representation Theory, vol. 9. Springer Science & Business Media (2012)
28. Knapp, A.W.: Lie Groups Beyond an Introduction, vol. 140. Springer (1996)
29. Sepanski, M.R.: Compact Lie Groups. Springer (2007)
30. Simon, B.: Representations of Finite and Compact Groups. Graduate Studies in Mathematics, vol. 10. American Mathematical Society (1996)
31. Vicsek, T., Czirók, A., Ben-Jacob, E., Cohen, I., Shochet, O.: Novel type of phase transition in a system of self-driven particles. Phys. Rev. Lett. **75**(6), 1226 (1995)
32. Warner, F.W.: Foundations Of Differentiable Manifolds and Lie Groups. Springer (1983)
33. Weyl, H.: The Classical Groups: Their Invariants and Representations, vol. 1. Princeton University Press (1946)

Some Remarks About the Link Between the Fisher Information and Landau or Landau-Fermi-Dirac Entropy Dissipation

Laurent Desvillettes

Abstract We present in this work variants of existing estimates for the Landau or Landau-Fermi-Dirac entropy dissipation, in terms of Fisher information, in the hard potential case. The specificity of those variants is that the entropy is never used in the estimates (in order to control possible concentrations on a zero measure set). The proofs are significantly simplified with respect to previous papers on the subject.

Keywords Landau equation · Landau operator · Entropy dissipation · Fisher information

1 Introduction

1.1 Description of the Landau Operator and Its Entropy Dissipation

We write down the Landau operator appearing in kinetic theory (cf. [9, 21]), defined by

$$Q_\gamma(f, f)(v) = \nabla_v \cdot \left\{ \int_{\mathbb{R}^3} |v-w|^{2+\gamma}\, \Pi(v-w) \Big(f(w)\, \nabla f(v) - f(v)\, \nabla f(w) \Big)\, dw \right\}, \tag{1}$$

where

$$\Pi_{ij}(z) = \delta_{ij} - \frac{z_i z_j}{|z|^2} \tag{2}$$

is the i, j-component of the orthogonal projection Π onto $z^\perp := \{y \,/\, y \cdot z = 0\}$.

L. Desvillettes (✉)
Université Paris Cité and Sorbonne Université, CNRS, IUF, Institut de Mathématiques de Jussieu-Paris Rive Gauche, Paris, France
e-mail: desvillettes@math.univ-paris-diderot.fr

E. Carlen et al. (eds.), *From Particle Systems to Partial Differential Equations*, Springer Proceedings in Mathematics & Statistics 465,
https://doi.org/10.1007/978-3-031-65195-3_3

In this paper, we focus on the case of so-called Maxwell molecules, hard spheres or hard potentials, that is

$$\gamma \in [0, 1]. \tag{3}$$

The Landau operator satisfies (at the formal level) the conservation of mass, momentum and kinetic energy, that is, for any $f := f(v) \geq 0$:

$$\int_{\mathbb{R}^3} Q_\gamma(f, f)(v) \begin{pmatrix} 1 \\ v_i \\ |v|^2/2 \end{pmatrix} dv = \begin{pmatrix} 0 \\ 0 \\ 0 \end{pmatrix}. \tag{4}$$

We also have the following property of nonnegativity of the so-called entropy dissipation associated to the Landau operator, namely (at the formal level), for any $f := f(v) > 0$,

$$D_\gamma(f) := -\int_{\mathbb{R}^3} Q_\gamma(f, f)(v) \ln f(v)\, dv \tag{5}$$

$$= \frac{1}{2} \iint_{\mathbb{R}^3\times\mathbb{R}^3} f(v)\, f(w)\, |v-w|^{2+\gamma}\, \Pi(v-w) \left(\frac{\nabla f}{f}(v) - \frac{\nabla f}{f}(w) \right)$$

$$\cdot \left(\frac{\nabla f}{f}(v) - \frac{\nabla f}{f}(w) \right) dvdw \geq 0.$$

In the last years, several works (cf. [11–14, 18]) have been devoted to the proof of estimates from below of the quantity $D_\gamma(f)$ in terms of quantities like the Fisher information $F(f) := \int_{\mathbb{R}^3} \frac{|\nabla f(v)|^2}{f(v)}\, dv$, or the relative (with respect to a Maxwellian function $M := M(v)$) Fisher information $F_{rel}(f) := \int_{\mathbb{R}^3} \left| \frac{\nabla f}{f}(v) - \frac{\nabla M}{M}(v) \right|^2 f(v)\, dv$.

Those estimates (typically the ones related to $F(f)$) yield results of regularity for the solutions to the Landau equation (cf. [11, 15, 19]). When they are complemented with the logarithmic Sobolev inequality of Gross (cf. [20]), they also (typically the ones related to $F_{rel}(f)$) yield results for the large time behavior of the Landau equation (cf. [8, 18]), following the lines of the entropy-entropy dissipation method introduced in kinetic theory in the 90s by [6, 7, 23] (cf. [16] for more historical details and an explanation of the general context).

We write down a typical estimate, taken from Corollary 2.5 of [12], which holds when γ satisfies (3), and $f > 0$,

$$\int_{\mathbb{R}^3} f(v) \begin{pmatrix} 1 \\ v \\ |v|^2 \end{pmatrix} dv = \begin{pmatrix} 1 \\ 0 \\ 3 \end{pmatrix}$$

(and when the terms appearing in the inequality are defined):

$$\int_{\mathbb{R}^3} \frac{|\nabla f(v)|^2}{f(v)}\, dv \le 3072\, \Delta(f)^{-2} \left\{ 8448 + 48\sqrt{1+\pi}\, D_\gamma(f)\, ||f||_{L^2(\mathbb{R}^3)} \right\}, \tag{6}$$

where

$$\Delta(f) := \text{Det}\left(\int_{\mathbb{R}^3} f(w)\,(1+|w|^2)^{-1/2} \begin{bmatrix} 1 & w_i & w_j \\ w_i & w_i^2 & w_i\, w_j \\ w_j & w_i\, w_j & w_j^2 \end{bmatrix} dw \right). \tag{7}$$

The purpose of this paper is to show that it is possible to obtain estimates like (6) with the following improvements:

- No terms involving a determinant like (7) appears;
- The numerical constants are much smaller than the ones appearing in (6);
- The proof is short;
- These new estimates extend somewhat to the case of the Landau-Fermi-Dirac entropy dissipation.

The price to pay in order to get those improvements is the following restriction: the estimates that we will write down hold only when $D_\gamma(f)$ is not too large. We will see however that this is not a problem when one wishes to prove the exponential decay towards equilibrium with explicit rate for the solutions to the Landau equation with hard potentials or hard spheres.

We notice that the statement and computations presented below have the same flavour as the material presented in Chap. 5 of [18] (more generally, the methodology for getting estimates relating D_γ and F or F_{rel} was introduced in [18], recalling some of the computations of [10]). The conditions on the cross section are however more general in the present paper, and the proofs are simplified a lot with respect to what is presented in [18].

Our main result writes:

Theorem 1 *We consider γ satisfying (3). Then for all $f := f(v) > 0$ (sufficiently smooth for the terms in the estimate to make sense) satisfying*

$$\int_{\mathbb{R}^3} f(v) \begin{pmatrix} 1 \\ v \\ |v|^2 \end{pmatrix} dv = \begin{pmatrix} 1 \\ 0 \\ 3 \end{pmatrix}, \tag{8}$$

and such that

$$\left(\int_{\mathbb{R}^3} |v|^4\, f(v)\, dv \right) \left(1 + 2\sqrt{\pi}\, ||f||_{L^2(\mathbb{R}^3)}\right)\, D_\gamma(f) \le \frac{27}{32},$$

the following estimate holds:

$$\int_{\mathbb{R}^3} \left| \frac{\nabla f(v)}{f(v)} + v \right|^2 f(v)\, dv \le \left[\frac{64}{9} \left(\int_{\mathbb{R}^3} |v|^4 f(v)\, dv \right) \left(1 + 2\sqrt{\pi}\, ||f||_{L^2(\mathbb{R}^3)}\right) \right.$$
$$\left. + 48 + 32\sqrt{\pi}\, ||\, |\cdot|^2 f||_{L^2(\mathbb{R}^3)} \right] D_\gamma(f). \tag{9}$$

Remark 1 Inequality (9) can be further simplified, and be transformed for example in

$$\int_{\mathbb{R}^3} \left| \frac{\nabla f(v)}{f(v)} + v \right|^2 f(v)\, dv \le 200\, ||f||^2_{L^2_6(\mathbb{R}^3)}\, D_\gamma(f),$$

as soon as

$$||f||^2_{L^2_6(\mathbb{R}^3)}\, D_\gamma(f) \le 0.062,$$

where we denote from now on (for any $p \in [1, \infty[$ and $q \in \mathbb{R}$)

$$||f||^p_{L^p_q(\mathbb{R}^3)} := \int_{\mathbb{R}^3} (1 + |v|^2)^{pq/2}\, |f(v)|^p\, dv.$$

Indeed, one can check thanks to Hölder's inequality that (when f satisfies (8))

$$\left(\int_{\mathbb{R}^3} |v|^4 f(v)\, dv \right) \left(1 + 2\sqrt{\pi}\, ||f||_{L^2(\mathbb{R}^3)}\right) \le \left(2\pi\, \sqrt{\pi} + \frac{\pi^2}{4} \right) ||f||^2_{L^2_6(\mathbb{R}^3)}.$$

We see at this point that the numerical constant (that is, 200) is indeed much smaller than the ones appearing in inequality (6).

Section 2 is devoted to the proof of Theorem 1. Then, we show in Sect. 3 how this result can be used to obtain in a quick way the explicit exponential decay towards equilibrium for the solutions to the Landau equation with hard potentials or hard spheres. Finally, we show in Sect. 4 that Theorem 1 can be extended somewhat in the case of the Landau-Fermi-Dirac equation.

2 Proof of Theorem 1

We start here the

Proof of Theorem 1: We denote in this proof, for $i = 1, 2, 3$,

$$I_i(f) := \int_{\mathbb{R}^3} f(v)\, v_i^2\, dv, \tag{10}$$

so that $I_1(f) + I_2(f) + I_3(f) = 3$.

Up to the use of a linear isometry of $\mathbb{R}^3$, we assume like in [18] without loss of generality that, for $i, j = 1, 2, 3$ such that $i \neq j$,

$$\int_{\mathbb{R}^3} f(v)\, v_i\, v_j\, dv = 0. \tag{11}$$

Note that the Landau operator, like the Boltzmann operator (from which it is a singular limit: the so called grazing collisions limit) is indeed built to be Galilean-invariant.

We first observe that

$$D_\gamma(f) = \frac{1}{2} \sum_{i=1}^{3} \int\int_{\mathbb{R}^3 \times \mathbb{R}^3} f(v)\, f(w)\, |v-w|^\gamma\, [q^f_{i,i+1}(v,w)]^2\, dv dw,$$

where for $i, j = 1, 2, 3$,

$$q^f_{i,j}(v,w) := (v_i - w_i) \left(\frac{\partial_j f}{f}(v) - \frac{\partial_j f}{f}(w) \right) - (v_j - w_j) \left(\frac{\partial_i f}{f}(v) - \frac{\partial_i f}{f}(w) \right), \tag{12}$$

and by definition $q^f_{i,4} := q^f_{i,1}$.

Using the normalizations (8) and (11), we multiply (12) by $f(w)$ and $w_i\, f(w)$ and integrate on $\mathbb{R}^3$ in the w-variable. We get for $i \neq j$ the identities:

$$v_i\, \frac{\partial_j f}{f}(v) - v_j\, \frac{\partial_i f}{f}(v) = \int_{\mathbb{R}^3} f(w)\, q^f_{i,j}(v,w)\, dw, \tag{13}$$

$$I_i(f)\, \frac{\partial_j f}{f}(v) + v_j = - \int_{\mathbb{R}^3} w_i\, f(w)\, q^f_{i,j}(v,w)\, dw. \tag{14}$$

It is convenient to write (14) under the form

$$I_i(f) \left(\frac{\partial_j f}{f}(v) + v_j \right) = (I_i(f) - 1)\, v_j - \int_{\mathbb{R}^3} w_i\, f(w)\, q^f_{i,j}(v,w)\, dw. \tag{15}$$

As a consequence, using the Cauchy-Schwarz inequality,

$$\begin{aligned} I_i(f)^2 \int_{\mathbb{R}^3} \left| \frac{\partial_j f}{f}(v) + v_j \right|^2 f(v)\, dv &\le 2\, (I_i(f) - 1)^2\, I_j(f) \\ &\quad + 2 \int_{\mathbb{R}^3} f(v) \left| \int_{\mathbb{R}^3} w_i\, f(w)\, q^f_{i,j}(v,w)\, dw \right|^2 dv \\ \le 2\, (I_i(f) - 1)^2\, I_j(f) + 2 \int_{\mathbb{R}^3} f(v) &\left[\int_{\mathbb{R}^3} f(w)\, |v-w|^\gamma\, |q^f_{i,j}(v,w)|^2\, dw \right] \\ &\times \left[\int_{\mathbb{R}^3} f(w)\, |v-w|^{-\gamma}\, w_i^2\, dw \right] dv \end{aligned}$$

$$\leq 2\,(I_i(f)-1)^2\, I_j(f) + 2\int_{\mathbb{R}^3} f(v)\left[\int_{\mathbb{R}^3} f(w)\,|v-w|^{\gamma}\,|q^f_{i,j}(v,w)|^2\,dw\right] dv\,\sigma_\gamma(f), \tag{16}$$

where

$$\sigma_\gamma(f) := \sup_{v\in\mathbb{R}^3}\int_{\mathbb{R}^3} f(w)\,|v-w|^{-\gamma}\,|w|^2\,dw. \tag{17}$$

Multiplying identity (13) by $v_i\, v_j\, f(v)$ and integrating on $\mathbb{R}^3$ in the v-variable, we see that

$$I_i(f) - I_j(f) = \int\int_{\mathbb{R}^3\times\mathbb{R}^3} f(v)\, f(w)\, v_i\, v_j\, q^f_{i,j}(v,w)\, dvdw$$

so that, using again the Cauchy-Schwarz inequality,

$$|I_j(f) - I_i(f)| \leq \left[\int\int_{\mathbb{R}^3\times\mathbb{R}^3} f(v)\, f(w)\,|v-w|^{\gamma}\,|q^f_{i,j}(v,w)|^2\, dvdw\right]^{1/2}$$

$$\times\left[\int\int_{\mathbb{R}^3\times\mathbb{R}^3} f(v)\, f(w)\,|v-w|^{-\gamma}\, v_i^2\, v_j^2\, dvdw\right]^{1/2}$$

$$\leq \sqrt{\frac{2}{3}}\, D_\gamma(f)^{1/2}\, s_\gamma(f)^{1/2},$$

where

$$s_\gamma(f) := \int\int_{\mathbb{R}^3\times\mathbb{R}^3} f(v)\, f(w)\,|v-w|^{-\gamma}\,|v|^4\, dvdw. \tag{18}$$

As a consequence,

$$|I_i(f) - 1|^2 = \left(\frac{2}{3} I_i(f) - \frac{1}{3} I_{i+1}(f) - \frac{1}{3} I_{i+2}(f)\right)^2$$

$$\leq \frac{2}{9}(I_i(f) - I_{i+1}(f))^2 + \frac{2}{9}(I_i(f) - I_{i+2}(f))^2 \leq \frac{8}{27}\, D_\gamma(f)\, s_\gamma(f).$$

We deduce from this estimate that $I_i(f) \geq \frac{1}{2}$ as soon as

$$D_\gamma(f)\, s_\gamma(f) \leq \frac{27}{32}, \tag{19}$$

since (19) implies that $|I_i(f) - 1| \leq \frac{1}{2}$.

Under condition (19), we see therefore that (16) leads to

$$\int_{\mathbb{R}^3} \left| \frac{\partial_j f}{f}(v) + v_j \right|^2 f(v) \, \mathrm{dv} \le 8\,(I_i(f) - 1)^2\, I_j(f)$$

$$+ 8 \int_{\mathbb{R}^3} f(v) \left[\int_{\mathbb{R}^3} f(w)\, |v-w|^\gamma \, |q^f_{i,j}(v,w)|^2 \, dw \right] \sigma_\gamma(f),$$

so that finally, summing from $j = 1$ to 3 (and choosing $i = j + 1$)

$$\int_{\mathbb{R}^3} \left| \frac{\nabla f}{f}(v) + v \right|^2 f(v)\, dv \le \left(\frac{64}{9}\, s_\gamma(f) + 16\,\sigma_\gamma(f) \right) D_\gamma(f). \qquad (20)$$

Then we observe that

$$s_\gamma(f) \le \int\int_{|v-w|\ge 1} |v|^4\, f(v)\, f(w)\, dv dw + \int\int_{|v-w|\le 1} |v|^4\, f(v)\, f(w)\, |v-w|^{-\gamma}\, dv dw$$

$$\le \left(\int_{\mathbb{R}^3} |v|^4\, f(v)\, dv \right) \left(1 + \sup_{v\in\mathbb{R}^3} \int_{|v-w|\le 1} f(w)\, |v-w|^{-\gamma}\, dw \right)$$

$$\le \left(\int_{\mathbb{R}^3} |v|^4\, f(v)\, dv \right) \left(1 + 2\sqrt{\pi}\, ||f||_{L^2(\mathbb{R}^3)} \right), \qquad (21)$$

and that

$$\sigma_\gamma(f) \le \sup_{v\in\mathbb{R}^3} \int_{|v-w|\ge 1} |w|^2\, f(w)\, dw$$

$$+ \sup_{v\in\mathbb{R}^3} \int_{|v-w|\le 1} |w|^2\, |v-w|^{-\gamma}\, f(w)\, dw$$

$$\le 3 + 2\sqrt{\pi}\, ||\,|\cdot|^2 f\,||_{L^2(\mathbb{R}^3)}. \qquad (22)$$

We conclude the proof of Theorem 1 by using estimates (21) and (22) in estimate (20). □

3 Application to the Large Time Behavior of the Landau Equation

We show in this section that Theorem 1 quickly leads to a result of exponential convergence towards equilibrium with explicit rate, for the Landau equation in the case of hard potentials or hard spheres.

Note that in [18], such a result is proven in the so-called overMaxwelian case, which includes the case of Maxwell molecules, but not the case of hard potentials or hard spheres, because in those cases, the cross section cancels at some point. Those cases are treated as a special case (when the quantum parameter ε tends to 0) of the results obtained in [2] for the Landau-Fermi-Dirac equation. What we now present is a new (somewhat simpler) proof.

We start with a simple Lemma which replaces here Proposition 6 and its consequences in [18]:

Lemma 1 *Suppose that $H \in C^1(\mathbb{R}_+;\mathbb{R}_+)$ and $D \in C(\mathbb{R}_+;\mathbb{R}_+)$ are such that*

$$- H' = D.$$

We also assume that for some real numbers $q, c_0 > 0$

$$D \le q \qquad \Rightarrow \qquad D \ge c_0\, H. \tag{23}$$

Then, for all $t \ge H(0)/q$, the following estimate holds:

$$H(t) \le H(0)\, \exp(c_0\, H(0)/q)\, e^{-c_0\, t}. \tag{24}$$

In the sequel, H and D will be the respective relative entropy and entropy dissipation of the Landau and the Landau-Fermi-Dirac equations.

Proof of Lemma 1: We first observe that if for some $\bar{t} \ge 0$, one has $H(\bar{t}) \le q/c_0$, then for all $t \ge \bar{t}$,

$$H(t) \le q/c_0 \tag{25}$$

since H is nonincreasing.

Then either $D(t) \le q$ and thanks to (23), $D(t) \ge c_0\, H(t)$; or $D(t) \ge q \ge c_0\, H(t)$ thanks to (25).

In both cases, $-H'(t) \ge c_0\, H(t)$ for $t \ge \bar{t}$. Consequently,

$$\forall t \ge \bar{t}, \qquad H(t) \le H(\bar{t})\, e^{-c_0\,(t-\bar{t})}. \tag{26}$$

We apply estimate (26) for

$$\bar{t} := \inf\{\tau \ge 0,\, H(\tau) \le q/c_0\}.$$

We see that $H(s) \ge q/c_0$ for $s \in]0, \bar{t}]$ (this interval is empty if $\bar{t} = 0$, and might be of infinite length). Then, for $s \in]0, \bar{t}]$, if $D(s) \le q$ we see that $D(s) \ge c_0\, H(s) \ge q$ (because of (23)). Therefore $D(s) \ge q$ on $[0, \bar{t}]$, so that $H'(s) \le -q$ on the same interval, and finally

$$0 \le H(\bar{t}) \le H(0) - q\, \bar{t},$$

so that $\bar{t} \le H(0)/q$.

Coming back to (26), we see that for $t \ge H(0)/q$ $(\ge \bar{t})$,

$$H(t) \le H(\bar{t}) \exp(c_0 H(0)/q) e^{-c_0 t}.$$

We conclude that estimate (24) holds (using the fact that H is a nonincreasing function). □

We now explain how this Lemma (together with Theorem 1) yields a result of exponential convergence towards equilibrium with explicit rate for the Landau equation.

Using Theorem 5 of [17], we see that when $\gamma \in]0, 1]$ (that is, in the case of hard potentials or hard spheres), and when the initial datum $f_{in} \ge 0$ lies in $L^2_3(\mathbb{R}^3) \cap L^1_3(\mathbb{R}^3)$ (beware that the definition of L^p_q is different here and in [17]), then there exists a solution $f := f(t, v) \ge 0$ to the (spatially homogeneous) Landau equation $\partial_t f = Q_\gamma(f, f)$ (with initial datum f_{in}) lying in $L^\infty([1, \infty[; L^2_6(\mathbb{R}^3))$, and satisfying the entropy identity in the strong sense (this is actually true if the initial datum is bounded below by a Maxwellian and has a sufficient number of moments; if the initial datum does not fulfill those requirements, one has to approximate it by initial data fulfilling it):

$$-\frac{d}{dt} H(f(t, \cdot)|M) = D_\gamma(f(t, \cdot)),$$

where

$$H(f|M) := \int_{\mathbb{R}^3} f(v) \log f(v) dv - \int_{\mathbb{R}^3} M(v) \log M(v) dv,$$

and $M(v) = (2\pi)^{-3/2} e^{-\frac{|v|^2}{2}}$.

At this point we assume the normalization $\int_{\mathbb{R}^3} f_{in}(v) \begin{pmatrix} 1 \\ v \\ |v|^2 \end{pmatrix} dv = \begin{pmatrix} 1 \\ 0 \\ 3 \end{pmatrix}$, and apply Theorem 1 (more precisely Remark 1) with $f := f(t, \cdot)$ (the considered solution to the Landau equation).

It is then possible to use Lemma 1 (with a time translation), using

$$H := H(f(t, \cdot)|M), \qquad D := D_\gamma(f(t, \cdot)),$$

$$q := 0.22 \, ||f||^{-2}_{L^\infty([1,\infty[; L^2_6(\mathbb{R}^3))}, \qquad c_0 := c_1 \, 200^{-1} \, ||f||^{-2}_{L^\infty([1,\infty[; L^2_6(\mathbb{R}^3))},$$

where c_1 is the constant in the logarithmic Sobolev inequality of Gross [20].

We obtain thanks to Lemma 1 that this solution decays exponentially fast, with an explicit rate, towards the equilibrium M.

It is possible to obtain the same result without assuming the normalization (but with a different Maxwellian equilibrium M) thanks to the change of unknown $f \to a\, f(b\,(\cdot - u))$, where $a, b > 0$, and $u \in \mathbb{R}^3$.

Remark 2 A similar result can be obtained for the case of Maxwell molecules ($\gamma = 0$), but in that case, direct explicit computations can be exploited in order to get the exact rate of exponential convergence.

The case of soft potentials $\gamma \in [-3, 0[$ is somewhat different. Moments estimates still exist (cf. [11] and [8]), as well as smoothness estimates when $\gamma \in [-2, 0[$ (case of moderately soft potentials, cf. for example [25] and [11]), but they are not uniform in time. It is nevertheless possible (at least in the Coulomb case $\gamma = -3$) to obtain a stretched exponential estimate of convergence towards equilibrium (cf. [8]), using methods related to those of [24]. However, it is not obvious to know if such a result can be recovered with a simplified proof using arguments similar to those used in this paper.

4 Extension to the Landau-Fermi-Dirac Equation

The Landau Fermi Dirac operator (cf. [1–4]) is defined by

$$Q_{\gamma,LFD}(f,f)(v) = \nabla_v \cdot \left\{ \int_{\mathbb{R}^3} |v-w|^{2+\gamma}\, \Pi(v-w) \Big(f(w)\,(1-\varepsilon\, f(w))\, \nabla f(v) \right.$$

$$\left. - f(v)\,(1-\varepsilon\, f(v)) \nabla f(w) \Big)\, dw \right\}, \tag{27}$$

where Π is defined by (2), and $\varepsilon > 0$ is a parameter which mesures the amount of quantum effects (due to the Pauli exclusion principle).

For this operator, we still have (at the formal level) the conservation of mass, momentum and kinetic energy, that is, for any $f := f(v) \ge 0$:

$$\int_{\mathbb{R}^3} Q_{\gamma,LFD}(f,f)(v) \begin{pmatrix} 1 \\ v_i \\ |v|^2/2 \end{pmatrix} dv = \begin{pmatrix} 0 \\ 0 \\ 0 \end{pmatrix}. \tag{28}$$

The entropy dissipation associated to the Landau-Fermi-Dirac operator also remains a nonnegative quantity, but it is not identical to the entropy dissipation associated to the Landau operator. Note that the entropy itself is different in the classical and quantum case (cf. (43)). More precisely (at the formal level), for any $f := f(v) > 0$, it writes

$$D_{\gamma,LFD}(f) := - \int_{\mathbb{R}^3} Q_{\gamma,LFD}(f,f)(v)\, [\ln f(v) - \ln(1-\varepsilon\, f(v))]\, dv \tag{29}$$

$$= \frac{1}{2} \iint_{\mathbb{R}^3 \times \mathbb{R}^3} f(v)\, (1-\varepsilon\, f(v))\, f(w)\, (1-\varepsilon\, f(w)) |v-w|^{2+\gamma}\, \Pi(v-w)$$

$$\left(\frac{\nabla f}{f\,(1-\varepsilon\, f)}(v) - \frac{\nabla f}{f\,(1-\varepsilon\, f)}(w) \right) \cdot \left(\frac{\nabla f}{f\,(1-\varepsilon\, f)}(v) - \frac{\nabla f}{f\,(1-\varepsilon f)}(w) \right) dv dw \ge 0.$$

In [2], it is shown that this quantity can be related to a quantity close to the Fisher information (more precisely, a quantum variant of F or F_{rel}). More precisely, the following estimate is presented (Proposition 2.12 of [2]): for γ satisfying assumption (3) and for $f := f(v) > 0$ such that (8) holds and

$$\forall v \in \mathbb{R}^3 \qquad 1 - \varepsilon\, f(v) \geq \kappa_0, \tag{30}$$

one has (when the quantities below are well defined)

$$\int_{\mathbb{R}^3} \left| \frac{\nabla f(v)}{f(v)(1-\varepsilon f(v))} - K v \right|^2 f(v)\, dv \leq 510\, (\min_i I_i(f))^{-3}\, \kappa_0^{-2} \max(1, B_\gamma(f))$$

$$\times \max\left(1, \int f(v)\,(1+|v|^2)^{1+\frac{\gamma}{2}}\, dv\right) J_\gamma(f)\, D_{\gamma,LFD}(f), \tag{31}$$

where $K := \frac{1}{\varepsilon}\int_{\mathbb{R}^3} \ln(1-\varepsilon f(v))\, dv$, the directional temperature $I_i(f)$ is defined in (10),

$$B_\gamma(f) := \left(\min_{i\neq j} \inf_{\sigma\in S^1} \int_{\mathbb{R}^3} \left| \sigma_1 \frac{v_i}{\langle v\rangle} - \sigma_2 \frac{v_j}{\langle v\rangle} \right|^2 f(v)\, dv \right)^{-1},$$

and

$$J_\gamma(f) := \sup_{v\in\mathbb{R}^3} \langle v\rangle^\gamma \int_{\mathbb{R}^3} f(w)\, |w-v|^{-\gamma}\, (1+|w|^2)\, dw.$$

In the same spirit as in Sect. 2, we propose an estimate which is close to (5), but somewhat simpler, and much easier to prove. It writes:

Proposition 2 *We consider γ satisfying (3). Then for all $f := f(v) > 0$ (sufficiently smooth for the terms in the estimate to make sense) satisfying the normalization (8), such that*

$$\forall v \in \mathbb{R}^3 \qquad 1 - \varepsilon\, f(v) \geq \kappa_0, \tag{32}$$

and such that

$$\left(\int_{\mathbb{R}^3} |v|^4\, f(v)\, dv\right) \left(1 + 2\sqrt{\pi}\, ||f||_{L^2(\mathbb{R}^3)}\right) \kappa_0^{-1}\, D_{\gamma,LFD}(f) \leq \frac{27}{32},$$

the following estimate holds:

$$\int_{\mathbb{R}^3} \left| \frac{\nabla f(v)}{f(v)\,(1-\varepsilon f(v))} - K\, v \right|^2 f(v)\, dv \le \left[\frac{32}{3}\, \kappa_0^{-1} \left(\int_{\mathbb{R}^3} |v|^4\, f(v)\, dv \right) \left(1 + 2\sqrt{\pi}\, ||f||_{L^2(\mathbb{R}^3)}\right) \right.$$

$$\left. + 216\, \left(1 + 2\sqrt{\pi}\, ||f||_{L^2(\mathbb{R}^3)}\right) + 24\,(3 + 2\sqrt{\pi}\, ||\, |\cdot|^2\, f||_{L^2(\mathbb{R}^3)}) \right] \kappa_0^{-2}\, D_{\gamma,LFD}(f), \tag{33}$$

where we recall that $K := \frac{1}{\varepsilon} \int_{\mathbb{R}^3} \ln(1 - \varepsilon f(v))\, dv.$

Remark 3 An important feature of the above estimate is that it is uniform with respect to ε (for $\varepsilon > 0$ such that (32) holds, like all the estimates provided in [2]). One can check that when $\varepsilon \to 0$, Proposition 2 becomes Theorem 1, up to the numerical constants. Remark 1 can also be adapted to the case of the Landau-Fermi-Dirac operator.

Proof of Proposition 2 : We first observe that

$$D_{\gamma,LFD}(f) = \frac{1}{2} \sum_{i=1}^{3} \int\!\!\int_{\mathbb{R}^3 \times \mathbb{R}^3} f(v)\,(1 - \varepsilon f(v))\, f(w)\,(1 - \varepsilon f(w))$$

$$\times\ |v - w|^{\gamma}\, [r^f_{i,i+1}(v, w)]^2\, dv dw,$$

where for $i, j = 1, 2, 3$,

$$r^f_{i,j}(v, w) := (v_i - w_i) \left(\frac{\partial_j f}{f\,(1 - \varepsilon f)}(v) - \frac{\partial_j f}{f\,(1 - \varepsilon f)}(w) \right)$$

$$- (v_j - w_j) \left(\frac{\partial_i f}{f\,(1 - \varepsilon f)}(v) - \frac{\partial_i f}{f\,(1 - \varepsilon f)}(w) \right), \tag{34}$$

and, by definition, $r^f_{i,4} := r^f_{i,1}$.

Using the normalization (8) and assuming without loss of generality that (11) holds, we multiply (34) by $f(w)$ and $w_i\, f(w)$ and integrate on $\mathbb{R}^3$ in the w-variable. We get the identities, when $i \neq j$,

$$v_i\, \frac{\partial_j f}{f\,(1 - \varepsilon f)}(v) - v_j\, \frac{\partial_i f}{f\,(1 - \varepsilon f)}(v) = \int_{\mathbb{R}^3} f(w)\, r^f_{i,j}(v, w)\, dw, \tag{35}$$

$$I_i(f)\, \frac{\partial_j f}{f\,(1 - \varepsilon f)}(v) - K\, v_j + \frac{1}{\varepsilon} \int_{\mathbb{R}^3} \ln(1 - \varepsilon f(w))\, w_j\, dw = - \int_{\mathbb{R}^3} w_i\, f(w)\, r^f_{i,j}(v, w)\, dw. \tag{36}$$

This last identity can be rewritten as

$$I_i(f)\left(\frac{\partial_j f}{f\,(1-\varepsilon f)}(v) - K\,v_j\right) = -(I_i(f)-1)\,K\,v_j$$

$$-\frac{1}{\varepsilon}\int_{\mathbb{R}^3} \ln(1-\varepsilon f(w))\,w_j\,dw - \int_{\mathbb{R}^3} w_i\,f(w)\,r^f_{i,j}(v,w)\,dw. \tag{37}$$

Multiplying identity (35) by $v_i\,v_j\,f(v)\,(1-\varepsilon f(v))$ and integrating on $\mathbb{R}^3$ in the v-variable, we see that

$$I_j(f) - I_i(f) = \int\int_{\mathbb{R}^3\times\mathbb{R}^3} f(v)\,(1-\varepsilon f(v))\,f(w)\,v_i\,v_j\,r^f_{i,j}(v,w)\,dvdw$$

so that, using again the Cauchy-Schwarz inequality as in the case of the Landau equation,

$$|I_j(f) - I_i(f)| \le \sqrt{\frac{2}{3}}\,\kappa_0^{-1/2}\,D_{\gamma,LFD}(f)^{1/2}\,s_\gamma(f)^{1/2},$$

where $s_\gamma(f)$ is defined by (18). As a consequence, once again as in the case of the Landau equation,

$$|I_i(f) - 1|^2 \le \frac{8}{27}\,\kappa_0^{-1}\,D_{\gamma,LFD}(f)\,s_\gamma(f).$$

We deduce from this estimate that $I_i(f) \ge \frac{1}{2}$ as soon as

$$D_{\gamma,LFD}(f)\,s_\gamma(f)\,\kappa_0^{-1} \le \frac{27}{32}. \tag{38}$$

Then, multiplying (36) by $f(v)$ and integrating on $\mathbb{R}^3$ with respect to the variable v, we see that

$$\frac{1}{\varepsilon}\int_{\mathbb{R}^3} \ln(1-\varepsilon f(w))\,w_j\,dw = \int\int_{\mathbb{R}^3\times\mathbb{R}^3} f(v)\,f(w)\,w_i\,r^f_{i,j}(v,w)\,dvdw,$$

so that

$$\left|\frac{1}{\varepsilon}\int_{\mathbb{R}^3} \ln(1-\varepsilon f(w))\,w_j\,dw\right| \le \left(\int\int_{\mathbb{R}^3\times\mathbb{R}^3} f(v)\,f(w)\,|v-w|^\gamma\,|r^f_{i,j}(v,w)|^2\,dvdw\right)^{1/2}$$

$$\times\left(\int\int_{\mathbb{R}^3\times\mathbb{R}^3} f(v)\,f(w)\,|v-w|^{-\gamma}\,|w_i|^2\,dvdw\right)^{1/2}$$

$$\le \sqrt{2}\,\kappa_0^{-1}\,D_{\gamma,LFD}(f)^{1/2}\,\Sigma_\gamma(f)^{1/2},$$

where

$$\Sigma_\gamma(f) := \int\int_{\mathbb{R}^3\times\mathbb{R}^3} f(v)\, f(w)\, |v-w|^{-\gamma}\, |w|^2\, dvdw.$$

Under condition (38), we see that estimate (37) leads to

$$\int_{\mathbb{R}^3} \left| \frac{\partial_j f}{f\,(1-\varepsilon f)}(v) - K\, v_j \right|^2 f(v)\ \mathrm{dv} \le 12\,(I_i(f)-1)^2\, K^2\, I_j(f) + 24\,\kappa_0^{-2}\, D_{\gamma,LFD}(f)\, \Sigma_\gamma(f)$$

$$+\ 12 \int_{\mathbb{R}^3} f(v) \left[\int_{\mathbb{R}^3} f(w)\, |v-w|^\gamma\, |r^f_{i,j}(v,w)|^2\, dw \right] dv\ \sigma_\gamma(f),$$

so that finally, summing from $j=1$ to 3 (and choosing $i=j+1$)

$$\int_{\mathbb{R}^3} \left| \frac{\nabla f}{f\,(1-\varepsilon\, f)}(v) - K\, v \right|^2 f(v)\, dv \le \left[\frac{32}{3}\,\kappa_0^{-1}\, K^2\, s_\gamma(f) + 72\,\kappa_0^{-2}\, \Sigma_\gamma(f) \right.$$

$$\left. +\ 24\,\kappa_0^{-2}\, \sigma_\gamma(f) \right] D_{\gamma,LFD}(f).$$

At this level, we observe that (working as for establishing (21)),

$$\Sigma_\gamma(f) \le 3\, \left(1 + 2\sqrt{\pi}\, ||f||_{L^2(\mathbb{R}^3)}\right). \tag{39}$$

Observing that

$$|K| \le \int_{\mathbb{R}^3} \frac{f(v)}{1-\varepsilon\, f(v)}\, dv \le \kappa_0^{-1},$$

and remembering (21), (22) and (39), we can conclude the proof of Proposition 2. □

We end up this section by briefly explaining how Proposition 2 leads to a result of exponential convergence with an explicit and ε-independent (for ε not too large) rate towards equilibrium for the solutions to the (spatially homogeneous) Landau-Fermi-Dirac equations with hard potentials or hard spheres (that is $\gamma\in]0,1]$). Since the proof closely follows the computations and estimates of [2], we do not present all details.

First, observing that $\int_{\mathbb{R}^3} f(v)\, |v|^4\, dv$, $||f||_{L^2(\mathbb{R}^3)}$ and $||\,|\cdot|^2\, f||_{L^2(\mathbb{R}^3)}$ can be estimated by $||f||_{L^2_6(\mathbb{R}^3)}$ (see Remark 1), we see thanks to Proposition 2 that for some $A_f, B_f > 0$ depending only on $||f||_{L^2_6(\mathbb{R}^3)}$ and κ_0, the estimate

$$\int_{\mathbb{R}^3} \left| \frac{\nabla f(v)}{f(v)\,(1-\varepsilon f(v))} - K\, v \right|^2 f(v)\, dv \le B_f\, D_{\gamma,LFD}(f) \tag{40}$$

holds for all normalized $f>0$ such that (32) is satisfied, and such that

$$A_f \, D_{\gamma,LFD}(f) \le 1.$$

Estimate (40) can be rewritten (see [2], Sect. 3.1) as

$$\int_{\mathbb{R}^3} \left| \frac{\nabla f(v)}{f(v)\,(1-\varepsilon f(v))} + 2b_\varepsilon \, v \right|^2 f(v)\, dv - 3\,(K + 2b_\varepsilon)^2 \le B_f \, D_{\gamma,LFD}(f), \tag{41}$$

where b_ε is defined (for ε not too large) in [2], as the unique real number such that

$$\int_{\mathbb{R}^3} \frac{a_\varepsilon \, e^{-b_\varepsilon |v|^2}}{1 + \varepsilon \, a_\varepsilon \, e^{-b_\varepsilon |v|^2}} \begin{pmatrix} 1 \\ |v|^2 \end{pmatrix} dv = \begin{pmatrix} 1 \\ 3 \end{pmatrix}.$$

Using the estimate (extracted from [2], Lemma 3.1)

$$|K + 2b_\varepsilon| \le \frac{2\varepsilon}{\kappa_0^2} \max(||f||_\infty, ||\mathscr{M}_\varepsilon||_\infty) \, ||f - \mathscr{M}_\varepsilon||_{L^1(\mathbb{R}^3)},$$

where

$$\mathscr{M}_\varepsilon := \frac{a_\varepsilon \, e^{-b_\varepsilon |v|^2}}{1 + \varepsilon \, a_\varepsilon \, e^{-b_\varepsilon |v|^2}},$$

we obtain the estimate

$$\int_{\mathbb{R}^3} \left| \frac{\nabla f(v)}{f(v)\,(1-\varepsilon f(v))} + 2b_\varepsilon \, v \right|^2 f(v)\, dv - 12 \, \frac{\varepsilon^2}{\kappa_0^4} \max(||f||_\infty^2, ||\mathscr{M}_\varepsilon||_\infty^2) \, ||f - \mathscr{M}_\varepsilon||_{L^1(\mathbb{R}^3)}^2$$

$$\le B_f \, D_{\gamma,LFD}(f). \tag{42}$$

Defining

$$H(f|\mathscr{M}_\varepsilon) := S_\varepsilon(\mathscr{M}_\varepsilon) - S_\varepsilon(f), \tag{43}$$

where

$$S_\varepsilon(f) := -\frac{1}{\varepsilon} \int_{\mathbb{R}^3} \Big[\varepsilon \, f \, \log(\varepsilon f) + (1 - \varepsilon f) \, \log(1 - \varepsilon f) \Big] \, dv,$$

and using the Csiszar-Kullback inequality for Fermi-Dirac relative entropy (cf. [22])

$$||f - \mathscr{M}_\varepsilon||_{L^1(\mathbb{R}^3)}^2 \le 2 \, H(f|\mathscr{M}_\varepsilon),$$

which holds for normalized f, and the corresponding logarithmic Sobolev-like inequality (cf. [5])

$$\int_{\mathbb{R}^3} \left| \frac{\nabla f(v)}{f(v)\,(1-\varepsilon f(v))} + 2b_\varepsilon \, v \right|^2 f(v)\, dv \ge 2b_\varepsilon \, H(f|\mathscr{M}_\varepsilon),$$

we end up with

$$\left[2b_\varepsilon - 24\,\frac{\varepsilon^2}{\kappa_0^4}\,\max(||f||_\infty^2, ||\mathscr{M}_\varepsilon||_\infty^2)\right] H(f|\mathscr{M}_\varepsilon) \le B_f\, D_{\gamma,LFD}(f). \tag{44}$$

This last estimate is a functional estimate which holds for all normalized $f > 0$ such that (32) is satisfied, under the condition $A_f\, D_{\gamma,LFD}(f) \le 1$.

We now consider a normalized initial datum $f_{in} \ge 0$ in $L^1_4(\mathbb{R}^3)$, and a solution $f := f(t,v) \ge 0$ of the (spatially homogeneous) Landau-Fermi-Dirac $\partial_t f = Q_{\gamma,LFD}(f,f)$ (for $\gamma \in]0,1]$), known to satisfy the estimate

$$\sup_{t\ge t_0} ||f(t,\cdot)||_{L^2_6(\mathbb{R}^3)} + \sup_{t\ge t_0} ||f(t,\cdot)||_{L^\infty(\mathbb{R}^3)} \le C_{t_0},$$

for some $C_{t_0} > 0$ which depends on $t_0 > 0$ (cf. Propositions 4.1 and 4.2 of [2]).

This ensures that for $\varepsilon \in]0,\varepsilon^*]$, where $\varepsilon^* > 0$ is given (and can be made explicit), the quantities $A_{f(t,\cdot)}$, $B_{f(t,\cdot)}$, $||f(t,\cdot)||_\infty$ are bounded when $t \ge t_0$, and (32) holds for $f(t,\cdot)$.

At this level, it is possible to use Lemma 1 since the entropy identity

$$-\frac{d}{dt} H(f(t,\cdot)|\mathscr{M}_\varepsilon) = D_{\gamma,LFD}(f(t,\cdot))$$

holds for the considered solutions of the Landau-Fermi-Dirac equation, and to conclude.

Acknowledgements This paper is dedicated to the memory of Maria Conceicao Carvalho, who played a pionneering role in the establishment of the entropy-entropy dissipation method in kinetic theory.

References

1. Alonso, R., Bagland, V., Lods, B.: Long time dynamics for the Landau-Fermi-Dirac equation with hard potentials. J. Differential Equations **270**, 596–663 (2021)
2. R. Alonso, V. Bagland, L. Desvillettes, and B. Lods: About the use of entropy production for the Landau-Fermi-Dirac equation. *Journal of Statistical Physics* **183**, (2021)
3. Alonso, R., Bagland, V., Desvillettes, L., Lods, B.: About the Landau-Fermi-Dirac equation with moderately soft potentials. Arch. Rational Mech. Anal **244**, 779–875 (2022)
4. Bagland, V.: Well-posedness for the spatially homogeneous Landau-Fermi-Dirac equation for hard potentials. Proc. Roy. Soc. Edinburgh Sect. A. **134**, 415–447 (2004)
5. Carrillo, J.A., Laurençot, Ph., Rosado, J.: Fermi-Dirac-Fokker-Planck equation: well-posedness and long-time asymptotics. J. Differential Equations **247**, 2209–2234 (2009)
6. Carlen, E.A., Carvalho, M.C.: Strict entropy production bounds and stability of the rate of convergence to equilibrium for the Boltzmann equation. J. Stat. Phys. **67**(3–4), 575–608 (1992)

7. Carlen, E.A., Carvalho, M.C.: Entropy production estimates for Boltzmann equations with physically realistic collision kernels. J. Stat. Phys. **74**(3–4), 743–782 (1994)
8. Carrapatoso, K., Desvillettes, L., He, L.: Estimates for the large time behavior of the Landau equation in the Coulomb case. Arch. Rational Mech. Anal. **224**, 381–420 (2017)
9. Chapman, S., Cowling, T.G.: The mathematical theory of non-uniform gases. Cambridge Univ. Press, London (1952)
10. Desvillettes, L.: Entropy dissipation rate and convergence in kinetic equations. Commun. Math. Phys. **123**(4), 687–702 (1989)
11. Desvillettes, L.: Entropy dissipation estimates for the Landau equation in the Coulomb case and applications. J. Funct. Anal. **269**, 1359–1403 (2015)
12. L. Desvillettes. Entropy dissipation estimates for the Landau equation: General cross sections. Proceedings of the conference PSPDE III, Braga, 2014
13. L. Desvillettes. Structure entropique du noyau de collision de Landau, Proceedings of the seminar Laurent Schwartz - EDP et applications (2014-2015), Exp. No. 14
14. L. Desvillettes. About Boltzmann's H Theorem for Landau Equation (Autour du Théorème H de Boltzmann), Proceedings of the seminar Laurent Schwartz - EDP et applications (2019-2020), Exp. No. 9
15. L. Desvillettes, L.-B. He and J.-C. Jiang. A new monotonicity formula for the spatially homogeneous Landau equation with Coulomb potential and its applications. Preprint arXiv:2011.00386. Accepted for publication in the J. European Math. Society
16. Desvillettes, L., Mouhot, C., Villani, C.: Celebrating Cercignani's conjecture for the Boltzmann equation. Kinet. Relat. Models **4**, 277–294 (2011)
17. L. Desvillettes and C. Villani. On the spatially homogeneous Landau equation for hard potentials. Part I. Existence, uniqueness and smoothness. *Commun. Partial Differential Equations*, **25**, n.1-2 (2000), 179-259
18. L. Desvillettes and C. Villani. On the spatially homogeneous Landau equation for hard potentials. Part II. H-Theorem and applications. *Commun. Partial Differential Equations*, **25**, n.1-2 (2000), 261-298
19. Golse, F., Gualdani, M.-P., Imbert, C., Vasseur, A.: Partial Regularity in Time for the Space-Homogeneous Landau equation with Coulomb Potential. Ann. Sci. Ecole Normale Supérieure **55**(6), 1575–1611 (2022)
20. Gross, L.: Logarithmic Sobolev inequalities. Amer. J. Math. **97**, 1061–1083 (1975)
21. Lifschitz, E.M., Pitaevskii, L.P.: Physical kinetics. Perg. Press, Oxford (1981)
22. Lu, X., Wennberg, B.: On stability and strong convergence for the spatially homogeneous Boltzmann equation for Fermi-Dirac particles. Arch. Ration. Mech. Anal. **168**, 1–34 (2003)
23. Toscani, G.: Entropy production and the rate of convergence to equilibrium for the Fokker-Planck equation. Quart. Appl. Math. **57**(3), 521–541 (1999)
24. Toscani, G., Villani, C.: On the trend to equilibrium for some dissipative systems with slowly increasing a priori bounds. J. Statist. Phys. **98**(5–6), 1279–1309 (2000)
25. Wu, K.-C.: Global in time estimates for the spatially homogeneous Landau equation with soft potentials. J. Funct. Anal. **266**, 3134–3155 (2014)

The Entropic Journey of Kac's Model

Amit Einav

Abstract The goal of this paper is to review the advances that were made during the last few decades in the study of the entropy, and in particular the entropy method, for Kac's many particle system.

Keywords Kac's model · Mean field limits · Convergence to equilibrium · Spectral gap · The entropy method

1 Introduction

Our fascination with a mathematical description for real life phenomena can be viewed as a cornerstone in the birth of the mathematical subject known as *(modern) Analysis* back in the 17th century. Our desire to understand such phenomena has not lessened over the course of the centuries—if anything, our expanding familiarity with the universe and its mysteries only deepened it. Particular interest, especially in the last century, is given to phenomena which involve many elements.

Systems that revolve around many particles, substances, animals or agents are all around us—from the air we breath to the galaxies around us, herds of buffaloes roaming the African plains and red blood cells coursing through our bodies. Even our attempts to navigate the crowd in a busy shopping centre is another example of a system of many elements. As common as such systems are, their mathematical investigation is far from simple.

The goal of our paper is to give a short overview on how we mathematically treat such systems and to focus our attention on one particular important model: Kac's many particle model. Kac's model, introduced in 1956 by Mark Kac (see [15]), was

---

In loving memory of Maria Conceição Carvalho (São), who is always with us in our memories and our hearts.

A. Einav (✉)
School of Mathematical Sciences, Durham University, Upper Mountjoy Campus, Stockton Road, DH1 3LE Durham, United Kingdom
e-mail: amit.einav@durham.ac.uk

E. Carlen et al. (eds.), *From Particle Systems to Partial Differential Equations*,
Springer Proceedings in Mathematics & Statistics 465,
https://doi.org/10.1007/978-3-031-65195-3_4

conceived to give a probabilistic justification to the famous Boltzmann equation[1] and is one of the first examples for the *mean field limit* procedure.

Our paper will almost solely concern itself with the *functional* study of this model and its convergence to equilibrium via the so-called *entropy method*.[2] Many contribution to this study, as well as to the Boltzmann equation which is intimately connected to Kac's model, have been made over the last 70 years. For the sake of brevity we will only mention works that are connected to the topic we present here, and even that will be mostly limited to those papers that are directly related to the results we will present.

Further information about the history and study of Kac's model and the Boltzmann equation can be found in [9, 17, 19] and references within.

1.1 Many Elements Systems—The Microscopic, Macroscopic and Mesoscopic Framework

When one considers systems of many elements there are usually three frameworks one can adopt: *microscopic, macroscopic and mesoscopic.*

Microscopic Approach

This approach is, in a sense, the oldest and the most accurate approach. In the microscopic framework we view each element in the system as an individual and consider, and attempt to solve, the phase space trajectorial equations associated to these elements. In the common case where the interaction between the elements of the system is binary, i.e. happens only between two elements, a typical microscopic description of a system is given by

$$\begin{aligned} \frac{d}{dt}x_i(t) &= v_i(t), \\ \frac{d}{dt}v_i(t) &= \frac{1}{N}\sum_{j\neq i} K\left(x_i(t)-x_j(t)\right), \end{aligned} \tag{1}$$

where the function K represents the force of the interaction between the ith and jth elements and is assumed to depend only on the relative distance between them. The factor of $1/N$ appearing next to the sum of the second equation (corresponding to Newton's second law) makes sure that the force acting on each element is of order of magnitude 1.

[1] As opposed to the rigorous derivation of it from Newtonian mechanics, which is known as Hilbert's 6th problem.

[2] We will be remiss if we won't mention that there are many other studies on the convergence to equilibrium, such as the impressive [17].

While extremely accurate in theory, solving a system such as (1) is not really feasible in most cases—even when we have theorems that assure us that there exists a solution. Our significant computational power is not enough to solve even relatively simple systems of this form. However, in most cases we don't really need to.

Macroscopic Approach

Faced with the seemingly herculean task of trying to solve microscopic systems of equations, people have realised that for all intent and purposes we do *not* need to understand the individual behaviour of each element of the system in order to understand its behaviour. Indeed, as most early systems that were investigated revolved around particles of gases or fluids, systems that comprise of elements that are not perceptible to the human eye, it was understood that the behaviour we are looking for is more *statistical*, or *probabilistic*, in nature. The core idea of the macroscopic approach is to consider a part of a system as a "unit" whose evolution we would like to investigate. This part of the system should be sufficiently large to include enough elements to facilitate probabilistic considerations, yet sufficiently small with respect to the entire system to be considered as "point mass". Macroscopic approaches are the realm of *Fluid Dynamics* and typical equations in it include the Euler equation

$$\frac{\partial u}{\partial t} + (u \cdot \nabla)\, u = -\frac{1}{\rho}\nabla p + F, \tag{2}$$

and the Navier-Stokes equation[3]

$$\frac{\partial u}{\partial t} + (u \cdot \nabla)\, u = -\frac{1}{\rho}\nabla p + \nu \Delta u, \tag{3}$$

where u is the velocity field of the fluid, ρ is the density of it, p is the pressure, F is the external force, and ν is the viscosity of the fluid.

Mesoscopic Approach

The mesoscopic approach appeared around the end of the 19th century in the mathematical study of the kinetic theory of gasses. It was spearheaded by figures such as Maxwell and Boltzmann. This approach is an "in between" approach between the microscopic and macroscopic. The object of investigation of mesoscopic equations is an *average element* of the system. Consequently, mesoscopic equations usually consider the evolution of probability density functions.

An extremely important (and one of the first) mesoscopic equation is the so-called Boltzmann equation, given by

[3] Equation (3) represents the Navier-Stokes equation without external force.

$$\partial_t f(t,x,v) + v \cdot \nabla_x f(t,x,v) = Q_B(f,f)(t,x,v), \tag{4}$$

with

$$Q_B(f,f)(x,v) = \int_{\mathbb{R}^d \times \mathbb{S}^{d-1}} B(v, v_*, \sigma) \left(f(x, v') f(x, v'_*) - f(x,v) f(x, v_*) \right) dv_* d\sigma,$$

where

$$\begin{aligned} v' &= \frac{v+v_*}{2} + \frac{|v-v_*|\,\sigma}{2}, \\ v'_* &= \frac{v+v_*}{2} - \frac{|v-v_*|\,\sigma}{2}, \end{aligned} \tag{5}$$

and $B(v, v_*, \sigma)$ being the collision kernel (more on which shortly).

The Boltzmann equation describes the evolution of an average particle in dilute (or rarefied) gas with no external forces which is governed by two processes:

1. The left hand side of (4), known as the *transport part* of the equation, represents the free motion of the average particle prior to its collision with another particle. It is based on the simple transport equation

$$\begin{aligned} &\partial_t f(t,x,v) + v \cdot \nabla_x f(t,x,v) = 0, \\ &f(0,x,v) = f_0(x,v), \end{aligned}$$

   whose solution

$$f(t,x,v) = f_0(x - vt, v)$$

   is constant on trajectories of paths of constant velocities. The transport part is very common in many physically relevant mesoscopic equations.
2. The right hand side of (4), known as the *collision or interaction* part, represents the effect of the interactions, collisions in the case of gases, on the evolution of the probability density. In the case of the Boltzmann equation, and many other mesoscopic equations, these collisions happen when two (average) particles reach the same spatial position. How will this affect our "possibility" to find a particle at time t in this position x and velocity v? We need to consider two scenarios:

- The particles arrived at velocities v' and v'_* to the position x and collided in a way that preserves the energy and momentum. Consequently, they have scattered in accordance to a unit vector σ and ended up with velocities v and v_*, thus *increasing* the possibility to find an average particle at position x and velocity v. The connection between v', v'_*, v, v_*, and the scattering angle σ is given by (5)[4] and this process is encoded in the *gain* term of Q_B

[4] One can show that this relation is the only viable one if one assumes conservation of energy and momentum.

$$\int_{\mathbb{R}^d \times \mathbb{S}^{d-1}} B\left(v, v_*, \sigma\right) f\left(x, v'\right) f\left(x, v'_*\right) dv_* d\sigma.$$

The function $B\left(v, v_*, \sigma\right)$, the *collision kernel*, gives us the intensity of such collision and depends only on the physics of the problem. The integration over σ and v_* is present to take into account all possible scattering and resulting additional (less relevant) velocity. There is, however, some subtle but fundamental assumption here: the collision process we've described speaks about two (average) particles being at the same position x with velocities v' and v'_* respectively. This information is encoded in the function $f_2\left(x, x, v', v'_*\right)$ where f_2 is the *joint probability density* of these average particles. The gain term, however, contains the expression $f\left(x, v'\right) f\left(x, v'_*\right)$ instead. This implies an assumption of *independence* between these particles prior to the collision. This is known as Boltzmann's *molecular chaos* or *Stosszahlansatz*.

- The same consideration as above, together with Boltzmann's molecular chaos assumption, gives us the *loss* term in Q_B

$$-\int_{\mathbb{R}^d \times \mathbb{S}^{d-1}} B\left(v, v_*, \sigma\right) f\left(x, v\right) f\left(x, v_*\right) dv_* d\sigma$$

which expresses the *decrease* in the possibility to find an average particle at position x and velocity v and corresponds to the process that the two particles arrived to the point x at velocities v and v_* and their resulting post collision velocities do not include (up to a set of measure zero) the velocity v.

It is worth to mention that in most situations the collision kernel B is given by

$$B\left(v, v_*, \sigma\right) = \left|v - v_*\right|^\gamma b\left(\cos\theta\right). \tag{6}$$

with γ, known as the *hardness of the potential*, being in $[-d, 1]$.

Nowadays we consider many equations to be "kinetic equations", even when they do not have a clear mesoscopic interpretation, and use common tools to investigate them. Examples include Fokker-Planck equations, coagulation and fragmentation equations, and reversible and irreversible reaction-diffusion equations.

1.2 Moving Between the Microscopic, Macroscopic, and Mesoscopic Frameworks

It is evident from the description in Sect. 1.1 that the different frameworks one has for equations and models that pertain to many element phenomena come in hierarchical order. It shouldn't come as a big surprise then that we can try and "move" between these approaches. The process of inferring an equation in one framework from another framework requires some limiting procedure. The two common procedures we employ are expressed in the following diagram:

$$\text{Microscopic} \xrightarrow[\text{Mean Field Limits}]{} \text{Mesoscopic} \xrightarrow[\text{Hydrodynamical Limits}]{} \text{Macroscopic}$$

We will mostly focus on mean field limits, but would mention hydrodynamical limits for the sake of completion.

From Meso to Macro—Hydrodynamical Limits

The mesoscopic framework "lives" in the microscopic world of individual elements yet concerns itself only with average "individual" behaviours. These behaviours, we imagine, are enough for us to understand how the macroscopic "unit" evolves. In order for us to be able to achieve this goal, however, we must scale our spatial and time variables since these "units" of the macroscopic world contain a large quantity of mesoscopic elements, and the time scale where phenomena occur in these levels is not of the same order of magnitude.

This scaling is usually done by introducing a new "smallness" parameter, pertaining to the "zooming out" in space and time and the impact of the scaling on the strength of interactions between the elements. One can use such approach, which known as *hydrodynamical limits*, to find connection between fluid equations, such as the Euler equation, and the Boltzmann equation. As this topic is outside the remit of this review paper, we will not elaborate more on it.

From Micro to Meso—Mean Field Limits

It shouldn't come as a shock that we should expect to be able to get information on the behaviour of an average element of a system, i.e. on the mesoscopic framework, from the complete deterministic behaviour of every element in it, i.e. the microscopic framework.

The first step one must undertake to achieve such a goal is to recast the trajectorial microscopic equations, expressed in (1), in a more probabilistic way. Let us, thus, consider a system of N elements, each of which lives in the phase space $\mathscr{X} \times \mathscr{V}$. In order to consider the problem from a probabilistic perspective we need to explore the *probability density function* of the entire ensemble, which we will denote by $F_N(\boldsymbol{X}_N, \boldsymbol{V}_N)$ with $(\boldsymbol{X}_N, \boldsymbol{V}_N) \in (\mathscr{X} \times \mathscr{V})^N$. As F_N remains constants on the trajectories of the individual elements of the system we find that the system of Eq. (1) is equivalent, at least formally, to the following equation

$$\begin{aligned} &\partial_t F_N(\boldsymbol{X}_N, \boldsymbol{V}_N) + \boldsymbol{V}_N \cdot \nabla_{\boldsymbol{X}_N} F_N(\boldsymbol{X}_N, \boldsymbol{V}_N) \\ &\qquad + \frac{1}{N} \sum_{i=1}^{N} \sum_{i \neq j} K\left(x_i - x_j\right) \cdot \nabla_{v_i} F_N(\boldsymbol{X}_N, \boldsymbol{V}_N) = 0 \end{aligned} \tag{7}$$

which is known as the *master equation* or the *Liouville equation*.

As the mesoscopic approach concerns itself with only one average element, we need to reduce (7) to an equation that relates to such element. Before we do so we notice that in order for the above to make sense, and indeed in order for us to even be able to consider a probabilistic approach here, we shouldn't be able to distinguish between the elements of the system. Functionally, what this means is that the probability density function must be *symmetric* under permutation of its phase-space variables, i.e. for any σ the group of permutation of $\{1, \ldots, N\}$, we must have that

$$F_N(x_1, \ldots, x_N, v_1, \ldots, v_N) = F_N\left(x_{\sigma(1)}, \ldots, x_{\sigma(N)}, v_{\sigma(1)}, \ldots, v_{\sigma(N)}\right).$$

The evolution of one average element of the system, therefore, is determined by an equation for the *first marginal* of F_N which we will denote by $F_{N,1}$. Since F_N is symmetric, it doesn't matter which variable we consider. To find the evolution of this marginal we need to integrate out the remaining variables in (7). This, however, will not yield a closed equation since the term that involves the function $K\left(x_i - x_j\right)$ causes correlations that can't be integrated out. Indeed, using our symmetry assumption we find that the evolution of $F_{N,1}$ is given by

$$\begin{aligned} &\partial_t F_{N,1}(x_1, v_1) + v_1 \cdot \nabla_{x_1} F_{N,1}(x_1, v_1) \\ &+ \frac{N-1}{N} \int_{\mathcal{X} \times \mathcal{V}} K(x_1 - x_2) \cdot \nabla_{v_1} F_{N,2}(x_1, x_2, v_1, v_2)\, dx_2 dv_2 = 0. \end{aligned} \tag{8}$$

An attempt to find the evolution of the second marginal, $F_{N,2}$, in order to close the above equation will result in an equation that will depend (via an integration just like above) on $F_{N,3}$. In general one can form a hierarchy of N equations, known as the *BBGKY*[5] hierarchy, which, up to the Nth level, details the evolution of $F_{N,k}$ with respect to $F_{N,k+1}$.

As can be seen from the above, we can't expect to be able to close our hierarchy as it stands. At this point there are two crucial observation we would like to make:

(1) In order for us to be able to justify considering an average element of the system we must have an enormous amount of elements in it. Mathematically speaking, this means that we would like to consider the limit $N \to \infty$.
(2) Most system that we consider in our investigations do not evolve "randomly" (for lack of a better word) and we expect some phenomenon to emergence, a sort of *asymptotic correlation* between the elements that becomes more and more pronounced as the number of elements in the system increases. One example, which came from the study of dilute gases, is that since in such gases particles hardly see each other, one could expect that any *fixed* group of k particles will become more and more *independent* as the number of gas particle increases. In other words,

[5] Bogoliubov-Born-Green-Kirkwood-Yvon.

$$\begin{aligned} F_{N,1}(x_1, v_1) &\underset{N \text{ large}}{\approx} f(x_1, v_1), \\ F_{N,2}(x_1, x_2, v_1, v_2) &\underset{N \text{ large}}{\approx} f(x_1, v_1) f(x_2, v_2), \\ &\vdots \\ F_{N,k}(\boldsymbol{X}_k, \boldsymbol{V}_k) &\underset{N \text{ large}}{\approx} f^{\otimes k}(\boldsymbol{X}_k, \boldsymbol{V}_k). \end{aligned} \tag{9}$$

The condition expressed in (9), which we will get back to in the next section, is known as *chaos* or *chaoticity*. Assuming that this condition holds under the evolution of the master equation we find that taking the limit of the number of elements to infinity in (8) gives the equation

$$\partial_t f(x, v) + v \cdot \nabla_x f(x, v) + (K * \rho)(x) \nabla_v f(x, v) = 0,$$

where $\rho(x) = \int_{\mathcal{V}} f(x, v) dv$. The above represents the equation of a *limiting* average element.

The process of using asymptotic correlation on a marginal equation to attain a limit equation is known as the *mean field limit* procedure. The solution to the limiting equation is known as the *mean field limit*. There are a few things we should emphasise at this point:

- The true starting point of the mean field limit procedure was not the microscopic framework and the set of Eq. (1)—it was the master equation (7). This observation allows us to start our mean field study *not* from a deterministic trajectorial system but from so-called average model, expressed via a master equation for a probability density of an ensemble of elements (which is usually governed by a Poisson jump processes). This line of investigation and the use of mean field limit approaches is an extremely active field and includes models that range from dilute gases to swarming of animals and decision making.
- Asymptotic correlations are *functional properties* that involve no evolution. As such, to properly apply a mean field limit approach one *must* show that the appropriate asymptotic correlation *propagates* with the evolution of the master equation.
- To this day there is only *one* asymptotic correlation we use in our mean field approach—chaoticity. The notion of chaos in the form presented here was conceived by Kac in his 1956 work on his particle model (see [15]).

We have now covered the background and framework for the rest of our work—the study of Kac's original particle model.

2 Kac's Model and the Study of the Convergence to Equilibrium

2.1 Kac's Model

Showing the validity of the Boltzmann equation from Newtonian mechanics is an old problem which was formally articulated in Hilbert's famous symposium at 1900. It is commonly known as *Hilbert's 6th problem*. While some progress has been made, most notable by Lanford in 1975 (for some history and a modern take on Landord's ideas see [12]), the problem remains open. In his 1956 work, [15], Kac introduced an average model (in the sense presented in the end of the previous section) to try and give a probabilistic justification to the Boltzmann equation.

Kac's model considered N gas particles which interact via a binary collision process arriving in a Poisson stream. The jump process associated to these collisions replaces the trajectorial role of the spatial variable in the original equation and consequently Kac's model is spatially homogenous. To simplify things further, Kac assumed that each particle has a one dimensional velocity. The price of this simplification is the fact that unlike the Boltzmann equation, the collisional process in Kac's model does not preserve both energy and momentum.[6]

The process that governs the evolution of Kac's model is the following: when the Poisson clock chimes, two particles who are chosen uniformly from the ensemble, say the $i$the and jth particle, collide and scatter with a random scattering angle, θ, which is distributed uniformly on $[0, 2\pi]$. The resulting new velocities of the particles are given by

$$\begin{aligned} v_i(\theta) &= v_i \cos(\theta) + v_j \sin(\theta), \\ v_j(\theta) &= -v_i \sin(\theta) + v_j \cos(\theta). \end{aligned} \tag{10}$$

Kac's evolution, i.e. his master equation, is constructed in a similar way to Boltzmann equation and includes a gain and loss terms. It is given by

$$\partial_t F_N(t, \boldsymbol{V}_N) = \mathscr{L}_N F_N = N(\mathscr{Q} - I) F_N(t, \boldsymbol{V}_N), \quad \boldsymbol{V}_N \in \mathbb{S}^{N-1}\left(\sqrt{N}\right) \tag{11}$$

where

$$\mathscr{Q}F(\boldsymbol{V}_N) = \frac{1}{\binom{N}{2}} \sum_{i<j} \frac{1}{2\pi} \int_0^{2\pi} F_N\left(R_{i,j,\theta}(\boldsymbol{V}_N)\right) d\theta,$$

[6] Kac's model can be (and have been) extended to a higher dimensional model where this has been rectified (see [16]).

with

$$\left(R_{i,j,\theta}\left(\boldsymbol{V}_N\right)\right)_l = \begin{cases} v_l & l \neq i, j, \\ v_i\left(\theta\right) & l = i, \\ v_j\left(\theta\right) & l = j. \end{cases}$$

A couple of remarks:

- The factor N, appearing before the gain and loss term (represented by $\mathscr{Q}$ and I), is there to make sure that the average time between two collisions is of order of magnitude 1 and *is independent in* N. This is crucial if we want to achieve a meaningful limit when N goes to infinity.
- The phase space in Kac's model is $\mathbb{S}^{N-1}\left(\sqrt{N}\right)$ and *not* $\mathbb{R}^N$. The reason behind this is that the collision process described above conserves the energy of the ensemble (but not the momentum). As the particles in this model are indistinguishable we know that if one of them has kinetic energy E then all of them must have the same energy. Consequently, the total energy of the ensemble is NE. Choosing E to be $1/2$ under the assumption that the mass of all particles is 1, gives us that $\boldsymbol{V}_N \in \mathbb{S}^{N-1}\left(\sqrt{N}\right)$. The fact that we are restricting ourselves to a sphere will play an important role, and add some difficulties, in the study of the long time behaviour of the solutions to our master equation. In this work we will sometimes refer to $\mathbb{S}^{N-1}\left(\sqrt{N}\right)$ as *Kac's sphere*.

The Connection Between Kac's Model and the Boltzmann Equation

Following on the ideas presented in Sect. 1.2, one can find the first equation in the BBGKY hierarchy for Kac's master equation and attempt to perform the mean field limit procedure. A straight forward calculation (which is a bit more complicated since we are working on Kac's sphere) shows that the equation for $F_{N,1}$ is given by

$$\partial_t F_{N,1}(v) = \frac{1}{\pi}\int_0^{2\pi}\int_{\mathbb{R}}\left(F_{N,2}(v(\theta), (w(\theta)) - F_{N,2}(v, w)\right) d\vartheta dw$$

where $v\left(\theta\right)$ and $w\left(\theta\right)$ are given by the same formula as $v_i\left(\theta\right)$ and $v_j\left(\theta\right)$ in (10). In order to proceed with the mean filed limit approach on the above we require two things: an asymptotic correlation and a proof of its propagation.

The notion of *chaos*, mentioned in Sect. 2.1, was conceived by Kac for this purpose and was presented for the first time in the same paper, [15]. The formal definition of this notion, extended to general probability measures, is given by:

Definition 1 Let $\mathscr{X}$ be a Polish space. We say that a sequence of symmetric probability measures, $\mu_N \in \mathscr{P}\left(\mathscr{X}^N\right)$ with $N \in \mathbb{N}$, is μ_0-chaotic for some probability measure $\mu_0 \in \mathscr{P}\left(\mathscr{X}\right)$ if

$$\Pi_k \mu_N \underset{N\to\infty}{\overset{\text{weak}}{\longrightarrow}} \mu_0^{\otimes k}$$

where $\Pi_k \mu_N$ is the k-th marginal of μ_N.[7]

In the case of Kac's model we say that a sequence of symmetric probability density functions on $\mathbb{S}^{N-1}\left(\sqrt{N}\right)$, $\{F_N\}_{N\in\mathbb{N}}$, is f-chaotic, where f is a probability density on $\mathbb{R}$, if $F_N d\sigma_N$ is fdx-chaotic, where $d\sigma_N$ is the uniform probability measure on $\mathbb{S}^{N-1}\left(\sqrt{N}\right)$.

Following on his definition of chaoticity, Kac has shown that this notion is propagated by his master equation and concluded his model's mean field limit equation

$$\partial_t f(v) = \frac{1}{\pi}\int \left(f(v(\theta))f(w(\theta)) - f(v_1)f(v_2)\right) d\theta dw, \tag{12}$$

which is known as the *Boltzmann-Kac equation*. When comparing the above to (4), one can view it as a 1-dimensional version of the Boltzmann equation for Maxwell molecules, i.e. for the case where $\gamma = 0$ and b is a constant in (6). With that, Kac has managed to give a probabilistic justification to the validity of the Boltzmann equation.

2.2 *Convergence to Equilibrium—The Spectral Gap*

At the time of conception of Kac's model, the study of non-linear equations was not as developed as it is today. Besides giving a validation to Boltzmann equation (and establishing the mean field limit approach with his notion of chaoticity), Kac was interested in trying to use his limiting procedure to "push down" information from his many particle model, which is governed by a simple yet dimensional dependent operator, to its limit equation. In particular, a question that was, and is to this day, of great interest to the community was that of the rate of convergence to equilibrium for this equation.

A closer look at the operator that governs Kac's master equation, $\mathcal{L}_N$, reveals the following:

- $\mathcal{L}_N$ is composed of 2-dimensional rotations and averaging. As such it is a bounded linear operator from $L^p\left(\mathbb{S}^{N-1}\left(\sqrt{N}\right), d\sigma_N\right)$ to $L^p\left(\mathbb{S}^{N-1}\left(\sqrt{N}\right), d\sigma_N\right)$ for all $p \in [1,\infty]$.

[7] This condition can be reformulated using empirical measures. $\{\mu_N\}_{N\in\mathbb{N}}$ will be μ_0-chaotic if and only if for any random vector $(X_1, \ldots, X_N)$ with law μ_N we have that the random *empirical measure*

$$\mu_N = \frac{1}{N}\sum_{i=1}^{N} \delta_{X_i}$$

converges in law towards the deterministic measure μ_0.

- Continuing on the above, a simple calculation shows that $\mathscr{L}_N$ is also self-adjoint on $L^2\left(\mathbb{S}^{N-1}\left(\sqrt{N}\right), d\sigma_N\right)$.
- $\mathscr{L}_N F = 0$ if and only if F is a radial function. As our underlying space is a sphere we conclude that a normalised basis to the kernel of $\mathscr{L}_N$ in $\left(L^2\left(\mathbb{S}^{N-1}\left(\sqrt{N}\right)\right), d\sigma_N\right)$, and the only steady state to the equation with a unit mass, is $F_N \equiv 1$. The fact that we have a unique normalised steady state is another reason why one needs to restrict the ensemble to a sphere in Kac's model.

The fact that $\mathscr{L}_N$ is self-adjoint on $L^2\left(\mathbb{S}^{N-1}\left(\sqrt{N}\right), d\sigma_N\right)$ and has a one dimensional kernel implies that the convergence to the equilibrium in Kac's model, which is the probability density $F_N \equiv 1$, is determined by the *spectral gap* of the operator $\mathscr{L}_N$:

$$\Delta_N = \inf_{F_N \perp 1,\ \|F_N\|_{L^2\left(\mathbb{S}^{N-1}\left(\sqrt{N}\right), d\sigma_N\right)}=1} \langle F_N, \mathscr{L}_N F_N \rangle .$$

Kac has realised that in order to be able to utilise this information for his mean field limit he needed to find a lower bound on Δ_N that is *uniform in* N. He conjectured in his work that

$$\Delta = \inf_{N\in\mathbb{N}} \Delta_N > 0.$$

It took 44 years to resolve this conjecture. The first proof was given by Janvresse in [14] and later, in their work [7], Carlen, Carvalho and Loss showed that

$$\Delta_N = \frac{N+2}{2(N-1)}. \tag{13}$$

It is worth the note that the study of spectral gaps in many particle models, and general Markov processes that relate to the notion of chaoticity, remains a fascinating and relevant topic. We would like to mention in particular the recent work of Carlen, Carvalho and Loss [8].

While equality (13) seems very encouraging, it was known long before the conjecture was resolved that the spectral gap is not the way to go forward. The reason behind this is the fact that the distance that is associated to the spectral gap, the L^2 norm, does not adhere to the notion of chaoticity.

Intuitively speaking, if $\{F_N\}_{N\in\mathbb{N}}$ is f-chaotic then one could imagine that $F_N \approx f^{\otimes N}$. This is quite untrue as it neglects correlation that appear, but it epitomises the underlying problem of the L^2 distance as the L^2 norm of a tensorisation (at least on a tensorised space) becomes the *product* of the L^2 norms of the "generator" of this tensorisation. On Kac's sphere one can find a sequence $\{F_N\}_{N\in\mathbb{N}}$ of probability densities such that

$$\|F_N\|_{L^2(\mathbb{S}^{N-1}(\sqrt{N}), d\sigma_N)} \geq C^N$$

for some $C > 1$. Combining the above with (13) yields the sharp estimate

$$\|F_N(t) - 1\|_{L^2(\mathbb{S}^{N-1}(\sqrt{N}), d\sigma_N)} \leq e^{-\frac{(N+2)t}{2(N-1)}} \|F_N(0) - 1\|_{L^2(\mathbb{S}^{N-1}(\sqrt{N}), d\sigma_N)}$$

showing the problem we face: as the right hand side of the above dominates an expression of the form $C^N e^{-\frac{t}{2}}$ we see that we won't be able to take N to infinity and get something meaningful. Intuitively, the above shows that the time we need to wait before seeing significant decay of our solution is of order N, i.e. the number of particles.

2.3 Convergence to Equilibrium—The Entropy

The failure of the spectral gap method due to its underlying distance, as described above, made people realise that we need to consider a different type of distance with which we measure how far a solution is from equilibrium. It didn't take very long to realise that perhaps we should take a leaf from Boltzmann's book and consider the (non-linear) "distance" that is *the entropy*. Modelled after the Boltzmann relative entropy,[8] given by

$$\mathscr{H}(f|f_\infty) = \int_{\mathbb{R}} h\left(f(x)|f_\infty(x)\right) dx$$

where $h\left(x|y\right) = x \log\left(\frac{x}{y}\right) - x + y$ and f_∞ is the equilibrium of the equation, Kac's entropy is given by

$$\mathscr{H}_N(F_N) = \int_{\mathbb{S}^{N-1}(\sqrt{N})} F_N \log F_N d\sigma_N. \tag{14}$$

An indication that we have indeed chosen the right "distance" comes from its adherence to the notion of chaoticity, at least intuitively. Assuming that $F_N \approx f^{\otimes N}$ one can show that

$$\mathscr{H}_N\left(F_N\right) \approx \int_{\mathbb{S}^{N-1}\left(\sqrt{N}\right)} f^{\otimes N}\left(\boldsymbol{V}_N\right) \log\left(f^{\otimes N}\left(\boldsymbol{V}_N\right)\right) d\sigma_N = \sum_{i=1}^{N} \int_{\mathbb{S}^{N-1}\left(\sqrt{N}\right)} f^{\otimes N}\left(\boldsymbol{V}_N\right) \log\left(f\left(v_i\right)\right) d\sigma_N$$

$$\underset{\text{symmetry}}{=} N \int_{\mathbb{S}^{N-1}\left(\sqrt{N}\right)} f^{\otimes N}\left(\boldsymbol{V}_N\right) \log\left(f\left(v_1\right)\right) d\sigma_N \approx N \mathscr{H}\left(f|\mathscr{M}\right)$$

with $\mathscr{M}(v) = (2\pi)^{-1/2} e^{-v^2/2}$ appearing due to the integration of $\prod_{j\neq 1} f(v_j)$ against $d\sigma_N$. The above tells us, at least informally, that if $\mathscr{H}_N\left(F_N(t)\right)$ converges exponentially to zero with a rate that is independent of N, then a simple rescaling by N

[8] We call is a *relative* entropy as it measures a function relative to another, in our case the equilibrium. There are many possibilities for entropies and we refer the reader to [1] where they can find some common examples used in other kinetic equations.

of the functional should give us an exponential convergence to equilibrium for the functional $\mathscr{H}\left(f(t)|\mathscr{M}\right)$ in the limit equation.

The next natural question is, thus, *how can we estimate quantitatively the convergence of* $\mathscr{H}_N$ *to zero?*

The method we'll use to explore this question is known as *the entropy method.* Let us motivate the ideas of this method by going back to spectral gaps.
Consider the equation

$$\partial_t u = Lu,$$

where L is a semi-definite negative operator whose kernel is one dimensional. Let u_∞ be a normalised basis for this kernel which corresponds to the unique normalised equilibrium of the system. The evolution of the natural distance, the L^2 norm, between the solution to the above, $u(t)$, and the equilibrium u_∞ is given by

$$\frac{d}{dt}\|u(t)-u_\infty\|^2 = 2\left\langle u(t)-u_\infty, L\left(u(t)-u_\infty\right)\right\rangle.$$

The right hand side of the above expression brings to mind the fact that (in certain settings) we can find the eigenvalues of a self-adjoint operator L by looking at $\langle u, Lu\rangle$ with $\|u\| = 1$. The spectral gap of L, λ, can be shown to equal

$$\lambda = \inf_{f\perp u_\infty,\ \|f\|=1} \langle f, -Lf\rangle$$

and consequently we see that as long as $\langle u(t), u_\infty\rangle = 1$ for $t \geq 0$, which is guaranteed if and only if $\langle u(0), u_\infty\rangle = 1$,[9] we have that[10]

$$\frac{d}{dt}\|u(t)-u_\infty\|^2 = 2\left\langle u(t)-u_\infty, L\left(u(t)-u_\infty\right)\right\rangle \leq -2\lambda\|u(t)-u_\infty\|^2.$$

This differential inequality results in an exponential convergence to equilibrium with explicit rate which is the spectral gap

$$\|u(t)-u_\infty\| \leq e^{-\lambda t}\|u(0)-u_\infty\|.$$

At this point we notice that we managed to achieve a concrete rate of convergence by finding a *functional inequality* (in our case based on a spectral property) that

---

[9]

$$\frac{d}{dt}\langle u(t), u_\infty\rangle = \langle Lu(t), u_\infty\rangle = \langle u(t), Lu_\infty\rangle = 0.$$

This is the conservation of mass property for the equation.

[10] We require this condition since only then do we have that

$$u(t)-u_\infty = u(t) - \langle u(t), u_\infty\rangle u_\infty \in \{u_\infty\}^\perp.$$

connected between the dissipation of our proposed distance and the distance itself. This is the key idea of the entropy method—we just need to perform it on the entropy.

Given an entropy for our system (or more precisely—a relative entropy), i.e. a Lyapnov functional $\mathscr{E}(f|f_\infty)$[11] such that $\mathscr{E}(f|f_\infty) = 0$ if and only if $f = f_\infty$, the entropy method proceeds as follows:

- Differentiating the entropy along the flow of the evolution equation we find that its dissipation can be written as $-\mathscr{D}(f(t))$, where the new *non-negative* functional $\mathscr{D}$ is known as the *entropy production*.
- Putting the evolution of our system aside[12] we search for an explicit function $\Phi : \mathbb{R}_+ \to \mathbb{R}_+$ such that

$$\mathscr{D}(f) \geq \Phi(\mathscr{E}(f|f_\infty)) \tag{15}$$

  for all appropriate functions.
- Once the above functional inequality has been found, we recall that $\mathscr{D}$ is connected to the dissipation of $\mathscr{E}$ and conclude the following differential inequality

$$\frac{d}{dt}\mathscr{E}(f(t)|f_\infty) \leq -\Phi(\mathscr{E}(f(t)|f_\infty))$$

  from which we conclude a *quantitative expression* of the convergence to zero of $\mathscr{E}(f(t)|f_\infty)$.

The most desirable situation in the application of the entropy method, and in a sense the most physical one, is when Φ in (15) is given by $\Phi(x) = \lambda x$ for some $\lambda > 0$. We can think about this case as an "entropic" spectral gap as it implies the functional inequality

$$\mathscr{D}(f) \geq \lambda \mathscr{E}(f|f_\infty)$$

which results in an exponential convergence to equilibrium.

Applying the entropy method to our entropy $\mathscr{H}_N$ under the flow of Kac's master equation (7) we find that the entropy production in our case is given by

$$\mathscr{D}(F_N) = -\langle \log F_N, \mathscr{L}_N F_N \rangle_{L^2(\mathbb{S}^{N-1}(\sqrt{N}), d\sigma_N)}.$$

In order to achieve our desired exponential convergence of $\mathscr{H}_N(F_N)$[13] we consider the entropic spectral gap

$$\Gamma_N = \inf_{F_N d\sigma_N \in \mathscr{P}(\mathbb{S}^{N-1}(\sqrt{N}))} \frac{\mathscr{D}_N(F_N)}{\mathscr{H}_N(F_N)}$$

[11] I.e. a functional such that $\frac{d}{dt}\mathscr{E}(f(t)|f_\infty) \leq 0$ for any solution, $f(t)$, of the equation.

[12] In a sense we hope that our entropy and its production have truly captured the geometry of the problem and the relevant properties of the evolution.

[13] It is worth to note that $\mathscr{H}_N$ is indeed a *relative* functional. It measures the entropic distance of $F_N d\sigma_N$ from $1 d\sigma_N$.

and, much like Kac's spectral gap, ask whether or not

$$\Gamma = \inf_N \Gamma_N > 0.$$

In his 2003 paper, [20], Villani has shown that[14]

$$\Gamma_N \geq \frac{2}{N-1}$$

and conjectured that it is of an order of magnitude $1/N$. If true, since by definition

$$\mathscr{H}_N\left(F_N(t)\right) \leq e^{-\Gamma_N t}\mathscr{H}_N\left(F_N(0)\right),$$

Villani's conjecture would imply that in order to see significant decay in the entropy on Kac's sphere we must wait a time that is proportional to N—exactly like our spectral gap instance.

As we will see, the problem we're facing in this setting, unlike the spectral gap, is not the underlying distance—it is the production itself.

An important step in the investigation of Villani's conjecture was made in the 2010 work of Carlen, Carvahlo, Le-Roux, Loss and Villlani, [6] where the authors have managed to show that

$$\Gamma = 0.$$

Expanding on the ideas presented in that paper, Villani's conjecture has been shown to be essentially true in the work of the author [11] where he proved the following:

Theorem 1 *For any* $0 < \eta < 1$ *there exists an explicit constant* C_η, *that blows up as* η *goes to* 1, *such that*

$$\Gamma_N \leq \frac{C_\eta}{N^\eta}.$$

The intuition behind the proof of these results can be explained relatively simply: How can we create an ensemble that converges to equilibrium slowly? Consider an system in which most of the particles are fairly stable but there is a small amount of particles that is extremely energetic—carrying half the energy of the *entire* system, for instance. If the ensemble is chaotic, i.e. "almost independent", it will take a very long time for these energetic particles to meet the slow moving particles and transfer some of their energy so that the entire system can equilibrate.

Showing this precisely relies heavily on understanding chaoticity on Kac's sphere and on constructing a very natural type of chaotic state—the so-called *conditioned tensorisation*.

[14] The formula presented here was achieved by some simplification of the argument of [20] in the case of the Kac model we present here (see [6] for more details).

3 Conditioned Tensorisation—The Notion of "Almost Independence"

3.1 Conditioned Tensorisations

Looking back at the description of the mean field limit process, our definition of chaos, and its propagation, one question which we have yet to address becomes apparent: Are there chaotic states at all?

Had our phase state been of the form $\mathscr{X}^N$, for some Polish space $\mathscr{X}$ (for instance $\mathbb{R}^N$) then chaoticity, which seem to mean "almost independence",[15] could be attained by considering an independent state, i.e. a tensorisation of a given probability density f on $\mathscr{X}$, $F_N = f^{\otimes N}$. This, however, is impossible on $\mathbb{S}^{N-1}\left(\sqrt{N}\right)$ since the velocity variables can not be independent. Can we, though, find a chaotic state on Kac's sphere that looks as if it is almost a tensorised state?

One natural approach presents itself readily: Given a probability density function on $\mathbb{R}$, f, we can tensorise it to get an independent state on $\mathbb{R}^N$ and then *restrict it to Kac's sphere* in hope that not too many correlations will manifest. In other words, given a probability density function f on $\mathbb{R}$ we define

$$F_N\left(\boldsymbol{V}_N\right) = \frac{f^{\otimes N}\left(\boldsymbol{V}_N\right)}{\mathscr{Z}_N\left(f, \sqrt{N}\right)}, \quad \text{where} \quad \mathscr{Z}_N\left(f, \sqrt{N}\right) = \int_{\mathbb{S}^{N-1}\left(\sqrt{N}\right)} f^{\otimes N}\left(\boldsymbol{V}_N\right) d\sigma_N. \tag{16}$$

F_N of the above form is called the *conditioned tensorisation of* f and the function $\mathscr{Z}_N$ is known as *the normalisation function.*

A few immediate questions come to mind when looking at the definition of these states:

- Are they well defined? Clearly if f is in addition continuous then $\mathscr{Z}_N\left(f, \sqrt{N}\right)$ is well defined and F_N is indeed a probability density on $\mathbb{S}^{N-1}\left(\sqrt{N}\right)$ with respect to $d\sigma_N$. What happens, however, if f is only measurable? $\mathscr{Z}_N\left(f, \sqrt{N}\right)$ is an integration over a set of measure zero with respect to dx and as such we require some sort of trace theorem.
- Are there any restriction on f? From our set up of Kac's model, and in particular our choice of $\mathbb{S}^{N-1}\left(\sqrt{N}\right)$ as the underlying space, we know that the random variable that is associated to f, say $\mathscr{V}$, is supposed to have a unit kinetic energy. This means that to be consistent we must have that

$$\int_{\mathbb{R}} v^2 f(v) dv = 1.$$

[15] We do need to be quite careful here. The independence that chaos guarantees is always "capped" at a *given* finite marginal. We have no idea what happens with the correlations between a number of elements that is of order of N.

Since the above implies that the independent random ensemble $(\mathscr{V}_1, \ldots, \mathscr{V}_N)$ (whose values are in $\mathbb{R}^N$) has a kinetic energy of value N we have a chance, according to the law of large numbers, that the ensemble is concentrated enough on $\mathbb{S}^{N-1}\left(\sqrt{N}\right)$ for its restriction to Kac's sphere to make sense.

- Is F_N indeed chaotic? It is possible to show (see [6, 11] and [10] for more information) that

$$F_{N,k}(v_1, \ldots, v_k) = \frac{\left|\mathbb{S}^{N-k-1}\right|}{\left|\mathbb{S}^{N-1}\right|} \frac{1}{N^{\frac{N-2}{2}}} \left(N - \sum_{i=1}^{k} v_i^2\right)_+^{\frac{N-k-2}{2}} \frac{\mathscr{Z}_{N-k}\left(f, \sqrt{\left(N - \sum_{i=1}^{k} v_i^2\right)_+}\right)}{\mathscr{Z}_N\left(f, \sqrt{N}\right)} f^{\otimes k}(v_1, \ldots, v_k) \tag{17}$$

with

$$\mathscr{Z}_n(f, r) = \int_{\mathbb{S}^{m-1}(r)} f^{\otimes n}(\boldsymbol{V}_n)\, d\sigma_{n,r} \tag{18}$$

and where $d\sigma_{n,r}$ is the uniform probability measure on $\mathbb{S}^{n-1}(r)$. In particular $d\sigma_N = d\sigma_{N,\sqrt{N}}$.

Conditioned tensorisation are not a new concept. In fact, Kac mentioned them in his work [15]. However, the conditions that he needed to impose on f in his discussion were very stringent. A better understanding of these states, on how one can consider the trace operation which we seem to employ by defining them, and on the question of chaoticity was achieved in the work of Carlen, Carvalho, Le Roux, Loss and Villani, [6], which included some beautiful probabilistic observations and ideas.

To truly understand conditioned tensorisation we must understand the behaviour of the normalisation function $\mathscr{Z}_N(f, r)$ for all $r \in \left[0, \sqrt{N}\right]$.

By its definition, we expect that $\mathscr{Z}_n(f, r)$ will tell us, in some sense, how concentrated $f^{\otimes n}$ is on $\mathbb{S}^{n-1}(r)$—up to some geometric factors. A more careful look, however, reveals that this concentration is intimately connected to the probability density of $\sum_{i=1}^{n} \mathscr{V}_i^2$—the random variable of the kinetic energy of the ensemble.

Indeed, if h is the probability density of the variable $\mathscr{V}^2$, attained from the probability density of $\mathscr{V}$, f, by

$$h(r) = \frac{f\left(\sqrt{r}\right) + f\left(-\sqrt{r}\right)}{2r}, \qquad \forall r > 0 \tag{19}$$

we have that for any radial function on $\mathbb{R}^n$, $\phi(\boldsymbol{X}_n) = \varphi\left(|\boldsymbol{X}_n|\right)$

$$\int_{\mathbb{R}^n} \phi\left(\boldsymbol{V}_n\right) f^{\otimes n}\left(\boldsymbol{V}_n\right) d\boldsymbol{V}_n = \int_{r=0}^{\infty} \varphi(r) \left|\mathbb{S}^{n-1}\right| r^{n-1} \underbrace{\int_{\mathbb{S}^{n-1}(r)} f^{\otimes n}\left(\boldsymbol{V}_n\right) d\sigma_{n,r}}_{=\mathcal{Z}_n(f,r)} .$$

On the other hand

$$\int_{\mathbb{R}^n} \phi\left(\boldsymbol{V}_n\right) f^{\otimes n}\left(\boldsymbol{V}_n\right) d\boldsymbol{V}_n = \int_0^{\infty} \varphi\left(\sqrt{t}\right) h^{\otimes n}(t)\, dr = \int_0^{\infty} \varphi(r) 2r h^{\otimes n}\left(r^2\right) dr,$$

where we have used the independence of $\left\{\mathcal{V}_i^2\right\}_{i=1,\ldots,n}$ which follows from that of $\{\mathcal{V}_i\}_{i=1,\ldots,n}$. From all the above we can conclude that

$$\mathcal{Z}_n\left(f, \sqrt{r}\right) = \frac{2h^{\otimes n}(r)}{r^{\frac{n-2}{2}}\left|\mathbb{S}^{n-1}\right|}. \tag{20}$$

Formula (20) gives us a way to go forward in investigating $\mathcal{Z}_N$—we would like to find a *local Central Limit Theorem*[16] for the tensorisation of h. One version of such theorem, expressed via f, was shown in [6]:

Theorem 2 *Let f be a probability density function on $\mathbb{R}$ such that $f \in L^p(\mathbb{R})$ for some $p > 1$. Assume in addition that*

$$\int_{\mathbb{R}} v^2 f(v) dv = 1 \quad \textit{and} \quad \int_{\mathbb{R}} v^4 f(v) dv < \infty.$$

Then $\mathcal{Z}_N(f, \sqrt{r})$ is well defined for any $r \in [0, N]$, and

$$\mathcal{Z}_N(f, \sqrt{r}) = \frac{2}{\sqrt{N}\Sigma\left|\mathbb{S}^{N-1}\right| r^{\frac{N-2}{2}}}\left(\frac{e^{-\frac{(r-N)^2}{2N\Sigma^2}}}{\sqrt{2\pi}} + \lambda_N(r)\right), \tag{21}$$

with

$$\sup_{r\in[0,N]} |\lambda_N(r)| \underset{N\to\infty}{\longrightarrow} 0,$$

and where $\Sigma^2 = \int_{\mathbb{R}} v^4 f(v)dv - 1$.

Besides the fact that the above theorem gives us relatively natural conditions under which conditioned tensorisation of a probability density f makes sense, plugging (21) in (17) shows almost immediately the such states are f-chaotic.

The usefulness of this local CLT doesn't end there—it shows that conditioned tensorisation are indeed very natural in Kac's setting as their entropy and production scale linearly in N. This, in turn, will be the crucial step we need to show Carlen et. al's result and its extension to Theorem 1.

[16] I.e. a central limit theorem for the probability density of the independent sum.

3.2 Conditioned Tensorisation and Entropic Convergence to Equilibrium

Consider the conditioned tensorisarion of f, F_N, where f satisfies the conditions of Theorem 2. Then, as can be seen in [6, 11],

$$\mathscr{H}_N\left(F_N\right)=\frac{N\left|\mathbb{S}^{N-2}\right|}{N^{\frac{N-2}{2}}\left|\mathbb{S}^{N-1}\right|\mathscr{Z}_N\left(f,\sqrt{N}\right)}\left[\int_{-\sqrt{N}}^{N} f(v)\log f(v)\left(N-v^2\right)^{\frac{N-3}{2}}\right.$$
$$\left.\mathscr{Z}_{N-1}\left(f,\sqrt{N-v^2}\right)dv\right]-\log\left(\mathscr{Z}_N\left(f,\sqrt{N}\right)\right). \tag{22}$$

and

$$\mathscr{D}_N\left(F_N\right)=\frac{N\left|\mathbb{S}^{N-3}\right|}{4\pi N^{\frac{N-2}{2}}\left|\mathbb{S}^{N-1}\right|\mathscr{Z}_N\left(f,\sqrt{N}\right)}$$
$$\int_0^{2\pi}d\theta\int_{v_1^2+v_2^2\leq N}\psi\left(f(v_1)f(v_2),f\left(v_1\left(\theta\right)\right)f\left(v_2\left(\theta\right)\right)\right) \tag{23}$$
$$\left(N-v_1^2-v_2^2\right)^{\frac{N-4}{2}}\mathscr{Z}_{N-2}\left(f,\sqrt{N-v_1^2-v_2^2}\right)dv_1dv_2.$$

where

$$\psi\left(x,y\right)=\left(x-y\right)\log\left(\frac{x}{y}\right). \tag{24}$$

Utilising Theorem 2 we conclude that

Theorem 3 *Let f be a probability density on $\mathbb{R}$. Assume that the conditions of Theorem 2 hold and let F_N be the conditioned tensorisation of f. Then*

$$\lim_{N\to\infty}\frac{\mathscr{H}_N\left(F_N\right)}{N}=\int_{\mathbb{R}}f(v)\log f(v)dv+\frac{1}{2}+\frac{\log 2\pi}{2}=\mathscr{H}\left(f|\mathscr{M}\right)$$

and

$$\lim_{N\to\infty}\frac{\mathscr{D}_N\left(F_N\right)}{N}=\frac{1}{4\pi}\int_{\mathbb{R}^2\times[0,2\pi]}\psi\left(f(v_1)f(v_2),f\left(v_1\left(\theta\right)\right)f\left(v_2\left(\theta\right)\right)\right)dv_1dv_2d\theta=\frac{\mathscr{D}\left(f\right)}{2}$$

where

$$\mathscr{D}(f)=\frac{1}{2\pi}\int_{\mathbb{R}\times[0,2\pi]}\psi\left(f(v)f(v_*),f\left(v\left(\theta\right)\right)f\left(v_*\left(\theta\right)\right)\right)dvdv_*d\theta, \tag{25}$$

with $v_1\left(\theta\right)$ and $v_2\left(\theta\right)$, as well as $v\left(\theta\right)$ and $v_*\left(\theta\right)$, given by the same formula as (10).

Besides further cementing the status of conditioned tensorisation as a worthy and natural family to explore in the setting of Kac's model, Theorem 3 is the key to showing Carlen et. al. claim that

$$\inf_{N\in\mathbb{N}} \Gamma_N = \inf_{N\in\mathbb{N}} \inf_{F_N d\sigma_N \in \mathscr{P}\left(\mathbb{S}^{N-1}(\sqrt{N})\right)} \frac{\mathscr{D}_N\left(F_N\right)}{\mathscr{H}_N\left(F_N\right)} = 0.$$

Indeed, as long as f satisfies the condition of Theorem 2, we find that by considering the family of condition tensorisation of f, which we will denote by $\left[f^{\otimes N}\right]$, that

$$\inf_{N\in\mathbb{N}} \Gamma_N \leq \lim_{N\to\infty} \frac{\mathscr{D}_N\left(\left[f^{\otimes N}\right]\right)}{\mathscr{H}_N\left(\left[f^{\otimes N}\right]\right)} = \frac{\mathscr{D}(f)}{2\mathscr{H}(f|\mathscr{M})}. \tag{26}$$

Following up on the intuition we presented in the end of Sect. 2.3 we consider a "generator" f that somehow gives an average particle very low probability to be very energetic and almost full probability to be stable. A natural choice is given by[17]

$$f_\delta(v) = \delta \mathscr{M}_{\frac{1}{2\delta}}(v) + (1-\delta)\mathscr{M}_{\frac{1}{2(1-\delta)}}(v) \tag{27}$$

where $\mathscr{M}_a(v) = (2\pi a)^{-1/2} e^{-\frac{v^2}{2a}}$ and $0 < \delta < 1$. Plugging it in (26) yields the estimate

$$\inf_{N\in\mathbb{N}} \Gamma_N \leq -C\delta \log(\delta),$$

where C is a fixed bounded constant. Taking δ to zero gives the desired result.

The above idea, however, is not enough to give an explicit bound on every Γ_N as is expressed in Theorem 1. The reason behind this lies in the choice (27). While it captures the intuition that an average particle has small probability to be very energetic, this probability—and consequently the (average) portion of the particles in the ensemble that will be very energetic, is *fixed* and involves a non-negligible amount of particles. This is why we are getting a uniform in N bound on Γ_N. In order to utilise the same idea and get an N-dependent bound for Γ_N at the same time, we need to allow the number of particles that carry the high energy to be significantly smaller than N. In other words, we need to let the probability that an average particle will be energetic to decrease with N. We will, thus, consider a generator of the form

$$f_{\delta_N}(v) = \delta_N \mathscr{M}_{\frac{1}{2\delta_N}}(v) + (1-\delta_N)\mathscr{M}_{\frac{1}{2(1-\delta_N)}}(v) \tag{28}$$

for some choice of $\{\delta_N\}_{N\in\mathbb{N}}$ that goes to zero as N goes to infinity.

While this seems not too different to what we've done before, there is a crucial ingredient in being able to use the above idea that is new. The result of

[17] Note that $\int_{\mathbb{R}} v^2 \mathscr{M}_a(v)dv = a$ which shows that each part of f_δ carries the same kinetic energy.

Carlen et. al.'s relied heavily on the local-CLT theorem which is expressed in Theorem 2. It is unclear if the same result holds in this case due to the following two reasons:

(1) Theorem 2 has considered the expression (20). Since our function f_N depends on N we need to be careful when we tensorise it (or the associated h_N) a number of times that is of order N. Such orders are particularly important to us as the rescaled entropy and productions depend on $\mathscr{Z}_N$, $\mathscr{Z}_{N-1}$ and $\mathscr{Z}_{N-2}$.
(2) A straight forward calculation shows that

$$\int_{\mathbb{R}} v^4 f_{\delta_N}(v)dv = \frac{3}{4\delta_N\left(1-\delta_N\right)}$$

which is not bounded as N goes to infinity. This implies that the conditions of Theorem 2 do not hold uniformly in N.

The resolution to the above issues was given in [11] where the author developed, under certain conditions, a new local-CLT that allowed for dependence in N in the probability density for the generator of the conditioned tensorisation. The theorem allowed for some blow up in N for the fourth moment. We will not state it here, as it is quite technical, and refer the interested reader to the aforementioned paper.

With this local-CLT the author has managed to show the following concentration estimate for the normalisation function of f_{δ_N}, defined by (28):

Theorem 4 *Let $\{\delta_N\}_{N\in\mathbb{N}}$ be a sequence of positive numbers that satisfies*

- *δ_N is dominated by a some power of N.*
- *There exists $\beta > 0$ such that*

$$\delta_N^{1+2\beta} N \underset{N\to\infty}{\longrightarrow} \infty$$

and

$$\delta_N^{1+3\beta} N \underset{N\to\infty}{\longrightarrow} 0.$$

Then, for any fixed j

$$Z_{N-j}\left(f_{\delta_N}, \sqrt{u}\right) = \frac{2}{\sqrt{N-j}\cdot\Sigma_{\delta_N}\cdot\left|\mathbb{S}^{N-j-1}\right|u^{\frac{N-j}{2}-1}} \left(\frac{e^{-\frac{(u-N+j)^2}{2(N-j)\Sigma_{\delta_N}^2}}}{\sqrt{2\pi}} + \lambda_j(N-j,u)\right) \tag{29}$$

with

$$\lim_{N\to\infty}\sup_{u\in\mathbb{R}}\left|\lambda_j(N-j,u)\right| = 0,$$

where

$$\Sigma^2_{\delta_N} = \int_{\mathbb{R}} v^4 \delta_N(v)dv - 1 = \frac{3}{4\delta_N(1-\delta_N)} - 1.$$

Using the above theorem for the conditioned tensorisation of f_{δ_N} with $\delta_N = N^{2\beta-1}$ and $0 < \beta < 1/6$ chosen arbitrarily, together with (22) and (23) gives us the proof of Theorem 1

Remark 1 We would like to mention at this point that additional extensions of the local-CLT presented in Theorem 2 have been explored. In particular, in their work [10] the authors have developed a similar concentration result for the normalisation function of a probability density function f such that $f \in L^p(\mathbb{R})$ for some $p > 1$, $\int_{\mathbb{R}} v^2 f(v)dv = 1$ and

$$\int_{-\sqrt{x}}^{\sqrt{x}} v^4 f(v) \underset{x\to\infty}{\sim} Cx^{2-\alpha}$$

for some $C > 0$ and $1 < \alpha < 2$. In this case the concentration of $\mathscr{Z}_N$ is measured not with a Gaussian, as in Theorems 2 and 4, but with a the attractor that is associated to an α-stable Lévi process.

While our attempt to utilise the mean field limit approach with our entropic study of convergence to equilibrium in Kac's model has not been successful, we have found intuitive families of chaotic states—conditioned tensorisations. These families are a crucial ingredient in what comes next—the (somewhat) fulfilment of Kac's vision.

4 Kac's Vision Fulfilled

4.1 New Notions of Chaoticity

When we discussed the entropic study of the convergence to equilibrium in Kac's model back in Sect. 2.3, we motivated our decision to choose the entropy as our "distance" with the intuitive reasoning that the entropy of chaotic states should scale linearly in N. At no point, however, did we prove that it is true in general. We did show in our previous section that conditioned tensorisations do satisfy this property. Moreover—their entropy production *also* scales linearly in N. These observations motivate the following definitions

Definition 2 Let $F_N \in \mathscr{P}\left(\mathbb{S}^{N-1}\left(\sqrt{N}\right), d\sigma_N\right)$ with $N \in \mathbb{N}$ be an $f-$chaotic family of probability densities.

(i) We say that $\{F_N\}_{N\in\mathbb{N}}$ is *f-entropically chaotic* if

$$\lim_{N\to\infty} \frac{\mathscr{H}_N(F_N)}{N} = \mathscr{H}(f|\mathscr{M}). \tag{30}$$

(ii) We say that $\{F_N\}_{N\in\mathbb{N}}$ is *f-strongly entropic chaotic* if in addition to the above

$$\lim_{N\to\infty} \frac{\mathscr{D}_N(F_N)}{N} = \frac{\mathscr{D}(f)}{2}, \tag{31}$$

where $\mathscr{D}(f)$ is given in (25).

The notion of entropic chaoticity was originally defined and investigated in [6]. Further work, such that of Hauray and Mischler [13], have expanded on this study and even showed that under certain conditions the chaoticity of a family $\{F_N\}_{N\in\mathbb{N}}$ can follow from condition (30).

The reason why we identified and defined these new stronger notion of chaoticity is that on closer examination of our approach to the investigation of the convergence to equilibrium, such notions offer a different avenue for attempting to deduce information for the Boltzmann-Kac equation (12) from Kac's model.

In Sect. 2 we tried to use the entropy method to find an explicit convergence rate for $\mathscr{H}_N(F_N(t))$ in hope that we can then pass it to $\mathscr{H}(f(t)|\mathscr{M})$ by rescaling Kac's entropy and taking N to infinity—implicitly assuming entropic chaoticity. By doing so, however, we have lost some of the geometry of the process as we've "thrown away" the functional inequality we've found and only kept its conseqeunce.

The idea that "nice enough" chaotic states, like strongly entropic ones, scale linearly in N both in entropy and its production give rise to the intuition that we should look for a functional inequality of the form

$$\frac{\mathscr{D}_N(F_N)}{N} \geq \Phi\left(\frac{\mathscr{H}_N(F_N)}{N}\right),$$

for some function Φ.

4.2 What Do We Expect—The Entropy Method for the Boltzmann Equation

Since the time when Kac presented his model and expressed his hope to infer information on the Boltzmann equation from it, many techniques to deal with non-linear equations have been developed. In fact, nowadays many questions that pertain to various aspects of Kac's model are more complicated to solve than similar questions for the Boltzmann equation.

Before we delve into exploring functional inequalities of rescaled entropy and its production, we want to pause and reflect on the study of the entropy method in the Boltzmann equation.

The paper where Villani's managed to show that the entropic spectral gap Γ_N is bounded from below by $2/(N-1)$, [20], was in fact mostly dedicated to the study of the entropy method for the Boltzmann equation. In their work [2], Bobylev and Cercignani have managed to show that when the collision kernel B is given by

$$B(v, v_*, \sigma) = 1 + |v - v_*|^\gamma$$

with $\gamma \in [0, 2)$ then one finds that for the spatially homoegenous Boltzmann equation

$$\inf_{f dv \in \mathscr{P}(\mathbb{R}^d)} \frac{\mathfrak{D}_\gamma(f)}{\mathscr{H}(f|\mathscr{M})} = 0,$$

where

$$\mathfrak{D}_\gamma(f) = \frac{1}{4}\int_{\mathbb{R}^d \times \mathbb{R}^d \times \mathbb{S}^{d-1}} (1 + |v - v_*|^\gamma)\, \psi\left(f(v) f(v_*), f\left(v', v_*'\right)\right) dv dv_* d\sigma,$$

with ψ defined as in (24). Note that this result is consistent with the study on Kac's sphere.

Inspired by the works of Carlen and Carvahlo [3, 4], Toscani and Villani managed to show in [18] that if one assumes that f is smooth enough and has a Maxwellian-type lower bound (we leave the precise conditions to [18]) we have that for any $\epsilon > 0$ there exists a constant $C_\epsilon(f)$ that depends on f such that

$$\mathfrak{D}_\gamma(f) \geq C_\epsilon(f) \mathscr{H}(f|\mathscr{M})^{1+\epsilon} \tag{32}$$

giving an "almost" entropic gap result. In his work [20], Villani has extended the result to various other collusion kernels as well as have shown the existence of an entropic spectral gap for the super quadratic collision kernel (i.e. when $\gamma = 2$).

Following on the above, and on the intimate interplay between Kac's model and the Boltzmann equation, it is reasonable for us to seek inequalities of the form (32) on the level of Kac's sphere.

4.3 General Kac's Models and the "Correct" Inequalities

In the context of Boltzmann equation, Kac's original model fitted into the framework of Maxwellian molecules (i.e., the case where the collision kernel B is constant). Extensions of this model can be made, at least formally, to the case where the collisions in Kac's process occurs with an intensity that depends on the scattering (i.e. on θ) and the velocities of the colliding particles (i.e. on v_i and v_j if the particles that collided were the ith and jth ones).[18]

The natural extension to Kac's master equation which we will consider here is given by

$$\partial_t F_N(t, \boldsymbol{V}_N) = \mathscr{L}_{N,\gamma} F_N(t, \boldsymbol{V}_N), \qquad \boldsymbol{V}_N \in \mathbb{S}^{N-1}\left(\sqrt{N}\right) \tag{33}$$

[18] In terms of a jump processes, the dependence on v_i and v_j is slightly problematic but we continue as if no additional justification for the proposed model is needed.

where

$$\mathscr{L}_{N,\gamma} F(\boldsymbol{V}_N) = -\frac{N}{2\pi} \frac{1}{\binom{N}{2}} \sum_{i<j} \left(1 + \left(v_i^2 + v_j^2\right)\right)^\gamma \int_0^{2\pi} \left(F_N(\boldsymbol{V}_N) - F_N\left(R_{i,j,\theta}(\boldsymbol{V}_N)\right)\right) d\theta,$$

and $0 \leq \gamma \leq 1$.

Under the assumption of the propagation of chaos we can repeat the mean field approach in this model to find the limit equation

$$\partial_t f(v) = \frac{1}{\pi} \int_{-\pi}^{\pi} \left(1 + (v^2 + w^2)\right)^\gamma \left(f(v(\theta)) f(w(\theta)) - f(v) f(w)\right) d\theta dw, \tag{34}$$

where $v(\theta)$ and $w(\theta)$ are given by formula (10).

Much like in Kac's original model we can try and explore entropic convergence to equilibrium of (33) by finding a connection between the entropy of the system, $\mathscr{H}_N(F_N)$, and its production

$$\mathscr{D}_{N,\gamma} = -\left\langle \log F_N, \mathscr{L}_{N,\gamma} F_N \right\rangle_{L^2\left(\mathbb{S}^{N-1}(\sqrt{N}), d\sigma_N\right)}.$$

A straight forward calculation finds that

$$\mathscr{D}_{N,\gamma}(F_N) = \frac{1}{4\pi} \frac{N}{\binom{N}{2}} \sum_{i<j} \int_{\mathbb{S}^{N-1}(\sqrt{N})} \int_0^{2\pi} \left(1 + \left(v_i^2 + v_j^2\right)\right)^\gamma \psi\left(F_N(\boldsymbol{V}_N), F_N\left(R_{i,j,\theta}(\boldsymbol{V}_N)\right)\right) d\sigma_N d\theta. \tag{35}$$

where $\psi(x, y) = (x - y) \log(x/y)$ (as defined in (24)).

Following on the discussion in Sect. 4.2 we are encouraged to search for a functional inequalities of the form

$$\frac{\mathscr{D}_{N,\gamma}(F_N)}{N} \geq C_\epsilon \left(\frac{\mathscr{H}_N(F_N)}{N}\right)^{1+\epsilon} \tag{36}$$

for a given, hopefully arbitrary, $\epsilon > 0$.

We naturally don't expect this to hold for all chaotic states. However, something can still be said in the general case: In his work [20], Villani has shown the following inequality:

$$\mathscr{D}_{N,\gamma}(F_N) \geq C_\gamma \frac{\mathscr{H}_N(F_N)}{N^{1-\gamma}}. \tag{37}$$

This inequality, however, is not of the form (36) unless $\gamma = 1$.

To try and find conditions under which we expect to have an inequality of the form (36), let us look again at the intuitive idea that chaotic states F_N behaves like

$f^{\otimes N}$. Since we've already given an intuitive outlook in Sect. 2.3 to why $\mathscr{H}_N(F_N) \approx N\mathscr{H}(f|\mathscr{M})$ we turn our attention to $\mathscr{D}_{N,\gamma}$.

Following on (35) and using symmetry we see that

$$\frac{\mathscr{D}_{N,\gamma}(F_N)}{N} = \frac{1}{4\pi}\int_{\mathbb{S}^{N-1}\left(\sqrt{N}\right)}\int_0^{2\pi}\left(1+\left(v_1^2+v_2^2\right)\right)^{\gamma}\psi\left(F_N\left(\boldsymbol{V}_N\right),F_N\left(R_{1,2,\theta}\left(\boldsymbol{V}_N\right)\right)\right)d\sigma_N d\theta.$$

Plugging in $F_N \approx f^{\otimes N}$ we get

$$\begin{aligned}&\psi\left(F_N\left(\boldsymbol{V}_N\right),F_N\left(R_{1,2,\theta}\left(\boldsymbol{V}_N\right)\right)\right)\\&\quad\approx\left(F_N\left(\boldsymbol{V}_N\right)-F_N\left(R_{1,2,\theta}\left(\boldsymbol{V}_N\right)\right)\right)\log\left(\frac{f\left(v_1(\theta)\right)f\left(v_2(\theta)\right)}{f\left(v_1\right)f\left(v_2\right)}\right)\end{aligned}$$

and notice that the dependence in the dimension is completely cancelled in the logarithm, leaving only an $L^1\left(\mathbb{S}^{N-1}\left(\sqrt{N}\right), d\sigma_N\right)$ like term. This, in a sense, is exactly why the scaling works for such states. As F_N can't be expected to look like $f^{\otimes N}$, besides for conditioned tensorisation, we might need to impose stringer control on the integral that appears above. This motivates the following definition:

Definition 3 We say that a family of symmetric probability densities $F_N \in \mathscr{P}\left(\mathbb{S}^{N-1}\left(\sqrt{N}\right), d\sigma_N\right)$ with $N \in \mathbb{N}$ has the log-power property of order $\beta > 0$, or is log-power of order β in short, if there exists $\mathscr{C}_\beta > 0$ such that

$$\sup_{N\in\mathbb{N}}\frac{1}{2\pi}\int_0^{2\pi}\int_{\mathbb{S}^{N-1}(\sqrt{N})}\psi_\beta\left(F_N\left(\boldsymbol{V}_N\right),F_N\left(R_{1,2,\theta}\left(\boldsymbol{V}_N\right)\right)\right)d\sigma^N d\theta \leq \mathscr{C}_\beta^{1+\beta} \tag{38}$$

where $\psi_\beta(x,y) = |x-y|\left|\log\left(\frac{x}{y}\right)\right|^{1+\beta}$. We will call the constant $\mathscr{C}_\beta$ the *log power* constant.

The log power property is indeed the right ingredient we needed to achieve an inequality of the form (36). On its own it is not enough, but by adding a natural requirement of a uniform in N moment control for $F_{N,1}$, Carlen, Carvalho and the author of this paper have managed to show the following in [5]:

Theorem 5 *Let $F_N \in \mathscr{P}\left(\mathbb{S}^{N-1}\left(\sqrt{N}\right), d\sigma_N\right)$ with $N \in \mathbb{N}$ be family of symmetric density functions. Assume in addition that there exists some $\beta > 0$ such that $\{F_N\}_{N\in\mathbb{N}}$ is log-power of order β and that there exists $k > 1 + 1/\beta$ such that*

$$M_{2k} = \sup_{N\in\mathbb{N}} M_{2k}\left(F_{N,1}\right) < \infty, \tag{39}$$

where

$$M_{2k}\left(F_{N,1}\right) = \int_{\mathbb{R}} v^{2k}F_{N,1}(v)dv = \int_{\mathbb{S}^{N-1}(\sqrt{N})} v_1^{2k}F_N\left(\boldsymbol{V}_N\right)d\sigma_N. \tag{40}$$

Then, there exists an explicit $C_{k,\gamma,\beta} > 0$ *that depends only on* β*, the log power constant of* $\{F_N\}_{N\in\mathbb{N}}$, $\mathscr{C}_\beta$, *and* M_{2k} *such that*

$$\frac{\mathscr{D}_{N,\gamma}(F_N)}{N} \geq C_{k,\gamma,\beta}\left(\frac{\mathscr{H}_N(F_N)}{N}\right)^{1+\frac{(1-\gamma)(1+\beta)}{k\beta-(1+\beta)}}. \tag{41}$$

A few observation on the above theorem:

- The condition $k > 1 + 1/\beta$ gives us an interplay between our moment control, k, and the log-power parameter, β. The tighter control we have on the integral (38), the less moment we need.
- Moments conditions are very natural in the setting of the Boltzmann equation and many other kinetic equations so we shouldn't be too surprised by the appearance of $M_{2k}\left(F_{N,1}\right)$.
- For a given $\epsilon > 0$ we see that if F_N is log-power of order β and we have moment control of high enough order then we can recover the inequality

$$\frac{\mathscr{D}_{N,\gamma}(F_N)}{N} \geq C_\epsilon\left(\frac{\mathscr{H}_N(F_N)}{N}\right)^{1+\epsilon}.$$

The difficulty in showing Theorem 5 was mostly in finding the right condition under which it holds, i.e. the notion of log-power of some order. Once we have convinced ourselves that it is the right ingredient for the desired functional inequality, the proof is a fairly straight forward combination of (37) for $\gamma = 1$ and interpolation techniques. We would like to mention at this point that using a similar technique for the Boltzmann equation is exactly how Villani has shown (32) in [20]. However, this, at least in the author's opinion, is much more involved than the proof of Theorem 5.

Due to its relative simplicity, let us give the proof to this theorem here:

Proof (*Proof of Theorem* 5) For any $\lambda > 0$ we define the set

$$A_\lambda = \left\{v_1, v_2 \in \mathbb{R} \;\middle|\; 1 + v_1^2 + v_2^2 > \lambda\right\}.$$

Using the symmetry of F_N we find that

$$\begin{aligned}\mathscr{D}_{N,1}(F_N) \leq& \lambda^{1-\gamma} D_{N,\gamma}(F_N)\\ &+ \frac{N}{4\pi}\int_{\mathbb{S}^{N-1}\left(\sqrt{N}\right)\cap A_\lambda}\int_0^{2\pi}\left(1+v_1^2+v_2^2\right)\psi\left(F_N\left(V_N\right), F_N\left(R_{1,2,\theta}\left(V_N\right)\right)\right) d\sigma_N d\theta\end{aligned}$$

$$\leq \lambda^{1-\gamma} D_{N,\gamma}(F_N)$$

$$+\frac{N}{2}\left(\frac{1}{2\pi}\int_0^{2\pi}\int_{\mathbb{S}^{N-1}\left(\sqrt{N}\right)}\left|\log\left(\frac{F_N\left(R_{1,2,\theta}\left(V_N\right)\right)}{F_N\left(V_N\right)}\right)\right|^{1+\beta}\left|F_N\left(V_N\right)-F_N\left(R_{1,2,\theta}\left(V_N\right)\right)\right|d\sigma_N d\theta\right)^{\frac{1}{1+\beta}}$$

$$\left(2\int_{\mathbb{S}^{N-1}\left(\sqrt{N}\right)\cap A_\lambda}\left(1+v_1^2+v_2^2\right)^{\frac{1+\beta}{\beta}} F_N\left(V_N\right)d\sigma^N\right)^{\frac{\beta}{1+\beta}}.$$

Using the log-power property and the fact that for all $\eta > 0$

$$\left(1+v_1^2+v_2^2\right)^\eta \leq 3^\eta\left(1+|v_1|^{2\eta}+|v_2|^{2\eta}\right)$$

we find that

$$\frac{D_{N,1}(F_N)}{N} \leq \lambda^{1-\gamma}\frac{D_{N,\gamma}(F_N)}{N}+\frac{2^{\frac{\beta}{1+\beta}}3^{\frac{k\beta}{1+\beta}}\mathscr{C}_\beta}{2}\frac{1}{\lambda^{\frac{k\beta}{1+\beta}-1}}\left(1+2M_{2k}\right)^{\frac{\beta}{1+\beta}}.$$

Optimising over λ yields

$$\frac{D_{N,1}(F_N)}{N} \leq \frac{k\beta-\gamma(1+\beta)}{(1+\beta)(1-\gamma)}\left(\frac{(1+\beta)(1-\gamma)}{k\beta-(1+\beta)}\right)^{\frac{k\beta-(1+\beta)}{k\beta-\gamma(1+\beta)}}$$
$$\frac{3^{\frac{k\beta(1-\gamma)}{k\beta-\gamma(1+\beta)}}\mathscr{C}_\beta^{\frac{(1+\beta)(1-\gamma)}{k\beta-\gamma(1+\beta)}}}{2^{\frac{1-\gamma}{k\beta-\gamma(1+\beta)}}}\left(1+2M_{2k}\right)^{\frac{\beta(1-\gamma)}{k\beta-\gamma(1+\beta)}}\left(\frac{D_{N,\gamma}(F_N)}{N}\right)^{\frac{k\beta-(1+\beta)}{k\beta-\gamma(1+\beta)}},$$

which together with the fact that $\mathscr{D}_{N,1}\left(F_N\right) \geq C_1\mathscr{H}_N\left(F_N\right)$ concludes the proof.

While we may want to pat ourselves on the back for a job well done, a few questions do require our attention:

- In order to be able to use inequality (41) to find a functional inequality for the Boltzmann equation by using the mean field method we need to consider functions that are strongly entropic chaotic and have the log-power property. Are there any such functions at all?
- Much like when we discussed the notion of chaoticity in Sect. 2.1, we wonder if the log-power property, as well as the notion of strongly entropic chaoticity, propagate with the master equation (33)?

While we will shortly show the existence of log-power states, we still don't know the answer to the second question above. Initial attempts to explore the propagation of the log-power property were met with surprising resistance and the author is still unsure which ingredient is missing or was overlooked. What we did achieve, however, is still enough for us to be able to fulfil Kac's original vision of "pushing down" information from the many particle system to its limit equation.

4.4 Kac's Vision Fulfilled

As was mentioned in Sect. 2.2, Kac viewed his model as more than just means to justify the Boltzmann equation. He wanted to use it and the mean field limit approach to gain information about the Boltzmann equation—in particular about its convergence to equilibrium. His spectral gap approach, and later on the entropic study of Kac's model, relied on finding functionals that, up to a simple rescaling, converge to zero as time goes to infinity uniformly in N. Consequently, the propagation of associated notions under Kac's master equation was essential to this process.

The approach presented in this section, and in particular inequality (41), gives us the means to not only transfer rates of convergence, but to "push down" a functional inequality, giving us information about the geometry of the problem and the setting. Indeed, assuming that we found an f-strongly entropic family, $\{F_N\}_{N\in\mathbb{N}}$, with respect to the generalised Kac master equation (33), i.e. a family such that

$$\lim_{N\to\infty}\frac{\mathscr{H}_N(F_N)}{N}=\mathscr{H}(f|\mathscr{M})$$

and

$$\lim_{N\to\infty}\frac{\mathscr{D}_{N,\gamma}(f)}{N}=\frac{\mathscr{D}_\gamma(f)}{2}$$

with

$$\mathscr{D}_\gamma(f)=\frac{1}{2\pi}\int_{\mathbb{R}\times[0,2\pi]}\left(1+v^2+v_*^2\right)^\gamma \psi\left(f(v)f(v_*),\, f(v(\theta))\,f(v_*(\theta))\right)dvdv_*d\theta, \tag{42}$$

and assuming that $\{F_N\}_{N\in\mathbb{N}}$ is log-power of order β than (41) would immediately imply that

$$\mathscr{D}_\gamma(f)\geq 2C_{k,\gamma,\beta}\left(\mathscr{H}(f|\mathscr{M})\right)^{1+\frac{(1-\gamma)(1+\beta)}{k\beta-(1+\beta)}}. \tag{43}$$

If we can find conditions on f and $\{F_N\}_{N\in\mathbb{N}}$ *which propagate with respect to the generalised Boltzmann-Kac equation* (34) such that the associated requirement of Theorem 5 remain true for all time, we would be able to upgrade (43) to a differential inequality that will give us an *explicit* rate of convergence to equilibrium in the Boltzmann-Kac equation.

Can we, then, find f-strongly entropic chaotic families that are also log-power of some order? Conditioned tensorisations rear their head again. The strong chaoticity of these families will easily follow from (22) and a γ-modified variation of (23)[19] as long as the conditions of Theorem 2 hold. The question of the log-power property is answered in the next theorem, taken from [5].

---

[19] We just need to add $\left(1+v_1^2+v_2^2\right)^\gamma$ inside the integral.

Theorem 6 *Let f be a probability density on $\mathbb{R}$ that satisfies the conditions of Theorem 2. Assume in addition that $\|f\|_\infty < \infty$ and that there exist $\beta > 0$ and a positive function Φ such that*

$$f(v) \geq e^{-\Phi(v)},$$

$$M_{\Phi,\beta} = \int_{\mathbb{R}} \Phi(v)^{1+\beta} f(v)dv < \infty,$$

and

$$M_{avg,\Phi,\beta} = \int_{\mathbb{R}^2} \left(\int_0^{2\pi} \Phi(v_1(\theta))^{1+\beta} d\theta \right) f(v_1)f(v_2)dv_1dv_2 < \infty.$$

Then the conditioned tensorisation of f, $\{F_N\}_{N\in\mathbb{N}}$, has the log-power property of order β for $N \geq 3$. Moreover, the log power constant, $\mathscr{C}_\beta$, can be chosen to be

$$\mathscr{C}_\beta = \left[2^{1+2\beta}\sqrt{3}\frac{1+\sqrt{2\pi}\sup_{N\in\mathbb{N}}\sup_u |\lambda_{N-1}(u)|}{1-\sqrt{2\pi}\sup_{N\in\mathbb{N}}\sup_u |\lambda_N|(u)} \mathfrak{M}_{\epsilon,\phi,\beta} \right]^{\frac{1}{1+\beta}}, \tag{44}$$

where λ_N is the same as in Theorem 2 and

$$\mathfrak{M}_{\epsilon,\phi,\beta} = 2\left(\sup_{x\geq 1} \frac{\log x}{x^\epsilon}\right)^{1+\beta} \|f\|_\infty^{\epsilon(1+\beta)} + M_{\Phi,\beta} + M_{avg,\Phi,\beta}.$$

We can further restrict the conditions on our generator f to ones that satisfy the requirements of Theorem 6 *and* propagate with the generalised Boltzmann-Kac equation, resulting in a *time dependent* family of conditioned tensorisation, $\left[f(t)^{\otimes N}\right]_{N\in\mathbb{N}}$, which is $f(t)$-strongly entropic chaotic *and* log-power of an order which is independent of N. Consequently, plugging this family in (41) and taking N to infinity we can obtain an explicit rate of convergence to equilibrium. We conclude this section with the following theorem from [5]:

Theorem 7 *Let f be a probability density on $\mathbb{R}$ that satisfies the conditions of Theorem 2. Assume in addition that*

- *There exists $\beta > 0$ and $k > 1 + 1/\beta$ such that*

$$M_{\max(2k,k(1+\beta),4)}(f) = \int_{\mathbb{R}} |v|^{\max(2k,k(1+\beta),4)} f(v)dv < \infty.$$

- $$I(f) = \int_{\mathbb{R}} \frac{\left(f'(x)\right)^2}{f(x)}dx < \infty.$$

- *$f(v) \geq Ce^{-|v|^2}$ for all $v \in \mathbb{R}$.*

Then, there exists an explicit constant, $\mathscr{C}$, depending only on f, its moments, and β, such that

$$D_\gamma(f) \geq \mathscr{C} H\left(f|M\right)^{1+\frac{(1-\gamma)(1+\beta)}{k\beta-(1+\beta)}}. \tag{45}$$

Moreover, if $f(t)$ is the solution to (34) *then* (45) *holds for all t with constants that are independent of time.*

5 Final Remarks

The study and ideas that govern mean field limits, chaoticity, and Kac's model have a long and illustrious history. In our review we have mostly focused on the *functional* aspect of this study—in particular for Kac's particle system. We have tried to illustrate where issues arise when one studies the explicit convergence to equilibrium of the system—and tried to show that sometimes finding the right inequality and (family of) function(s) can circumvent some of the dynamical issues in the former study. There is still much to explore within and outside the setting of this model and the author expects that the best is yet to come.

References

1. Arnold, A., Markowich, P., Toscani, G., Unterreiter, A.: On convex Sobolev inequalities and the rate of convergence to equilibrium for Fokker-Planck type equations. Commun. Partial Differ. Equ. **26**(1–2), 43–100 (2001)
2. Bobylev, A., Cercignani, C.: On the rate of entropy production for the Boltzmann equation. J. Stat. Phys. **94**(3–4), 603–618 (1999)
3. Carlen, E.A., Carvalho, M.C.: Strict entropy production bounds and stability of the rate of convergence to equilibrium for the Boltzmann equation. J. Stat. Phys. **67**(3–4), 575–608 (1992)
4. Carlen, E.A., Carvalho, M.C.: Entropy production estimates for Boltzmann equations with physically realistic collision kernels. J. Stat. Phys. **74**(3–4), 743–782 (1994)
5. Carlen, E.A., Carvalho, M.C., Einav, A.: Entropy production inequalities for the Kac walk. Kinet. Relat. Models **11**(2), 219–238 (2018)
6. Carlen, E.A., Carvalho, M.C., Le Roux, J., Loss, M., Villani, C.: Entropy and chaos in the Kac model. Kinet. Relat. Models **3**(1), 85–122 (2010)
7. Carlen, E.A., Carvalho, M.C., Loss, M.: Determination of the spectral gap for Kac's master equation and related stochastic evolution. Acta Math. **191**(1), 1–54 (2003)
8. Carlen, E.A., Carvalho, M.C., Loss, M.: Spectral gaps for reversible Markov processes with chaotic invariant measures: the Kac process with hard sphere collisions in three dimensions. Ann. Prob. **48**(6), 2807–2844 (2020)
9. Carlen, E.A., Carvalho, M.C., Loss M.: Kinetic theory and the Kac master equation. In *Entropy and the quantum II*, Contemporary Mathematics, vol. 552, pp. 1–20. American Mathematical Society, Providence, RI (2011)
10. Carrapatoso, K., Einav, A.: Chaos and entropic chaos in Kac's model without high moments. Electron. J. Probab. **18**(78), 38 (2013)
11. Einav, A.: On Villani's conjecture concerning entropy production for the Kac master equation. Kinet. Relat. Models **4**(2), 479–497 (2011)

12. Gallagher, I., Saint-Raymond, L., Texier, B.: From Newton to Boltzmann: hard spheres and short-range potentials. Zurich Lectures in Advanced Mathematics. European Mathematical Society (EMS), Zürich (2013)
13. Hauray, M., Mischler, S.: On Kac's chaos and related problems. J. Funct. Anal. **266**(10), 6055–6157 (2014)
14. Janvresse, E.: Spectral gap for Kac's model of Boltzmann equation. Ann. Prob. **29**(1), 288–304 (2001)
15. Kac, M.: Foundations of kinetic theory. In: Proceedings of the Third Berkeley Symposium on Mathematical Statistics and Probability, 1954–1955, vol. III, pp. 171–197. University of California Press, Berkeley-Los Angeles, Calif. (1956)
16. Jr, McKean, H.P.: An exponential formula for solving Boltmann's equation for a Maxwellian gas. J. Comb. Theory **2**, 358–382 (1967)
17. Mischler, S., Mouhot, C.: Kac's program in kinetic theory. Invent. Math. **193**(1), 1–147 (2013)
18. Toscani, G., Villani, C.: Sharp entropy dissipation bounds and explicit rate of trend to equilibrium for the spatially homogeneous Boltzmann equation. Commun. Math. Phys. **203**(3), 667–706 (1999)
19. Villani, C.: A review of mathematical topics in collisional kinetic theory. Handbook of Mathematical Fluid Dynamics, vol. 1, pp. 71–74. North-Holland, 2002
20. Villani, C.: Cercignani's conjecture is sometimes true and always almost true. Commun. Math. Phys. **234**(3), 455–490 (2003)

Entropy Production Bounds for the Kac Model are Uniform in the Number of Particles

Luís Simão Ferreira

Abstract In this paper, we prove fully quantitative entropy power law bounds for Kac's collision model, uniform in the number of particles, by making use of the HWI inequality and Wasserstein distance power laws obtained by Rousset [34] via a Markovian coupling method. In turn, these bounds allow us to derive entropy power laws for the spatially homogeneous Boltzmann equation, akin to the results by Villani [40], by relatively elementary methods and requiring only moment and mild integrability conditions on the initial distribution.

Keywords Particle systems · Kac's process · Boltzmann equation · Convergence to equilibrium · Entropy production · Power law · Coupling · HWI inequality · Propagation of chaos

1 Introduction

Kickstarted by the 1956 seminal paper of Mark Kac [25], particle systems and approach to equilibrium has proved itself to be an extremely fruitful topic of research in the recent decades, due to its wide range of available techniques.

In this work, the concept of chaos was introduced as a formalization of asymptotic independence in order to connect microscopic quantities with the macroscopic description of the evolution of a rarefied gas.

To make use of this new tool, Kac described a caricature stochastic process modelling the binary collisions in a gas made of N identical molecules, now known as *Kac's process* or *Kac's walk*, with particular interest in its convergence to equilibrium.

L. S. Ferreira (✉)
Business Research Unit ISCTE-IUL, Lisboa, Portugal
e-mail: lsafa1@iscte-iul.pt

E. Carlen et al. (eds.), *From Particle Systems to Partial Differential Equations*,
Springer Proceedings in Mathematics & Statistics 465,
https://doi.org/10.1007/978-3-031-65195-3_5

Later on, in 1959 [26], he moved on to a more realistic version of the toy model, with 3 dimensional velocities undergoing momentum and energy conserving collisions, whose connection with the spatially homogeneous Boltzmann equation became clear with mean-field theory. With appropriate scaling, the normalized empirical measure converges to a solution of the PDE as $N \to +\infty$, giving hope to the idea that discrete probabilistic methods could lead to strong results on the mean-field statistics.

The study of relaxation to equilibrium has since been largely reliant on analytic and functional techniques. The historically preferred method for such estimates is known as the *spectral gap* method. The central object of this approach is the generator of the Markov semi-group of the process, $\mathscr{L}_N$. Since initial distribution F_0^N of Kac's process evolves according to

$$\frac{\partial}{\partial t} F_t^N = -\mathscr{L}_N F_t^N \,, \tag{1}$$

a lower bound Δ_N on the spectrum of $\mathscr{L}_N$ implies exponential convergence to equilibrium by the spectral theorem,

$$\|e^{-\mathscr{L}_N t} F_0^N - 1\|_2 \leq e^{-\Delta_N t} \|F_0^N - 1\|_2 \,. \tag{2}$$

While the existence of the spectral gap did not pose an obstacle, the persistence of the gap as $N \to +\infty$ proved itself to be a remarkable problem. For Kac's caricature with one dimensional velocities, he conjectured a spectral gap Δ_N bounded away from 0 uniformly in N, that is

$$\liminf_{N \to +\infty} \Delta_N > 0. \tag{3}$$

The solution was finally found in 2000 by Janvresse [24] and the exact values of Δ_N were found later on by Carlen, Carvalho and Loss [8].

The existence of a spectral gap for the physically realistic model with 3 dimensional velocities was tackled over a series of papers [5, 6, 9, 10] by the same authors, together with Geronimo, and the remaining computation of the 3 particle case was carried out in [20]. Furthermore, the authors of [29] have used uniform in time propagation of chaos to pull back the long time stability of the mean field limit to prove convergence to equilibrium of the N-particle case.

One other interesting technique gained relevance by the works and conjectures of Carlo Cercignani [13]. Stronger than spectral gap estimates, the study of entropy production inequalities has been at the focus of the functional analytic community.

Relative entropy, defined between two probability measures μ and ν on a measurable space Ω, with μ absolutely continuous relative to ν, as

$$H(\mu|\nu) := \int_\Omega \frac{d\mu}{d\nu} \log \frac{d\mu}{d\nu} d\nu \,, \tag{4}$$

can also be used as a stronger measure of deviation from equilibrium, by virtue of the Csiszar-Kullnack-Leibler-Pinsker inequality,

$$H(\mu|\nu) \geq \frac{1}{2}\|\mu - \nu\|_{TV}^2, \tag{5}$$

with $\|\cdot\|_{TV}$ denoting the total variation norm.

Its logarithmic scaling makes it a natural candidate as an approach for Kac's problem when the solutions show chaotic behaviour and asymptotically factorize, i.e., they are well approximated by a tensorization,

$$F_t^N \approx f_t^{\otimes N}. \tag{6}$$

Thus, one would be interested in obtaining entropy production estimates of the type

$$\frac{d}{dt}H(F_t^N|\sigma^N) \leq -C_N H(F_t^N|\sigma^N), \tag{7}$$

where σ^N denotes the uniform equilibrium measure on the N particle state space, in particular to extend such bounds to solutions of the Boltzmann equation.

As it turned out, the computation of the entropy production constant is a much harder problem when compared to the computation of the spectral gap, leading to many new developments in the theory of log-Sobolev inequalities.

Moreover, previously obtained estimates indicating that

$$C_N \sim \frac{1}{N} \tag{8}$$

for Kac's toy model appear sharp [19, 40], meaning that a general result on exponential convergence to equilibrium would disappear in the $N \to +\infty$ limit.

Nonetheless, a series of results by Toscani and Villani [37–39], with Villani later settling Cercignani's conjecture [40], has shown that, while this method may not be able to directly derive exponential convergence to equilibrium for solutions of the Boltzmann equation, a rate of $o(t^{-\infty})$ can be achieved, provided there is enough regularity.

In terms of probability, the Kac process posed an interesting challenge by itself. While much work has been done regarding equilibration and convergence of Markov chains in discrete spaces in search of Doeblin [17] type results (see, for instance, [14, 22]), a less studied case is that of Markov processes in continuous state spaces of large dimension, such as Kac's walk, and the use of explicit probabilistic methods for weak convergence.

Recently, Rousset [34] was the first to obtain convergence results by the means of *coupling* different realizations of Kac's process, whose work our results largely rely on. With this method, he proved a N-uniform quantitative version of Tanaka's theorem [36], that originally showed the quadratic Wasserstein distance between solutions of the Boltzmann-Kac equation and equilibrium was decreasing, but without estimates on the rate.

Rousset's estimates require only reasonable moment conditions on initial data. In this paper, we are able to make use of these estimates to deduce power laws for entropy of solutions of the spatially homogeneous Boltzmann equation for Maxwellian molecules, requiring just the same moment bounds, improving previous results as no regularity or differentiability assumptions are needed. Similarly, these imply the same rates of convergence in L_2.

In order to state the main results, we will now fix notation and go over the main definitions of the paper.

1.1 Kac's Collision Process

Let $N \in \mathbb{N}$ denote the number of particles in our system, $0 \leq \alpha \leq 2$ a physical parameter and $\mathscr{S}_N$ the set of N-tuples $\vec{v} = (v_1, ..., v_N)$ of velocities v_j in $\mathbb{R}^d$ such that

$$\frac{1}{N}\sum_{j=1}^{N} |v_j|^2 = 1, \tag{9}$$

$$\frac{1}{N}\sum_{j=1}^{N} v_j = 0. \tag{10}$$

We refer to Eqs. (9) and (10) respectively as the energy and momentum conditions, meaning that, without loss of generality, we consider a normalized system of particles of mass 2 with average energy equal to 1 in the center of mass reference frame. For the most part, we will be considering the physically relevant case of 3-dimensional velocities.

The system, denoted $V_t^N \in \mathscr{S}_N$, then undergoes momentum and energy conserving collisions in a stochastic manner, whose dynamics, called *Kac's process*, can be described as follows.

To each pair of particles $\{i, j\}$, we associate an exponential random variable $T_{i,j}$ modelling the time until the first collision of the pair, with rate given by

$$\lambda_{i,j} = \binom{N}{2}^{-1} |v_i - v_j|^{\alpha} \tag{11}$$

and take the *first collision time* T as the first among these,

$$T = \min_{i<j} T_{i,j}, \tag{12}$$

after which the minimizing pair of particles (i_*, j_*) collides and the process jumps from

$$V_{T^-}^N = (v_1, ..., v_i, ..., v_j, ..., v_N) \in \mathscr{S}_N$$

to

$$V_T^N = (v_1, ..., v_i^*, ..., v_j^*, .., v_N) \in \mathscr{S}_N,$$

starting anew. Given the constrictions imposed, we can parametrize these collisions on the sphere of radius $\sqrt{v_i^2 + v_j^2}$ in $\mathbb{R}^d$ as

$$\begin{aligned} v_i^*(\sigma) &= \frac{v_i + v_j}{2} + \frac{|v_i - v_j|}{2}\sigma, \\ v_j^*(\sigma) &= \frac{v_i + v_j}{2} - \frac{|v_i - v_j|}{2}\sigma, \end{aligned} \tag{13}$$

with $\sigma \in \mathbb{S}^{d-1}$.

The unit vector σ is randomly drawn according to the considered collision kernel, which determines any directional preferences of the setting or the collisional process itself.

It may be specified by any non-negative even function b on the interval $[-1, 1]$ such that, for any fixed $\sigma' \in \mathbb{S}^2$,

$$\int_{\mathbb{S}^2} b(\sigma \cdot \sigma')d\sigma = 1. \tag{14}$$

In this paper, we are particularly interested in the case where it is chosen uniformly in $\mathbb{S}^2$, i.e., taking $b(x) = 1$.

This parametrization is also particularly useful in our probabilistic context, relative to alternative ones such as the n parametrization, where the post-collisional velocities are obtained by reflecting along a plane with normal n and whose Jacobian is non-constant.

The parameter α describes the interaction range. The case $\alpha < 0$ is called the *soft potential*, which becomes unbounded for particles with close velocities, whereas $\alpha > 0$ is called the *hard potential*, which instead is unbounded as $|v - v_*| \to +\infty$.

The case $\alpha = 1$ is the most physically relevant, describing *hard sphere* collisions, and $\alpha = 2$, the *super hard sphere* case, while not realistic, is of interest as well in the study of physical behaviours [10, 40].

In this paper, however, we will focus on *Maxwell molecules*, with parameter $\alpha = 0$, whose interactions do not depend on their velocities.

Finally, notice that the infinitesimal generator of V_t^N can be written as

$$L_{N,\alpha} f = -N\binom{N}{2}^{-1} \sum_{i<j} |v_i - v_j|^\alpha \left(f - [f]^{(i,j)}\right), \tag{15}$$

with

$$[f]^{(i,j)}(\vec{v}) := \int_{\mathbb{S}^{d-1}} f(R_{i,j,\sigma}\vec{v})d\sigma \tag{16}$$

and $R_{i,j,\sigma}$ the rotation operator describing parametrization (13).

Thus, if V_0^N has a density f_0, the density for times $t > 0$ satisfies Kac's master equation,

$$\frac{\partial}{\partial t} f_t = L_{N,\alpha} f_t\,, \tag{17}$$

subject to the boundary condition $\lim_{t\to 0} f_t = f_0$.

1.2 Wasserstein Distance

In order to study convergence to equilibrium, we will fix a reference metric, for which there are many possible options. In order to employ probabilistic tools, the most useful metric is the *Wasserstein distance*.

The Wasserstein p-distance between two probability measures π_1, π_2 is defined as

$$W_p^p(\pi_1, \pi_2) = \inf_{\pi\in\Pi(\pi_1,\pi_2)} \int_{\Omega\times\Omega} d_p(u, v)^p \pi(du, dv), \tag{18}$$

with the infimum being taken over all possible couplings of the pair π_1, π_2.

In this context, a coupling is simply a probability measure π on the product space $\Omega \times \Omega$ with π_1 and π_2 as its marginals.

Alternatively, by Kantorovich duality, it can be also defined as a supremum instead:

$$W_p^p(\pi_1, \pi_2) = \sup_{\varphi,\psi} \left(\int_\Omega \varphi d\pi_1 - \int_\Omega \psi d\pi_2 \right), \tag{19}$$

with $\varphi \in L^1(d\pi_1)$, $\psi \in L^1(d\pi_2)$ satisfying

$$\varphi(u) - \psi(v) \le d_p(u, v)^p. \tag{20}$$

Equation (18) can be controlled by constructing particular couplings, while the dual definition give us some analytical versatility to prove convergence results. In this paper, we will consider the Wasserstein 2-distance, due to its geometric properties.

In practice, we will also be interested in the *symmetric* Wasserstein 2-distance, $W_{2,\mathrm{sym}}(\pi_1, \pi_2)$, where π_1, π_2 are instead defined on the quotient space $\mathbb{R}^n/\mathrm{Sym}$. One alternative way to define it on the original state space is instead taking the infimum over all particle permutations.

This choice is due to the symmetric Wasserstein distance admitting a random variable representation with almost surely positively correlated variables. This will allow us to couple these variables in a pathwise fashion and the positive correlation condition enables the use of geometric tools to prove convergence.

These two distances are equal for symmetric, or exchangeable, distributions,

$$W_2(\pi_1, \pi_2) = W_{2,\text{sym}}(\pi_1, \pi_2), \forall \pi_1, \pi_2 \in \mathscr{P}_{\text{sym}}((\mathbb{R}^d)^N). \tag{21}$$

This does not pose an obstacle to us, as the symmetric sector needs a more careful analysis and is of most physical relevance, due to its relation with the many particle limit.

We also note that, by the symmetry of the generator of Kac's process, particle exchangeability is propagated in time for individual copies of the process, but this will not be globally true for the coupled process.

Furthermore, we shall consider a normalized version of the Wasserstein distance, taking

$$d_2(u, v)^2 = \langle |u - v|^2 \rangle_N = \frac{1}{N} \sum_{k=1}^{N} |u_k - v_k|^2, \tag{22}$$

using $\langle \cdot \rangle_N := \frac{1}{N} \sum_{k=1}^{N}$ to denote the particle average operator.

1.3 Entropy, Entropy Production and Fisher Information

Let H denote the *entropy functional relative to the uniform measure* defined for densities in $\mathscr{S}_N$:

$$H(f) := \int_{\mathscr{S}_N} f \log f d\sigma^N. \tag{23}$$

The *entropy production functional* of a stochastic process is defined as minus the rate of change of entropy under the evolution generated by the process. In our context, we find that the entropy production functional of Kac's process is given by

$$D_{L_{N,\alpha}}(f) := -\frac{d}{dt} H(e^{tL_{N,\alpha}} f) \bigg|_{t=0} \geq 0. \tag{24}$$

One final quantity to consider is the *Fisher information*, defined as

$$I(f) := \int_{\mathscr{S}_N} \frac{|\nabla_{\mathscr{S}_N} f|^2}{f} d\sigma^N = \int_{\mathscr{S}_N} f |\nabla_{\mathscr{S}_N} \log f|^2 d\sigma^N, \tag{25}$$

which may be usefully interpreted as the entropy production of diffusion (i.e., of the Laplacian operator) [4]. To simplify the notation, we will denote the tangential gradient of $\mathscr{S}_N$ simply as ∇. The following lemma reminds us of two properties of particular usefulness.

Lemma 1 *The Fisher information functional is convex and sub-additive.*

Proof The function

$$g(v,t) = \frac{\langle v, v\rangle}{t} \tag{26}$$

is jointly convex in $(v,t) \in \mathbb{R}^n \times \mathbb{R}^+$. Taking $v = \nabla f$ and $t = f$, by the linearity of the gradient, we deduce the first part of the lemma.

The second half of the statement follows from convexity and homogeneity of degree 1: for any $\lambda \in \mathbb{R}$,

$$I(\lambda f) = \lambda I(f) \tag{27}$$

and

$$I(f+g) = 2I(f/2+g/2) \leq 2(I(f)/2 + I(g)/2) = I(f) + I(g). \tag{28}$$

1.4 Main Results

With the previous concepts in mind, we are able to state our main results.

Theorem 1 (Entropy power law for Kac's process) *Let $(V_t^N)_{t\geq 0}$ be an N particle Kac's process with Maxwell molecules and constant angular collision kernel, symmetric under particle permutations, and denote its law by $Law(V_t^N) = \pi_t^N$. Let $\delta > 0$ and $p, q > 1$ such that $1/p + 1/q = 1$.*

Furthermore, assume that the initial condition V_0^N has a finite moment of order $2q(1+\delta)$, i.e.,

$$\mathbb{E}\Big\langle |V_0^N|^{2q(1+\delta)} \Big\rangle_N < +\infty. \tag{29}$$

Then, we have that

$$\frac{H(\pi_t^N|\sigma^N)}{N} \leq \sqrt{\frac{I(\pi_0^N|\sigma^N)}{N}} \left(W_{2,sym}(\pi_0^N, \sigma^N)^{-1/\delta} + \frac{c_{\delta,q,N}}{\delta} t \right)^{-\delta}, \tag{30}$$

with

$$c_{\delta,q,N} := \frac{\kappa_{\delta,q}}{\left(\beta_{\delta,q} \mathbb{E}\Big\langle |V_0^N|^{2q(1+\delta)} \Big\rangle_N \right)^{1/2q\delta}}, \tag{31}$$

$$\kappa_{\delta,q} := 2^{-9/2-2/\delta} \frac{d-2}{d-1} \frac{(1+\delta)^{1+1/\delta}}{(1+2\delta)^{1+2/\delta}} \left(\frac{d-1}{d}\right)^{1+1/2\delta} \mathbb{E}\left[|G_d|^{2p(1+\delta)} \right]^{-1/2p\delta}, \tag{32}$$

G_d being a d-dimensional standard Gaussian random variable, and $\beta_{\delta,q}$ an explicitly computable constant.

Theorem 2 (Entropy power law for the Boltzmann equation) *Let $\delta > 0,\ p, q > 1$ such that $1/p + 1/q = 1$, $f_0 \in L^s(\mathbb{R}^3)$, $s > 1$, a probability density with finite sixth moment and entropy such that*

$$\int_{\mathbb{R}^3} \begin{bmatrix} 1 \\ v \\ v^2 \end{bmatrix} f_0(v)dv = \begin{bmatrix} 1 \\ 0 \\ 1 \end{bmatrix}. \tag{33}$$

Additionally, suppose that f_0 has a finite moment of order $2q(1+\delta)$, i.e.,

$$\int_{\mathbb{R}^3} |v|^{2q(1+\delta)} f_0(v)dv < +\infty\,. \tag{34}$$

Let $f_t = f(v, t)$ denote the solution of the initial value problem for the spatially homogeneous Boltzmann equation with Maxwell molecules and constant angular collision kernel,

$$\frac{\partial f_t}{\partial t} = \int_{\mathbb{S}^2}\int_{\mathbb{R}^3} \Big(f(v^*, t) f(w^*, t) - f(v, t) f(w, t) \Big) dw d\sigma\,, \tag{35}$$

$$\lim_{t \to 0^+} f_t = f_0\,. \tag{36}$$

Then, we have that

$$H(f_t|\gamma) \le \sqrt{I(f_0)} \left(2^{-1/\delta} + \frac{c_{\delta,q}}{\delta} t \right)^{-\delta}, \tag{37}$$

with γ being the standard Gaussian measure on $\mathbb{R}^3$,

$$c_{\delta,q} = c_{\delta,q}(f_0) := \frac{\kappa_{\delta,q}}{\left(\beta_{\delta,q} \int_{\mathbb{R}^3} |v|^{2q(1+\delta)} f_0(v)dv \right)^{1/2q\delta}}, \tag{38}$$

$\kappa_{\delta,q}$ defined by (32) *and $\beta_{\delta,q}$ also explicitly computable.*

1.5 Structure of the Paper

The paper and proofs are then organized as follows.

In Sect. 2, we describe the approach of Rousset [34] to proving Wasserstein distance power laws through Markov coupling. Here, we show the construction of the coupling providing an upper bound for the power law, adapting notation and bounds as needed. We also take the time to show some of the shortcomings of the method and how some stronger results are still inaccessible.

In Sect. 3, we introduce our main tool, the HWI inequality, and apply it to our setting, proving Theorem 1. While geometrical in its nature, it nonetheless allows us to obtain analytical bounds on entropy using probabilistic objects. In particular, Wasserstein distance power laws obtained via coupling imply an entropy power law.

To conclude, in Sect. 4 we make use of propagation of chaos and the idea of conditioned product measures to prove Theorem 2, translating the finite particle result into an analogous one for solutions of the spatially homogeneous Boltzmann equation.

2 Markovian Coupling Method

The main technique on which our results build on relies on introducing dependence between an arbitrary Kac's process and another one starting already in equilibrium, forcing them to converge in some sense to the same state at a quantifiable rate.

This idea is captured by the concept of a *Markovian coupling*, consisting of a pair of copies of a Markov process, $t \to (U_t, V_t) \in E \times E$, which is itself a Markov process on the product state space.

In this section, we explain the theorems obtained by Rousset [34] on convergence to equilibrium in the Wasserstein distance, derived with the aforementioned technique, which lay the groundwork for our results.

Before we move on to our context with a continuous time jump process in a continuous state space, let us illustrate this procedure with a card magic trick, known as the *Kruskal Count*. We refer to [27] for an exposition of the mathematical details.

Suppose a magician has a deck of shuffled cards. First, you pick a number between one and ten and the magician starts showing the cards one by one. When as many cards as the number you picked have passed, your new number becomes the one on the current card (face cards are worth 5), and so on until the deck is finished.

The magician's goal is to guess the last card you landed on. While this seems impossible without knowing the starting seed number, the best possible guess is obtained by playing the game along himself, starting with his own seed. Due to the Markovian nature of a player's card sequence, only the final state matters, and if eventually his sequence matches the audience's, they will remain coupled afterwards.

In fact, with this strategy, there is a probability of about 85% that the magician guesses right. In Markovian coupling terms, the probability of two copies of the Kruscal count process becoming coupled increases over time, independently of the initial conditions.

In this toy model, we essentially couple the two Kruscal card sequences by playing with the same deck. Next, we will see how we can do a similar procedure to two copies of Kac's process. For an exposition on general Markovian coupling theory and convergence to equilibrium, we refer to [3].

2.1 Convergence to Equilibrium

Now, suppose that U_t is a Kac process with velocities initially distributed according to the equilibrium distribution, so that

$$\mathrm{Law}(U_t) = \pi_\infty^N = \sigma^N \, , \forall t \geq 0 \, , \tag{39}$$

and let the initial condition of the process V_t be any symmetric distribution π_0^N, with π_t^N being its distribution at time t under Kac's process.

From Eq. (18), we recall that the coupling imposes a bound on information distance

$$W_2^2(\pi_t^N, \sigma^N) \leq \mathbb{E}\left[\left\langle |U_t - V_t|^2 \right\rangle_N\right] \left(= \int_{\mathscr{S}_N \times \mathscr{S}_N} d_2(u, v)^2 \pi(du, dv) \right). \tag{40}$$

A necessary condition to eventually prove a power law convergence is that our coupling is *weakly contracting*. In other words, the existence of an almost surely decreasing coupling for Kac's process such that, for any initial conditions and $0 \leq t \leq t + h$,

$$\left\langle |U_{t+h} - V_{t+h}|^2 \right\rangle_N \leq \left\langle |U_t - V_t|^2 \right\rangle_N \quad \text{a.s.}, \tag{41}$$

implies that the expected quadratic coupling distance is non-increasing,

$$\mathbb{E}\left[\left\langle |U_{t+h} - V_{t+h}|^2 \right\rangle_N\right] \leq \mathbb{E}\left[\left\langle |U_t - V_t|^2 \right\rangle_N\right]. \tag{42}$$

The starting step for obtaining such objects is through *simultaneous coupling*, where the corresponding pairs of particles have the same collision times on both copies. However, this is only possible if the jump rates are state independent, limiting the scope of this technique to the case of Maxwell molecules.

When coupling the two copies of Kac's process, we must also pay special attention that the introduced dependencies don't skew the distributions of any of them or change the law of the process.

For simplicity, take $N = 3$ and let $t \geq 0$, $h > 0$. Suppose that a jump occurs once during the interval $(t, t + h]$ between particles 2 and 3, without loss of generality.

Initially, we have $U = (u_1, u, u_*)$, $V = (v_1, v, v_*)$, which we can write in terms of the pre-collisional direction $\sigma_u = (u - u_*)/|u - u_*|$,

$$u = \frac{u + u_*}{2} + \frac{|u - u_*|}{2}\sigma_u \, , \tag{43}$$

$$u_* = \frac{u + u_*}{2} - \frac{|u - u_*|}{2}\sigma_u \tag{44}$$

and analogously for v, v_*.

Recalling our parametrization of the collisions, the post-collisional velocities are given by a similar expression, with the pre-collisional directions σ_u, σ_v simply being replaced with new random directions σ_u', σ_v'.

Thus, the coupling distances before and after the collisions are given by

$$d_2(U_t, V_t)^2 = |u_1 - v_1|^2 + 2\left|\frac{u + u_*}{2} - \frac{v + v_*}{2}\right|^2 + 2\left|\frac{|u - u_*|}{2}\sigma_u - \frac{|v - v_*|}{2}\sigma_v\right|^2 \tag{45}$$

and

$$d_2(U_{t+h}, V_{t+h})^2 = |u_1 - v_1|^2 + 2\left|\frac{u + u_*}{2} - \frac{v + v_*}{2}\right|^2 + 2\left|\frac{|u - u_*|}{2}\sigma_u' - \frac{|v - v_*|}{2}\sigma_v'\right|^2, \tag{46}$$

and we have the equality

$$d_2(U_{t+h}, V_{t+h})^2 - d_2(U_t, V_t)^2 = -|u - u_*||v - v_*|\left(\sigma_u' \cdot \sigma_v' - \sigma_u \cdot \sigma_v\right), \tag{47}$$

which is also valid for the general N particle process.

We are left with discussing on how to introduce dependence between σ_v' and σ_u', in order to make this coupling distance a.s. non-increasing.

In the work of Rousset [34], the choice of parallel transport based coupling leads to both copies having the same scattering angles. That is,

$$\sigma_v' \cdot \sigma_v = \sigma_u' \cdot \sigma_u =: \cos\theta. \tag{48}$$

In order to make it a contractive coupling, the rotation in the remaining degrees of freedom ensures the "alignment" of post-collisional velocities. This allows for the treatment of scattering distributions with arbitrary angular cut-off, and the results also hold for the no angular cut-off case by a limiting procedure.

2.2 Wasserstein Distance Power Law

Rousset's proof then relies on the computation of the *quadratic coupling creation*,

$$\frac{d}{dt}\mathbb{E}\left[\left\langle |U_t - V_t|^2 \right\rangle_N\right]\Big|_{t=0} = -\mathbb{E}\left[\mathscr{C}_2(U_t, V_t)\right] \leq 0, \tag{49}$$

where $\mathscr{C}_2$ is the *two-body coupling creation functional*,

$$\mathscr{C}_2(u, v) = \frac{d-2}{2d-2}\langle |u - u_*||v - v_*| - (u - u_*)\cdot(v - v_*)\rangle_N \geq 0, \tag{50}$$

can be interpreted as the alignment between the differences $v - v_*$ and $u - u_*$.

A general sharp inequality from linear algebra then allows for the two quantities to be related, provided there is positive correlation between the two initial processes.

The following theorem ensures that, due to exchangeability, there is a random variable representation of the two initial distributions such that this assumption is satisfied.

Lemma 2 ([34]) *Let $\pi_1, \pi_2 \in \mathscr{P}_{sym}((\mathbb{R}^d)^N)$ be two exchangeable probabilities. Then, there exists a random variable representation $(U_0, V_0, \Sigma) \in (\mathbb{R}^d \times \mathbb{R}^d)^N \times Sym_N$ with $Law(U_0) = \pi_1$, $Law(V_0) = \pi_2$, such that:*

1. $W^2_{2,sym}(\pi_1, \pi_2) = \mathbb{E}\left\langle |U_0 - V_{0,\Sigma(\cdot)}|^2 \right\rangle_N,$
   where $V_{0,\Sigma(\cdot)}$ denotes the random variable V_0 with particles rearranged according to Σ;
2. *If V_0 is almost surely centered ($\langle V_0\rangle_N = 0$ a.s.), then $\langle U_0 \cdot V_{0,\Sigma(.)}\rangle_N \geq 0$ a.s..*

Before the coupled process begins, the coupled distributions are completely exchangeable, meaning that there is no natural ordering or pairing of the particles between the two copies yet, but this property is broken for $t > 0$.

Nevertheless, the particle pairs are rearranged initially so that the positive correlation assumption is satisfied and the constructed pairing shows a power law coupling creation, proving:

Theorem 3 ([34]) *Let $t \mapsto V_t \in (\mathbb{R}^d)^N$ be any Kac's conservative particle system with Maxwell molecules and denote $\pi_t^N := Law(V_t)$. Then, for any $\delta > 0$, $p, q > 1$ such that $1/p + 1/q = 1$, the following trend to the equilibrium π_∞^N holds:*

$$\frac{d^+}{dt} W_{2,sym}(\pi_t^N, \sigma^N) \leq -c_{\delta,q,N}(\pi_t) W_{2,sym}(\pi_t^N, \sigma^N)^{1+1/\delta}, \tag{51}$$

with

$$c_{\delta,q,N}(\pi_t) := k_{\delta,q,N}\mathbb{E}\left[\left\langle |V_t|^{2q(1+\delta)}\right\rangle_N\right]^{-1/2q\delta}, \tag{52}$$

and $k_{\delta,q,N} > 0$ an explicit numerical constant:

$$k_{\delta,q,N} = 2^{-9/2-2/\delta}\frac{d-2}{d-1}\frac{(1+\delta)^{1+1/\delta}}{(1+2\delta)^{1+2/\delta}} \times \mathbb{E}\left[\left\langle \kappa^{p(1+\delta)}_{\langle U\otimes U\rangle_N}\left|\frac{U - U_*}{\sqrt{2}}\right|^{2p(1+\delta)}\right\rangle_N\right]^{-1/2p\delta}. \tag{53}$$

Here, U, U_* are i.i.d. uniform random variables on $\mathscr{S}_N$ and κ_S denotes the spectral quantity

$$\kappa_S := (1 - \lambda_{max}(S))^{-1} \in \left[\frac{d}{d-1}, +\infty\right], \tag{54}$$

with $\lambda_{max}(S) \leq 1$ the spectral radius of a trace 1 positive symmetric matrix S.

One can equivalently write the differential inequality (51) as

$$W_{2,\mathrm{sym}}(\pi_t^N, \sigma^N) \leq \left(W_{2,\mathrm{sym}}(\pi_0^N, \sigma^N)^{-1/\delta} + \frac{1}{\delta} \int_0^t c_{\delta,q,N}(\pi_s^N) ds \right)^{-\delta}, \tag{55}$$

by solving the differential equation given by the equality and comparing initial conditions.

Moreover, the dependency in N can be removed from $k_{\delta,q,N}$ by the following monotonous limit, based on a central limit-type argument by Rousset,

$$\begin{aligned} &\lim_{N\to+\infty} \mathbb{E}\left[\left\langle \kappa_{\langle U \otimes U \rangle_N}^{p(1+\delta)} \left|\frac{U-U_*}{\sqrt{2}}\right|^{2p(1+\delta)} \right\rangle_N\right]^{-1/2p\delta} \\ &= \left(\frac{d-1}{d}\right)^{1+1/2\delta} \mathbb{E}\left[|G_d|^{2p(1+\delta)}\right]^{-1/2p\delta}, \end{aligned} \tag{56}$$

G_d a d-dimensional Gaussian random variable with $\mathbb{E}|G_d|^2 = 1$, so that we have

$$\begin{aligned} k_{\delta,q,N} \geq k_{\delta,q} :=& 2^{-9/2-2/\delta} \frac{d-2}{d-1} \frac{(1+\delta)^{1+1/\delta}}{(1+2\delta)^{1+2/\delta}} \\ &\times \left(\frac{d-1}{d}\right)^{1+1/2\delta} \mathbb{E}\left[|G_d|^{2p(1+\delta)}\right]^{-1/2p\delta}. \end{aligned} \tag{57}$$

2.3 *Moment Bounds*

The next step towards a more useful form of inequality (55) is bounding the integrand from below, independently from time, with such bound depending on $(\pi_s^N)_{0\leq s\leq t}$ only through the initial condition π_0^N.

For this purpose, first recall the inequality known as *Povzner's identity* [31]:

Theorem 4 (Povzner's identity) *For all $p \in (2, +\infty)$, there is an explicit constant $\beta(B, p)$, with B the collision kernel, such that, for all $v, v_* \in \mathbb{R}^3$ and $u = (v - v_*)/|v - v_*|$,*

$$\begin{aligned} &\int_{\mathbb{S}^{d-1}} (|v'|^p + |v_*'|^p - |v|^p - |v_*|^p) B(u, d\sigma) \\ &\leq -\beta(|v|^p + |v_*|^p) + \beta^{-1}(|v||v_*|^{p-1} + |v|^{p-1}|v_*|). \end{aligned} \tag{58}$$

This identity is a powerful tool through which Norris [31] derived moment bounds for Kac's process. Adapting his results to our context and notation, we have the following theorem:

Theorem 5 ([31]) *Let $(V_t^N)_{t\geq 0}$ be Kac's process, $p \in [2, +\infty)$ and $q \in (2, +\infty)$, with $p \leq q$. There exists an explicit constant $C(p,q)$, such that, for all $t \geq 0$,*

$$\mathbb{E}\Big\langle |V_t|^q \Big\rangle_N \leq C(p,q)(1+t^{p-q})\mathbb{E}\Big\langle |V_0|^p \Big\rangle_N. \tag{59}$$

Moreover, there is a constant $C(q,q)$ such that, for all $t \geq 0$,

$$\mathbb{E}\left(\sup_{s\leq t} \Big\langle |V_s|^q \Big\rangle_N \right) \leq C(q,q)(1+t)\mathbb{E}\Big\langle |V_0|^q \Big\rangle_N. \tag{60}$$

While this result also allows us to require lower moment bounds at the cost of time growth, we make the most out of Eq. (59) by taking $p = q$, giving us

$$\mathbb{E}\Big\langle |V_t|^q \Big\rangle_N \leq 2C(q,q)\mathbb{E}\Big\langle |V_0|^q \Big\rangle_N. \tag{61}$$

All together, plugging inequalities (57) and (61) back into definition (52), we get

$$c_{\delta,q,N}(\pi_t^N) \geq \frac{k_{\delta,q}}{\left(2C\mathbb{E}\Big\langle |V_0|^{2q(1+\delta)} \Big\rangle_N\right)^{1/2q\delta}} := c_{\delta,q,N}, \tag{62}$$

independent of time, with $C := C\Big(2q(1+\delta), 2q(1+\delta)\Big)$ given by Norris's constant, and the power law given by (55) becomes

$$W_{2,\mathrm{sym}}(\pi_t^N, \sigma^N) \leq \left(W_{2,\mathrm{sym}}(\pi_0^N, \sigma^N)^{-1/\delta} + \frac{c_{\delta,q,N}}{\delta} t \right)^{-\delta}. \tag{63}$$

Before moving on to the proofs of the main results, let us explain the necessity of sub-exponential rates.

We begin by computing the asymptotics of $k_{\delta,q}$ as $\delta \to +\infty$, which if finite and positive would yield exponential convergence.

Note that, as $\mathbb{E}|G_d|^2 = 1$,

$$d|G_d|^2 = \sum_{i=1}^{d} Z_i^2 \sim \chi_d^2,$$

with the Z_i's being i.i.d. one-dimensional standard Gaussian random variables and χ_d^2 denoting the Chi-squared distribution with d degrees of freedom. With this in mind, we can compute these moments exactly:

$$\mathbb{E}\left[|G_d|^{2p(1+\delta)}\right] = d^{-p(1+\delta)}\mathbb{E}\left[\chi_d^{p(1+\delta)}\right] = \left(\frac{2}{d}\right)^{p(1+\delta)} \frac{\Gamma\left(\frac{d}{2}+p(1+\delta)\right)}{\Gamma\left(\frac{d}{2}\right)} \tag{64}$$

and we have

$$\begin{aligned}\lim_{N\to+\infty} k_{\delta,q,N} =& 2^{-9/2-2/\delta}\frac{d-2}{d-1}\frac{(1+\delta)^{1+1/\delta}}{(1+2\delta)^{1+2/\delta}}\left(\frac{d-1}{d}\right)^{1+1/2\delta} \\ &\times \left(\frac{2}{d}\right)^{-\frac{1+\delta}{\delta}} \frac{\Gamma\left(\frac{d}{2}+p(1+\delta)\right)^{-1/2p\delta}}{\Gamma\left(\frac{d}{2}\right)^{-1/2p\delta}}.\end{aligned} \tag{65}$$

Unfortunately, we find that

$$\mathbb{E}\left[|G_d|^{2p(1+\delta)}\right]^{-1/2q\delta} \xrightarrow[\delta\to+\infty]{} 0$$

polynomially, and exponential convergence is unreachable through this method. Note that, as $q \to 1$, $p \to +\infty$ and $k_{\delta,q,N} \xrightarrow[q\to 1]{} 0$ for the same reasons.

Even if exponential convergence for fixed N was possible, there would be no way to translate this result into an analogous one for the Boltzmann equation. However, it is also not possible to obtain such a result for finite particles using this method, as a counter-example can be constructed as a continuous perturbation of the identity coupling at equilibrium, as shown in Rousset's work.

3 HWI Inequality

We now introduce the main tool of this paper, the so called HWI inequality. First obtained by Otto and Villani [32] in the context of logarithmic Sobolev inequalities and optimal transport inequalities, such as the one by Talagrand [35], this is a powerful functional inequality relating entropy, information and transport, which has allowed for significant developments in the context of chaos estimates [12, 18, 23]

Moreover, it has deep geometrical meaning, as it can be formally derived by inducing a Riemannian-like structure in the set of probability measures of a manifold with the Wasserstein distance as its geodesic distance. However, these arguments suffer from a shortfall in regularity and alternative rigorous proofs were given. We refer to [21]for an extensive treatment and proof, from which we cite the relevant setting and inequalities.

Let $\mathscr{P}_2(X)$ denote the set of probability measures with finite second moment on a metric space (X, d). Let $(M, g, \mathbf{m})$ be a smooth Riemannian manifold without boundary of dimension n, with reference measure $\mathbf{m} := e^{-V}\mathrm{Vol} \in \mathscr{P}_2(M)$, Vol the volume measure on M and $V : M \to [0, +\infty)$ a potential belonging to $C^\infty(M)$.

We define $\mathscr{P}_2(M)$ by understanding M as a metric space endowed with the geodesic distance.

For fixed $m \in \mathbb{N}$, the *Bakry-Émery Ricci tensor* is then defined as

$$\mathrm{Ric}_{V,m} := \mathrm{Ric}_g - (m-n)\frac{\mathrm{Hess}(e^{-\frac{V}{m-n}})}{e^{-\frac{V}{m-n}}} \tag{66}$$

Additionally, suppose that for some $\kappa \in \mathbb{R}$ and $m \geq n$ it satisfies

$$\mathrm{Ric}_{V,m} \geq \kappa g\,. \tag{67}$$

Theorem 6 (HWI Inequality) *Under the previous conditions, the following inequalities are satisfied* $\forall \nu \in \mathscr{P}_2(M)$:

1. *HWI inequality:*

$$H(\nu|\mathbf{m}) \leq W_2(\nu, \mathbf{m})\sqrt{I(\nu|\mathbf{m})} - \frac{\kappa}{2}W_2^2(\nu, \mathbf{m})\,; \tag{68}$$

2. *Talagrand inequality:*

$$\frac{\kappa}{2}W_2^2(\nu, \mathbf{m}) \leq H(\nu|\mathbf{m}); \tag{69}$$

3. *Logarithmic Sobolev inequality: assume that* $\kappa > 0$*, then*

$$H(\nu|\mathbf{m}) \leq \frac{1}{2\kappa}I(\nu|\mathbf{m}). \tag{70}$$

Note that the Wasserstein distance in Eq. (68) is not the N-rescaled version we consider previously.

3.1 *Proof of the Entropy Power Law for Kac's Process*

We will now prove Theorem 1.

Proof Firstly, since the state space of Kac's process, $\mathscr{S}_N$, is isometric to a sphere of radius $\sqrt{N}$ in $\mathbb{R}^{3(N-1)}$, we see that the Ricci tensor satisfies

$$R_{i,j} = \left(3 - \frac{4}{N}\right)g_{i,j}. \tag{71}$$

Taking the potential V to be constant, we conclude that the conditions for the HWI inequality setting are trivially verified.

Thus, let π_t^N denote the measure of Kac's process at time t with initial condition π_0^N, and $\pi_\infty^N = \sigma^N$ the equilibrium uniform measure.

Since $\kappa > 0$, we may drop the extra negative term in the HWI inequality. Considering the rescaled inequality, with an extra $\sqrt{N}$ factor to account for the particle-averaged Wasserstein distance, we obtain

$$\begin{aligned} H(\pi_t^N|\sigma^N) &\le W_2(\pi_t^N,\sigma^N)\sqrt{N\cdot I(\pi_t^N|\sigma^N)} - \frac{\kappa N}{2}W_2^2(\pi_t^N,\sigma^N) \\ &\le W_2(\pi_t^N,\sigma^N)\sqrt{N\cdot I(\pi_t^N|\sigma^N)}. \end{aligned} \tag{72}$$

The power law for entropy is then easily obtained after combining Eqs. (72) and (63), provided that we can bound in time the Fisher information of the solutions.

Fortunately, solutions of Kac's evolution equation have non-increasing Fisher information.

To see this, let

$$Q(f) := \binom{N}{2}^{-1}\sum_{i<j}[f]^{(i,j)}(v) = \binom{N}{2}^{-1}\sum_{i<j}\int_{\mathbb{S}^2} f(R_{i,j,\sigma}v)d\sigma \tag{73}$$

be the one step generator of Kac's walk, and define Kac's semi group P_t as

$$P_t f := e^{-tN}\sum_{k=0}^{+\infty}\frac{(tN)^k}{k!}Q^k(f). \tag{74}$$

One can verify that $P_t f$ is indeed a solution of Kac's evolution equation with initial condition f.

By the convexity of Fisher information (Lemma 1), we have that

$$I(P_t f) \le e^{-tN}\sum_{k=0}^{+\infty}\frac{(tN)^k}{k!}I(Q^k(f)). \tag{75}$$

Furthermore, again by convexity, we see that

$$\begin{aligned} I([f]^{i,j}) &= \int_{\mathscr{S}_N}\frac{\left|\nabla\left(\int_{\mathbb{S}^2} f(R_{i,j,\sigma}v)d\sigma\right)\right|^2}{\int_{\mathbb{S}^2} f(R_{i,j,\sigma}v)d\sigma}d\sigma^N \\ &\le \int_{\mathscr{S}_N}\int_{\mathbb{S}^2}\frac{|\nabla f(R_{i,j,\sigma}v)|^2}{f(R_{i,j,\sigma}v)}d\sigma d\sigma^N \\ &= \int_{\mathscr{S}_N}\frac{|\nabla f(v)|^2}{f(v)}d\sigma^N = I(f), \end{aligned} \tag{76}$$

and therefore $I(Q(f)) \le I(f)$, as Q is an average of the $[f]^{(i,j)}$'s.

We conclude that $I(Q^n(f)) \leq I(Q^{n-1}(f)) \leq ... \leq I(f)$, giving us

$$I(P_t f) \leq e^{-tN} \sum_{k=0}^{+\infty} \frac{(tN)^k}{k!} I(f) = I(f). \tag{77}$$

Thus, we obtain the inequality

$$H(\pi_t^N|\sigma^N) \leq W_2(\pi_t^N, \sigma^N)\sqrt{N \cdot I(\pi_0^N|\sigma^N)}, \tag{78}$$

and consequently

$$\frac{H(\pi_t^N|\sigma^N)}{N} \leq \sqrt{\frac{I(\pi_0^N|\sigma^N)}{N}} \left(W_{2,\text{sym}}(\pi_0^N, \sigma^N)^{-1/\delta} + \frac{c_{\delta,q,N}}{\delta} t \right)^{-\delta}, \tag{79}$$

proving Theorem 1.

4 Propagation of Chaos

Kac's process is governed by a linear equation describing the dynamics of N particles. Nonetheless, by taking "suitable" initial data, such that it asymptotically factorizes as the number of interacting particles becomes very large, the linearity is lost and the mean field limit leads to a strongly nonlinear equation instead, the spatially homogeneous Boltzmann equation:

$$\frac{\partial f_t}{\partial t} = \int_{\mathbb{S}^2} \int_{\mathbb{R}^3} \Big(f(v^*, t) f(w^*, t) - f(v, t) f(w, t) \Big) dw d\sigma. \tag{80}$$

We refer to [1, 30, 33] for treatments on existence results.

This asymptotic independence is captured by the notion of *chaos*.

Definition 1 **(Chaos)** Let μ be a given Borel probability measure on $\mathbb{R}$. Let μ^N be a probability measure on $\mathscr{S}_N$. Then the sequence of probability measures $\{\mu^N\}_{N\in\mathbb{N}}$ is said to be μ-chaotic if:

- Each μ^N is symmetric under interchange of particles;
- For each $k \in \mathbb{N}$, the first k particles marginal $P_k\mu^N$ converges to the k-fold tensor product $\mu^{\otimes k}$ as $N \to +\infty$, in the sense of weak convergence against bounded continuous functions.

One classic example of chaoticity is the uniform distribution on $\mathscr{S}_N$ and the Gaussian measure; that is, the sequence $\{\sigma^N\}_{N\in\mathbb{N}}$ is γ-chaotic.

For the simpler case of Kac's walk on the sphere with $d = 1$, this fact has been well established since the 19th century [28], observing that

$$d\gamma = \frac{1}{\sqrt{2\pi}} e^{-v^2/2} dv \tag{81}$$

and

$$P_1 \sigma^N dv = \left(1 - \frac{v^2}{N}\right)^{\frac{N-3}{2}} \frac{\Gamma(N/2)}{(N\pi)^{1/2}\Gamma(\frac{N-1}{2})} dv . \tag{82}$$

The analogous result for $\mathscr{S}_N$ is rather involved however, and the relevant marginals and proof can be found in Carrapatoso's work [12].

So, if we wish to derive a meaningful inequality valid for solutions of the Boltzmann equation, analogous to (79), by taking the limit in the number of particles ($N \to +\infty$), an f_0-chaotic sequence is necessary, for f_0 initial data for the Boltzmann equation.

Ideally, our sequence needs to have some kind of product structure, so that variables decouple as N gets large. However, the extent to how well our product measure approximates a measure on the sphere depends on how much of the energy is concentrated around it: if most of the mass concentrates away from radius $\sqrt{N}$, then too much information is lost and we do not get chaoticity.

Carlen, Carvalho, Le Roux, Loss and Villani [7] have given a precise construction for such sequences, called *conditioned tensor products*, or *conditioned product measures*, on the Kac sphere. The extension to $\mathscr{S}_N$ was studied by Carrapatoso [12].

Definition 2 (Conditioned product measures) Given a probability density f_0 on $\mathbb{R}^3$, the family of conditioned product measures defined by

$$[f_0^{\otimes N}]_{\mathscr{S}_N} = \frac{\Pi_{j=1}^N f_0(v_j)}{Z_N(f_0, \sqrt{N})} \sigma^N = \frac{\Pi_{j=1}^N (f_0(v_j)/\gamma(v_j))}{Z'_N(f_0, \sqrt{N})} \sigma^N , \tag{83}$$

with

$$Z_N(f_0, \sqrt{N}) := \int_{\mathscr{S}_N} f_0^{\otimes N} d\sigma^N , \tag{84}$$

$$Z'_N(f_0, \sqrt{N}) := \int_{\mathscr{S}_N} \left(\frac{f_0^{\otimes N}}{\gamma^{\otimes N}}\right) d\sigma^N , \tag{85}$$

is called the conditioned tensorization of f_0 to $\mathscr{S}_N$.

In order to simplify notation, we shall denote $[f_0^{\otimes N}]_{\mathscr{S}_N}$ as $f_0^{\otimes N}$ whenever it is clear from context. Furthermore, when we write $f_t^{\otimes N}$ we mean the density of Kac's process at time t with initial condition $f_0^{\otimes N}$, rather than the conditioned tensorization of f_t.

While we may define such families of measures for a large class of densities, requiring mild conditions on moments and integrability ensures highly desired properties. One such property, due to the product construction and the behavior of the normalization function, is that $f_0^{\otimes N}$ is f_0-chaotic. In fact, it can be shown that it is actually entropically f_0-chaotic:

Definition 3 (Entropic chaos) The sequence $\{\mu^N\}_{N\in\mathbb{N}}$ is said to be entropically μ-chaotic if it is μ-chaotic and satisfies

$$\lim_{N\to+\infty} \frac{H(\mu^N|\sigma^N)}{N} = H(\mu|\gamma). \tag{86}$$

Theorem 7 [12] *For any probability density $f \in L^p(\mathbb{R}^d)$, for $p > 1$, and with finite moment of order 6, the conditioned N-fold tensorization of f, $[f^{\otimes N}]_{\mathscr{S}_N}$ is entropically f-chaotic.*

While entropic chaos is a much stronger condition, we will make use of a property know as the *asymptotic upper semi extensivity of entropy* [7].

Lemma 3 *If μ^N is μ-chaotic, then the following holds:*

$$H(\mu|\gamma) \leq \liminf_{N\to+\infty} \frac{H(\mu^N|\sigma^N)}{N}. \tag{87}$$

The final link we need between solutions of Kac's process and solutions of the Boltzmann equation is Kac's classical theorem on *propagation of chaos* [29]:

Theorem 8 (Propagation of chaos) *Let f_0 be a probability density on $\mathbb{R}^3$, and $\{f_0^N\}_{N\in\mathbb{N}}$ be an $f_0(v)dv$-chaotic family of probability densities in $\mathscr{S}_N$, taken as initial distributions for N-particle Kac's process.*

Then, the sequence $\{f_t^N\}_{N\in\mathbb{N}}$ is $f(t,v)dv$-chaotic, where $f(t,v)$ is a solution to the Boltzmann equation (35) *with initial data f_0.*

Thus, the properties previously derived from the careful construction of the initial data are propagated in time.

4.1 Proof of the Entropy Power Law for the Boltzmann Equation

We are now in conditions of proving Theorem 2:

Proof By Kac's theorem, chaos is propagated under Kac's process and we have that $f_t^{\otimes N}$ is $f_t dv$-chaotic, with f_t being the solution to the spatially homogeneous Boltzmann equation (35) with initial condition f_0, and by Lemma 3

$$H(f_t|\gamma) \leq \liminf_{N\to+\infty} \frac{H(f_t^{\otimes N}|\sigma^N)}{N}. \tag{88}$$

Analysing the right hand side of Eq. (79), we first control the asymptotics of Fisher information.

In Carrapatoso's work [12], the first marginal of $f_0^{\otimes N}$ is determined to be

$$f_{0;v_1}^{\otimes N}(v_1) = f_0(v_1)\theta^N(v_1)\mathbb{1}_{|v_1|^2 \leq N-1}\,, \tag{89}$$

with

$$\theta^N(v) = \left(e^{\frac{-|v|^2}{2N}-\frac{-|v|^4}{2N}} + O(N^{-\frac{1}{2}})\right)\left(1 + O(N^{-1})\right), \tag{90}$$

and this fact allows him to show that

$$\frac{1}{N}I\left(f_0^{\otimes N}\middle|\sigma^N\right) \leq \int_{\mathbb{R}^3}\frac{|\nabla_{v_1}f_0(v_1)|^2}{f_0(v_1)}\theta^N(v_1)dv_1 \leq C < +\infty\,. \tag{91}$$

Noting that

$$\theta^N(v) \leq (1 + O(N^{-\frac{1}{2}}))(1 + O(N^{-1})) \tag{92}$$

and taking limits, we conclude

$$\liminf_{N\to+\infty}\frac{1}{N}I\left(f_0^{\otimes N}\middle|\sigma^N\right) \leq \int_{\mathbb{R}^3}\frac{|\nabla_v f_0(v)|^2}{f_0(v)}dv = I(f_0)\,. \tag{93}$$

With regards to the Wasserstein term, while it seems likely that

$$W_2^2(f_0^{\otimes N}, \sigma^N) \xrightarrow[N\to+\infty]{} W_2^2(f_0, \gamma)$$

holds, we do not need this result, since we have following inequalities:

$$W_2^2(f_0, \gamma) \leq \lim_{N\to+\infty} W_2^2(f_0^{\otimes N}, \sigma^N) \leq 2. \tag{94}$$

These can be deduced by either considering the trivial product coupling $f_0^{\otimes N} \otimes \sigma^N$, or by taking the upper bound $|U - V|^2 \leq 2N$ on the infimum definition of the Wasserstein distance.

We are left with computing a bound on the limit of $c_{\delta,q,N}$ as $N \to +\infty$, which follows again by the invariance of particle permutations and Eq. (89), observing that

$$\begin{aligned}
\mathbb{E}\left[\left\langle |V_0|^{2q(1+\delta)}\right\rangle_N\right] &= \frac{1}{N}\sum_{k=1}^{N}\int_{\mathscr{S}_N}|v_k|^{2q(1+\delta)}f_0^{\otimes N}d\sigma^N \\
&= \int_{\mathscr{S}_N}|v_1|^{2q(1+\delta)}f_0^{\otimes N}d\sigma^N \\
&\leq \int_{\mathbb{R}^3}|v|^{2q(1+\delta)}f_0(v)\theta^N(v)dv
\end{aligned}$$

$$\leq \left(\int_{\mathbb{R}^3} |v|^{2q(1+\delta)} f_0(v) dv \right)$$
$$\times \left(1 + O(N^{-\frac{1}{2}})\right)\left(1 + O(N^{-1})\right)$$
$$\xrightarrow[N\to+\infty]{} \int_{\mathbb{R}^3} |v|^{2q(1+\delta)} f_0(v) dv \,. \tag{95}$$

Note that $k_{\delta,q,N} \xrightarrow[N\to+\infty]{} k_{\delta,q}$, as seen in (57).

Combining all terms, we deduce

$$H(f_t|\gamma) \leq \sqrt{I(f_0)}\left(2^{-1/\delta} + \frac{c_{\delta,q}}{\delta}t\right)^{-\delta}, \tag{96}$$

concluding the proof of Theorem 2.

Acknowledgements I thank professor Eric Carlen for his guidance in the project over the past two years and for his suggestions and corrections. I also wish to thank CMAFcIO for their continued support.

References

1. Bobylev, A.V., Gamba, I.M.: Boltzmann equations for mixtures of maxwell gases: exact solutions and power like tails. J. Stat. Phys. **124**, 497–516 (2006)
2. Bobylev, A.V., Cercignani, C.: On the rate of entropy production for the Boltzmann equation. J. Stat. Phys. **94** (3–4), 603–618 (1999)
3. Burdzy, K., Kendall, W.S.: Efficient Markovian couplings: examples and counterexamples. Ann. Appl. Probab. **2**, 362–409 (2000)
4. Carlen, E.A., Lieb, E.H., Loss, M.: A sharp analog of young's inequality on S^N and related entropy inequalities. J. Geom. Anal. **14** (3), 487–520 (2004)
5. Carlen, E.A., Carvalho, M.C., Loss, M.: Determination of the spectral gap for Kac's master equation and related stochastic evolution. Acta Mathematica **191** (1), 1–54 (2003)
6. Carlen, E.A., Geronimo, J., Loss, M.: Determination of the spectral gap in the KAC model for physical momentum and energy conserving collisions. SIAM J. Math. Anal. **40** (1), 327–364 (2008)
7. Carlen, E.A., Carvalho, M.C., Le Roux, J., Loss, M., Villani, C.: Entropy and chaos in the KAC model. Kinet. Relat. Model. **3**(1), 85–122 (2010)
8. Carlen, E.A., Carvalho, M.C., Loss, M.: Many body aspects of approach to equilibrium. Journes Equations aux derivees partielles, Nantes, 5–9 Juin 2000
9. Carlen, E.A., Carvalho, M.C., Loss, M.: Spectral gap for the KAC model with hard sphere collisions. J. Funct. Anal. **266** (3), 1787–1832 (2014)
10. Carlen, E., Carvalho, M., Loss, M.: Spectral gaps for reversible Markov processes with chaotic invariant measures: the KAC process with hard sphere collisions in three dimensions. Ann. Probab. **48** (6), 2807–2844 (2020)
11. Carlen, E.A., Carvalho, M.C.: Strict entropy production bounds and stability of the rate of convergence to equilibrium for the Boltzmann equation. J. Stat. Phys. **67**(3–4), 575–608 (1992)

12. Carrapatoso, K.: Quantitative and qualitative KAC's Chaos on the Boltzmann's Sphere. Ann. Inst. H. Poincaré Prob. Stat. **51**(3), 993–1039 (2015)
13. Cercignani, C.: H-theorem and trend to equilibrium in the kinetic theory of gases. Arch. Mech Archiwum Mechaniki Stosowanej **34**, 231–241 (1982)
14. Cohn, H.: On a Paper by Doeblin on Non-Homogeneous Markov Chains. Adv. Appl. Probab. **13**(2), 388–401 (1981)
15. Desvillettes, L., Mouhot, C., Villani, C.: Celebrating Cercignani's conjecture for the Boltzmann equation. Kinet. Relat. Model. **4**(1), 277–294 (2011)
16. Desvillettes, L., Villani, C.: On the trend to global equilibrium for spatially inhomogeneous kinetic systems: the Boltzmann equation. Invent. Math. **159**, 245–316 (2005)
17. Le cas discontinu de probabilités en chaîne.: W. Doeblin. Publ. Fac. Sci. Univ. Masaryk (Brno) **236**, 3–13 (1937)
18. Einav, A.: On the subadditivity of the entropy on the sphere. J. Geom. Anal. **26**, 3098–3128 (2016)
19. Einav, A.: On Villani's conjecture concerning entropy production for the KAC master equation. Kinet. Relat. Models **4** (2), 479–497 (2011)
20. Ferreira, L.S.: A lower bound for the spectral gap of the conjugate KAC process with 3 interacting particles. Kinet. Relat. Model. **15** (1), 91–117 (2022)
21. Gentil, I., Léonard, C., Ripani, L., Tamanini, L.: An entropic interpolation proof of the HWI inequality. Stoch. Proc. Appl. **130**(2), 907–923 (2020)
22. Hairer, M.: Convergence of Markov Processes (2021). https://hairer.org/notes/Convergence.pdf. Accessed 3 Jan 2023
23. Hauray, M., Mischler, S.: On Kac's chaos and related problems. J. Funct. Anal. **266**(10), 6055–6157 (2014)
24. Janvresse, E.: Spectral gap for KAC's model of Boltzmann Equation. Ann. Prob. **29**, 288–304 (2001)
25. Kac, M.: Foundations of kinetic theory. In: Neyman, J. (ed.) Proceedings of the 3rd Berkeley Symposium on Mathematical Statistics and Probability, pp. 171–197. University of California (1956)
26. Kac, M.: Probability and Related Topics in Physical Sciences. Interscience Publication LTD., London, New York (1959)
27. Lagarias, J., Rains, E., Vanderbei, R.: The Kruskal Count. arXiv:math/0110143v1 [math.PR]. Accessed 13 Oct. 2001
28. Mehler, F.G.: Ueber die Entwicklung einer Function von beliebig vielen Variablen nach Laplaschen Functionen höherer Ordnungn. Crelle's J. **66**, 161–176 (1866)
29. Mischler, S., Mouhot, C.: About KAC's program in kinetic theory. Comptes Rendus Mathematique **349**(23), 1245–1250 (2011)
30. Mischler, S., Wennberg, B.: On the spatially homogeneous Boltzmann equation. Ann. Inst. Henri Poincaré **16**(4), 467–501 (1999)
31. Norris, J.: A consistency estimate for KAC's model with ellastic collisions in a dilute gas. Ann. Appl. Prob. **26**(2), 1029–1081 (2016)
32. Otto, F., Villani, C.: Generalization of an inequality by Talagrand, and links with the logarithmic Sobolev inequality. J. Funct. Anal. **173**(2), 361–400 (2000)
33. Pulvirenti, A., Toscani, G.: The theory of the nonlinear Boltzmann equation for Maxwell molecules in Fourier representation. Annal. Matem. Pura Appl. **171** (4), 181–204 (1996)
34. Rousset, M.: A N-uniform quantitative Tanaka's theorem for the conservative KAC's N-particle system with Maxwell molecules (2014). arxiv:1407.1965 [math.PR]
35. Talagrand, M.: Transportation cost for Gaussian and other product measures. Geom. Funct. Anal. **6**(3), 587–600 (1996)
36. Tanaka, H.: Probabilistic treatment of the Boltzmann equation of Maxwellian molecules. Probab. Theory Relat. Fields **46**(1), 67–105 (1978)
37. Toscani, G., Villani, C.: Probability metrics and uniqueness of the solution to the Boltzmann equation for a Maxwell gas. J. Stat. Phys. **94**(3–4), 619–637 (1999)

38. Toscani, G., Villani, C.: Sharp entropy dissipation bounds and explicit rate of trend to equilibrium for the spatially homogeneous Boltzmann equation. Commun. Math. Phys. **203**(3), 667–706 (1999)
39. Villani, C.: A review of mathematical topics in collisional kinetic theory. Handbook Math. Fluid Dyn. **1**, 71–74 (2002)
40. Villani, C.: Cercignani's conjecture is sometimes true and always almost true. Commun. Math. Phys. **234**, 455–490 (2003)

Quantum Wasserstein and Observability for Quantum Dynamics

François Golse

Abstract In recent years, there have been several extensions of various tools and methods of optimal transport to the quantum setting. In particular, a pseudometric analogous to the Wasserstein distance of exponent 2 has been defined in [F. Golse, T. Paul, Arch. Rational Mech. Anal. **223** (2017) 57–94] for the purpose of comparing probability densities defined on $\mathbf{R}^d \times \mathbf{R}^d$ with density operators on $L^2(\mathbf{R}^d)$. This pseudometric is particularly convenient if one seeks a quantitative error estimate for the classical limit of quantum dynamics. In this talk, we explain how to use this tool in order to study the observability problem for the Schrödinger or the von Neumann equations. Our analysis of this problem uses the quantum analogue of the Wasserstein distance, together with a geometric condition on the classical trajectories corresponding to the quantum dynamics under a condition analogous to the Bardos-Lebeau-Rauch geometric condition for the exact controllability of the wave equation [C. Bardos, G. Lebeau, J. Rauch, SIAM J. Control Opti. **30** (1992) 1024–1065]. The material presented in this paper is a review of a series of joint works with T. Paul, especially [Math. Models Methods Appl. Sciences, **32** (2022) 941–963].

Keywords Schrödinger equation · Von Neumann equation · Observability · Lions' Hilbert Uniqueness Method (HUM) · Wasserstein distance

1 Quantum Dynamics

This is a survey article reporting on the motivations, methods and results of a joint work with Paul [13]. The problem studied in this paper finds its origin in a discussion with C. Bardos on the problem of exact controllability for the Schrödinger equation.

In this talk, quantum dynamics refers to either the Schrödinger equation, or the von Neumann equation—also known as the quantum Liouville equation.

In memory of our dear friend and colleague, Maria Conceição Carvalho

F. Golse (✉)
École polytechnique, CMLS, Route de Saclay, 91128 Palaiseau Cedex, France
e-mail: francois.golse@polytechnique.edu

E. Carlen et al. (eds.), *From Particle Systems to Partial Differential Equations*, Springer Proceedings in Mathematics & Statistics 465,
https://doi.org/10.1007/978-3-031-65195-3_6

In quantum mechanics, the state at time t of a point particle of mass m at the position $x \in \mathbf{R}^d$ is defined in terms of its wave function $\psi(t,x) \in \mathbf{C}$, which satisfies

$$\int_{\mathbf{R}^d} |\psi(t,x)|^2 dx = 1\,, \qquad t \in \mathbf{R}\,.$$

Thus $|\psi(t,x)|^2 dx$ can be thought of as the probability of finding the particle at time t in an infinitesimal volume dx centered at $x \in \mathbf{R}^d$. Assuming that the particle is subject to a force deriving from some external potential $V \equiv V(x) \in \mathbf{R}$, the wave function satisfies the Schrödinger equation, which is the quantum analogue of Newton's second law of motion in classical mechanics:

$$i\hbar\partial_t\psi(t,x) = \left(-\tfrac{\hbar^2}{2m}\Delta_x + V(x)\right)\psi(t,x)\,, \qquad x \in \mathbf{R}^d\,.$$

The potential V is chosen so that the quantum Hamiltonian

$$\mathscr{H} := -\tfrac{\hbar^2}{2m}\Delta_x + V(x)\,,$$

which is defined a priori on $C_c^2(\mathbf{R}^d)$, has an extension that is an unbounded self-adjoint operator on $\mathfrak{H} = L^2(\mathbf{R}^d)$, so that $U(t) := e^{-it\mathscr{H}/\hbar}$ is a unitary group on $\mathfrak{H}$ by Stone's theorem. (For instance, one can assume that $d = 3$ and that the potential $V \in L^2(\mathbf{R}^3) + L^\infty(\mathbf{R}^3)$, in which case the domain of the self-adjoint extension of $\mathscr{H}$ is $H^2(\mathbf{R}^3)$: see the Kato-Rellich Theorem—Theorem 5.12 of [25]; another possibility is to assume that $V \in L^\infty_{loc}(\mathbf{R}^3)$ and that $V \geq -M$ for some constant $M > 0$—see Theorem 8.15 of [24]. One can also combine both results, and assume that $V = V_1 + V_2$ is the sum of a potential in $V_1 \in L^2(\mathbf{R}^3) + L^\infty(\mathbf{R}^3)$ and of a confining potential $V_2 \geq 0$ such that $V_2(x) \to +\infty$ as $|x| \to \infty$.) In that case,

$$\psi(t,\cdot) = e^{-it\mathscr{H}/\hbar}\psi(0,\cdot)\,, \qquad t \in \mathbf{R}\,,$$

so that the constraint $\|\psi(t,\cdot)\|_{L^2(\mathbf{R}^d)} = 1$ is satisfied for all $t \geq 0$ if it is satisfied initially, i.e. for $t = 0$.

The case of a wave function discussed above corresponds to a pure quantum state. One can consider instead the dynamics of mixed states, defined by a time-dependent density operator $R(t)$. We recall that a density operator on $\mathfrak{H}$ is

$$R \in \mathscr{L}(\mathfrak{H}) \quad \text{s.t.} \quad R = R^* \geq 0 \quad \text{and} \quad \operatorname{trace}_{\mathfrak{H}}(R) = 1\,.$$

Henceforth, we denote by $\mathscr{D}(\mathfrak{H})$ the set of density operators on $\mathfrak{H}$. We shall also need a notion of "finite energy" density operator (i.e. with finite energy for the harmonic oscillator). The set of finite energy density operators is denoted by

$$\mathscr{D}_2(\mathfrak{H}) := \left\{R \in \mathscr{D}(\mathfrak{H}) \text{ s.t. } \operatorname{trace}_{\mathfrak{H}}\left(R^{1/2}(|x|^2 - \hbar^2\Delta_x)R^{1/2}\right) < \infty\right\}.$$

The evolution of a density operator analogous to that of a wave function as above is

$$R(t) = e^{-it\mathscr{H}/\hbar} R(0) e^{+it\mathscr{H}/\hbar},$$

and one easily checks that the operator-valued function $t \mapsto R(t)$ satisfies the von Neumann equation

$$i\hbar\partial_t R(t) = [\mathscr{H}, R(t)] =: \mathscr{H}R(t) - R(t)\mathscr{H}.$$

(For instance, one can easily check that, if $\psi(t, \cdot) = e^{-it\mathscr{H}/\hbar}\psi(0, \cdot)$ is a solution of the Schrödinger equation, the orthogonal projection $P(t)$ on $\mathbf{C}\psi(t, \cdot)$ in $\mathfrak{H}$ is an example of density operator, and that $P(t) = e^{-it\mathscr{H}/\hbar} P(0) e^{+it\mathscr{H}/\hbar}$.)

2 Observability

In this section, we briefly explain how the notions of observability and controllability are intimately related. (The material in this section does not come from [13].)

A class $\mathscr{K}$ of quantum states at $t = 0$ is "observable" in a domain $\Omega \subset \mathbf{R}^d$ during a time $T > 0$ if there exists a constant $C_{OBS} > 0$ (which may depend on the class $\mathscr{K}$, on the observation domain Ω and the observation time T) such that

$$1(= \|\psi(0, \cdot)\|_{\mathfrak{H}}^2) \leq C_{OBS} \int_0^T \int_\Omega |\psi(t, x)|^2 dx dt$$

for all initial wave function in $\psi^{in} := \psi(0, \cdot) \in \mathscr{K}$, if the quantum dynamics is defined in terms of wave functions solutions of the Schrödinger equation. If the quantum dynamics is given in terms of density operators governed by the von Neumann equation, the condition above should be replaced with

$$1(= \|R(0)\|_1) \leq C_{OBS} \int_0^T \operatorname{trace}_{\mathfrak{H}}(\mathbf{1}_\Omega R(t)) dt \quad \text{for all } R^{in} = R(0) \in \mathscr{K},$$

where $\mathbf{1}_\Omega$ is the multiplication operator associated to the indicator function $\mathbf{1}_\Omega$ of the observation domain Ω.

Since the quantum dynamical equations (be it defined by the Schrödinger, or the von Neumann equation) is linear in terms of the wave function or of the density operator, observability as defined above implies that one can discriminate between two solutions by using only the partial information on these two solutions corresponding to observing them on the subdomain Ω of $\mathbf{R}^d$.

We invite the interested reader to think of the following example: let U be a domain (connected open set) in $\mathbf{C}$, and let $\Omega \subset U$ be a nonempty open subset of U. Let $\bar{\partial}$ be the Cauchy-Riemann operator, and let f be a distribution on U such that

$\bar{\partial} f = 0$. Assume that $f\big|_{\Omega} = 0$. By elliptic regularity, it is well known that f is a holomorphic function on U, so that the additional condition $f\big|_{\Omega} = 0$ implies that $f = 0$ on U by analytic continuation since U is connected. Notice however that in general, the analytic continuation theorem does not provide an observation constant —i.e. $C_{OBS} > 0$ such that, for each f holomorphic on U,

$$\sup_{z\in U} |f(z)| \leq C_{OBS} \sup_{z\in\Omega} |f(z)| \,.$$

We shall see below an application of the existence of $C_{OBS} > 0$ in the observation property for quantum dynamics.

The interest for this notion of observability is to be found in the community of mathematicians working on the controllability of PDEs, and comes from the well known (J.-L.) Lions Hilbert Uniqueness Method (HUM) (see also [7] for an earlier version). Here is how it works for a class of abstract differential equations including the Schrödinger equation.

Let H be a (possibly unbounded) self-adjoint operator on $\mathfrak{H} = L^2(\mathbf{R}^d)$ with domain $D(H)$. Consider the control problem (where f is the control)

$$\begin{cases} i\partial_t \phi = H\phi + \mathbf{1}_{(0,T)\times\Omega} f \,, \\ \phi\big|_{t=T} = 0 \,. \end{cases}$$

Can one drive any solution $\phi \in C_b(\mathbf{R}; L^2(\mathbf{R}^d))$ to 0 at time T by acting on the domain Ω via the control f only? In other words, is the control operator

$$\mathscr{C} : L^2((0,T)\times\Omega) \ni f \mapsto -i\phi\big|_{t=0} \in L^2(\mathbf{R}^d)$$

onto? To answer this question, one considers instead the observation problem

$$\begin{cases} i\partial_t \psi = H\psi \,, \\ \psi\big|_{t=0} = \psi^{in} \,, \end{cases}$$

and the observation operator

$$\mathscr{O} : L^2(\mathbf{R}^d) \ni \psi^{in} \mapsto \psi\big|_{(0,T)\times\Omega} \in L^2((0,T)\times\Omega) \,.$$

If the full class $C_b(\mathbf{R}; L^2(\mathbf{R}^d))$ of solutions of the Schrödinger equation is observable on $(0, T) \times \Omega$ in the sense of the definition above, the operator $\mathscr{O}$ is in particular one-to-one.

Since H is self-adjoint, it generates a unitary group $U(t) = e^{-itH}$ on $\mathfrak{H}$, and one easily checks that

$$\|\mathscr{O}\psi^{in}\|^2_{L^2(0,T)\times\Omega)} \leq \int_0^T \|U(t)\psi^{in}\|^2_{L^2(\mathbf{R}^d)} dt = T\|\psi^{in}\|^2_{L^2(\mathbf{R}^d)} \,,$$

so that $\mathscr{O}$ is a bounded operator with $\|\mathscr{O}\|^2 \leq T$. On the other hand, by the Duhamel formula

$$\mathscr{C} f = i \int_0^T U(-t)(\mathbf{1}_\Omega f(t))dt\,,$$

so that

$$\begin{aligned}\|\mathscr{C} f\|_{L^2(\mathbf{R}^d)} &\leq \int_0^T \|U(-t)(\mathbf{1}_\Omega f(t,\cdot))\|_{L^2(\mathbf{R}^d)}dt \\ &= \int_0^T \|f(t,\cdot))\|_{L^2(\Omega)}dt \leq T^{1/2}\|f\|_{L^2((0,T)\times\Omega)}\,.\end{aligned}$$

Thus $\mathscr{C}$ is a bounded operator with $\|\mathscr{C}\| \leq \sqrt{T}$. Lions' key observation is that $\mathscr{C}$ and $\mathscr{O}$ are adjoint of each other. Indeed

$$\begin{aligned}(\mathscr{C} f|\psi^{in})_{L^2(\mathbf{R}^d)} &= i\int_{\mathbf{R}^d} \overline{\phi(0,x)}\psi^{in}(x)dx = -\int_0^T i\partial_t\left(\int_{\mathbf{R}^d}\overline{\phi(t,x)}\psi(t,x)dx\right)dt \\ &= -\int_0^T\int_{\mathbf{R}^d}\overline{\phi(t,x)}H\psi(t,x)dxdt + \int_0^T\int_{\mathbf{R}^d}\overline{H\phi(t,x)}\psi(t,x)dxdt \\ &+ \int_0^T\int_\Omega \overline{f(t,x)}\psi(t,x)dxdt = (f|\mathscr{O}\psi^{in})_{L^2((0,T)\times\Omega)}\end{aligned}$$

since H is self-adjoint (glossing over issues involving the domain of H). Therefore (see for instance Corollary 2.17 in [4])

$$\overline{\operatorname{ran}(\mathscr{C})} = \ker(\mathscr{C}^*)^\perp = \ker(\mathscr{O})^\perp\,.$$

In other words, if $\ker(\mathscr{O}) = \{0\}$, the range of $\mathscr{C}$ is dense, meaning that one can control (i.e. drive to 0 in time T) a dense set of initial data $\psi^{in} = \psi(0,\cdot)$.

But in fact, the observability condition above says more. Consider the composed map $\mathscr{C} \circ \mathscr{O}$: this is a bounded linear operator from $L^2(\mathbf{R}^d)$ to itself, which is self-adjoint since

$$(\mathscr{C} \circ \mathscr{O})^* = \mathscr{O}^* \circ \mathscr{C}^* = \mathscr{C} \circ \mathscr{O}\,.$$

Besides,

$$(\psi^{in}|\mathscr{C} \circ \mathscr{O}\psi^{in})_{L^2(\mathbf{R}^d)} = \|\mathscr{O}\psi^{in}\|^2_{L^2((0,T)\times\Omega)} \geq C_{OBS}\|\psi^{in}\|^2_{L^2(\mathbf{R}^d)}\,,$$

so that $\ker(\mathscr{C} \circ \mathscr{O}) = \{0\}$ (since $\mathscr{C} \circ \mathscr{O}\psi^{in} = 0$ obviously implies $\psi^{in} = 0$), and $\overline{\operatorname{ran}(\mathscr{C} \circ \mathscr{O})} = L^2(\mathbf{R}^d)$ (again by Corollary 2.17 in [4]). But the inequality above and the Cauchy-Schwarz inequality obviously imply that

$$\|\mathscr{C} \circ \mathscr{O}\psi^{in}\|_{L^2(\mathbf{R}^d)} \geq C_{OBS}\|\psi^{in}\|_{L^2(\mathbf{R}^d)}\,,$$

and this implies in turn that $\operatorname{ran}(\mathscr{C} \circ \mathscr{O})$ is closed (see Theorem 2.20 in [4]). Hence

$$L^2(\mathbf{R}^d) = \operatorname{ran}(\mathscr{C} \circ \mathscr{O}) \subset \operatorname{ran}(\mathscr{C}), \quad \text{so that } \operatorname{ran}(\mathscr{C}) = L^2(\mathbf{R}^d).$$

Summarizing, the fact that the observation operator $\mathscr{O}$ is one-to-one, i.e. that one has $\ker \mathscr{O} = \{0\}$, is not sufficient to conclude that $\mathscr{C}$ is onto, i.e. that *every* solution of

$$i\partial_t \psi = H\psi + \mathbf{1}_\Omega f$$

can be driven to 0 at time T, a condition which can be written as

$$\psi(t, \cdot)\big|_{t=T} = 0,$$

by a control f localized in Ω. That $\ker \mathscr{O} = \{0\}$ implies that one can drive to 0 at time T only a *dense subset* of all solutions to the problem above. However, the observability inequality with a constant $C_{OBS} > 0$ implies that *every* solution of the equation above can be driven to 0 at time T by a control f localized in Ω. We hope that this discussion justifies the search for an observation inequality as above, with a positive observation constant C_{OBS}, in the case of quantum dynamics.

One of the reasons for which J.-L. Lions came up with his approach to controllability based on the HUM was the problem of controlling vibrations in large solid structures. His original formulation dealt with the wave equation [19], which is more involved than the almost trivial discussion of the Schrödinger equation presented above. The case of linear elasticity is of course very similar to the case of the wave equation.

3 Main Results

Another issue seriously complicates the controllability problem for vibrations in solid structures: in practice, vibrations are canceled by motors acting in some regions of the boundary, and not in the core of the structure. The question is to find the optimal way of positioning the motors at the boundary to do so.

In 1992, C. Bardos, G. Lebeau and J. Rauch came up with a rather universal approach to such controllability problems; their key idea was that it is enough to control only high frequency waves. Since the propagation of such waves follows the rules of geometric optics, they gave a geometric condition under which any solution of the wave equation can be driven to zero by some control at the region of the boundary satisfying their geometric condition. The mathematical vehicle for this kind of study is microlocal analysis, and specifically the propagation of the wave front set for solutions of the wave equation. While the propagation of singularities in some open domain is a rather simple matter, their reflection in the presence of boundaries requires more technical machinery (typically the work of Melrose-Sjöstrand [20, 21]). This is another way in which the observability problem mentioned in the previous section is much simpler than the problem considered in [2].

However, we shall retain the key idea of using geometric optics in [2]. In the context considered here, the analogue of geometric optics viewed as the high frequency limit of wave propagation is the classical limit of quantum dynamics. The analogue of the Bardos-Lebeau-Rauch geometric condition in that case is as follows.

3.1 The Geometric Observability Condition

First we consider the classical Hamiltonian corresponding to $\mathscr{H}$, i.e.

$$\mathbf{H}(x,\xi) := \tfrac{1}{2m}|\xi|^2 + V(x)\,.$$

Denote by $(X,\Xi)(t;x,\xi) \in \mathbf{R}^{2d}$ the flow of this classical Hamiltonian, assumed to be globally defined. (Later, we shall make assumptions on V implying the existence and uniqueness of such a global flow.) In other words, $t \mapsto (X,\Xi)(t;x,\xi)$ is the solution at time t of

$$\begin{cases} \dot{X} = \frac{\partial \mathbf{H}}{\partial \Xi}(X,\Xi) = \frac{1}{m}\Xi\,, \\ \dot{\Xi} = -\frac{\partial \mathbf{H}}{\partial X}(X,\Xi) = -\nabla V(X)\,, \end{cases} \qquad (X,\Xi)\big|_{t=0} = (x,\xi)\,.$$

Geometric condition. A triple (K,Ω,T) with $K \subset \mathbf{R}^{2d}$ compact, while $\Omega \subset \mathbf{R}^d$ is open and $T > 0$ is said to satisfies the geometric observability condition if

$$(GC) \qquad \text{for each } (x,\xi) \in K\,, \qquad \text{there exists } t \in (0,T) \text{ such that } X(t;x,\xi) \in \Omega\,.$$

In other, words, any classical trajectory leaving from the phase-space set K can be observed in Ω in the time interval $(0,T)$ (see Fig. 1). Here is an important implication of the geometric condition (GC).

Lemma 1 *For all $K \subset \mathbf{R}^d \times \mathbf{R}^d$ compact, all $\Omega \subset \mathbf{R}^d$ open and all $T > 0$,*

$$(K,\Omega,T) \textit{ satisfies } (GC) \implies C[K,\Omega,T] := \inf_{(x,\xi)\in K} \int_0^T \mathbf{1}_\Omega(X(t;x,\xi))dt > 0\,.$$

Here is a quick proof of this lemma.

Proof Since Ω is open in $\mathbf{R}^d$, its indicator function $\mathbf{1}_\Omega$ is lower semicontinuous (l.s.c.) on $\mathbf{R}^d$. By Fatou's lemma, the function

$$(x,\xi) \mapsto \int_0^T \mathbf{1}_\Omega(X(t;x,\xi))dt \in (0,+\infty) \text{ is l.s.c. on } \mathbf{R}^d \times \mathbf{R}^d\,.$$

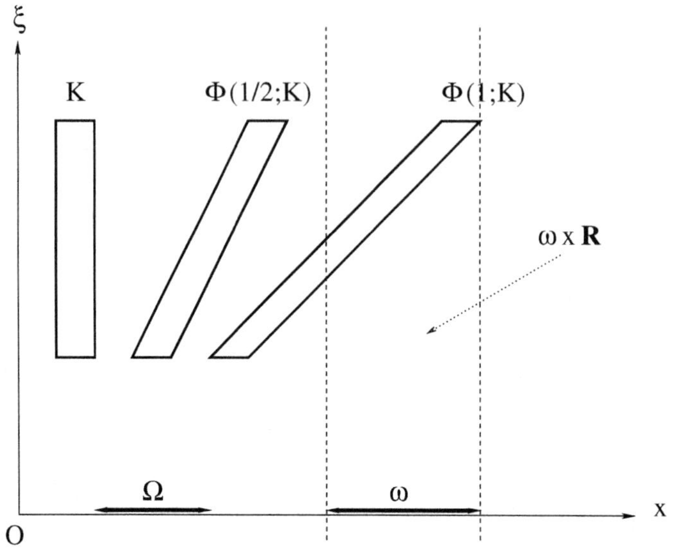

Fig. 1 The geometric condition in space dimension $d = 1$, with $V \equiv 0$. The classical free flow is $\Phi(t; x, \xi) := (X(t; x, \xi), \Xi(t; x, \xi)) = (x + t\xi, \xi)$. The picture represents the image of the closed phase-space rectangle K by the map $(x, \xi) \mapsto \Phi(t; x, \xi)$ at time $t = \frac{1}{2}$ and $t = 1$. The interval Ω satisfies the geometric condition with $T = 1$, at variance with ω. Indeed, phase-space points on the bottom side of K stay out of the strip $\omega \times \mathbf{R}$ for all $t \in [0, 1]$

Since K is compact in $\mathbf{R}^d \times \mathbf{R}^d$, the infimum of the nonnegative l.s.c. function above is attained at some point $(x^*, \xi^*) \in K$:

$$C[K, \Omega, T] = \int_0^T \mathbf{1}_\Omega(X(t; x^*, \xi^*))dt\,.$$

Applying (GC) to the phase space point $(x^*, \xi^*) \in K$ shows that there exists some time $t^* \in (0, T)$ such that

$$\mathbf{1}_\Omega(X(t^*; x^*, \xi^*)) = 1\,.$$

Since $t \mapsto \mathbf{1}_\Omega(X(t; x^*, \xi^*))$ is l.s.c. on $(0, T)$, there exists $\eta \in (0, T)$ such that

$$|t - t^*| < \eta \implies \mathbf{1}_\Omega(X(t; x^*, \xi^*)) = 1\,.$$

Hence

$$C[K, \Omega, T] = \int_0^T \mathbf{1}_\Omega(X(t; x^*, \xi^*))dt \geq \eta > 0\,.$$

This concludes the proof of Lemma 1. □

How the geometric condition can be used in the quantum setting may not be immediately obvious. In the first place, there is no obvious notion of "quantum trajectory" that would be shadowed by classical trajectories in phase space, and which would satisfy some variant of the geometric condition.

Nevertheless, under the geometric condition (GC) *à la* Bardos-Lebeau-Rauch, one arrives at a rather precise observability statement where the "geometric" constant $C[K, \Omega, T]$ obtained in Lemma 1 appears as an approximation of the observation constant.

Before stating this observability result, we need to introduce another mathematical object, which is a (partial) quantum analogue of the Wasserstein distance of exponent 2 (see Chap. 7 of [27]).

3.2 A Quantum Wasserstein Pseudometric

The material in this section is taken from [11]—see also the survey [10] presented at an earlier conference of this series.

By analogy with the notion of Wasserstein distance defined in Chap. 7 of [27], the pseudometric used in this work is based on a notion of "coupling" (or transport plan) between a probability density on phase space, i.e. $\mathbf{R}^d \times \mathbf{R}^d$, and a density operator on $\mathfrak{H} = L^2(\mathbf{R}^d)$.

At variance with the Wasserstein distance, which is defined on the set of probability measures, the pseudometric defined below is used to compare very different objects, viz. probability densities, i.e. functions defined on $\mathbf{R}^d \times \mathbf{R}^d$, with density operators defined on $L^2(\mathbf{R}^d)$. One could think of an analogy with the Gromov-Hausdorff metric which is used to compare compact sets which do not belong to the same ambient space, except that it is somehow important for our pseudometric to refrain from mapping the density operator to the set of probability densities, or the probability density to the set of density operators. This important point will be discussed later.

Definition 1 (*Couplings*) Let f be a probability density on $\mathbf{R}^{2d}$ and let $R \in \mathcal{D}(\mathfrak{H})$. A coupling of f and R is an operator-valued map

$$\mathbf{R}^{2d} \ni (x, \xi) \mapsto Q(x, \xi) = Q(x, \xi)^* \in \mathcal{L}(\mathfrak{H}) \text{ s.t. } Q(x, \xi) \geq 0 \text{ a.e.}$$
$$\text{s.t. } \operatorname{trace}_{\mathfrak{H}}(Q(x, \xi)) = f(x, \xi) \text{ a.e., and } \iint_{\mathbf{R}^{2d}} Q(x, \xi) dx d\xi = R\,.$$

The set of couplings of f and R is denoted $\mathcal{C}(f, R)$.

Notice that $\mathcal{C}(f, R) \neq \varnothing$ since the product map defined as follows

$$f \otimes_{\mathbf{C}} R : (x, \xi) \mapsto f(x, \xi) R\,,$$

always belongs to $\mathscr{C}(f, R)$. In the case of pure quantum states, the structure of couplings is particularly simple.

Lemma 2 *Assume that* $R \in \mathscr{D}(\mathfrak{H})$ *is a rank-1 orthogonal projection. Then, for each probability density* f *on* $\mathbf{R}^{2d}$*, one has*

$$\mathscr{C}(f, R) = \{f \otimes_{\mathbf{C}} R\}.$$

(This result is Lemma 3.6 in [14].)

Next we define a "cost of transport" from the phase-space point $(x, \xi) \in \mathbf{R}^d \times \mathbf{R}^d$ to the quantum position y and the quantum momentum $-i\hbar\nabla_y$:

$$c_{\lambda,\hbar}(x, \xi) := \lambda^2|x - y|^2 + |\xi + i\hbar\nabla_y|^2 .$$

Observe that $c_{\lambda,\hbar}(x, \xi)$ is the Hamiltonian of a harmonic oscillator on $\mathfrak{H} = L^2(\mathbf{R}^d_y)$ shifted by (x, ξ). In particular

$$c_{\lambda,\hbar}(x, \xi) = c_{\lambda,\hbar}(x, \xi)^* \geq \lambda d\hbar I_{\mathfrak{H}} ,$$

an inequality which can be seen as a variant of Heisenberg's uncertainty principle. Notice also that the domain of $c_{\lambda,\hbar}(x, \xi)$ is

$$D(c_{\lambda,\hbar}(x, \xi)) = H^2(\mathbf{R}^d) \cap L^2(\mathbf{R}^d; |y|^4 dy) .$$

This definition of the cost of transport is precisely what allows comparing apples with pears—here, probability densities on $\mathbf{R}^d_x \times \mathbf{R}^d_\xi$ with density operators on $L^2(\mathbf{R}^d_y)$.

With this, we can define a notion of pseudometric analogous to the Wasserstein distance of exponent 2, which will be used to compare probability densities on phase spaced and density operators.

Definition 2 (*Pseudometric*) Let $R \in \mathscr{D}_2(\mathfrak{H})$ and let f be a probability density on $\mathbf{R}^d \times \mathbf{R}^d$ with finite second order moments. Then

$$\mathfrak{d}_\lambda(f, R) := \inf_{Q \in \mathscr{C}(f,R)} \left(\iint_{\mathbf{R}^{2d}} \operatorname{trace}_{\mathfrak{H}}(Q(x, \xi)^{\frac{1}{2}} c_{\lambda,\hbar}(x, \xi) Q(x, \xi)^{\frac{1}{2}}) dx d\xi \right)^{1/2} .$$

(See Definition 2.2 of [11] in the case $\lambda = 1$.)

Because of the inequality on $c_{\lambda,\hbar}(x, \xi)$ analogous to Heisenberg's uncertainty principle recalled above, the pseudometric $\mathfrak{d}_\lambda$ satisfies the bound

$$\mathfrak{d}_\lambda(f, R)^2 \geq \lambda d\hbar$$

for all $R \in \mathscr{D}_2(R)$ and all probability density f on $\mathbf{R}^d \times \mathbf{R}^d$.

In the sequel, we shall use Dirac's bra-ket notation: if $\psi \equiv \psi(x) \in \mathbf{C}$ is a square-integrable measurable function such that

$$\int_{\mathbf{R}^d} |\psi(x)|^2 dx = 1\,,$$

the notation $|\psi\rangle$ (the ket $|\psi\rangle$) designates the function ψ viewed as a vector of $\mathfrak{H}$, while the notation $\langle\psi|$ (the bra $\langle\psi|$) designates the linear functional

$$\langle\psi| : \mathfrak{H} \ni \phi \mapsto \int_{\mathbf{R}^d} \overline{\psi(x)}\phi(x)dx \in \mathbf{C}\,.$$

If T is an operator on $\mathfrak{H}$, the notation $\langle\psi|T|\phi\rangle$ designates is the complex number

$$\langle\psi|T|\phi\rangle := \int_{\mathbf{R}^d} \overline{\psi(x)}(T\phi)(x)dx\,.$$

On the other hand, the notation $|\psi\rangle\langle\psi|$, where $\psi \in \mathfrak{H}$ is such that $\|\psi\|_{\mathfrak{H}} = 1$, designates the orthogonal projection on $\mathbf{C}\psi$, in other words the linear map

$$\mathfrak{H} \ni \phi \mapsto \langle\psi|\phi\rangle\psi \in \mathfrak{H}\,.$$

There are two classes of density operators R for which the pseudometric $\mathfrak{d}_\lambda(f, R)$ can be computed, or estimated explicitly.

First, if $R \in \mathcal{D}_2(\mathfrak{H})$ is a rank-1 operator, i.e. if

$$R = |\psi\rangle\langle\psi| \quad \text{with } \psi \in H^1(\mathbf{R}^d) \cap L^2(\mathbf{R}^d; |y|^2 dy)\,,$$

then

$$\mathfrak{d}_\lambda(f, R) = \left(\iint_{\mathbf{R}^{2d}} \langle\psi|c_{\lambda,\hbar}(x,\xi)|\psi\rangle f(x,\xi)dxd\xi \right)^{1/2}\,,$$

since $\mathcal{C}(f, R) = \{f \otimes_{\mathbf{C}} R\}$ by Lemma 2.

Another interesting example is the case of Toeplitz density operators. First we recall the notion of Schrödinger coherent states: for all $q, p \in \mathbf{R}^d$, set

$$|q, p\rangle(x) := (\pi\hbar)^{-d/4} e^{-|x-q|^2/2\hbar} e^{ip\cdot(x-q/2)/\hbar}\,.$$

Definition 3 (*Toeplitz density operators*) If f is a probability density on $\mathbf{R}^d \times \mathbf{R}^d$, the operator

$$\mathcal{T}[f] := \iint_{\mathbf{R}^d\times\mathbf{R}^d} |q, p\rangle\langle q, p| f(q, p)dqdp \text{ belongs to } \mathcal{D}(\mathfrak{H})\,.$$

One easily checks that $\mathscr{T}[f] \in \mathscr{D}_2(\mathfrak{H})$ if f has finite second-order moments. It has been proved in [11] (see also Proposition 2.5 of [13]) that, for all f, g, probability densities on $\mathbf{R}^d \times \mathbf{R}^d$ with finite second order moments,

$$\mathfrak{d}_\lambda(f, \mathscr{T}[g])^2 \leq \max(1, \lambda^2)\mathscr{W}_2(f, g)^2 + \tfrac{1}{2}(1+\lambda^2)d\hbar\,,$$

where $\mathscr{W}_2$ is the Wasserstein distance of exponent 2 defined in Chap. 7 of [27] (see Sect. 7.1, and especially Theorem 7.3 in that reference).

3.3 *The Observability Result*

With the material already introduced in the sections above, we can state our observability inequality for quantum dynamics. This is the main result in [13] (stated there as Theorem 1.1).

Theorem 1 *Assume that $V \in C^{1,1}(\mathbf{R}^d)$ is such that $V(x) \to +\infty$ as $|x| \to \infty$, and let $K \subset \mathbf{R}^d \times \mathbf{R}^d$ compact, $\Omega \subset \mathbf{R}^d$ open and $T > 0$ satisfy the Bardos-Lebeau-Rauch geometric condition (GC). Let $U(t)$ be the quantum dynamics (unitary group on $\mathfrak{H}$) defined as*

$$U(t) := e^{-it\mathscr{H}/\hbar} \quad \text{where} \quad \mathscr{H} := -\tfrac{\hbar^2}{2m}\Delta + V\,.$$

Then, for all $R^{in} \in \mathscr{D}_2(\mathfrak{H})$ and all $\delta > 0$, denoting $\Omega_\delta := \Omega + B(0,\delta)$, one has

$$\int_0^T \operatorname{trace}_{\mathfrak{H}}(\mathbf{1}_{\Omega_\delta} U(t) R^{in} U(t)^*)dt \geq \underbrace{C[K,\Omega,T]}_{\text{geometric constant}} \underbrace{-\frac{1}{\delta}\inf_{\lambda>0}\frac{T}{\lambda}E\left(\tfrac{T}{2}\left(\tfrac{\lambda}{m}+\tfrac{L}{\lambda}\right)\right)\inf_{\operatorname{supp}(f^{in})\subset K}\mathfrak{d}_\lambda(f^{in}, R^{in})}_{\text{semiclassical correction}}\,,$$

where

$$L := \operatorname{Lip}(\nabla V) \quad \text{and} \quad E(z) := \frac{e^z - 1}{z}\,.$$

Before going further, let us give two examples of solutions (characterized through their initial data) for which the semiclassical correction in the inequality above is known explicitly. In general, it is quite complicated to compute $\mathfrak{d}_\lambda(f^{in}, R^{in})$—of course, since the semiclassical correction involves an infimum, and since $\mathfrak{d}_\lambda$ itself is defined as an infimum, one can always choose some initial data f^{in} and some coupling in $\mathscr{C}(f^{in}, R^{in})$ to compute an upper bound of this semiclassical correction. However, such a procedure will most likely not give a very accurate estimate of this semiclassical correction.

The two examples below are based on those cases (already mentioned in the preceding section) for which the pseudometric $\mathfrak{d}_\lambda$ can be computed explicitly, or at least estimated rather precisely.

Example 1 (*the case of Toeplitz initial data*) Assume that $R^{in} = \mathscr{T}[g^{in}]$ where g^{in} is a probability density with support in the compact $K \subset \mathbf{R}^d \times \mathbf{R}^d$ (so that in particular $\mathscr{T}[g] \in \mathscr{D}_2(\mathfrak{H})$). Then

$$\inf_{\mathrm{supp}(f^{in})\subset K} \mathfrak{d}_\lambda(f^{in}, \mathscr{T}[g^{in}])^2 = \mathfrak{d}_\lambda(g^{in}, \mathscr{T}[g^{in}])^2 \le \tfrac{1}{2}(\lambda^2+1)d\hbar,$$

so that, according to the observability inequality in Theorem 1

$$\int_0^T \mathrm{trace}_{\mathfrak{H}}(\mathbf{1}_{\Omega_\delta} U(t)\mathscr{T}[g^{in}]U(t)^*)dt \ge C[K,\Omega,T] \\ - \frac{1}{\delta} \inf_{\lambda>0} \tfrac{T}{\lambda} E\left(\tfrac{T}{2}\left(\tfrac{\lambda}{m}+\tfrac{L}{\lambda}\right)\right)\sqrt{\tfrac{1}{2}(\lambda^2+1)d\hbar}.$$

In that case, the quantum correction is completely explicit, and of order $O(\hbar^{1/2})$. Therefore, it is possible to obtain an explicit observation constant by adjusting $\delta > 0$ so that

$$\inf_{\lambda>0} \tfrac{T}{\lambda} E\left(\tfrac{T}{2}\left(\tfrac{\lambda}{m}+\tfrac{L}{\lambda}\right)\right)\sqrt{\tfrac{1}{2}(\lambda^2+1)d\hbar} < \delta C[K,\Omega,T].$$

(Observe that there is no need to assume that $\hbar \to 0$ for this inequality to hold.)

Example 2 (*the case of pure states*) Let $R^{in} = |\psi^{in}\rangle\langle\psi^{in}|$ with $\psi^{in} \in H^1(\mathbf{R}^d) \cap L^2(\mathbf{R}^d; |y|^2 dy)$ and $\|\psi^{in}\|_{\mathfrak{H}} = 1$. Choose $\lambda = 1$ and

$$f^{in}(q,p) := \frac{|\langle q,p|\psi^{in}\rangle|^2}{(2\pi\hbar)^d}.$$

(The function f^{in} is referred to as the "Husimi transform" of the operator $|\psi^{in}\rangle\langle\psi^{in}|$.) Assume that

$$\delta C[K,\Omega,T] \iint_K |\langle q,p|\psi^{in}\rangle|^2 \tfrac{dqdp}{(2\pi\hbar)^d} > D[T,\mathrm{Lip}(\nabla V)]\Sigma[\psi^{in}],$$

where

$$D[T,L] := 4\frac{e^{(1+L)T/2}-1}{1+L},$$
$$\Sigma[\psi^{in}]^2 := \langle\psi^{in}|\,|x|^2|\psi^{in}\rangle - |\langle\psi^{in}|x|\psi^{in}\rangle|^2 \\ + \langle\psi^{in}| - \hbar^2\Delta_x|\psi^{in}\rangle \quad |\langle\psi^{in}| \quad i\hbar\nabla_x|\psi^{in}\rangle|^2.$$

We have used here the bound

$$\mathfrak{d}_1(f^{in}, |\psi^{in}\rangle\langle\psi^{in}|) \leq 2\Sigma[\psi^{in}]$$

from Proposition 9.1 in [12]). In that case, one has the observation inequality

$$\|\psi^{in}\|_{\mathfrak{H}}^2 = 1 \leq C_{OBS} \int_0^T \int_{\Omega} |U(t)\psi^{in}(x)|^2 dx dt\,,$$

where

$$\frac{1}{C_{OBS}} = C[K, \Omega, T] \iint_K |\langle q, p|\psi^{in}\rangle|^2 \tfrac{dqdp}{(2\pi\hbar)^d} - D[T, \mathrm{Lip}(\nabla V)] \tfrac{\Sigma[\psi^{in}]}{\delta}\,.$$

Remarks. Several remarks are in order before explaining how this result is proved.

(1) There is an abundant literature on the observation and control problems for the Schrödinger equation: see for instance [1, 3, 5, 8, 9, 15–18, 22, 23, 26] (this list being of course by no means exhaustive). All these works are based on mathematical techniques somewhat different from the ones used in this talk, involving mostly harmonic analysis, or microlocal analysis.

(2) Notice that the regularity assumptions used on the potential V in Theorem 1 are exactly those needed for the quantum Hamiltonian $\mathscr{H}$ to have a self-adjoint extension to $\mathfrak{H}$, and for the classical Hamiltonian $\mathbf{H}(x, \xi) := \frac{1}{2m}|\xi|^2 + V(x)$ to generate a global flow on $\mathbf{R}^d \times \mathbf{R}^d$ (by the Cauchy-Lipshitz theorem). These assumptions are therefore the minimal requirements for all the mathematical objects considered in this statement to be well defined. Microlocal techniques, or semiclassical analysis usually require a lot more regularity on coefficients of the PDEs considered, especially in order to construct Fourier integral parametrices.

(3) Even though the classical limit of the Schrödinger (in the case of pure states), of the von Neumann equation (for mixed states) is the key argument connecting the quantum dynamics with the Bardos-Lebeau-Rauch type geometric condition, which is formulated in terms of the corresponding classical dynamics, there is no need to assume that $\hbar \to 0$ in the inequality stated in Theorem 1 in order to obtain an observability constant. Assuming that (K, Ω, T) satisfies (GC), solutions of the von Neumann or Schrödinger equations can be observed in an enlarged observation domain $\Omega_\delta = \Omega + B(0, \delta)$ provided that $\delta > 0$ is chosen large enough in terms of the geometric constant $C[K, \Omega, T]$ and of the semiclassical correction. The reason for considering the enlarged observation domain is that the quantum dynamics is not concentrated on the classical trajectories which appear in the geometric condition (GC). Since the condition (GC) is precisely defined in terms of the classical trajectories, it is natural to expect an observation inequality in the quantum case on a slightly larger domain. How large $\delta > 0$ needs to be for such an observation inequality to hold is given by completely explicit formulas involving $C[K, \Omega, T]$ together with the quantum Wasserstein pseudometric $\mathfrak{d}_\lambda(f^{in}, R^{in})$.

4 Proofs of the Main Results

4.1 Propagation of $\mathfrak{d}_\lambda$ by the Quantum and Classical Dynamics

A key ingredient in the proof of Theorem 1 is an upper bound $\mathfrak{d}_\lambda(f(t,\cdot,\cdot), R(t))$ in terms of $\mathfrak{d}_\lambda(f(0,\cdot,\cdot), R(0))$, when f is a solution of the classical Liouville equation

$$\partial_t f + \{\tfrac{1}{2m}|\xi|^2 + V(x), f\} = 0\,,$$

while R is a solution of the von Neumann equation

$$\partial_t R = -\tfrac{i}{\hbar}[-\tfrac{\hbar^2}{2m}\Delta + V, R(t)]\,.$$

Lemma 3 *Assume $V \in C^{1,1}(\mathbf{R}^d)$ is such that $\mathscr{H} := -\frac{1}{2}\hbar^2\Delta_x + V(x)$ has a self-adjoint extension to $\mathfrak{H}$, and let $U(t) := e^{-it\mathscr{H}/\hbar}$, while $\Phi(t; x, \xi) = (X, \Xi)(t; x, \xi)$ is the flow of the classical Hamiltonian $\mathbf{H}(x, \xi) := \frac{1}{2}|\xi|^2 + V(x)$. Then, for each probability density f^{in} on $\mathbf{R}^d \times \mathbf{R}^d$ with finite second order moments and each initial density operator $R^{in} \in \mathscr{D}_2(\mathfrak{H})$, it holds*

$$\mathfrak{d}_\lambda(f^{in} \circ \Phi(t,\cdot,\cdot), U(t)R^{in}U(t)^*) \le \mathfrak{d}_\lambda(f^{in}, R^{in})e^{L|t|}\,, \qquad t \ge 0\,,$$

with

$$L := \frac{1}{2}\left(\frac{\lambda}{m} + \frac{\mathrm{Lip}(\nabla V)}{\lambda}\right).$$

(For instance, if one assumes that $V(x) \to +\infty$ as $|x| \to \infty$, the quantum Hamiltonian has a self-adjoint extension to $\mathfrak{H}$ by Theorem 8.15 of [24], and the classical Hamiltonian flow Φ is globally defined on $\mathbf{R} \times \mathbf{R}^d \times \mathbf{R}^d$ according to statement (10.5.5) of [6].)

This result is a (slightly) simpler variant of Theorem 2.5 in [11], a complete proof of which can be fund in Sect. 3 of [13] (see Theorem 3.1 and its proof in that reference). We shall not reproduce it here, but, as a motivation, we recall a simple computation on the problem of pair dispersion[1] in classical mechanics.

Consider two trajectories of the classical Hamiltonian

$$\mathbf{H}(x, \xi) = \tfrac{1}{2m}|\xi|^2 + V(x)\,.$$

In other words, let $t \mapsto (X_t, \Xi_t) \in \mathbf{R}^d \times \mathbf{R}^d$ and $t \mapsto (Y_t, H_t) \in \mathbf{R}^d \times \mathbf{R}^d$ satisfy

$$\begin{cases} \dot{X}_t = \frac{1}{m}\Xi_t\,, \\ \dot{\Xi}_t = -\nabla V(X_t)\,, \end{cases} \qquad \begin{cases} \dot{Y}_t = \frac{1}{m}H_t\,, \\ \dot{H}_t = -\nabla V(Y_t)\,. \end{cases}$$

[1] This terminology is borrowed from the dynamics of vortices in fluid mechanics.

Since

$$\dot{X}_t - \dot{Y}_t = \tfrac{1}{m}(\Xi_t - H_t)\,, \quad \text{and} \quad \dot{\Xi}_t - \dot{H}_t = -(\nabla V(X_t) - \nabla V(Y_t))\,,$$

one has

$$\begin{aligned} &\tfrac{d}{dt}(\lambda^2|X_t - Y_t|^2 + |\Xi_t - H_t|^2) \\ &= 2\lambda^2(X_t - Y_t)\cdot\tfrac{1}{m}(\Xi_t - H_t) - 2(\Xi_t - H_t)\cdot(\nabla V(X_t) - \nabla V(Y_t)) \\ &\le 2\tfrac{\lambda^2}{m}|X_t - Y_t||\Xi_t - H_t| + 2|\Xi_t - H_t||\nabla V(X_t) - \nabla V(Y_t)| \\ &\le 2\tfrac{\lambda^2}{m}|X_t - Y_t||\Xi_t - H_t| + 2\tfrac{\mathrm{Lip}(\nabla V)}{\lambda}|\Xi_t - H_t|\lambda|X_t - Y_t| \\ &\le \left(\tfrac{\lambda}{m} + \tfrac{\mathrm{Lip}(\nabla V)}{\lambda}\right)(\lambda^2|X_t - Y_t|^2 + |\Xi_t - H_t|^2)\,. \end{aligned}$$

By Gronwall's inequality, we conclude that, for each $t \ge 0$, it holds

$$\lambda^2|X_t - Y_t|^2 + |\Xi_t - H_t|^2 \le \left(\lambda^2|X_0 - Y_0|^2 + |\Xi_0 - H_0|^2\right)\exp\left(\left(\tfrac{\lambda}{m} + \tfrac{\mathrm{Lip}(\nabla V)}{\lambda}\right)t\right)\,.$$

In the special case $\lambda = 1$, this inequality can be understood in terms of the Wasserstein distance of exponent 2 (see Sect. 7.1 in Chap. 7 of [27] for the definition of this distance) between Dirac measures:

$$\begin{aligned} \mathscr{W}_2(\delta_{X_t,\Xi_t}, \delta_{Y_t,H_t}) &= \lambda^2|X_t - Y_t|^2 + |\Xi_t - H_t|^2 \\ &= \mathscr{W}_2(\Phi(t;\cdot,\cdot)\#\delta_{X_0,\Xi_0}, \Phi(t;\cdot,\cdot)\#\delta_{Y_0,H_0}) \\ &\le \mathscr{W}_2(\delta_{X_0,\Xi_0}, \delta_{Y_0,H_0})\exp\left(\left(\tfrac{1}{m} + \mathrm{Lip}(\nabla V)\right)t\right)\,. \end{aligned}$$

This inequality is easily generalized in turn to any pair of probability densities f^{in} and g^{in} on $\mathbf{R}^d \times \mathbf{R}^d$ with finite second order moments:

$$\mathscr{W}_2(f^{in}\circ\Phi(t;\cdot,\cdot), g^{in}\circ\Phi(t;\cdot,\cdot)) \le \mathscr{W}_2(f^{in}, g^{in})\exp\left(\left(\tfrac{1}{m} + \mathrm{Lip}(\nabla V)\right)t\right)\,.$$

The proof of Lemma 3 is done by analogy with the elementary estimates above. The idea is to propagate the transport cost $c_{\lambda,\hbar}(x,\xi)$ as follows

$$c_{\lambda,\hbar}(x,\xi) \mapsto U(t)^* c_{\lambda,\hbar}(\Phi(t;x,\xi))U(t)\,,$$

and to evaluate this operator with an observable that is an arbitrary coupling Q^{in} of f^{in} and R^{in}. In other words

$$\iint_{\mathbf{R}^d\times\mathbf{R}^d} \operatorname{trace}_{\mathfrak{H}}\left(Q^{in}(x,\xi)^{\frac{1}{2}}U(t)^*c_{\lambda,\hbar}(\Phi(t;x,\xi))U(t)Q^{in}(x,\xi)^{\frac{1}{2}}\right)dxd\xi$$
$$=\iint_{\mathbf{R}^d\times\mathbf{R}^d} \operatorname{trace}_{\mathfrak{H}}\left(Q(t,x,\xi)^{\frac{1}{2}}c_{\lambda,\hbar}(x,\xi)Q(t,x,\xi)^{\frac{1}{2}}\right)dxd\xi$$
$$\geq \mathfrak{d}_\lambda(f(t,\cdot,\cdot),R(t))^2\,,$$

since one easily checks that

$$t\mapsto Q(t,x,\xi):=U(t)Q^{in}(\Phi(-t;x,\xi))^{\frac{1}{2}}U(t)^*\text{ belongs to }\mathscr{C}(f(t,\cdot,\cdot),R(t))\,.$$

On the other hand, a formal computation shows that

$$\partial_t\left(U(t)^*c_{\lambda,\hbar}(\Phi(t;x,\xi))U(t)\right)\Big|_{t=0}$$
$$=\{\tfrac{1}{2m}|\xi|^2+V(x),c_{\lambda,\hbar}(x,\xi)\}+\tfrac{i}{\hbar}[-\tfrac{\hbar^2}{2m}\Delta+V,c_{\lambda,\hbar}(x,\xi)\}$$
$$=\tfrac{\lambda^2}{m}(\xi+i\hbar\nabla_y)\cdot(x-y)+\tfrac{\lambda^2}{m}(x-y)\cdot(\xi+i\hbar\nabla_y)$$
$$-(\nabla V(x)-\nabla V(y))\cdot(\xi+i\hbar\nabla_y)-(\xi+i\hbar\nabla_y)\cdot(\nabla V(x)-\nabla\text{ V(y)})$$
$$\leq\tfrac{\lambda}{m}c_{\lambda,\hbar}(x,\xi)+\tfrac{\mathrm{Lip}(\nabla V)}{\lambda}\left(\tfrac{\lambda^2}{\mathrm{Lip}(\nabla V)^2}|\nabla V(x)-\nabla V(y)|^2+|\xi+i\hbar\nabla_y|^2\right)$$
$$\leq\left(\tfrac{\lambda}{m}+\tfrac{\mathrm{Lip}(\nabla V)}{\lambda}\right)c_{\lambda,\hbar}(x,\xi)\,.$$

With this observation, one easily checks that

$$\frac{d}{dt}\iint_{\mathbf{R}^d\times\mathbf{R}^d} \operatorname{trace}_{\mathfrak{H}}\left(Q(t,x,\xi)^{\frac{1}{2}}c_{\lambda,\hbar}(x,\xi)Q(t,x,\xi)^{\frac{1}{2}}\right)dxd\xi$$
$$\leq\left(\tfrac{\lambda}{m}+\tfrac{\mathrm{Lip}(\nabla V)}{\lambda}\right)\iint_{\mathbf{R}^d\times\mathbf{R}^d} \operatorname{trace}_{\mathfrak{H}}\left(Q(t,x,\xi)^{\frac{1}{2}}c_{\lambda,\hbar}(x,\xi)Q(t,x,\xi)^{\frac{1}{2}}\right)dxd\xi\,,$$

and this implies Lemma 3 by choosing a minimizing sequence $Q_n^{in}\in\mathscr{C}(f^{in},R^{in})$ so that

$$\iint_{\mathbf{R}^d\times\mathbf{R}^d} \operatorname{trace}_{\mathfrak{H}}\left(Q^{in}(x,\xi)^{\frac{1}{2}}c_{\lambda,\hbar}(x,\xi)Q^{in}(x,\xi)^{\frac{1}{2}}\right)dxd\xi\to\mathfrak{d}_\lambda(f^{in},R^{in})^2$$

as $n\to\infty$. □

4.2 *Proof of Theorem 1*

Consider $f(t,\cdot,\cdot):=f^{in}\circ\Phi(t;\cdot,\cdot)$ (where Φ is the classical Hamiltonian flow on $\mathbf{R}^d\times\mathbf{R}^d$ generated by the classical Hamiltonian $\mathbf{H}(x,\xi):=\frac{1}{2m}|\xi|^2+V(x)$), and $R(t):=U(t)R^{in}U(t)^*$ (where $U(t)=e^{-it\mathscr{H}/\hbar}$ is the Schrödinger group generated by the quantum Hamiltonian $\mathscr{H}=-\frac{\hbar^2}{2m}\Delta+V$).

For each $Q(t) \in \mathscr{C}(f(t,\cdot,\cdot), R(t))$ and each Lipschitz continuous function χ on $\mathbf{R}^d$, it holds

$$\begin{aligned}
&\left|\operatorname{trace}_{\mathfrak{H}}(\chi R(t)) - \iint_{\mathbf{R}^{2d}} \chi(x) f(t,x,\xi)dxd\xi\right| \\
&= \left|\iint_{\mathbf{R}^{2d}} \operatorname{trace}_{\mathfrak{H}}((\chi(y)-\chi(x))Q(t,x,\xi))dxd\xi\right| \\
&\le \tfrac{\operatorname{Lip}(\chi)}{\lambda}\left(\iint_{\mathbf{R}^{2d}} \operatorname{trace}_{\mathfrak{H}}(Q_t^{\frac12}(\lambda^2|x-y|^2+|\xi+i\hbar\nabla_y|^2)Q_t^{\frac12})dxd\xi\right)^{\frac12},
\end{aligned}$$

so that, by the propagation lemma (Lemma 3 above), one has

$$\begin{aligned}
\left|\operatorname{trace}_{\mathfrak{H}}(\chi R(t)) - \iint_{\mathbf{R}^{2d}} \chi(x) f(t,x,\xi)dxd\xi\right| &\le \tfrac{\operatorname{Lip}(\chi)}{\lambda}\mathfrak{d}_\lambda(f(t,\cdot,\cdot),R(t)) \\
&\le \tfrac{\operatorname{Lip}(\chi)}{\lambda}\mathfrak{d}_\lambda(f^{in},R^{in})\exp\left(\tfrac12 t\left(\tfrac{\lambda}{m}+\tfrac{\operatorname{Lip}(\nabla V)}{\lambda}\right)\right).
\end{aligned}$$

By Liouville's theorem, the phase space volume element $dxd\xi$ is invariant by any Hamiltonian flow, so that

$$\iint_{\mathbf{R}^{2d}} \chi(x) f(t,x,\xi)dxd\xi = \iint_{\mathbf{R}^{2d}} \chi(X(t;x,\xi)) f^{in}(x,\xi)dxd\xi\,.$$

Therefore, we conclude from the inequality above that

$$\begin{aligned}
\int_0^T \operatorname{trace}_{\mathfrak{H}}(\chi R(t))dt \ge &\inf_{(x,\xi)\in K}\int_0^T \chi(X(t;x,\xi))dt \iint_K f^{in}(x,\xi)dxd\xi \\
&- \frac{\operatorname{Lip}(\chi)}{\lambda}\frac{\exp\left(\tfrac12 T\left(\tfrac{\lambda}{m}+\tfrac{\operatorname{Lip}(\nabla V)}{\lambda}\right)\right)-1}{\tfrac12\left(\tfrac{\lambda}{m}+\tfrac{\operatorname{Lip}(\nabla V)}{\lambda}\right)}\mathfrak{d}_\lambda(f^{in},R^{in})\,.
\end{aligned}$$

The inequality in Theorem 1 follows from choosing

$$\chi(x) := \left(1-\frac{\operatorname{dist}(x,\Omega)}{\delta}\right)_+,$$

so that $\operatorname{Lip}(\chi) = \frac{1}{\delta}$. □

References

1. Anantharaman, N., Léautaud, M., Macià, F.: Wigner measures and observability for the Schrödinger equation on the disk. Invent. Math. **206**, 485–599 (2016)
2. Bardos, C., Lebeau, G., Rauch, J.: Sharp sufficient conditions for the observation, control and stabilization of waves from the boundary. SIAM J. Control Opti. **30**, 1024–1065 (1992)
3. Bourgain, J., Burq, N., Zworski, M.: Control for Schrödinger operators on 2-tori: rough potentials. J. Eur. Math. Soc. **15**, 1597–1628 (2013)
4. Brezis, H.: Functional Analysis, Sobolev Spaces and Partial Differential Equations. Springer, New York, Dordrecht, Heidelberg, London (2011)
5. Burq, N., Zworski, M.: Control for Schrödinger Operators on Tori. Math. Res. Lett. **19**, 309–324 (2012)
6. Dieudonné, J.: Foundations of Modern Analysis, vol. 1. Academic Press, New York, London, 19690
7. Dolecki, S., Russell, D.L.: A general theory of observation and control. SIAM J. Control Optim. **15**, 185–220 (1977)
8. Fabre, C., Quelques résultats de contrôlabilité exacte de l'équation de Schrödinger. Application à l'équation des plaques vibrantes. C. R. Acad. Sci. Paris Sér. I Math. **312**, 61–66 (1991)
9. Fabre, C.: Résultats de contrôlabilité exacte interne pour l'équation de Schrödinger et leurs lirnites asymptotique: applications à certaines équations de plaques vibrantes Asymptotic Anal. **5**, 343–379 (1992)
10. Golse, F.: From the N-body Schrödinger equation to the Vlasov equation. In: From particle systems to partial differential equations. PSPDE IV, Braga, Portugal, 16–18 Dec. 2015. Springer Proceedings in Mathematics & Statistics, vol. 209, pp. 199–219 (2017)
11. Golse, F., Paul, T.: The Schrödinger equation in the mean-field and semiclassical regime. Arch. Rational Mech. Anal. **223**, 57–94 (2017)
12. Golse, F., Paul, T.: Semiclassical evolution with low regularity. J. Math. Pures et Appl. **151**, 257–311 (2021)
13. Golse, F., Paul, T.: Quantitative observability for the Schrödinger and Heisenberg equations: an optimal transport approach. Math. Models Methods Appl. Sci. **32**, 941–963 (2022)
14. Golse, F., Paul, T.: Optimal transport pseudometrics for quantum and classical densities. J. Funct. Anal. **32**, 109417 (2022)
15. Lasiecka, I., Triggiani, R., Zhang, X.: Global Uniqueness, Observability and Stabilization of Nonconservative Schrödinger Equations Via Pointwise Carleman Estimates. Part I: $H^1(\Omega)$-Estimates. J. Inv. Ill-Posed Probl. **11**, 1–63 (2003)
16. Lasiecka, I., Triggiani, R., Zhang, X.: Global Uniqueness, observability and stabilization of nonconservative Schrödinger Equations Via Pointwise Carleman Estimates. Part II: $L^2(\Omega)$-Estimates. J. Inv. Ill-Posed Probl. **12**, 183–231 (2004)
17. Laurent, C.: Internal control of the Schrödinger equation. Math. Control Relat. Fields **4**, 161–186 (2014)
18. Lebeau, G.: Contrôle de l'équation de Schrödinger. J. Math. Pures et Appl. **71**, 267–291 (1992)
19. Lions, J.-L.: Contrôlabilité Exacte, Perturbation et Stabilisation de Systèmes Distribués. Tome 1: Contrôlabilité Exacte. Masson, Paris (1988)
20. Melrose, R.B., Sjöstrand, J.: Singularities of boundary value problems I. Commun. Pure Appl. Math. **31**, 593–617 (1978)
21. Melrose, R.B., Sjöstrand, J.: Singularities of boundary value Problems II. Commun. Pure Appl. Math. **35**, 129–168 (1982)
22. Phung, K.D.: Observability and Control of Schrödinger Equations. SIAM J. Control Optim. **40**, 211–230 (2001)
23. Privat, Y., Trélat, E., Zuazua, E.: Optimal observability of the multi-dimensional wave and Schrödinger equations in quantum ergodic domains. J. Eur. Math. Soc. **18**, 1043–1111 (2016)
24. Reed, M., Simon, B.: Methods of Modern Mathematical Physics. I: Functional Analysis. Revised edition. Academic Press, Inc. (1980)

25. Reed, M., Simon, B.: Methods of Modern Mathematical Physics. II: Fourier Analysis, Self-Adjointness. Academic Press, Inc. (1975)
26. Tenenbaum, G., Tucsnak, M.: Fast and strongly localized observation for the Schrödinger equation. Trans. Am. Math. Soc. **361**, 951–977 (2009)
27. Villani, C.: Topics in Optimal Transportation. American Mathematical Society, Providence RI (2003)

Results About the Free Kawasaki Dynamics of Continuous Particle Systems in Infinite Volume: Long-time Asymptotics and Hydrodynamic Limit

Yuri G. Kondratiev, Tobias Kuna, Maria João Oliveira, José Luís da Silva, and Ludwig Streit

In memory of our colleague and friend São

(Maria Conceição Carvalho)

Abstract An infinite particle system of independent jumping particles in infinite volume is considered. Their construction is recalled, further properties are derived, the relation with hierarchical equations, Poissonian analysis, and second quantization are discussed. The hydrodynamic limit for a general initial distribution satisfying a mixing condition is derived. The long-time asymptotics is computed under an

---

Yuri G. Kondratiev—Deceased.

Y. G. Kondratiev
Reading, UK

T. Kuna (✉)
Dipartimento di Ingegneria, Scienze dell'Informazione e Matematica,
Universitá degli Studi dell'Aquila, L'Aquila, Italy
e-mail: tobias.kuna@univaq.it

Department of Mathematics and Statistics, University of Reading, Reading, UK

M. J. Oliveira
DCeT, Universidade Aberta, 1269-001 Lisbon, Portugal
e-mail: mjoliveira@ciencias.ulisboa.pt

CMAFCIO, University of Lisbon, 1749-016 Lisbon, Portugal

J. L. da Silva
CCM, University of Madeira, P 9000-390 Funchal, Portugal
e-mail: joses@staff.uma.pt

J. L. da Silva · L. Streit
CIMA, Faculty of Exact Sciences and Engineering, University of Madeira,
Campus da Penteada, 9020-105 Funchal, Portugal
e-mail: streit@uma.pt

L. Streit
Forschungszentrum BiBoS, Universität Bielefeld, 33501 Bielefeld, Germany

E. Carlen et al. (eds.), *From Particle Systems to Partial Differential Equations*,
Springer Proceedings in Mathematics & Statistics 465,
https://doi.org/10.1007/978-3-031-65195-3_7

extra assumption. The relation with constructions based on infinite volume limits is discussed.

Keywords Infinite particle systems · Kawasaki dynamics · Hydrodynamic limit · Long-time asymptotics

1 Introduction

Here we study systems with a large number of particles where the dynamics consists of particles jumping at random times. To gain insights to these systems, one studies them at different scales which are relevant to the system, for us: the scale on which the individual movement is happening, the typical distance between particles, the scale on which we observe the system and finally the size of the system. The key feature of this paper is that we consider the regime where the system size is much larger than all the other scales and try to evaluate what kind of challenges that may bring. The most commonly studied case is where the scale on which the system is observed and the system size are of the same order, which simplifies the situation substantially. Most considerations in this paper become almost trivial in this case, see Sect. 5.3 for a detailed discussion. We are presenting hitherto unpublished results regarding infinite systems, to do so with their proofs would exceed the present framework. The latter can be found in (arXiv:0912.1312).

In this paper, we will walk through mathematical and some conceptual challenges which appear when the system size is much larger than the other scales. For simplicity we model this through an infinite system, as it means that all results derived are automatically independent of the system size. For example, relaxation rates, which depend on the system size, may easily give rise to so slow relaxation of large systems failing to capture the physically relevant relaxation mechanism. Infinite systems are very challenging, see [5, 49] for Hamiltonian dynamics and [13, 50] for gradient diffusions. Despite serious efforts in both cases, our understanding is far from complete and rigorous results are sparse. In order to get a better understanding of the challenges one has to face treating infinite systems, we will consider a non-interacting system where each particle follows a jump process. This allows us to have a very explicit representation of the dynamics and to treat the following three properties in great detail: construction of dynamics (Sects. 3 and 4), long-time asymptotics (Sects. 5.1 and 6.1), and hydrodynamic limits (Sects. 5.2 and 6.2).

The study of the dynamics of an infinite particle system without interaction can be reduced to the analysis of a type of one-particle dynamics. However, this one-particle dynamics being an effective description of the infinite particle system dynamics has to be considered with initial "distributions" which are not integrable, but merely bounded, as they actually describe the particle density of the infinite particle system. To stress this different interpretation, we will call it a pseudo-one-particle system in the following. Each constant initial "distribution" gives rise to a distinct invariant measure of the infinite particle system, in our case, Poisson point random fields with constant intensity. These are all mutually singular probability measures, and hence the system cannot be ergodic. If we want to understand the dynamics of the infinite particle system in full, we are forced to work with initial distributions which are singular towards each other. In this paper, we consider increasingly more general initial distributions corresponding to different physical realms. In Sect. 4 we consider system in equilibrium that is point random fields with density with respect to a Poisson point random field with constant intensity, in Sect. 5 locally equilibrium, that is Poisson point random fields with non-constant intensity and in Sect. 6 non-equilibrium, that is general point random fields with some moment conditions.

The reader may wonder whether, in the free case, not everything is totally obvious. We try to tackle all difficulties using the most straightforward explicit techniques to make the challenges transparent. All these difficulties are innate for large system size, see Sect. 5.3. Though most of the techniques will not extend to the interacting case, the challenges should be maintained if not increased.

Let us give a more technical introduction into the paper. In Sect. 2 we describe the generator of the infinite particle dynamics, the state space and associated notions. The dynamic considered here is a version in the continuum of the well established Kawasaki dynamics for lattice gas systems. These are random evolutions of particle systems in which individual particles jump in the space with rates leading to a Gibbs state in the continuum as an invariant measure [2, 17, 31]. Having the particle number as a conserved quantity these dynamics give rise to a continuous family of invariant measures which are important to study the so-called hydrodynamic limits, see e.g. [6, 22]. Continuous versions of Glauber dynamics were introduced in [3, 14, 16, 24, 27, 31, 43]. In Sect. 2.2 we introduce and recall properties of the associated (pseudo)-one-particle operator.

In Sect. 3 we discuss different ways to construct the infinite particle dynamics. In Sect. 3.1 we recall a probabilistic approach worked out in [32]. The free infinite particle dynamics is just the product of countably many copies of the one-particle dynamics. However, to consider this product as a stochastic movement in the configuration space as further step, one has to show that a.s. in any finite time interval only finitely many jump events are visible in a bounded observation window, i.e., the number of particles in the window stays finite and only finitely many particles pass the window. This does not hold for any initial configuration and the processes can only be started in a restricted set of initial configurations or in initial distributions supported therein. We add to the results of [32] the description of the path-space measure corresponding to the process.

This construction method cannot be extended to interacting systems. A probabilistic construction writing the system as a limit of finite particle systems will require a quite detailed control, see e.g. [13]. In Sect. 3.2 we present an analytic approach to construct the semi-group which may also work for interacting systems. The approach is based on writing the dynamics in terms of the system of correlation functions, see [2] for such a construction in the interacting case. If one only starts in probability measure with a density with respect to an invariant measure, that is in the equilibrium case, techniques from second quantisation become available, cf. Sect. 4.

In Sect. 4.1 we give a description of the dynamics in terms of Fock spaces which also allow a comparison with non-relativistic quantum fields type descriptions. In the case of symmetric dynamics, the powerful methods of Dirichlet forms may be applied, in particular, existence can be shown even for general interacting systems, cf. [31] and references therein. In the case of the free Kawasaki process, one can apply in addition second quantization techniques, which give a full description of the L^2-theory, also for non-symmetric jump rates, cf. Sect. 4.2.

In Sect. 5 we consider the case in which one starts the process in a Poisson point random field with non-constant intensity, that is the system is in local equilibrium. The path space measure is then a Poisson point random field on path space. L^p techniques break down already in this case and hence Dirichlet form techniques cannot be applied, see Remark 6. In Sect. 5.1 we consider the long-time asymptotics of the one-dimensional distribution in time and show that it converges to an invariant measure, that is a Poisson random field with constant intensity. The constant is the arithmetic mean of the initial intensity, see Definition 1. This is easily shown in Proposition 3 under the assumption that the Fourier transform of the intensity z is a signed measure. For such z the arithmetic mean exists, see Corollary 1, but not any bounded non-negative function z for which the arithmetic mean exists and the time asymptotics converges has a signed-measure as Fourier transform, see Remark 9. There also exist bounded non-negative functions for which the arithmetic mean does not exist, see Remark 8. In order to establish a general result, one needs better analytic control of the behaviour of z at infinity.

In Sect. 5.2 we scale space and time simultaneously to obtain a macroscopic description in terms of partial differential equation. When the Fourier transform of the activity is a signed measure, the proper choice of a scale and the computation of the limit is direct. For general bounded activities, one has to control higher derivatives, see Proposition 6.

In Sect. 5.3, we compare the results for systems where the system size is of the same order as the scale on which one observes the system and we derive a representation of the continuous Fourier transform in order to see the technical difference with the previous results. The finite system size induces a spectral gap of size L^{-2} which eliminates all regularity requirements on the jump rate and the activity z. We also show that for a mildly larger system size this effect essentially maintains. In [8] the authors showed the convergence to the hydrodynamic limit, with much more sophisticated techniques, uniformly in time for an infinite particle system but of independent Brownian particles. We also discuss why in the infinite volume one cannot give a comparison result between the free Kawasaki dynamics and the system of independent Brownian particles.

The results described above about long-time asymptotics and hydrodynamic limits are extended to general initial distributions in Sect. 6. We require some kind of mixing property for the initial distributions, the so-called decay of correlation, which we formulate in terms of the cumulants (or Ursell functions). For the hydrodynamic limit, we have to require in addition that the first correlation function of the initial measure converges reasonably under the scaling.

We prove in Sect. 6.3 that the conditions on the initial distributions required above are fulfilled, in particular, by Gibbs measures in the high-temperature low-activity regime.

This article is a testimony to the enthusiasm, deep scientific knowledge and commitment of our collaborator, co-author and friend Yuri Kondratiev. He passed away on the 5th of September of 2023.

2 Kawasaki Dynamics

2.1 *The Generator*

The configuration space $\Gamma := \Gamma_{\mathbb{R}^d}$ over $\mathbb{R}^d$, $d \in \mathbb{N}$, is defined as the set of all locally finite subsets of $\mathbb{R}^d$,

$$\Gamma := \left\{\gamma \subset \mathbb{R}^d : |\gamma_\Lambda| < \infty \text{ for every compact } \Lambda \subset \mathbb{R}^d\right\},$$

where $|\cdot|$ denotes the cardinality of a set and $\gamma_\Lambda := \gamma \cap \Lambda$. As usual we identify each $\gamma \in \Gamma$ with the non-negative Radon measure $\sum_{x\in\gamma} \delta_x \in \mathscr{M}(\mathbb{R}^d)$, where δ_x is the Dirac measure with mass at x, $\sum_{x\in\emptyset} \delta_x$ is, by definition, the zero measure, and $\mathscr{M}(\mathbb{R}^d)$ denotes the space of all non-negative Radon measures on the Borel σ-algebra $\mathscr{B}(\mathbb{R}^d)$. This identification allows to endow Γ with the topology induced by the vague topology on $\mathscr{M}(\mathbb{R}^d)$, i.e., the weakest topology on Γ with respect to which all mappings

$$\Gamma \ni \gamma \longmapsto \langle f, \gamma\rangle := \int_{\mathbb{R}^d} \gamma(dx)\, f(x) = \sum_{x\in\gamma} f(x), \quad f \in C_c(\mathbb{R}^d),$$

are continuous. Here $C_c(\mathbb{R}^d)$ denotes the set of all continuous functions on $\mathbb{R}^d$ with compact support. By $\mathscr{B}(\Gamma)$ we will denote the corresponding Borel σ-algebra on Γ.

Given a non-negative function $a \in L^1(\mathbb{R}^d, dx)$, the generator $L := L_a$ of the free Kawasaki dynamics for an infinite particle system is given by the informal expression

$$(LF)(\gamma) := \sum_{x\in\gamma} \int_{\mathbb{R}^d} dy\, a(x-y)\, (F(\gamma \setminus x \cup y) - F(\gamma))\,. \tag{1}$$

We proceed to give a rigorous meaning to the right-hand side of (1). Let $\mathscr{O}_c(\mathbb{R}^d)$ denote the set of all open sets in $\mathbb{R}^d$ with compact closure. A $\mathscr{B}(\Gamma)$-measurable function F is called cylinder and exponentially bounded whenever there is a $\Lambda \in \mathscr{O}_c(\mathbb{R}^d)$ such that $F(\gamma) = F(\gamma_\Lambda)$ for all $\gamma \in \Gamma$, and $|F(\gamma)| \leq Ce^{c|\gamma_\Lambda|}$, $\gamma \in \Gamma$, for some $C, c > 0$. For such a function F one has

$$\sum_{x\in\gamma}\int_{\mathbb{R}^d} dy\, a(x-y)\,|F(\gamma\setminus x\cup y)-F(\gamma)|$$
$$\leq \left[|\gamma_\Lambda|\int_{\mathbb{R}^d} dy\, a(y) + \int_\Lambda dy \sum_{x\in\gamma} a(x-y)\right] 2Ce^{c(|\gamma_\Lambda|+1)}, \tag{2}$$

which is finite provided the configuration γ is an element in $\Gamma_a \subset \Gamma$,

$$\Gamma_a := \left\{\gamma\in\Gamma : y \mapsto \sum_{x\in\gamma} a(x-y) \text{ is } L^1_{\text{loc}}(\mathbb{R}^d, dy)\right\}.$$

In this case, the sum and the integral in (1) are finite, and thus the operator L is well-defined on the space $\mathscr{F}L^0_{eb}(\Gamma_a)$ of all cylinder functions exponentially bounded on Γ restricted to Γ_a.

Concerning the set Γ_a, we note that $\mu(\Gamma_a) = 1$ for any probability measure μ on Γ with first correlation function $k_\mu^{(1)}$ bounded. This follows from the fact that for each closed ball $B(n) \subset \mathbb{R}^d$ centered at 0 and radius $n \in \mathbb{N}$ we have

$$\int_\Gamma \mu(d\gamma)\int_{B(n)} dy \sum_{x\in\gamma} a(x-y) = \int_{\mathbb{R}^d} dx\, k_\mu^{(1)}(x)\int_{B(n)} dy\, a(x-y)$$
$$\leq C\|a\|_{L^1(\mathbb{R}^d,dx)}\mathrm{vol}(B(n))$$

for any constant $C \geq |k_\mu^{(1)}|$. Here $\mathrm{vol}(B(n))$ denotes the volume of $B(n)$ with respect to the Lebesgue measure on $\mathbb{R}^d$. Clearly, the latter implies that $\mu(\Gamma\setminus\Gamma_a) = 0$.

In view of these considerations, throughout this work we shall restrict our setting to Γ_a.

Among the elements in $\mathscr{F}L^0_{eb}(\Gamma_a)$ we distinguish the functions $e_B(f)$, called Bogoliubov exponentials,

$$e_B(f,\gamma) := \prod_{x\in\gamma}(1+f(x)), \quad \gamma\in\Gamma, \tag{3}$$

for f being any bounded $\mathscr{B}(\mathbb{R}^d)$-measurable function with compact support ($f \in B_c(\mathbb{R}^d)$). An important reason to single out these functions is due to the especially simple form for the action of L on them, namely, for all $f \in B_c(\mathbb{R}^d)$ and all $\gamma \in \Gamma_a$,

$$(Le_B(f))(\gamma) = \sum_{x\in\gamma}(Af)(x)e_B(f,\gamma\setminus x), \tag{4}$$

where

$$Af(x) := \int_{\mathbb{R}^d} dy\, a(x-y)\,(f(y)-f(x)). \tag{5}$$

In the sequel we call the linear operator A the pseudo-one-particle operator, though we will explain in which way it differs from the genuine one-particle operator. Due to its special role throughout this work, its properties will be studied in more detail in the next subsection.

Given a locally integrable function $z \geq 0$, we remind that the Poisson measure π_z with intensity z is the unique probability measure on Γ for which the Laplace transform is given by

$$\int_\Gamma \pi_z(d\gamma)\, e^{\langle\varphi,\gamma\rangle} = \exp\left(\int_{\mathbb{R}^d} dx\, \left(e^{\varphi(x)} - 1\right) z(x)\right), \tag{6}$$

for all φ in the Schwartz space $\mathscr{D}(\mathbb{R}^d) := C_c^\infty(\mathbb{R}^d)$ of all infinitely differentiable functions with compact support. We recall that a measure μ on Γ_a is called infinitesimally reversible with respect to an operator L, whenever L is symmetric in $L^2(\Gamma_a, \mu)$.

Lemma 1 *Assume that a is an even function. Then for any real number $z > 0$ the Poisson measure π_z with constant intensity z is an infinitesimally reversible measure with respect to L.*

Remark 1 It is clear that the linear space spanned by the class of functions $e_B(f)$, $f \in B_c(\mathbb{R}^d)$, is in $\mathscr{F}L^0_{eb}(\Gamma_a)$. The space $\mathscr{F}L^0_{eb}(\Gamma_a)$ also contains the class of coherent states $e_{\pi_z}(f)$ corresponding to $f \in B_c(\mathbb{R}^d)$,

$$e_{\pi_z}(f) := \exp\left(-\int_{\mathbb{R}^d} dx\, z(x) f(x)\right) e_B(f), \tag{7}$$

see Sect. 4.1 for a use of this relation.

2.2 The One-particle Operator

In the sequel the pseudo-one-particle operator A introduced in (5) will play an essential role. Because of this, in this subsection we shall collect its main properties used below. We observe that in stochastic analysis the operator A is known as a relatively simple example of a bounded generator of a Markov jump process on $\mathbb{R}^d$ (see e.g. [11, Sect. 4.2]).

In terms of the usual convolution $*$ of functions, we also note that

$$Af = a * f - a^{(0)} f, \quad a^{(0)} := \int_{\mathbb{R}^d} a(x)\, dx. \tag{8}$$

Therefore, the properties of convolution of functions (namely, Young's inequality) lead straightforwardly to the L^p results stated in the next proposition. There, we also consider the real Banach spaces $B(\mathbb{R}^d)$ and $C_\infty(\mathbb{R}^d)$, respectively, of all bounded measurable functions and of all continuous functions vanishing at infinity, both with the supremum norm $\|f\|_u := \sup_{x\in\mathbb{R}^d} |f(x)|$. We recall that a strongly continuous contraction semigroup $(T_t)_{t\geq 0}$ is called sub-Markovian whenever for $0 \leq f \leq 1$ it follows $0 \leq T_t f \leq 1$ for every $t \geq 0$. If, in addition, $T_t f_n \nearrow 1$ for some sequence $f_n \nearrow 1$, then $(T_t)_{t\geq 0}$ is called Markovian. Here, for the spaces $B(\mathbb{R}^d)$ and $C_\infty(\mathbb{R}^d)$ the convergence is pointwise, and for an L^p-space the convergence is almost everywhere. A strongly continuous contraction semigroup $(T_t)_{t\geq 0}$ is called positivity preserving, if $f \geq 0$ implies $T_t f \geq 0$ for every $t \geq 0$.

Proposition 1 *The linear operator A is a bounded operator on $L^p(\mathbb{R}^d, zdx)$, for $z > 0$ constant and $1 \leq p \leq \infty$ (on $B(\mathbb{R}^d)$ and on $C_\infty(\mathbb{R}^d)$). As a consequence, A is the generator of a uniformly continuous semigroup $(e^{tA})_{t\geq 0}$ on $L^p(\mathbb{R}^d, zdx)$, $1 \leq p \leq \infty$ (on $B(\mathbb{R}^d)$ and on $C_\infty(\mathbb{R}^d)$),*

$$e^{tA} := \sum_{n=0}^{\infty} \frac{(tA)^n}{n!}, \tag{9}$$

the sum converging in norm for every $t \geq 0$. Moreover, $(e^{tA})_{t\geq 0}$ is a positivity preserving (contraction) semigroup on each $L^p(\mathbb{R}^d, zdx)$ space, $1 \leq p \leq \infty$ (on $B(\mathbb{R}^d)$ and on $C_\infty(\mathbb{R}^d)$), Markovian on $L^p(\mathbb{R}^d, zdx)$ for $1 \leq p < \infty$.

For the proof see e.g. [11, Sect. 4.2, 21].

Due to (8), the operator A, as well as the semigroup $(e^{tA})_{t\geq 0}$, both either on a $L^p(\mathbb{R}^d, zdx)$ space, $1 \leq p < \infty$, or on $C_\infty(\mathbb{R}^d)$, may be expressed in terms of the Fourier transform,

$$\hat{f}(k) := \frac{1}{(2\pi)^{d/2}} \int_{\mathbb{R}^d} dx\, e^{-i\langle x,k\rangle} f(x),$$

see e.g. [21].

Proposition 2 *For every function φ in the Schwartz space $\mathscr{S}(\mathbb{R}^d)$ of tempered test functions one has*

$$\widehat{A\varphi}(k) = (2\pi)^{d/2} \left(\hat{a}(k) - \hat{a}(0)\right) \hat{\varphi}(k), \quad k \in \mathbb{R}^d,$$

and

$$\left(e^{tA}\varphi\right)(x) = \frac{1}{(2\pi)^{d/2}} \int_{\mathbb{R}^d} dk\, e^{i\langle k,x\rangle} e^{t(2\pi)^{d/2}(\hat{a}(k)-\hat{a}(0))} \hat{\varphi}(k), \quad x \in \mathbb{R}^d. \tag{10}$$

Remark 2 Since a is non-negative, for all $k \in \mathbb{R}^d$ one has

$$\mathrm{Re}(\hat{a}(k) - \hat{a}(0)) = \frac{1}{(2\pi)^{d/2}} \int_{\mathbb{R}^d} dx\, (\cos(\langle k, x\rangle) - 1) a(x) \leq 0. \tag{11}$$

The equality in (11) holds only for $k = 0$. This follows from the fact that the set $\{x : \langle k, x\rangle = 2n\pi, n \in \mathbb{Z}\}$ has zero Lebesgue measure if and only if $k \neq 0$.

Remark 3 According to Proposition 2, the semigroup $(e^{tA})_{t\geq 0}$ is defined by a kernel μ_t, $t \geq 0$, which Fourier transform is given by

$$\hat{\mu}_t(k) = \frac{1}{(2\pi)^{d/2}} e^{t(2\pi)^{d/2}(\hat{a}(k)-\hat{a}(0))}.$$

Because a is non-negative, $\hat{a}$ and thus $\hat{\mu}_t$, $t \geq 0$, are positive definite functions. Bochner's theorem yields that each μ_t is a non-negative finite measure on $\mathbb{R}^d$. We note that the Markovian property of $(e^{tA})_{t\geq 0}$ means that each μ_t is actually a probability measure. For more details see e.g. [21].

Remark 4 It is easy to check that on the space $L^2(\mathbb{R}^d, zdx)$ the adjoint operator A^* of A is defined for all $f \in L^2(\mathbb{R}^d, zdx)$ by

$$(A^* f)(x) = \int_{\mathbb{R}^d} dy\, a(y-x)(f(y) - f(x)) + \int_{\mathbb{R}} dy\big(a(y) - a(-y)\big)\, f(x).$$

As a consequence, if a is an even function, then A is a bounded self-adjoint operator on $L^2(\mathbb{R}^d, zdx)$. In this case, it follows from (9) that the semigroup $(e^{tA})_{t\geq 0}$ is a self-adjoint contraction on $L^2(\mathbb{R}^d, zdx)$.

Remark 5 For a non-constant activity parameter z, the operator A is still bounded but, in general, the semigroup $(e^{tA})_{t\geq 0}$ is not any longer a contraction. For instance, on $\mathbb{R}$ consider $a(x) = e^{-x^2}$ and $z(x) = 1 + e^{-x^2+4x}$. A simple calculation shows that for $\varphi(x) = \pi^{-1/2}e^{-x^2}$ one has $\int_{\mathbb{R}} dx\, \varphi(x)(A\varphi)(x)z(x) > 0$, proving that the semigroup $(e^{tA})_{t\geq 0}$ cannot be a contraction in $L^2(\mathbb{R}^d, zdx)$.

3 Independent Infinite Particle Processes

3.1 A Probabilistic Approach

3.1.1 One-particle Process

As we mentioned at the beginning of Sect. 2.2, the operator A is the generator of a Markov jump process on $\mathbb{R}^d$. Following e.g. [11], we can explicitly construct this process as follows. Consider the Markov chain $(Y_k)_{k\in\mathbb{N}_0}$ on $\mathbb{R}^d$, with transition density function $\mu(x, y) = \frac{a(x-y)}{a^{(0)}}$. Let $(Z_t)_{t\geq 0}$ be a Poisson process with parameter $a^{(0)}$ independent of the Markov chain. We then define the Markov process $(X_t)_{t\geq 0}$ by $X_t := Y_{Z_t}, t \geq 0$. This process has A as generator and, by construction, it has *cadlag* paths in $\mathbb{R}^d$. We denote by $D([0, \infty), \mathbb{R}^d)$ the set of all *cadlag* paths from $[0, \infty)$ to $\mathbb{R}^d$ and by P^x the path-space measure corresponding to the process X starting at $x \in \mathbb{R}^d$. By E_x we denote the expectation w.r.t. this measure. Hence, for $\varphi \in \mathcal{D}(\mathbb{R}^d)$ we have that

$$E_x[\varphi(X_t)] = e^{tA}\varphi(x). \tag{12}$$

3.1.2 Construction

The process on Γ corresponding to L is the following random evolution: each particle evolves according to the above jump process, independently of the other particles, cf. Lemma 2 below. This independent infinite particle process was rigorously constructed in [32] (see also [25]). In the following we present the main results therein.

We need further assumptions on the jump kernel a to make this construction rigorous. It is sufficient for the construction done in [32] to assume that there exist an $\alpha > 2d$ and a $C > 0$ such that

$$0 \leq a(y) \leq \frac{C}{(1+|y|)^{\alpha}}, \qquad \text{for all } y \in \mathbb{R}^d. \tag{13}$$

Then [32] shows that there exists a Markov process $\big(D([0,\infty),\Theta), (\mathbf{X}_t)_{t\geq 0}, (\mathbf{P}_\gamma)_{\gamma\in\Theta}\big)$ on the set

$$\Theta := \left\{\gamma \in \Gamma \,:\, \exists m \in \mathbb{N} \text{ such that } |\gamma_{B(n)}| \leq m \operatorname{vol}(B(n)), \quad \forall n \in \mathbb{N}\right\}.$$

Here $D([0,\infty),\Theta)$ denotes the set of all *cadlag* paths from $[0,\infty)$ to Θ, the process $\mathbf{X}_t : D([0,\infty),\Theta) \to \Theta$ is the canonical one, that is, $\mathbf{X}_t(\omega) := \omega(t)$, $\omega \in D([0,\infty),\Theta)$, and each $\mathbf{P}_\gamma$ is the path-space measure of the process starting at a $\gamma \in \Theta$. By $\mathbf{E}_\gamma$ we denote the expectation w.r.t. $\mathbf{P}_\gamma$.

Note, that it is impossible to let the process start from an arbitrary initial configuration $\gamma \in \Gamma$, as follows implicitly from the discussion before Theorem 2.2 in [32]. One is obliged to restrict the set of possible initial configurations, in our case to Θ. Hence we can start the process from configurations in Θ and initial distributions μ supported on Θ. The restriction that $\mu(\Theta) = 1$ is not very severe. For example, it is sufficient that for all n the random variable $|\gamma_{B(n)}|$ has all exponential moments finite with respect to μ and $|\gamma_{B(n)}|/\operatorname{vol}(B(n))$ has a μ integral uniformly bounded in n. In particular, for any probability measure μ on Γ which correlation functions $k_\mu^{(n)}$, $n \in \mathbb{N}$, fulfill the so-called Ruelle bound, i.e., there is a $C > 0$ such that $k_\mu^{(n)} \leq C^n$ for every $n \in \mathbb{N}$. This holds, for instance, for Gibbs measures w.r.t. superstable, lower and upper regular potentials, cf. [48]. Trivially, it holds for Poisson measures which have as intensity a non-negative bounded measurable function. For a constant intensity this result was shown using ergodicity in [41]. However, finiteness of exponential moments or even finiteness of all moments are not necessary, one may assume instead decay of correlation type properties.

3.1.3 Starting from a Configuration and Transition Propabilities

Choosing, for a fixed initial configuration $\gamma \in \Theta$, an enumeration $\{x_n\}_{n\in\mathbb{N}}$, the infinite particle process can be described more explicitly. For each n let us consider an independent copy of the one-particle jump process $((X_t^{(n)})_{t\geq 0}, P^{x_n})$ introduced at the beginning of the section. In [32] it is shown that if one starts in a $\gamma = \{x_n\}_{n\in\mathbb{N}} \in \Theta$, then $\bigotimes_{n=1}^{\infty} P^{x_n}$-a.s. for all times the sequence $(X_t^{(n)})_{n\in\mathbb{N}}$ has neither accumulation points nor two entries of the sequence coincide and hence it can be identified with the configuration $\{X_t^{(n)}\}_{n\in\mathbb{N}} \in \Gamma$. Furthermore, it has been shown that this configuration lies in Θ. Hence $(\{X_t^{(n)}\}_{n\in\mathbb{N}})_{t\geq 0}$ is the corresponding process on the configuration space Θ and the path-space measure is the "symmetrization" of $\bigotimes_{n=1}^{\infty} P^{x_n}$. Moreover, the transition probability $(\mathbf{P}_t)_{t\geq 0}$ of the process $(\mathbf{X}_t)_{t\geq 0}$ is just the product of the one-particle transition probabilities $e^{tA}(x-y)dy$, i.e., $\prod_{n=1}^{\infty} e^{tA}(x_n - y_n)dy_n$.

As a consequence, we can give explicit formulas for a lot of expressions. The Fourier-Laplace transform of the distribution of the processes at a fixed time, is for all non-positive $\varphi \in \mathscr{D}(\mathbb{R}^d)$ given by

$$\mathbf{E}_\gamma \left[e^{\langle \varphi, \mathbf{X}_t\rangle}\right] = \int_\Theta \mathbf{P}_t(\gamma, d\xi)\, e^{\langle \varphi, \xi\rangle} = \prod_{x\in\gamma} E_x\left[e^{\varphi(X_t)}\right] = \prod_{x\in\gamma} e^{tA}\varphi(x), \tag{14}$$

see also [32] (and also Lemma 3 below for an extension to a wider class of φ), alternatively we can use the Bogoliubov exponentials introduced in (3) for $\varphi \in \mathscr{D}(\mathbb{R}^d)$ with $-1 < \varphi \leq 0$,

$$\mathbf{E}_\gamma \left[e_B(\varphi, \mathbf{X}_t)\right] = \prod_{x\in\gamma} E_x\left[\varphi(X_t) + 1\right] = e_B(e^{tA}\varphi, \gamma). \tag{15}$$

More generally, it follows that the Fourier-Laplace transform of the joint distribution of the processes at different times can be expressed as

$$\mathbf{E}_\gamma \left[e^{\langle \varphi_1, \mathbf{X}_{t_1}\rangle} \cdots e^{\langle \varphi_n, \mathbf{X}_{t_n}\rangle}\right] = \prod_{x\in\gamma} E_x\left[e^{\varphi_1(X_{t_1})} \cdots e^{\varphi_n(X_{t_n})}\right],$$

for all $0 \leq t_1 < \ldots < t_n$ and all non-positive $\varphi_1, \ldots, \varphi_n \in \mathscr{D}(\mathbb{R}^d)$, $n \geq 2$. By a monotone approximation procedure using Riemannian sums, this relation allows us to calculate the Laplace transform of the path-space itself measure $\mathbf{P}_\gamma$, i.e.,

$$\mathbf{E}_\gamma \left[e^{-\int dt \langle \varphi(t,\cdot), \mathbf{X}_t\rangle}\right] = \prod_{x\in\gamma} E_x\left[e^{-\int dt \varphi(t, X_t)}\right], \tag{16}$$

for all non-negative continuous functions φ from $[0,\infty)\times\mathbb{R}^d$ to $\mathbb{R}$ with compact support.

The next result gives a relation between the process $\mathbf{X}$ and the operator L.

Lemma 2 *For all $F \in \mathscr{F}L^0_{eb}(\Theta)$ and all $\gamma \in \Theta$ there holds*

$$\mathbf{E}_\gamma \left[F(\mathbf{X}_t) - F(\mathbf{X}_0) - \int_0^t ds L F(\mathbf{X}_s)\right] = 0. \tag{17}$$

Lemma 3 *Let* $\chi(y) = \frac{C'}{(1+|y|)^{\alpha/2}}$, *for some* $C' > 0$ *and* $\alpha > 0$ *as in (13). Then there exists a* $c > 0$ *such that* $A\chi \leq c\chi$ *and* $e^{tA}\chi \leq e^{ct}\chi$ *for all* $t \geq 0$. *Moreover, for all* $\gamma \in \Theta$ *and all measurable functions* φ *such that* $\frac{|\varphi|}{\chi}$ *is bounded, one has that the product* $\prod_{x\in\gamma}(1+\varphi(x))$ *is absolutely convergent and*

$$\mathbf{E}_\gamma\left[e^{\langle\varphi,\mathbf{X}_t\rangle}\right] < \infty.$$

3.1.4 Starting from a Distribution

In Sects. 5 and 6 one considers the one-dimensional in time distributions of processes starting with initial distributions μ which are probability measures on Θ. The path-space measure $\mathbf{P}^\mu$ corresponding to such a process is given by $\int_\Theta \mathbf{P}_\gamma \mu(d\gamma)$. Its one-dimensional distribution is a probability measure $P^{\mathbf{X}}_{\mu,t}$ on Θ defined for all non-negative measurable functions F by

$$\int_\Theta P^{\mathbf{X}}_{\mu,t}(d\gamma)F(\gamma) := \int_\Theta \mu(d\gamma)\mathbf{E}_\gamma\left[F(\mathbf{X}_t)\right]. \tag{18}$$

In particular, for $\mu = \delta_\gamma$, $\gamma \in \Theta$, the one-dimensional distribution coincides with the transition kernel $\mathbf{P}_t(\gamma, \cdot)$ described above.

For functions F being Bogoliubov exponentials, definition (18) leads to the so-called Bogoliubov functionals [4]. By definition, the Bogoliubov functional corresponding to a probability measure μ on Γ is defined by

$$B_\mu(\varphi) := \int_\Gamma e_B(\varphi,\gamma)\mu(d\gamma), \qquad \varphi \in \mathscr{D}(\mathbb{R}^d).$$

Such a functional is an analogue of the Fourier-Laplace transform on configuration spaces, cf. [36]. Due to (15) there is an interesting relation between the Bogoliubov functional corresponding to the initial distribution and the Bogoliubov functional corresponding to the one-dimensional distribution of the process at a time $t > 0$, namely,

$$\int_\Theta P^{\mathbf{X}}_{\mu,t}(d\gamma)e_B(\varphi,\gamma) = \int_\Theta \mu(d\gamma)e_B(e^{tA}\varphi,\gamma) = B_\mu(e^{tA}\varphi). \tag{19}$$

In particular, for $\mu = \pi_z$ for some bounded intensity function $z \geq 0$ one finds

$$\int_\Theta e_B(\varphi,\gamma)P^{\mathbf{X}}_{\pi_z,t}(d\gamma) = \int_\Theta e_B(e^{tA}\varphi,\gamma)\pi_z(d\gamma) = \exp\left(\int_{\mathbb{R}^d} e^{tA}\varphi(x)z(x)dx\right), \tag{20}$$

for all $\varphi \in \mathscr{D}(\mathbb{R}^d)$. In Sect. 5 we shall consider this special case in more detail.

One may ask why one does not work with the more standard Fourier-Laplace transform. Reconsidering (14) one sees that at least formally

$$\mathbf{E}_\gamma\left[e^{\langle\varphi,\mathbf{X}_t\rangle}\right] = e^{\langle \ln(e^{tA}e^{\varphi}),\gamma\rangle} \tag{21}$$

and notes that the pseudo-one-particle dynamic obtained in this way $\varphi \mapsto \ln(e^{tA}e^{\varphi})$ is non-linear.

3.1.5 Pseudo-one-Particle Operator

We call in this paper A the pseudo-one-particle operator. This is motivated by (19) which gives us the possibility to study time development of the infinite particle system in terms of the semi-group e^{tA}. Indeed, the Bogoliubov exponentials, like the usual exponentials, form a class of function large enough to describe the dynamics uniquely. The difference with the dynamics of a system of one particle is, beside the exponential in (19), the interpretation of z. In the one-particle system, z would describe the initial distribution of the particle and hence given by an $L^1(\mathbb{R}^d, dx)$ function. In the infinite particle system z describes the intensity of the Poisson measure, that is the density of particles and hence it is natural to consider $z \in L^\infty(\mathbb{R}^d, dx)$. The difference is in which space we consider the semi-group to act, and we will see in Sect. 5 and in Sect. 6 that the qualitative behaviour is quite different. Note that for a Poisson measure on the configuration space the number of particles is a.s. finite if and only if $z \in L^1(\mathbb{R}^d, dx)$.

3.2 An Analytic Approach

Within the framework of infinite dimensional analysis on configuration spaces [26] one may derive alternative representations and constructions of the dynamics. Instead of describing the infinite particle dynamics on Γ through the Kolmogorov equation $\frac{\partial}{\partial t}F_t = LF_t$, the so-called K-transform [37, 38] allows an alternative description for the action of L on $\mathcal{F}L^0_{eb}(\Gamma_a)$.

Given the space $\Gamma_0 := \{\gamma \in \Gamma : |\gamma| < \infty\}$ of finite configurations endowed with the metrizable topology as described in [26], let $B_{exp,ls}(\Gamma_0)$ be the space of all exponentially bounded Borel measurable functions G (i.e., $|G(\eta)| \le C_1 e^{C_2|\eta|}$ for some $C_1, C_2 > 0$) with local support (i.e., there is a $\Lambda \in \mathcal{O}_c(\mathbb{R}^d)$ such that $G(\eta) = 0$ for all $\eta \in \Gamma_0$ with $|\eta \cap (\mathbb{R}^d \setminus \Lambda)| \neq 0$). The K-transform of a $G \in B_{exp,ls}(\Gamma_0)$ is the mapping $KG : \Gamma \to \mathbb{R}$ defined for all $\gamma \in \Gamma$ by $(KG)(\gamma) := \sum_{\substack{\eta\subset\gamma\\ |\eta|<\infty}} G(\eta)$. Note that $K\left(B_{exp,ls}(\Gamma_0)\right) \subset \mathcal{F}L^0_{eb}(\Gamma_a)$. Given a $G \in B_{exp,ls}(\Gamma_0)$, then in terms of the operator L we obtain

$$\begin{aligned}(L(KG))(\gamma) = \sum_{x\in\gamma}\int_{\mathbb{R}^d\setminus(\gamma\setminus x)} dy\, a(x-y)\big[&(KG)(\gamma\setminus x) + (KG(\cdot\cup y))(\gamma\setminus x)\\ &-(KG)(\gamma\setminus x) - (KG(\cdot\cup x))(\gamma\setminus x)\big],\end{aligned}$$

which leads to the so-called symbol acting on quasi-observables $\hat{L}$,

$$(\hat{L}G)(\eta) := \sum_{x\in\eta} \int_{\mathbb{R}^d} dy\, a(x-y)\,(G(\eta\setminus x \cup y) - G(\eta)), \quad G \in B_{exp,ls}(\Gamma_0),$$

and the corresponding time evolution equation $\frac{\partial}{\partial t} G_t = \hat{L} G_t$. The transition kernel $\hat{P}_t$ corresponding to $\hat{L}$ is then for $\eta = \{x_1, \ldots, x_n\}$ given by

$$\int_{\Gamma_0} \hat{P}_t(\eta', d\eta)\, G(\eta) = \int_{\mathbb{R}^{dn}} dy_1 \ldots dy_n \prod_{i=1}^{n} e^{tA}(x_i - y_i) G(\{y_1, \ldots, y_n\}).$$

This allows us to extend the explicit formula for transition kernels of $\mathbf{X}$ to the class of all so-called observables of additive type

$$\int_{\Theta} \mathbf{P}_t(\gamma, d\xi)\,(KG)(\xi) = K\left(\int_{\Gamma_0} \hat{P}_t(\cdot, d\eta)\, G(\eta)\right)(\gamma).$$

By duality one may extend the dynamical description to correlation functions. For this purpose, on the space $\Gamma_0 = \cup_{n=0}^{\infty}\{\gamma \in \Gamma : |\gamma| = n\}$ let us consider the so-called Lebesgue-Poisson measure $\lambda_z := \sum_{n=0}^{\infty} \frac{1}{n!}(zm)^{(n)}$, where m denotes the Lebesgue measure and each $(zm)^{(n)}$, $n \in \mathbb{N}$, is the image measure on $\{\gamma \in \Gamma : |\gamma| = n\}$ of the product measure $z(x_1)dx_1 \cdots z(x_n)dx_n$ under the mapping $(\mathbb{R}^d)^n \ni (x_1, ..., x_n) \mapsto \{x_1, ..., x_n\}$. If for some probability measure μ on Γ there is a function k_μ on Γ_0 such that the equality $\int_\Gamma \mu(d\gamma)\,(KG)(\gamma) = \int_{\Gamma_0} \lambda(d\eta)\, G(\eta) k_\mu(\eta)$ holds for all $G \in B_{exp,ls}(\Gamma_0)$, then k_μ are the correlation function corresponding to μ. Here we abbreviate $\lambda := \lambda_1$. Denoting by $\hat{L}^*$ the dual operator of $\hat{L}$ in the sense

$$\int_{\Gamma_0} \lambda_z(d\eta)\,(\hat{L}G)(\eta) k(\eta) = \int_{\Gamma_0} \lambda_z(d\eta)\, G(\eta)(\hat{L}^* k)(\eta),$$

one obtains the following expression, cf. [12] for more details,

$$(\hat{L}^* k)(\eta) = \sum_{x\in\eta} \int_{\mathbb{R}^d} dy\, a(y-x) k(\eta \setminus x \cup y) - |\eta| k(\eta) a^{(0)}.$$

The corresponding time evolution equation for correlation functions, $\frac{\partial}{\partial t} k_t = \hat{L}^* k_t$, is the analogue of the BBGKY-hierarchy for the case of the free Kawasaki dynamics. In our case this equation can be explicitly solved, namely,

$$k_t(\{x_1, \ldots, x_n\}) = \int_{\mathbb{R}^{dn}} dy_1 \cdots dy_n\, k_\mu(\{y_1, \ldots, y_n\}) \prod_{i=1}^{n} e^{tA}(y_i - x_i) \tag{22}$$

for the initial condition k_μ. Let us note that if one assumes a Ruelle bound for the initial correlation function k_μ, then all the above considerations can be made rigorous. Indeed, e^{tA} is a contraction in $L^\infty(\mathbb{R}^d, dx)$ and hence by (22) the correlation functions retain the Ruelle bound with time. In order to show that these correlation functions correspond to a probability measure μ_t on $\Theta \subset \Gamma$ one can either use similar arguments as in [23] or consider it as a moment problem on configuration space and try to verify the criteria in [19] for a version of the infinite dimensional moment problem giving a localization on the configuration space based on techniques established in [20] and guarantee uniqueness using Carleman bounds, see e.g. [18].

According to the previous considerations, the time evolution of the particle system may also be described in terms of Bogoliubov functionals, cf. [29], through the time evolution equation

$$\frac{\partial}{\partial t} B_{\mu,t}(\varphi) = \int_{\mathbb{R}^d} dx \int_{\mathbb{R}^d} dy\, a(x-y)(\varphi(y)-\varphi(x)) \frac{\delta B_{\mu,t}(\varphi)}{\delta \varphi(x)}, \quad \varphi \in \mathcal{D}(\mathbb{R}^d).$$

Here $\frac{\delta B_{\mu,t}(\varphi)}{\delta \varphi(x)}$ denotes the first variational derivative of $B_{\mu,t}$ at φ. Actually, $B_{\mu,t}(\varphi)$ is the Bogoliubov functional corresponding to the one-dimensional distribution of the process starting in μ. Hence, the previous equation has an explicit solution given by (19), i.e., $B_{\mu,t}(\varphi) = B_\mu(e^{tA}\varphi)$. If one develops the Bogoliubov functional in powers of φ then one recover the equations for the correlation functions derived above.

4 Equilibrium Dynamics

In this section we are interested in the representation of the generator L and its semigroup in terms of the creation, annihilation and second quantization operators as analytic expressions. These representations are possible because there is a well-known canonical unitary isomorphism between the (symmetric) Fock space and the Poisson space. The operator theoretic side can be only made rigorous for constant activity z. When a is symmetric then π_z for constant activity is a reversible measure, which gave rise to the name of the section. We start by recalling the aforementioned canonical unitary isomorphism. Our approach is based on [1, 33], but see also [28] and references therein.

4.1 The $L^2(\pi_z)$ Space and the Fock Space Representation

Let z be a non-negative locally integrable function, which we denote by $z(\cdot)$ in order to distinguish it from the case of constant intensity. We will explain in this sub-section why in order to use the Fock space representation to establish existence on has to assume that z is constant.

We consider the complex Hilbert space $L^2(\pi_{z(\cdot)}) := L^2(\Gamma, \mathscr{B}(\Gamma), \pi_{z(\cdot)})$ of square integrable complex valued functions on Γ with respect to the Poisson measure $\pi_{z(\cdot)}$. The coherent states introduced in Remark 1 generate the system of so-called Charlier polynomials, namely,

$$e_{\pi_{z(\cdot)}}(\varphi,\gamma) = \exp\left(-\int_{\mathbb{R}^d} dx\, z(x) f(x)\right) e_B(f) = \sum_{n=0}^{\infty} \frac{1}{n!} \langle C^n_{z(\cdot)}(\gamma), \varphi^{\otimes n}\rangle,$$

where $C^n_{z(\cdot)}(\gamma) \in (\mathscr{D}')^{\hat{\otimes} n}$ and $(\mathscr{D}')^{\hat{\otimes} n}$ is the n-th symmetric tensor product of the Schwartz distributions space $\mathscr{D}' := \mathscr{D}'(\mathbb{R}^d)$. This system is orthogonal and any $F \in L^2(\pi_{z(\cdot)})$ can be expanded in terms of Charlier polynomials

$$F(\gamma) = \sum_{n=0}^{\infty} \langle C^n_{z(\cdot)}(\gamma), f^{(n)}\rangle, \qquad f^{(n)} \in L^2(z(\cdot)dx)^{\hat{\otimes} n}, \tag{23}$$

where $L^2(z(\cdot)dx) := L^2(\mathbb{R}^d, z(\cdot)dx)$. This yields a unitary isomorphism $I_{\pi_{z(\cdot)}}$ between $L^2(\pi_{z(\cdot)})$ and the so-called symmetric Fock space

$$\mathscr{F}(L^2(z(\cdot)dx)) := \bigoplus_{n=0}^{\infty} n! L^2(z(\cdot)dx)^{\hat{\otimes} n}, \qquad L^2(z(\cdot)dx)^{\hat{\otimes} 0} := \mathbb{C}.$$

More precisely, for each $F \in L^2(\pi_{z(\cdot)})$ of the form (23) one has $I_{\pi_{z(\cdot)}}(F) = (f^{(n)})_{n=0}^{\infty}$. Next we recall the definition of annihilation, creation and second quantization operators on the total subset of Fock space vectors of the form $f^{(n)} = f_1 \hat{\otimes} \ldots \hat{\otimes} f_n$. The action of the creation operator, denoted by $a^+(h)$, $h \in L^2(z(\cdot)dx)$, on elements $f^{(n)} \in L^2(z(\cdot)dx)^{\hat{\otimes} n}$ is given by $a^+(h) f^{(n)} = h \hat{\otimes} f^{(n)}$. The annihilation operator $a^-(h)$ is the adjoint operator.

Given a semigroup $(e^{tA})_{t\geq 0}$ on the Hilbert space $L^2(z(\cdot)dx)$ one can construct potentially unbounded operator $\mathrm{Exp}(e^{tA})$ on $\mathscr{F}(L^2(z(\cdot)dx))$ defined by $e^{tA} \otimes \ldots \otimes e^{tA}$ on each space $L^2(z(\cdot)dx)^{\hat{\otimes} n}$. Note that whenever the operator norm of e^{tA} is greater than one $\mathrm{Exp}(e^{tA})$ becomes an unbounded operator. Hence in order to be in the realm of semi-group theory we have to assume that $(e^{tA})_{t\geq 0}$ is a contraction semi-group. In our case, for non-constant functions $z(\cdot)$ the semi-group e^{tA} is in general not a contraction, see Remark 5 and indeed the previous construction gives not rise to a semi-group, cf. Remark 6. So the constructions below are not mathematically rigorous for non-constant activity as they stand but well-defined on the coherent states, Bogoliubov exponentials respectively.

The generator of $\mathrm{Exp}(e^{tA})$ is the so-called second quantization operator $d\mathrm{Exp}A$ corresponding to A. The image of the Fock coherent state $e(f) := (f^{\otimes n}/n!)_{n=0}^{\infty}$, $f \in L^2(z(\cdot)dx)$, under $\mathrm{Exp}(e^{tA})$ is given by

$$\mathrm{Exp}(e^{tA})(e(f)) = e(e^{tA} f). \tag{24}$$

Through the unitary isomorphism $I_{\pi_z(\cdot)}$ a contraction semigroup $(\mathrm{Exp}_{\pi_{z(\cdot)}}(e^{tA}))_{t\geq 0}$ on $L^2(\pi_{z(\cdot)})$ is obtained. In particular, since $I^{-1}_{\pi_z(\cdot)}e(f) = e_{\pi_{z(\cdot)}}(f)$, it follows from (7) that

$$\mathrm{Exp}_{\pi_{z(\cdot)}}\left(e^{tA}\right) e_B(f) = e_B(e^{tA} f) \exp\left(-\int_{\mathbb{R}} (e^{tA} f(x) - f(x)) z(x) dx\right). \tag{25}$$

The operator L, differentiating (25), can be expressed in terms of creation and annihilation operators, namely,

$$L = d\mathrm{Exp}_{\pi_{z(\cdot)}}(A) + a^-_{\pi_{z(\cdot)}}(A^* z(\cdot)), \tag{26}$$

where $d\mathrm{Exp}_{\pi_{z(\cdot)}}(\ A)$ and $a^-_{\pi_{z(\cdot)}}$ are, respectively, the image of $d\mathrm{Exp}(A)$ and a^- under $I_{\pi_z(\cdot)}$, and hence

$$a^-_{\pi_{z(\cdot)}}(h) F(\gamma) = \int dx z(x) \left(F(\gamma \cup x) - F(\gamma)\right). \tag{27}$$

For more details see e.g. [28].

Finally, we would like to present the "annihilation and creation operators" in the form more common in physical literature. For each $x \in \mathbb{R}^d$ we define an operator $a^-(x)$ acting on $\boldsymbol{f} = (f^{(n)})_n$ by

$$(a^-(x)\boldsymbol{f})^{(n)}(y_1, \ldots, y_n) = \sqrt{n+1} f^{(n+1)}(x, y_1, \ldots, y_n).$$

The adjoint of the operator $a^-(x)$ is formally given by

$$(a^+(x)\boldsymbol{f})^{(n)}(y_1, \ldots, y_n) = \frac{1}{\sqrt{n}} \sum_{k=1}^{n} \delta(x - y_k) f^{(n-1)}(y_1, \ldots, \hat{y}_k, \ldots, y_n),$$

where $\hat{y}_k$ means that the k-th coordinate is excluded. Actually, the expression for $a^+(x)$ is well-defined as a quadratic form. It is easy to check the relations $a^{\pm}(f) = \int_{\mathbb{R}^d} a^{\pm}(x) f(x) dx$ in the sense of quadratic forms. Rewritten in a style more similar to the common usage in physics, the right hand side of (26) takes the form

$$\int dx\, z(x) \int dy\, \left(a(x-y) - a^{(0)}\delta(x-y)\right) \left(a^+_\pi(x) a^-_\pi(y) - a^-_\pi(y)\right),$$

where $a^{\pm}_\pi(x)$ are the image of $a^{\pm}(x)$ under $I_{\pi_z(\cdot)}$ for more details see [28].

4.2 Second Quantization Operator Approach

4.2.1 The Symmetric Case

If a is an even function, then the generator L is symmetric (see Lemma 1) and one can straightforwardly give an alternative construction to the one presented in Sect. 3 for the infinite particle dynamics. As a matter of fact, for a an even function, and for each constant $z > 0$, the operator L defined in (1) gives rise to a Dirichlet form on $L^2(\Gamma, \pi_z)$,

$$\int_\Gamma \pi_z(d\gamma)(LF)(\gamma)F(\gamma) = -\frac{1}{2}\int_\Gamma \pi_z(d\gamma)\sum_{x\in\gamma}\int_{\mathbb{R}^d} dy\, a(x-y)\,|F(\gamma\backslash x\cup y)-F(\gamma)|^2.$$

This allows the use of Dirichlet forms techniques to derive a Markov process on Γ with *cadlag* paths and having π_z as an invariant measure [31]. Actually, one can show that in this situation L is the second quantization operator corresponding to the non-positive self-adjoint operator A on $L^2(\mathbb{R}^d, zdx)$ (Remark 4). Hence, L is a non-positive essentially self-adjoint operator on $L^2(\Gamma, \pi_z)$, and it is the generator of a contraction semigroup on $L^2(\Gamma, \pi_z)$.

4.2.2 The Asymmetric Case

According to (15) and (7) the action of the transition kernel on coherent states corresponding π_z is given by

$$\mathbf{E}_\gamma\left[e_{\pi_z}(\varphi, \mathbf{X}_t)\right] = e_{\pi_z}(e^{tA}\varphi, \gamma)\exp\left(\int_{\mathbb{R}^d} dx\, \left(e^{tA}\varphi(x) - \varphi(x)\right)z\right). \tag{28}$$

This allows to express the action of the transition probability on coherent states in terms of annihilation and creation operators

$$\mathrm{Exp}_{\pi_z}(e^{tA})e^{a^-_{\pi_z}(e^{tA^*}z-z)}. \tag{29}$$

Observe that the right-hand side of (28) gives the action of a semigroup which preserves coherent states. Lemma 4 shows that L is the generator of a strongly continuous Markov semigroup.

Lemma 4 *Let $\mathcal{D}_{\mathrm{coh}}$ be the vector space spanned by all functions $e_B(\varphi)$ with $\varphi \in L^1(\mathbb{R}^d, zdx) \cap L^2(\mathbb{R}^d, zdx)$. The operator L restricted to $\mathcal{D}_{\mathrm{coh}}$ is closable in $L^2(\Gamma, \pi_z)$ and its closure is an extension of the operator $(L, \mathcal{F}L^0_{\mathrm{eb}}(\Gamma_a))$ defined in Sect. 2. Moreover, it is the generator of a Markov semigroup and when A is symmetric then $e^{tL} = \mathrm{Exp}_{\pi_z}(e^{tA})$.*

The proof is an adaptation for a non-symmetric generator of the proof of Proposition 4.1 in [1] and see also [30] for details. Technically, in our case the proof is simpler, because A is a bounded operator and we consider functions of the type $e_B(\varphi)$.

5 Local Equilibrium Dynamics

The considerations in Sect. 4 were essentially for Poisson measures π_z which have as activity parameter a constant function z. According to Lemma 1 and Sect. 2, such measures are reversible measures for the free Kawasaki process. As a first step towards other initial distributions, in this section we consider the so-called local equilibrium case, that is, Poisson measures with a non-constant activity parameter. The choice of the name "local equilibrium" is motivated by the following situation: Assume the activity z is varying on a scale much larger than the scale of the range of the jump kernel a. One considers the dynamics in a part of the volume with an extension much larger than the range of a but much smaller than the scale of variation of z. Then the Poisson measure effectively has a constant activity and thus is effectively a reversible measure for the dynamics. Thus, in this situation, the system is in local equilibrium. However, in this section, we do not consider an extra length scale of the variation of z, but it will appear in Sect. 5.2.

Remark 6 Although for bounded non-constant functions z the process starting in π_z can be constructed explicitly, cf. Sect. 3.1, one cannot expect that the expression given in (19) can be extended to a semigroup either in the $L^1(\Gamma_a, \pi_z)$ or in the $L^2(\Gamma_a, \pi_z)$ sense. The existence of the semigroup in an L^p-sense can only be expected w.r.t. an invariant measure. Indeed, this can be seen from the following equality

$$\frac{\|e^{tL}e_B(f)\|_{L^p(\pi_z)}}{\|e_B(f)\|_{L^p(\pi_z)}} = \exp\left(\frac{1}{p}\int_{\mathbb{R}^d}\Big(|1+e^{tA}f(x)|^p - |1+f(x)|^p\Big)z(x)dx\right),$$

and the fact that the operator norm of e^{tL} from $L^p(\pi_z)$ to itself admits the bound

$$\|e^{tL}\|_{\text{Op}} \geq \sup_{f\in\mathcal{D}(\mathbb{R}^d),\, f\geq 0,} \exp\left(\frac{1}{p}\int_{\mathbb{R}^d}\Big((1+e^{tA}f(x))^p - (1+f(x))^p\Big)z(x)dx\right),$$

where the right-hand side is infinite if $(e^{tA})_{t\geq 0}$ is not a contraction semigroup (see Remark 5).

Let $z \geq 0$ be a bounded measurable function. According to (20), for the one-dimensional distribution of the free Kawasaki process $(\mathbf{X}_t)_{t\geq 0}$ with initial distribution π_z one finds

$$\int_\Theta e_B(\varphi, \gamma) P^{\mathbf{X}}_{\pi_z,t}(d\gamma) = \exp\left(\int_{\mathbb{R}^d} dx\; e^{tA}\varphi(x)z(x)\right), \tag{30}$$

for every $\varphi \in \mathscr{D}(\mathbb{R}^d)$ and every $t \geq 0$.

For each $t \geq 0$ fixed, let us now consider the linear functional defined on $L^1(\mathbb{R}^d, dx)$ by

$$L^1(\mathbb{R}^d, dx) \ni f \mapsto \int_{\mathbb{R}^d} dx\, (e^{tA} f)(x)z(x).$$

Due to the contractivity property of the semigroup e^{tA} in $L^1(\mathbb{R}^d, dx)$, see Proposition 1, and to the boundedness of z, this functional is bounded on $L^1(\mathbb{R}^d, dx)$, and thus it is defined by a kernel $z_t \in L^\infty(\mathbb{R}^d, dx)$, that is,

$$\int_{\mathbb{R}^d} dx\, e^{tA} f(x)z(x) = \int_{\mathbb{R}^d} dx\, f(x)z_t(x), \tag{31}$$

for all $f \in L^1(\mathbb{R}^d, dx)$. Moreover, since e^{tA} is positivity preserving in $L^1(\mathbb{R}^d, dx)$, it follows from (31) that $z_t \geq 0$.

This shows by (30) that the one-dimensional distribution $P^{\mathbf{X}}_{\pi_z,t}$ is just the Poisson measure π_{z_t} (see also [7, 9]).

Furthermore, the path-space measure $\mathbf{P}^{\pi_z}$ corresponding to the process is also Poissonian. It is a Poisson measure on $\Gamma_{D([0,\infty),\mathbb{R}^d)}$ with intensity P^z, where P^z is the path-space measure on $D([0, \infty), \mathbb{R}^d)$ of the one-particle jump process corresponding to A with initial distribution $z(x)dx$. Using the fact that the path-space measure $\mathbf{P}_\gamma$ is supported on $D([0, \infty), \Theta)$ one easily sees that this implies that $\mathbf{P}^{\pi_z}$ is actually supported on a subset of $\Gamma_{D([0,\infty),\mathbb{R}^d)}$ which can be naturally identified with $D([0, \infty), \Theta)$.

Lemma 5 *The path-space measure $\mathbf{P}^{\pi_z}$ of the process $\mathbf{X}$ starting with initial distribution π_z is the Poisson measure π_{P^z} on $\Gamma_{D([0,\infty),\mathbb{R}^d)}$. In particular, for any continuous function φ on $[0, \infty) \times \mathbb{R}^d$ with compact support we have*

$$\int_{\Gamma_{D([0,\infty),\mathbb{R}^d)}} \mathbf{P}^{\pi_z}(d\omega) e^{-\int_0^\infty \langle \varphi(t,\cdot),\omega(t)\rangle dt} = \exp\left(\int_{D([0,\infty),\mathbb{R}^d)} \left[e^{-\int_0^\infty \varphi(t,\omega(t))dt} - 1 \right] P^z(d\omega) \right).$$

5.1 Long-time Asymptotics

In this subsection we want to study the behavior of the one-dimensional distribution of the infinite particle dynamics for large times. As mentioned above (31), the free Kawasaki dynamic leaves the Poissonian structure of the initial distribution unchanged and only the underlying intensity develops with time. Thus, the analysis of the long-time asymptotics behavior reduces to the long-time asymptotics of the intensity, see Lemma 6 for details. In other words, we reduced the problem to a pseudo-one-particle problem where the intensity plays the role of the initial distribution.

Definition 1 One says that a function $z \in L^1_{\mathrm{loc}}(\mathbb{R}^d, dx)$ has arithmetic mean, denoted by $\mathrm{mean}(z)$, whenever the following limit exists

$$\lim_{R \to +\infty} \frac{1}{\mathrm{vol}(B(R))} \int_{B(R)} dx\, z(x). \tag{32}$$

The following proposition gives a condition on the activity under which the long-time asymptotics of the one time distribution of the infinite particle dynamics exists, it is still Poissonian with constant intensity; the constant being the arithmetic mean of the initial intensity.

Proposition 3 *Let $z \geq 0$ be a bounded measurable function which Fourier transform is a signed measure. Then z has arithmetic mean and the one-dimensional distribution $P^{\mathbf{X}}_{\pi_z,t}$ converges weakly to $\pi_{\mathrm{mean}(z)}$ when t goes to infinity.*

Under the assumptions of Proposition 3, the activity z has arithmetic mean, cf. Corollary 1. Then, using Fourier transform, one can easily derive the long-time asymptotics for the pseudo-one-particle dynamics, cf. Lemma 7. By Lemma 6 this implies the existence of the long-time asymptotics for the infinite particle dynamics and identifies the limit. Using the same technique one can show that the long-time asymptotics for the infinite particle system only depends on the space asymptotics of the initial intensity z, cf. Corollary 2.

In Proposition 3, the assumption concerning the Fourier transform is actually not very transparent, because we cannot reasonably restate it in terms of z in the position variables. Therefore, we give several examples for illustration, cf. Example 1, and we derive certain properties of the arithmetic mean. Some classes of activities with reasonable asymptotic behavior do not fulfill the aforementioned Fourier transform assumption, e.g. Example 1(v). Note that also not every bounded function has an arithmetic mean, cf. Remark 8.

Let us discuss the properties of the arithmetic mean in some more details before we prove the lemmas mentioned above.

Remark 7 If two functions $z_1, z_2 \in L^1_{\mathrm{loc}}(\mathbb{R}^d, dx)$ have arithmetic mean, then for every $\alpha_1, \alpha_2 \in \mathbb{R}$ the function $\alpha_1 z_1 + \alpha_2 z_2$ also has arithmetic mean and $\mathrm{mean}(\alpha_1 z_1 + \alpha_2 z_2) = \alpha_1 \mathrm{mean}(z_1) + \alpha_2 \mathrm{mean}(z_2)$.

Example 1 To be more concrete we give some examples:

(i) If z is a constant function then $\mathrm{mean}(z) = z$.
(ii) If z decays to zero, i.e., for every $\varepsilon > 0$ there exists an $R > 0$ such that $|z(x)| \leq \varepsilon$ for $x \notin B(R)$, then $\mathrm{mean}(z) = 0$.
(iii) If $z \in L^p(\mathbb{R}^d, dx)$, $p \in [1, \infty)$, then $\mathrm{mean}(z) = 0$.
(iv) Also trigonometric functions have no influence, e.g., for $d = 1$ and $z(x) = 1 + \varepsilon \sin(x)$ we have $\mathrm{mean}(z) = 1$.
(v) A less trivial example is the following one. Given $z_0, z_1 \geq 0$, let z be the function defined at each $x = (x_1, \ldots, x_d) \in \mathbb{R}^d$ by

$$z(x) = \begin{cases} z_1 \text{ if } x_1 \geq 0 \\ z_0 \text{ otherwise} \end{cases}.$$

In this case one finds $\mathrm{mean}(z) = \frac{z_0+z_1}{2}$. Note that in this case $\hat{z}$ is not a signed measure.

Remark 8 The arithmetic mean does not exist for all bounded non-negative functions.

1. For $d = 1$ and

$$z(x) = \begin{cases} 1 \text{ if } 2^{2k} \leq |x| \leq 2^{2k+1} \text{ for a } k \in \mathbb{N}_0 \\ 0 \text{ otherwise} \end{cases},$$

the integral in (32) oscillates between $1/3$ and $2/3$. Thus z does not have arithmetic mean.

2. Another example is given by the following slowly oscillating function

$$z(x) = \cos(\ln(1+|x|)) + z_0, \quad x \in \mathbb{R}^d,$$

where z_0 is a constant greater or equal to 1. Then for large R it holds that

$$\frac{1}{\mathrm{vol}(B(R))} \int_{B(R)} z(x)\, dx \sim \frac{d}{\sqrt{1+d^2}} \sin\Big(\ln(R+1) + \arctan(d) \Big) + z_0.$$

In general slowly varying functions will show in general spurious behavior.

First, we prove that one may reduce the proof of Proposition 3 to a pseudo-one-particle system. This follows from the independent movement of the particles discussed in Sect. 3.

Lemma 6 *Let* $0 \leq z \in L^1_{\mathrm{loc}}(\mathbb{R}^d, dx)$ *be such that*

$$\lim_{t\to+\infty} \int_{\mathbb{R}^d} dx\, e^{tA}\varphi(x)\, z(x) =: \int_{\mathbb{R}^d} z_\infty(dx)\varphi(x)$$

exists for all $\varphi \in \mathcal{D}(\mathbb{R}^d)$. *If* z_∞ *is a (non-negative) Radon measure on* $\mathbb{R}^d$, *then the one-dimensional distribution* $P^{\mathbf{X}}_{\pi_z,t}$ *converges weakly to* π_{z_∞} *when* t *tends to infinity.*

It remains to derive the long-time asymptotics for the pseudo-one-particle system.

Lemma 7 *Let* $z \geq 0$ *be a bounded measurable function such that its Fourier transform* $\hat{z}$ *is a signed measure. For all* $\varphi \in \mathcal{S}(\mathbb{R}^d)$ *one has*

$$\lim_{t\to+\infty} \int_{\mathbb{R}^d} dx\, e^{tA}\varphi(x)\, z(x) = (2\pi)^{-d/2}\, \hat{z}(\{0\}) \int_{\mathbb{R}^d} dx\, \varphi(x). \tag{33}$$

The next lemma shows that the existence of the arithmetic mean is stable under L^1-convergence. In particular, it yields that the condition assumed in Proposition 3 on the Fourier transform of the intensity is sufficient to ensure the existence of the arithmetic mean.

Lemma 8 *Let $z \geq 0$ be a bounded measurable function. Assume that there exists a total subset of $L^1(\mathbb{R}^d, dx)$ and a $C > 0$ such that for all φ in that total subset we have*

$$\lim_{R\to\infty} R^{-d} \int_{\mathbb{R}^d} dx\, \varphi(x/R)z(x) = C \int_{\mathbb{R}^d} \varphi(x)dx. \tag{34}$$

Then z has an arithmetic mean. In addition, equality (34) holds for $C = \mathrm{mean}(z)$ and for any $\varphi \in L^1(\mathbb{R}^d, dx)$.

Corollary 1 *Given a bounded measurable function $z \geq 0$ the following two results hold:*

1. *If z has arithmetic mean, then the limit in (34) exists for all $\varphi \in L^1(\mathbb{R}^d, dx)$ and $C = \mathrm{mean}(z)$.*
2. *If the Fourier transform of z is a signed measure, then z has arithmetic mean and*

$$\mathrm{mean}(z) = (2\pi)^{-d/2}\, \hat{z}(\{0\}).$$

Remark 9 Let us underline the difference between $\hat{z}(\{0\})$ and the evaluation of $\hat{z}$ as a function at zero. We demonstrate this difference for the case when z is a bounded L^1-function. The Fourier transform of z is then a continuous function which we denote for the moment by $\tilde{z}$. Evaluation at zero makes sense in this case. However, we want to consider the Fourier transform as a generalized function, i.e., as a linear form on a function space. We assume that this linear form is regular enough to be expressed by a signed measure. If, as in our case, z is an L^1-function, then its Fourier transform is the measure $\hat{z}(dk) = \tilde{z}(k)dk$. Thus $\hat{z}(\{0\}) = 0$, which is totally different from $\tilde{z}(0) = \int_{\mathbb{R}^d} dx z(x)$.

Below we list the Fourier transforms (interpreted as generalized functions, signed measures, respectively) of the examples given in Example 1, provided they exist and with the same enumeration:

(i) $(2\pi)^{d/2} z\delta_0(dk)$.
(ii) a continuous function for $p = 1$ or an $L^{p/(p-1)}$-function for $1 < p \leq 2$ multiplied in both cases by the Lebesgue measure.
(iii) $(2\pi)^{d/2}\,(\delta_0(dk) + i\varepsilon/2\delta_1(dk) - i\varepsilon/2\delta_{-1}(dk))$.
(iv) In the one-dimensional case the Fourier transform is the following generalized function $\hat{z}(k) = \sqrt{2\pi}(z_0 + z_1)/2\delta_0(k) + i(z_0 - z_1)/\sqrt{2\pi}\,\mathscr{P}(1/k)$, where $\mathscr{P}(1/k)$ denotes the Cauchy principal value of $1/k$. Using this explicit formula the conclusion of Proposition 3 can be shown although the assumptions of Proposition 3 are not fulfilled.

Applying the same technique as in Lemma 7 we can prove that the time asymptotics depends only on the behavior of z at infinity.

Corollary 2 *Let $z_1, z_2 \geq 0$ be two bounded measurable functions. If $z_1 - z_2 \in L^p(\mathbb{R}^d, dx)$ for some $1 \leq p \leq 2$, then the free Kawasaki dynamics with initial distribution π_{z_1} and the free Kawasaki dynamics with initial distribution π_{z_2} have the same long-time asymptotic limit.*

Throughout this subsection, the study of the time asymptotic behavior of the free Kawasaki process with an initial distribution π_z was based on the analysis of the Laplace transform

$$\int_\Gamma \pi_z(d\gamma)\,\mathbf{E}_\gamma[e^{\langle\varphi,\mathbf{X}_t\rangle}], \tag{35}$$

cf. Lemma 6. Another way to interpret this is as follows: we have studied the time asymptotic behavior of the so-called empirical field corresponding to a $\varphi \in \mathscr{D}(\mathbb{R}^d)$,

$$n_t(\varphi, \mathbf{X}) := \langle\varphi, \mathbf{X}_t\rangle = \sum_{x\in\mathbf{X}_t} \varphi(x)$$

and showed that the limit $t \to \infty$ exists and identified the limiting distribution. The special role of the empirical fields is essential in the next subsection.

5.2 Hydrodynamic Limits

In the sequel let $z \geq 0$ be a bounded measurable function. In order to obtain a macroscopic description of our system, we rescale simultaneously the empirical field $n_t(\varphi, \mathbf{X}) = \langle\varphi, \mathbf{X}_t\rangle$, $\varphi \in \mathscr{D}(\mathbb{R}^d)$, in space and in time. The scale transformation in space is given by $\langle\varphi, \gamma\rangle \to \varepsilon^d\langle\varphi(\varepsilon\cdot), \gamma\rangle$, and in time by $t \to \varepsilon^{-\kappa}t$ for some $\kappa > 0$. To obtain non-trivial macroscopic density profiles, one has to scale the initial intensity as well, $z \to z(\varepsilon\cdot)$. This scaling yields a scaling of the Laplace transform of the empirical field, in other words, the Laplace transform of the one-dimensional distribution of the scaled process. For each $t \geq 0$ and each $\varphi \in \mathscr{D}(\mathbb{R}^d)$ we obtain from (15) the following form

$$\begin{aligned}
&\int_\Gamma \pi_{z(\varepsilon\cdot)}(d\gamma)\,\mathbf{E}_\gamma\left[e^{\varepsilon^d\langle\varphi(\varepsilon\cdot),\mathbf{X}_{\varepsilon^{-\kappa}t}\rangle}\right]\\
&= \int_\Gamma \pi_{z(\varepsilon\cdot)}(d\gamma)\,e_B\left(e^{\varepsilon^{-\kappa}tA}\left(e^{\varepsilon^d\varphi(\varepsilon\cdot)}-1\right),\gamma\right)\\
&= \exp\left(\int_{\mathbb{R}^d} dx\,\left(e^{\varepsilon^{-\kappa}tA}\left(e^{\varepsilon^d\varphi(\varepsilon\cdot)}-1\right)\right)(x)z(\varepsilon x)\right).
\end{aligned} \tag{36}$$

In the sequel we denote the scaled empirical field by

$$n_t^{(\varepsilon)}(\varphi, \mathbf{X}) := n_{\varepsilon^{-\kappa}t}(\varepsilon^d \varphi(\varepsilon\cdot), \mathbf{X}) = \varepsilon^d \langle \varphi(\varepsilon\cdot), \mathbf{X}_{\varepsilon^{-\kappa}t} \rangle . \tag{37}$$

According to the independent movement of the particles, we are again able to reduce the study of the infinite particle system to an effective pseudo-one-particle system. Again technical difficulties arise from the fact that the scale of the system size is much larger than the scale of space and time considered in the empirical field. Actually, the system size is infinite. Under the additional assumption that the Fourier transform of the activity z is a signed measure, the hydrodynamic limit can be derived rather directly using Fourier techniques, cf. Propositions 4 and 5. The general case of just bounded activities requires more technical involved considerations (postponed to Proposition 6).

Proposition 4 *Let $z \geq 0$ be a bounded measurable function such that its Fourier transform is a signed measure. For each $t \geq 0$ the following limit exists for all $\varphi \in \mathcal{D}(\mathbb{R}^d)$*

$$\lim_{\varepsilon \to 0^+} \int_\Theta \pi_{z(\varepsilon\cdot)}(d\gamma)\, \mathbf{E}_\gamma \left[e^{n_t^{(\varepsilon)}(\varphi, \mathbf{X})} \right] = \int_{\mathcal{D}'(\mathbb{R}^d)} \delta_{\rho_t}(d\omega) e^{\langle \varphi, \omega \rangle} \tag{38}$$

whenever one of the following conditions is fulfilled:

1. *If*

$$a_i^{(1)} := \int_{\mathbb{R}^d} dx\, x_i a(x) < \infty, \quad \forall i = 1, \ldots, d,$$

and $a^{(1)} := (a_1^{(1)}, \ldots, a_d^{(1)}) \neq 0$, then for $\kappa = 1$ the limiting density ρ_t is given by

$$\int_{\mathbb{R}^d} dx\, \rho_t(x)\varphi(x) := \int_{\mathbb{R}^d} dx\, z(x + ta^{(1)})\varphi(x), \quad \varphi \in \mathcal{D}(\mathbb{R}^d);$$

2. *If $a^{(1)} = 0$, and*

$$a_{ij}^{(2)} := \int_{\mathbb{R}^d} dx\, x_i x_j a(x) < \infty, \quad \forall i, j = 1, \ldots, d,$$

then for $\kappa = 2$ the limiting density ρ_t is given, for any $\varphi \in \mathcal{D}(\mathbb{R}^d)$, by

$$\int_{\mathbb{R}^d} dx\, \rho_t(x)\varphi(x) := \frac{1}{(2\pi)^{d/2}} \int_{\mathbb{R}^d} dx\, z(x) \int_{\mathbb{R}^d} dk\, e^{i\langle k, x\rangle} e^{-\frac{t}{2}\langle a^{(2)}k, k\rangle} \hat{\varphi}(k),$$

where $a^{(2)}$ denotes the $d \times d$ matrix with coefficients $a_{ij}^{(2)}$.

Remark 10 The limiting density $\rho_t(x) = z(x + ta^{(1)})$ obtained in Proposition 4 is the solution of the linear partial differential equation $\frac{\partial}{\partial t}\rho_t(x) = \langle a^{(1)}, \nabla \rho_t(x) \rangle =$

$\mathrm{div}(a^{(1)}\rho_t(x))$ with the initial condition $\rho_0 = z$. In the same way, the second case stated in Proposition 4 yields a limiting density which is the solution of the heat equation

$$\frac{\partial}{\partial t}\rho_t(x) = \frac{1}{2}\sum_{i,j=1}^{d} a_{ij}^{(2)} \frac{\partial^2}{\partial x_i \partial x_j}\rho_t(x)$$

with the same initial condition.

Given the function a, we may decompose a into a sum of an even function p and an odd function q, $a = p + q$. Note that one always has $p^{(1)} = 0$. Submitting also the function a itself to a proper scale transformation (beyond the previous scale transformations in space and in time), one may complement the statement of Proposition 4. In the proof of Proposition 4 one has to control $\lim_{\varepsilon\to 0^+} \frac{\hat{a}(-\varepsilon k)-\hat{a}(0)}{\varepsilon^\kappa}$ which suggests the scaling $a_\varepsilon := p + \varepsilon q$ and $\kappa = 2$. Note that the assumptions on the function a (i.e., $0 \le a \in L^1(\mathbb{R}^d, dx)$), carry over to a_ε, $\varepsilon > 0$. Hence all considerations done in Sect. 2 and following sections for the one-particle operator A and its underlying dynamics still hold for the operator A_ε defined by (5) with a replaced by a_ε. Denote by

$$a^{(1)} := \int_{\mathbb{R}^d} dx\, x a(x) = \int_{\mathbb{R}^d} dx\, x q(x) = \varepsilon^{-1} \int_{\mathbb{R}^d} dx\, x a_\varepsilon(x).$$

Proposition 5 ("weak asymmetry") *Let $z \ge 0$ be a bounded measurable function such that its Fourier transform is a signed measure. Under the above conditions, if $0 \ne q^{(1)} = a^{(1)} \in \mathbb{R}^d$ and $p_{ij}^{(2)} < \infty$ for every $i, j = 1, \ldots, d$, then for each $t \ge 0$ and each $\varphi \in \mathscr{D}(\mathbb{R}^d)$ the following limit exists, and it is given by*

$$\lim_{\varepsilon\to 0^+} \int_\Gamma \pi_{z(\varepsilon\cdot)}(d\gamma)\, \mathbf{E}_\gamma[e^{n_t^{(\varepsilon)}(\varphi,\mathbf{X})}]$$
$$= \exp\left(\frac{1}{(2\pi)^{d/2}} \int_{\mathbb{R}^d} dx\, z(x) \int_{\mathbb{R}^d} dk\, e^{i\langle k,x\rangle} e^{-it\langle a^{(1)},k\rangle - \frac{t}{2}\langle a^{(2)}k,k\rangle} \hat{\varphi}(k)\right).$$

Remark 11 Similarly to Remark 10, one may then conclude that for continuous differentiable z Proposition 5 leads to a limiting density ρ_t which is a solution of the partial differential equation

$$\frac{\partial}{\partial t}\rho_t(x) = \mathrm{div}(a^{(1)}\rho_t(x)) + \frac{1}{2}\sum_{i,j=1}^{d} a_{ij}^{(2)} \frac{\partial^2}{\partial x_i \partial x_j}\rho_t(x)$$

with the initial condition $\rho_0 = z$.

Proposition 6 *Assume that a has all moments finite. Then the results stated in Propositions 4 and 5 hold for all non-negative bounded measurable functions z.*

5.3 Finite Volume System

The situation is much easier when one starts with a particle system on a compact manifold. Let us explain this in the case, when we consider particle configurations on $[-L/2, L/2]^d$ with periodic boundary conditions, that is we consider the system on the torus. Let us start by giving a description of the system in global coordinates coming from the covering space of the torus. For any sufficiently decaying function φ on $\mathbb{R}^d$, denote by $\varphi_L(x) := \sum_{j\in\mathbb{Z}^d} \varphi(jL)$. Any function on the torus can be written in this way. Consider the particle system $\Gamma_L := \{\gamma \subset [-L/2, L/2]^d : |\gamma| < \infty\}$ with the jump rate given by a_L. The probabilistic construction is the finite product and hence automatically a configuration at least if one allows coinciding points. The particle dynamics extend to this case and can also be reduced in the same way to a pseudo-one-particle dynamics. Denote by A_L the generator of the latter. An easy calculation show that $A_L\varphi_L = (A\varphi)_L$ and hence $e^{tA_L}\varphi_L = (e^{tA}\varphi)_L$. Therefore, the semi-group of the pseudo-one-particle on the torus can be described in the following way

$$\int_{[-L/2,L/2]^d} \left(e^{tA_L}\varphi_L\right)(x)\, z(x)dx = \int_{\mathbb{R}^d} (e^{tA}\varphi)(x)\, P_L z(x)dx, \tag{39}$$

where

$$P_L z(x) := \sum_{j\in\mathbb{Z}} 1\!\!1_{[-L/2,L/2]^d}(x - jL)z(x - jL) \tag{40}$$

is the extension of z to a $(L\mathbb{Z})^d$ periodic function. Note that if φ has compact support then $\varphi_L = \varphi$ for L large enough. In order to connect with our previous consideration we take the Fourier transform and obtain the expression

$$(\sqrt{2\pi}L)^{-d} \sum_{j\in\mathbb{Z}^d} e^{t(2\pi)^{d/2}(\hat{a}(-j/L)-\hat{a}(0))}\hat{\varphi}(-j/L)\widehat{1\!\!1_{[-L/2,L/2]^d}} * \hat{z}(j/L). \tag{41}$$

5.3.1 Long-time Asymptotics

The behaviour at large time for general z can be easily established in the finite volume. As $\widehat{1\!\!1_{[-L/2,L/2]^d}} * \hat{z}$ is a continuous function one sees directly that for fixed L, the above converges to

$$(\sqrt{2\pi}L)^{-d}\widehat{1\!\!1_{[-L/2,L/2]^d}} * \hat{z}(0) = \frac{1}{L^d}\int_{[-L/2,L/2]^d} z(x)dx$$

with an exponential rate. When it converges the latter in turn converges to $\mathrm{mean}(z)$ for $L \to \infty$. We only used that $a \in L^1$ and that $z \in L^\infty$ is such that the arithmetic mean exists in contrast to the extra regularity required in Lemma 7, namely $\hat{z}$ is a signed measure.

In other words, we showed that the convergence is equivalent to the existence of the arithmetic mean without any regularity assumption of z except being bounded. Whereas, when one first sends L to infinity then (39) converges to (31) and then considering the time asymptotics afterwards can become much more involved. In the finite volume case one has a spectral gap proportional to L^{-2}, which will disappear in the $L \to \infty$. Let us first consider what happens when one sends t and L simultaneously to infinity but t is sufficiently larger than L. We need first a result to control the spectral behaviour of the semi-group near $k = 0$.

Lemma 9 *Assume that a has finite second moments and the matrix* $(a^{(2)}_{i,j})_{i,j=1,\dots,d}$ *is positive definite in the sense that there exists a constant* $\sigma' > 0$ *such that for all* $k \in \mathbb{R}^d$ *holds that* $\sum_{i,j=1}^d k_i k_j a^{(2)}_{i,j} \geq (\sigma')^{-2}|k|^2$. *Then for each* $\sigma > \sigma'$ *there exists a constant* $c > 0$ *such that*

$$\left| e^{t(2\pi)^{d/2}(\hat{a}(-k)-\hat{a}(0))} \right| \leq \exp\left(-t\frac{\min\{|k|^2, c\}}{2\sigma^2}\right). \tag{42}$$

Let us use this lemma and the bound

$$\left|(\sqrt{2\pi}L)^{-d}\, \widehat{1\!\!1_{[-L/2,L/2]^d}} * \hat{z}(j/L)\right| \leq \frac{1}{L^d}\int_{[-L/2,L/2]^d} z(x)dx \leq \|z\|_\infty \tag{43}$$

in order to estimate

$$\begin{aligned}
&(\sqrt{2\pi}L)^{-d} \sum_{j\in\mathbb{Z}^d\,:\,j\neq 0} e^{t(2\pi)^{d/2}\mathrm{Re}(\hat{a}(-j/L)-\hat{a}(0))}|\hat{\varphi}(-j/L)|\,\widehat{1\!\!1_{[-L/2,L/2]^d}} * \hat{z}(j/L)\\
&\leq \|z\|_\infty \sum_{j\in\mathbb{Z}^d\,:\,j\neq 0} e^{t(2\pi)^{d/2}\mathrm{Re}(\hat{a}(-j/L)-\hat{a}(0))}|\hat{\varphi}(-j/L)|\\
&\leq C\|z\|_\infty L^d e^{-t\frac{\min\{L^{-2},c\}}{2\sigma}} \to 0,
\end{aligned}$$

for some constant C. The latter converges to zero whenever $L^2 \ln L/t \to 0$.

One may think that one can use Lemma 9 to show that the time asymptotics of the semi-group $e^{tA}\varphi$ is the same as the one of the heat semi-group also in the infinite particle system, that is $L = \infty$. However, even when one ignores that Lemma 9 gives different exponential rates, one sees that the difference between the kernels can at best be bounded near $k = 0$ by a term of the form $|k|t^M e^{-ct|k|^2}$, where M is the order of the derivative considered. In the limit $t \to \infty$ this can be uniformly bounded by a term $|k|^{-2M+1}$, which becomes worse when one uses higher derivatives.

5.3.2 Hydrodynamic Limit

In finite volume also the study of the hydrodynamic limit is rather easy. In Fourier coordinates the exponent in (36) takes the form

$$(\sqrt{2\pi}L)^{-d}\sum_{j\in\mathbb{Z}^d} e^{t(2\pi)^{d/2}\varepsilon^{-2}(\hat{a}_\varepsilon(-\varepsilon j/L)-\hat{a}(0))}\hat{\varphi}(-j/L)\widehat{1\!\!1_{[-L/2,L/2]^d}}*\hat{z}(j/L). \quad (44)$$

The only ε dependence is in the jump kernel. As a is integrable, $\hat{a}_\varepsilon(-\varepsilon x) - \hat{a}(0)$ converges locally uniformly and without loss of generality we can assume that $\hat{\varphi}$ has compact support, the hydrodynamic limit is established immediately without any further assumptions on z. Again because of the spectral gap we only need that a has second moment and no further assumptions on z beside boundedness in contrast to Propositions 4 and 5 for the infinite volume system.

In Proposition 6 we were able to establish the hydrodynamic limit in the infinite volume system for a general z, but only under the assumption that a has all higher moments finite with increasing requirements for larger dimension d of the one-particle system.

However, one can establish the hydrodynamic limit with only little stronger requirements on a, namely that the third moment is finite for example (any power larger than 2 should be sufficient as well), if L does not grow too much compared to ε^-1. Indeed, we have to improve the bound a bit in the following way. As all derivatives of $t(2\pi)^{d/2}\varepsilon^{-2}(\hat{a}_\varepsilon(-\varepsilon k) - \hat{a}(0))$ in k are polynomially bounded in k and uniformly bounded in ε, one can use the mean value theorem to control the difference between the Fourier transform of the transition group and the limiting heat kernel, that is, to bound

$$\left| e^{\varepsilon^{-\kappa}t(2\pi)^{d/2}(\hat{a}_\varepsilon(-\varepsilon k)-\hat{a}(0))} - e^{it\langle a^{(1)},k\rangle - \frac{0^{2-\kappa}}{2}t\langle a^{(2)}k,k\rangle}\right| \leq C(1+|k|^2)^{1/2}\varepsilon. \quad (45)$$

Using that the term $\widehat{1\!\!1_{[-L/2,L/2]^d}} * \hat{z}(j/\varepsilon L)$ can be bounded by $\|z\|_\infty L^d$ and assuming that $\varepsilon L^d \to 0$ we can replace the exponent in (44) by $it\langle a^{(1)},k\rangle - \frac{0^{2-\kappa}}{2}t\langle a^{(2)}k,k\rangle$. Taking the inverse Fourier transform we obtain

$$\int_{\mathbb{R}^d} e^{-t\langle a^{(1)},\nabla\rangle - t\langle\nabla,a^{(2)}\nabla\rangle/2}\varphi(x)P_L z(x)dx, \quad (46)$$

which converges for example for bounded z and $\varphi \in L^1$ to the same expression with $P_L z$ replaced by z.

In both cases, long-time asymptotics and hydrodynamic limit, the spectral gap coming from the finite volume is simplifying the situation a lot. The spectral gap being of order L^{-2} is too small to describe the physical relevant decay of the system. However, the infinite volume variant of the two aforementioned cases show that even for $L = \infty$ one has the convergence result. The mechanism it is based on, must be very different, as the finite volume results are based on a spectral gap that disapears for $L = \infty$.

6 Non-equilibrium Dynamics

In this section we widen the class of initial distributions to measures far from equilibrium, that is we consider all probability measures μ on Θ as initial distributions subject only to a mild mixing condition. This means that we consider the processes constructed as in Sect. 3 but not necessarily with a Poissonian initial distribution. Assuming enough mixing of the initial measure μ, namely (48), and incorporating ideas from [8] we are able to generalize Proposition 3, see Proposition 7 in Sect. 6.1 and Proposition 6, see Theorem 1 in Sect. 6.2.

We formulate the mixinig requirement in terms of the second Ursell function (factorial cumulant), which can be expressed in terms of the first and second correlation function (factorial moments) defined in Sect. 3.2, namely

$$u_\mu^{(2)}(x, y) := k_\mu(\{x, y\}) - k_\mu(\{x\})k_\mu(\{y\}).$$

We denote in the following the first correlation function $x \mapsto k_\mu(\{x\})$ by ρ_μ. The condition on the second Ursell function, (48), is a rather weak mixing or decay of correlation condition. In Sect. 6.3 we show that this condition is fulfilled, in particular, by Gibbs measures in the high-temperature low-activity regime. The condition will also hold beyond that regime even in the presence of a phase transition, cf. e.g. [44].

6.1 Long-time Asymptotics

Recalling (19), the Laplace transform of the one-dimensional distribution $P_{\mu,t}^{\mathbf{X}}$ can be expressed in terms of a one-particle system, i.e., for all non-negative $f \in \mathscr{S}(\mathbb{R}^d)$

$$\int_\Gamma e^{-\langle f,\gamma\rangle} P_{\mu,t}^{\mathbf{X}}(d\gamma) = \int_\Theta e_B(e^{tA}(e^{-f} - 1), \gamma)\mu(d\gamma) = \int_\Theta e^{\langle \ln(e^{tA}(e^{-f}-1)+1),\gamma\rangle}\mu(d\gamma). \quad (47)$$

Proposition 7 *Let μ be a measure on Θ which has first and second correlation function. Assume that μ fulfills the following mixing condition*

$$\sup_{x\in\mathbb{R}^d} \int_{\mathbb{R}^d} u_\mu^{(2)}(x, y)dy < \infty. \quad (48)$$

In addition, we assume that the Fourier transform of the first correlation function ρ_μ is a signed measure. Then, the one-dimensional distribution $P^{\mathbf{X}}_{\mu,t}$ converges weakly to $\pi_{\mathrm{mean}(\rho_\mu)}$ when t tends to infinity.

6.2 Hydrodynamic Limits

As in Sect. 5.2 we want to study the rescaled empirical field $n_t^{(\varepsilon)}(\varphi, \mathbf{X}) = \varepsilon^d \langle \varphi(\varepsilon\cdot), \mathbf{X}_{\varepsilon^{-\kappa}t}\rangle$. Since we do not have any longer a natural parameter associated to the initial measure, one cannot formulate something like slowly varying intensities. However, one sees that a possible framework is to work with a quite arbitrary sequence of initial measures $(\mu_\varepsilon)_{\varepsilon>0}$. The main restriction on this sequence is that one has to assume a particular convergence for the first correlation measure described below in more details and the mixing condtion (48) uniformly in ε. In Corollary 4 we prove that these conditions are fulfilled by Gibbs measures in the high-temperature low-activity regime for slowly varying intensity $z(\varepsilon\cdot)$. Furthermore, the limit is identified.

Theorem 1 *Assume that a has all moments finite. Let $(\mu_\varepsilon)_{\varepsilon>0}$ be a sequence of measures on Θ such that ρ_{μ_ε} is uniformly bounded in ε, for all $x \in \mathbb{R}^d$ the limit $\lim_{\varepsilon\to 0^+} \rho_{\mu_\varepsilon}(\{x/\varepsilon\}) =: \rho_0(x)$ exists, and the following mixing condition*

$$\sup_{x\in\mathbb{R}^d,\varepsilon>0} \int_{\mathbb{R}^d} u^{(2)}_{\mu_\varepsilon}(x, y)dy < \infty$$

holds. Then, for each $t \geq 0$, the following limit exists for all non-negative $\varphi \in \mathcal{D}(\mathbb{R}^d)$

$$\lim_{\varepsilon\to 0^+} \int_\Gamma \mu_\varepsilon(d\gamma)\, \mathbf{E}_\gamma \left[e^{-n_t^{(\varepsilon)}(\varphi,\mathbf{X})}\right] =: \int_{\mathcal{D}'(\mathbb{R}^d)} \delta_{\rho_t}(d\omega) e^{-\langle\varphi,\omega\rangle}, \tag{49}$$

and the conclusions of Sect. 5.2 continue to hold.

6.3 Application to Gibbs Measures

In this subsection we prove that the hypothesis for the results of the previous subsection are fulfilled for a concrete class of non-equilibrium measures, namely, for Gibbs measures in the high-temperature low-activity regime.

In order to recall the definition of a Gibbs measure, first we have to introduce a pair potential $V : \mathbb{R}^d \to \mathbb{R} \cup \{+\infty\}$, that is, a measurable function such that $V(-x) = V(x) \in \mathbb{R}$ for all $x \in \mathbb{R}^d \setminus \{0\}$. For $\gamma \in \Gamma$ and $x \in \mathbb{R}^d \setminus \gamma$ we define a relative energy of interaction between a particle located at x and the configuration γ by

$$E(x,\gamma) := \begin{cases} \sum_{y\in\gamma} V(x-y), \text{ if } \sum_{y\in\gamma} |V(x-y)| < \infty \\ +\infty, \qquad \text{otherwise} \end{cases}.$$

A probability measure μ on Γ is called a Gibbs measure corresponding to V, an intensity function $z \geq 0$, and an inverse of temperature β whenever it fulfills the Georgii-Nguyen-Zessin equation [42, Theorem 2]

$$\int_\Gamma \mu(d\gamma) \sum_{x\in\gamma} H(x,\gamma) = \int_\Gamma \mu(d\gamma) \int_{\mathbb{R}^d} dx\, z(x) H(x, \gamma \cup \{x\}) e^{-\beta E(x,\gamma)} \tag{50}$$

for all positive measurable functions $H : \mathbb{R}^d \times \Gamma \to \mathbb{R}$. This definition is equivalent to the definition via DLR-equation, see [15, 34, 42, 46]. We observe that for $V \equiv 0$ (50) reduces to the Mecke identity, which yields an equivalent definition of the Poisson measure π_z [40, Theorem 3.1]. We also note that for either $V \equiv 0$ and z not being a constant or $V \neq 0$, a Gibbs measure neither is a reversible nor an invariant initial distribution for the free Kawasaki dynamics under consideration. In order to have thermodynamical behavior we assume that V is stable, i.e., there exists a $B > 0$ such that $\sum_{\{x,y\}\subset\eta} V(x-y) \geq -B|\eta|$ for all configurations $\eta \in \Gamma_0$. Furthermore, we shall assume that the parameters β, z are small (high-temperature low-activity regime), i.e.,

$$\|z\|_u e^{2\beta B+1} C(\beta) < 1,$$

where $C(\beta) := \int_{\mathbb{R}^d} dx\, (1 - e^{-\beta|V(x)|})$. These conditions are, in particular, sufficient to insure the existence of Gibbs measures, cf. [45, 47]. Moreover, the correlation functions corresponding to such measures exist and fulfil a Ruelle bound defined in Sect. 3, and thus, as noted there, they are supported on Θ.

In the high-temperature low-activity regime one has rather detailed information about the Ursell functions (factorial cumulants) $u_\mu : \Gamma_0 \to \mathbb{R}$ corresponding to μ, which are the bounded measurable functions such that for all $f \in \mathcal{D}(\mathbb{R}^d)$ holds

$$\int_\Gamma e^{\langle f,\gamma\rangle} \mu(d\gamma) = \exp\left(\int_{\Gamma_0} \prod_{y\in\eta} (e^{f(y)} - 1) u_\mu(\eta) \lambda(d\eta) \right). \tag{51}$$

The function $x \to u_\mu(\{x\})$ coincides with the first correlation function of μ.

We present the results necessary for the following. For further details, see e.g. [10, 39] and see also [35].

The Ursell functions can be expressed in terms of a sum over all connected graphs weighted by the Meyer-functions

$$k(\xi) := \sum_{G\in\mathcal{G}_c(\xi)} \prod_{\{x,y\}\in G} (e^{-\beta V(x-y)} - 1), \tag{52}$$

where $\mathscr{G}_c(\xi)$ denotes the set of all connected graphs with vertex set ξ:

$$u_\mu(\eta) := \int_{\Gamma_0} \lambda_z(d\xi) k(\eta \cup \xi) \prod_{x \in \eta} z(x). \tag{53}$$

Actually, the following bound is the key result of the cluster expansion of Penrose–Ruelle type

$$|k(\xi)| \le e^{2\beta B|\xi|} \sum_{T \in \mathscr{T}(\xi)} \prod_{\{x,y\} \in G} (e^{-\beta|V(x-y)|} - 1), \tag{54}$$

where $\mathscr{T}(\xi)$ denotes the set of all trees with set of vertices ξ. This leads to the following integrability bound

$$\int_{\mathbb{R}^{dn}} |u_\mu(\{x, y_1, \ldots, y_n\})| z(y_1) dy_1 \ldots z(y_n) dy_n$$
$$\le e^{(2\beta B+1)(n+1)} (\|z\|_u C(\beta))^n \sum_{m=0}^{\infty} \frac{(n+m+1)!}{m!} \left(e^{2\beta B+1} \|z\|_u C(\beta)\right)^m. \tag{55}$$

In particular, the mixing condition (48) of Proposition 7 and Theorem 1 holds.

We show that Gibbs measures in the high-temperature low-activity regime with a translation invariant potential fulfill the assumptions of Proposition 7.

Corollary 3 *Let $z \ge 0$ be a bounded measurable function which Fourier transform is a bounded signed measure. Let μ be a Gibbs measure corresponding to a translation invariant potential V described above, inverse temperature β and activity z which are in the high-temperature low-activity regime. Then the first correlation function ρ_μ has as Fourier transform a measure and the arithmetic mean*

$$\text{mean}(\rho_\mu) = \frac{1}{(2\pi)^{d/2}} \sum_{n=0}^{\infty} \frac{1}{n!} \int_{\mathbb{R}^{dn}} \overline{\hat{k_r}(p_1, \ldots, p_n) \hat{z}(\{p_1 + \ldots + p_n\})} \hat{z}(dp_1) \cdot \ldots \cdot \hat{z}(dp_n),$$

where

$$k_r(y_1, \ldots, y_n) := \sum_{G \in \tilde{\mathscr{G}}_c} \prod_{\{x_1, x_2\} \in G} (e^{-\beta V(x_1 - x_2)} - 1)$$

and where $\tilde{\mathscr{G}}_c$ denotes the set of all connected graphs with vertex set $(0, y_1, \ldots, y_n)$. As a consequence, all the assumptions of Proposition 7 are fulfilled.

As in Sect. 5.2, we consider initial measures with a slowly varying intensity, i.e., Gibbs measures corresponding to β, V, and $z(\varepsilon \cdot)$, which we denote by μ_ε. As $\|z\|_u$ is unchanged, all scaled measures μ_ε remain in the high-temperature low-activity regime and the bound (55) holds uniformly in $\varepsilon > 0$. For Gibbs measures, which are not Poisson measures, the first correlation function is not any longer just the intensity. The function appearing as the initial value in the limiting partial differential equation is the scaling limit of the first correlation function and not just the unscaled activity. Let us describe what the scaling limit of the first correlation function is. Denote by ρ_c^{equi} the correlation function corresponding to the Gibbs measure with constant activity c, temperature β and potential V. Due to the translation invariance of V this correlation function is a constant function. Given a function $z \geq 0$ denote by $x \mapsto \rho_{z(x)}^{\text{equi}}$ the function that associates to each x the constant value of the first correlation function of the Gibbs measure with constant activity $z(x)$. This function is the scaling limit of the first correlation function of μ_ε. Note that $\rho_\mu(x) = u_\mu(\{x\})$.

Corollary 4 *Given a bounded measurable function $z \geq 0$, a potential V, and an inverse temperature β fulfilling the conditions of the high-temperature and low-activity regime, the corresponding Gibbs measures fulfill all the assumptions of Theorem 1. Moreover,*

$$\rho_0(x) := \lim_{\varepsilon \to 0^+} u_{\mu_\varepsilon}(\{x/\varepsilon\}) = \rho_{z(x)}^{\text{equi}}. \tag{56}$$

Acknowledgements We would like to thanks N. Jacob, E. Lytvynov, M. Röckner, W. Hoh for fruitful discussions, T. Hilberdink for mentioning the theory of slowly varying functions, and T. Funaki for making us aware of [8]. M. J. O. was supported by the Portuguese national funds through FCT-Fundação para a Ciência e a Tecnologia, within the project UIDB/04561/2020. J. L. S. was partially supported by the grant UIDB/MAT/04674/2020 of FCT-Fundação para a Ciência e a Tecnologia, Portugal. Tobias Kuna is member of GNAMPA group of INdAM.

Appendix

We give some estimates necessary to establish the hydrodynamic limit.

Lemma 10 *Let $\varphi \in \mathcal{S}(\mathbb{R}^d)$ and $0 \leq \varepsilon \leq 1$ be given. Then*

$$\begin{aligned}
\|e^{tA}(e^{\varepsilon^d \varphi(\varepsilon\cdot)} - 1)\|_u &\leq \varepsilon^d \|\varphi\|_u e^{\|\varphi\|_u} \\
\|e^{tA}(e^{\varepsilon^d \varphi(\varepsilon\cdot)} - 1)\|_{L^1(\mathbb{R}^d,dx)} &\leq \|\varphi\|_{L^1(\mathbb{R}^d,dx)} e^{\|\varphi\|_{L^1(\mathbb{R}^d,dx)}} \\
\|e^{tA}(e^{\varepsilon^d \varphi(\varepsilon\cdot)} - 1 - \varepsilon^d \varphi(\varepsilon\cdot))\|_{L^1(\mathbb{R}^d,dx)} &\leq \varepsilon^d \|\varphi\|^2_{L^1(\mathbb{R}^d,dx)} e^{\|\varphi\|_{L^1(\mathbb{R}^d,dx)}}
\end{aligned} \tag{57}$$

Let us introduce the following two equivalent systems of norms for the locally convex topological vector space $\mathcal{S}(\mathbb{R}^d)$ for $A \in \mathbb{N}$ and $M \geq 0$:

$$\|f\|_{A,M,u} := \sum_{\substack{\alpha \in \mathbb{N}_0^d \\ |\alpha| \leq A}} \sup_{x \in \mathbb{R}^d} |D^\alpha f(x)| \left(1 + |x|^2\right)^M, \tag{58}$$

$$\|f\|_{A,M,2} := \sum_{\substack{\alpha \in \mathbb{N}_0^d \\ |\alpha| \leq A}} \left(\int_{\mathbb{R}^d} |D^\alpha f(x)|^2 \left(1 + |x|^2\right)^M dx \right)^{1/2}. \tag{59}$$

Lemma 11 *Let $f_1, f_2 \in C^\infty(\mathbb{R}^d; \mathbb{C})$ be two C^∞-functions with non-positive real part such that for an $A \in \mathbb{N}$ and a $M \geq 0$ one has $\|f_i\|_{A,-M,u} < \infty$. Then there exists a constant C depending on A and on $\|f_i\|_{A,-M,u}$ such that*

$$\|e^{f_1} - e^{f_2}\|_{A,-(A+1)M,2} \leq C\|f_1 - f_2\|_{A,-M,2}. \tag{60}$$

References

1. Albeverio, S., Kondratiev, Yu. G., Röckner, M.: Analysis and geometry on configuration spaces. J. Funct. Anal. **154**(2), 444–500 (1998)
2. Berns, C., Kondratiev, Yu. G., Kutoviy, O.: Construction of a state evolution for Kawasaki dynamics in continuum. Anal. Math. Phys. **3**(2), 97–117 (2013)
3. Bertini, L., Cancrini, N., Cesi, F.: The spectral gap for a Glauber-type dynamics in a continuous gas. Ann. Inst. H. Poincaré Probab. Statist. **38**, 91–108 (2002)
4. Bogoliubov, N.N.: Problems of a dynamical theory in statistical physics. Gostekhisdat, Moskau, 1946. In: Russian. English translation in de Boer, J., Uhlenbeck, G.E. (ed.) Studies in Statistical Mechanics, vol. 1, pp. 1–118. North-Holland, Amsterdam (1962)
5. Cavallaro, G., Marchioro, C., Spitoni, C.: Dynamics of infinitely many particles mutually interacting in three dimensions via a bounded superstable long-range potential. J. Stat. Phys. **120**(1–2):367–416 (2005)
6. De Masi, A., Presutti, E.: Mathematical methods for hydrodynamic limits. Lecture Notes in Mathematics, vol. 1501. Springer, Berlin (1991)
7. Dobrushin, R.L.: On Poisson's law for distribution of particles in space. Ukrain. Mat. Zh. 27–134 (1956). (In Russian)
8. Dobrushin, R.L., Siegmund-Schultze, R.: The hydrodynamic limit for systems of particles with independent evolution. Math. Nachr. **105**, 199–224 (1982)
9. Doob, J.L.: Stochastic Processes. Wiley, New York, London (1953)
10. Duneau, M., Souillard, B., Iagolnitzer, D.: Decay of correlations for infinite-range interactions. J. Math. Phys. **16**(8), 1662–1666 (1975)
11. Ethier, S.N., Kurtz, T.G.: Markov Processes. Characterization and Convergence. Wiley Series in Probability and Mathematical Statistics. Wiley, NewYork, Chichester, Brisbane, Toronto, and Singapore (1986)
12. Finkelshtein, D., Kondratiev, Yu. G., Oliveira, M.J.: Markov evolutions and hierarchical equations in the continuum. I. One-component systems. J. Evol. Equ. **9**(2), 197–233 (2009)
13. Fritz, J.: Gradient dynamics of infinite point systems. Ann. Probab. **15**(2), 478–514 (1987)
14. Garcia, N., Kurtz, T.: Spatial birth and death processes as solutions of stochastic equations. ALEA Lat. Am. J. Probab. Math. Stat. **1**, 281–303 (2006)
15. Georgii, H.O.: Canonical and grand canonical Gibbs states for continuum systems. Commun. Math. Phys. **48**, 31–51 (1976)
16. Glötzl, E.: Time reversible and Gibbsian point processes. I. Markovian spatial birth and death processes on a general phase space. Math. Nachr. **102**, 217–222 (1981)
17. Glötzl, E.: Time reversible and Gibbsian point processes. II. Markovian particle jump processes on a general phase space. Math. Nachr. **106**, 62–71 (1982)
18. Infusino, M.: Quasi-analyticity and determinacy of the full moment problem from finite to infinite dimensions. In: Stochastic and Infinite Dimensional Analysis, Trends in Mathematics, pp. 161–194. Birkhäuser (2016)
19. Infusino, M., Kuna, T.: The full moment problem on subsets of probabilities and point configurations. J. Math. Anal. Appl. **483**(1), 123551 (2020)
20. Infusino, M., Kuna, T., Rota, A.: The full infinite dimensional moment problem on semi-algebraic sets of generalized functions. J. Funct. Anal. **267**(5), 1382–1418 (2014)

21. Jacob, N.: Pseudo-Differential Operators and Markov Processes. Fourier Analysis and Semigroups, vol. I. Imperial College Press, London (2001)
22. Kipnis, C., Landim, C.: Scaling Limits of Interacting Particle Systems. Grundlehren der Mathematischen Wissenschaften, Springer, Berlin (1999)
23. Kondratiev, Yu. G., Kutoviy, O., Zhizhina, E.: Non-equilibrium Glauber type dynamics in continuum. J. Math. Phys. **47**(11), 113501 (2006)
24. Kondratiev, Yu. G., Lytvynov, E.: Glauber dynamics of continuous particle systems. Ann. Inst. H. Poincaré Probab. Statist. **41**(4), 685–702 (2005)
25. Kondratiev, Yu. G., Lytvynov, E., Röckner, M.: The heat semigroup on configuration spaces. Publ. Res. Inst. Math. Sci. **39**(1), 1–48 (2003)
26. Kondratiev, Yu. G., Kuna, T.: Harmonic analysis on configuration space I. General theory. Infin. Dimens. Anal. Quantum Probab. Relat. Top. **5**(2), 201–233 (2002)
27. Kondratiev, Yu. G., Kuna, T., Ohlerich, N.: Spectral gap for Glauber type dynamics for a special class of potentials. Electron. J. Probab. **18**(42), 18pp (2013)
28. Kondratiev, Yu. G., Kuna, T., Oliveira, M.J.: Analytic aspects of Poissonian white noise analysis. Methods Funct. Anal. Topol. **8**(4), 15–48 (2002)
29. Kondratiev, Yu. G., Kuna, T., Oliveira, M.J.: Holomorphic Bogoliubov functionals for interacting particle systems in continuum. J. Funct. Anal. **238**(2), 375–404 (2006)
30. Kondratiev, Yu. G., Kuna, T., Oliveira, M.J.: Extension of explicit formulas in Poissonian white noise analysis using harmonic analysis on configuration spaces. Condens. Matter Phys. **11**(2), 237–246 (2008)
31. Kondratiev, Yu. G., Lytvynov, E., Röckner, M.: Equilibrium Kawasaki dynamics of continuous particle systems. Infin. Dimens. Anal. Quantum Probab. Relat. Top. **7**(4), 185–209 (2007)
32. Kondratiev, Yu. G., Lytvynov, E., Röckner, M.: Non-equilibrium stochastic dynamics in continuum: The free case. Condens. Matter Phys. **11**(4(56)), 701–721 (2009)
33. Kondratiev, Yu. G., Silva, J.L., Streit, L., Us, G.F.: Analysis on Poisson and Gamma spaces. Infin. Dimens. Anal. Quantum Probab. Relat. Top. **1**(1), 91–117 (1998)
34. Kuna, T.: Studies in Configuration Space Analysis and Applications. Ph.D. thesis, Bonner Mathematische Schriften Nr. 324, University of Bonn (1999)
35. Kuna, T.: Properties of marked Gibbs measures at high temperature regime. Methods Funct. Anal. Topol. **7**, 33–54 (2001)
36. Kuna, T.: Bochner's theorem for point processes via Bogoliubov functionals. In preparation (2023)
37. Lenard, A.: States of classical statistical mechanical systems of infinitely many particles I. Arch. Rational Mech. Anal. **59**, 219–239 (1975)
38. Lenard, A.: States of classical statistical mechanical systems of infinitely many particles II. Arch. Rational Mech. Anal. **59**, 241–256 (1975)
39. Malyshev, V.A., Minlos, R.A.: Gibbs Random Fields: Cluster Expansions. Kluwer Academic Publishers, Dordrecht, Boston, London (1991)
40. Mecke, J.: Stationäre zufällige Maße auf lokalkompakten Abelschen Gruppen. Z. Wahrsch. verw. Gebiete **9**, 36–58 (1967)
41. Nguyen, X.X., Zessin, H.: Martin-Dynkin boundary of mixed Poisson processes. Z. Wahrsch. verw. Gebiete **37**, 191–200 (1976)
42. Nguyen, X.X., Zessin, H.: Integral and differential characterizations of the Gibbs process. Math. Nachr. **88**, 105–115 (1979)
43. Penrose, M.: Existence and spatial limit theorems for lattice and continuum particle systems. Probab. Surv. **5**, 1–36 (2008)
44. Presutti, E.: Scaling Limits in Statistical Mechanics and Microstructures in Continuum mechanics. Springer, Berlin (2009)
45. Procacci, A., Yuhjtman, S.A.: Convergence of Mayer and virial expansions and the Penrose tree-graph identity. Lett. Math. Phys. **107**, 31–46 (2017)
46. Puzicha, J.: Existenz und Charakterisierung von Gibbsmaßen. Master thesis, University of Bielefeld (1981)
47. Ruelle, D.: Statistical Mechanics. Rigorous Results. Benjamin, New York, Amsterdam (1969)

48. Ruelle, D.: Superstable interactions in classical statistical mechanics. Commun. Math. Phys. **18**, 127–159 (1970)
49. Sinai, Y.G. ed.: Ergodic theory with applications to dynamical systems and statistical mechanics, volume II of Encyclopaedia of Mathematical Sciences, Chapter III. Dynamical Systems of Statistical Mechanics and Kinetic Equations. Springer, Berlin, Heidelberg (1989)
50. Spohn, H.: Equilibrum fluctuations for interacting Brownian particles. Commun. Math. Phys. **103**, 1–33 (1986)

Approach to Equilibrium for the Kac Model

Federico Bonetto, Eric A. Carlen, Lukas Hauger, and Michael Loss

Abstract This is a review on the Kac Master Equation. Various issues will be presented such as the resolution of Kac's conjecture about the gap for the three dimensional hard sphere gas, entropic propagation of chaos and other topics such as systems coupled to reservoirs and thermostats. The discussion is informal with few proofs and those who are presented are only sketched.

Keywords Kinetic theory · Master equation · Kac Model · Decay of entropy · Decay of information · Collision process · Boltzmann equation · Kac master equation · Kac-boltzmann equation · Quantum Kac model

1 The One Dimensional Kac Model

It is the aim of this article to give an overview of a program to which Maria Carvalho contributed greatly and to show that her work has instigated further development in this area. In much of her career she worked on fundamental problems like to formulate the notion of equilibrium and to describe how systems that are out of equilibrium

Eric A. Carlen—Partially supported by U.S. National Science Foundation grant DMS-2055282.
Michael Loss—Partially supported by U.S. National Science Foundation grant DMS-2154340.

F. Bonetto · M. Loss (✉)
Georgia Institute of Technology, Atlanta, GA 30332, USA
e-mail: loss@math.gatech.edu

F. Bonetto
e-mail: federico.bonetto@math.gatech.edu

E. A. Carlen
Rutgers University, 110 Frelinghuysen Road, Piscataway, NJ 08854, USA
e-mail: carlen@math.rutgers.edu

L. Hauger
Max-Planck-Institute for Mathematics in the Sciences, Inselstraße 22, 04103 Leipzig, Germany
e-mail: hauger@mis.mpg.de

E. Carlen et al. (eds.), *From Particle Systems to Partial Differential Equations*,
Springer Proceedings in Mathematics & Statistics 465,
https://doi.org/10.1007/978-3-031-65195-3_8

tend towards it. One particular fruitful avenue she pursued with collaborators is an approach to kinetic theory through master equations.

The standard description of a classical gas, going back to Boltzmann [2], is to replace the evolution equation of the interacting many body problem by an effective equation describing the evolution of marginals, known as the Boltzmann equation. Despite heroic efforts by Lanford and others [19, 25], there is no satisfactory derivation of this equation from microscopic dynamics. There was substantial progress by Bodineau, Gallagher and Saint Raimond [1] who gave a rigorous derivation of the linearized Boltzmann equation describing a hard sphere gas near equilibrium. This seems to be the first rigorous derivation of an effective equation from microscopic dynamics.

A less ambitious approach is to construct models that illuminate certain questions of interest. Many of the difficulties stemming from a fundamental description are ignored in such an approach. However it has brought some of the issues into a sharp focus and may lead to new ideas on how to proceed with the 'fundamentalist program'. The pedagogical aspect is an added bonus of this phenomenological approach. Maria Carvalho and Eric Carlen introduced one of the authors to these kind of questions, in particular to an approach based on the Kac model. This paper is a review of some of the work done over the years, but we also present some newer developments, albeit with a light touch.

The Kac Master Equation describing the evolutions of the probability density $F(\boldsymbol{v}_N, t)$ for the velocities of a one-dimensional spatially homogeneous gas of N particle is given by

$$\frac{\partial F(\boldsymbol{v}_N, t)}{\partial t} = \frac{N}{\binom{N}{2}} \sum_{i<j} \left[\frac{1}{2\pi} \int_0^{2\pi} F(\boldsymbol{v}_N(i, j; \theta), t) d\theta - F(\boldsymbol{v}_N, t) \right] \tag{1}$$

with initial condition $F(\boldsymbol{v}_N, 0) = F_0(\boldsymbol{v}_N)$, where $\boldsymbol{v}_N = (v_1, \ldots, v_N)$ is the velocity vector of the particles,

$$\boldsymbol{v}_N(i, j; \theta) = (v_1, \ldots, v_i^*(\theta), \ldots, v_j^*(\theta), \ldots, v_N) \tag{2}$$

and

$$v_i^*(\theta) = v_i \cos\theta + v_j \sin\theta, \quad v_j^*(\theta) = -v_i \sin\theta + v_j \cos\theta \tag{3}$$

are the post-collisional velocities of the pair i, j undergoing this collision. Thus, the collisions are described by two dimensional random rotation. We assume that the density F is symmetric in the particle labels which is preserved by the time evolution. For a more detailed probabilistic explanation see [7, 8, 22] and especially [23]. It is useful to keep in mind that the rate at which a fixed particle collides with *any* other particle is independent of $N$and equals 2 in our case. A good way of thinking about the evolution equation is to write it as

$$\partial_t F = -\mathcal{L}_N F \, , \text{ where } \mathcal{L}_N = N(I - Q) \, , \tag{4}$$

where Q is the average over all rotations in the two dimensional coordinate hyperplanes. The solution of the master equation is then

$$F(\boldsymbol{v}_N, t) = e^{-\mathcal{L}_N t} F_0(\boldsymbol{v}_N)\,. \tag{5}$$

Since the collisions are modeled by two dimensional rotations, the kinetic energy

$$\sum_{j=1}^{N} v_j^2$$

is preserved and it suffices to consider the master equation (1) for densities $f \in L^1(\mathcal{S}_N, d\sigma_N)$ where $\mathcal{S}_N = \sqrt{N}\mathbb{S}^{N-1}$ and $d\sigma_N$ is the uniform probability measure on the sphere. We fix the total kinetic energy to be N, so that the average kinetic energy per particle is 1. If one requires for a one-dimensional collision that in addition the energy and the momentum are preserved, the only possible collisions would be that either the particles keep their velocities and pass through each other or exchange them. This would leave the symmetric density F invariant and hence not interesting.

It is fairly obvious that for product densities, i.e. densities of the form

$$F_N(\boldsymbol{v}_N) := \frac{\prod_{j=1}^{N} f(v_j)}{\int_{\mathcal{S}_N} \prod_{j=1}^{N} f(v_j)\, d\sigma_N}\,, \tag{6}$$

the product structure is not strictly preserved under the evolution. It is, however, preserved asymptotically as the particle number tends to infinity, a fact known as propagation of chaos, a notion that goes back to Boltzmann [2]. Consider a sequence $\{F_N(\boldsymbol{v}_N)\}_{N=1}^{\infty}$. Assume that there exists a function $f : \mathbb{R} \to \mathbb{R}_+$ such that for any $k \in \mathbb{N}$ and any bounded continuous function $\phi_k : \mathbb{R}^k \to \mathbb{R}$ we have

$$\lim_{N\to\infty} \int_{\mathcal{S}_N} F_N(\boldsymbol{v}_N)\phi_k(v_1, \ldots, v_k)\, d\sigma_N = \int_{\mathbb{R}^k} \prod_{j=1}^{k} f(v_j)\phi_k(v_1, \ldots, v_k)\, dv_1 \cdots dv_k\,,$$

then we say that the sequence F_N is *chaotic with limiting marginal* f. The simplest example is the function 1 on the sphere $\mathcal{S}_N$ which is easily seen to be chaotic with limiting marginal

$$\gamma(v) := (2\pi)^{-1/2} e^{-v^2/2}\,, \tag{7}$$

by a calculation that goes back to the 19th century to Mehler, Maxwell and Poincaré.

Kac's theorem states that chaos is propagated.

Theorem 1 (Kac Theorem) *Let $F_N(\boldsymbol{v}_N)$ be a chaotic sequence with limiting marginal $f_0(v)$ and $F_N(\boldsymbol{v}_N, t)$ be the solution of the Kac Master Equation with* *initial condition* $F_N(\mathbf{v}_N)$. *Then* $F_N(\mathbf{v}_N, t)$ *is chaotic with limiting marginal* $f(v,t)$ *and this function satisfies the Kac-Boltzmann Equation*

$$\frac{\partial f(v,t)}{\partial t} = \frac{1}{\pi}\int_0^{2\pi}\int_{-\infty}^{\infty}[f(v^*(\theta),t)f(w^*(\theta),t) - f(v,t)f(w,t)]\,dw.. \tag{8}$$

with $f(v,0) = f_0(v)$.

This theorem is proved in [22] and is the antecedent of many results of similar type. It was shown in [12] that any distribution on $\mathbb{R}$ is the limit marginal of a chaotic sequence and hence this leads to an existence result for solutions of the Kac-Boltzmann equation.

It is not very difficult to see that the semi-group generated by the Kac Master Equation is ergodic in the sense that the only invariant distribution, i.e. *equilibrium*, is the function 1 on $\mathcal{S}_N$. Thus, 0 is a nondegenerate eigenvalue of $\mathcal{L}_N$. Moreover, one observes that the generator $\mathcal{L}_N$ of this semigroup is a selfadjoint non-negative operator on $L^2(\mathcal{S}_N, d\sigma_N)$ and according to what was said before, the function 1 spans the null space of this operator. Thus, 0 is a simple eigenvalue of $\mathcal{L}_N$. The next largest eigenvalue Δ_N is called the *gap*. A simple application of the spectral theorem shows that $F_N(\mathbf{v}_N, t)$ converges to 1 as $t \to \infty$ in $L^2(\mathcal{S}_N, d\sigma_N)$ with a rate that is given by the gap

$$\|F_N(\cdot,t) - 1(\cdot)\|_2 \le e^{-\Delta_N t}\|F_N(\cdot,0) - 1(\cdot)\|_2 .$$

Kac [22] posed the conjecture that there exists a positive constant $c > 0$ independent of N such that $\Delta_N \ge c$ for all N. This conjecture was eventually proved in [21]. In [7] it was further shown that

$$\Delta_N = \frac{1}{2}\frac{N+2}{N-1} > \frac{1}{2} .$$

The Kac Boltzmann equation has Gaussian functions such as (7) as equilibria and it is customary to linearize the Kac-Boltzmann equation about one of these equilibria. In this way one obtains some information about the speed of approach to equilibrium by computing the gap for the generator of this linear equation. Interestingly, this is furnished by the Kac master equation in the limit as $N \to \infty$. The linearized version of the Kac-Boltzmann equation has the gap eigenvalue $1/2$ which is the limit of Δ_N as $N \to \infty$. But more is true. The gap eigenvalue for the Kac master Equation is not degenerate and the eigenfunction (up to normalization) is given by

$$\sum_{j=1}^{N}\left[v_j^4 - \frac{3}{N(N+2)}\right] .$$

As $N \to \infty$, the first marginal of this function converges to a fourth order hermitian polynomial which is the gap eigenfunction of the linearized Kac-Boltzmann equation. For details the reader may consult [8].

2 Momentum Preserving Collisions

As mentioned before, two particle collisions that preserve the kinetic energy and momentum are not very interesting in one dimension; particles either exchange the velocity or pass through each other without interaction. One can certainly write a Kac-type master equation for realistic collisional models such as hard spheres in three dimensions. Not only are the collisions energy and momentum preserving but there is the added complexity that the scattering cross section may depend on the momentum transfer $|v_i - v_j|$. There are multiple ways to parametrize the collision process and one convenient version used here is the swapping map

$$\begin{aligned} v_i^*(\sigma) &= \frac{v_i + v_j}{2} + \frac{|v_i - v_j|}{2}\sigma\,, \\ v_j^*(\sigma) &= \frac{v_i + v_j}{2} - \frac{|v_i - v_j|}{2}\sigma\,, \end{aligned} \tag{9}$$

where $\sigma \in \mathbb{S}^2$. To describe the collisions in the one dimensional case, we assumed that the angle θ was uniformly distributed. In the three dimensional case, we assume that there is given a function $b : [-1, 1] \to \mathbb{R}_+$ with

$$\int_{\mathbb{S}^2} b(\sigma \cdot \sigma')\, d\sigma = \frac{1}{2}\int_{-1}^{1} b(t)\, dt = 1\,. \tag{10}$$

Here $d\sigma$ is the uniform probability measure on $\mathbb{S}^2$. For hard spheres, $b = 1$, see [11] for an explanation how the various parametrizations are connected. The model analogous to (1) is given by

$$\begin{aligned} \frac{\partial F}{\partial t}(\boldsymbol{v}_N, t) &= -L_{N,\alpha} F(\boldsymbol{v}_N, t)) \\ &= -\frac{N}{\binom{N}{2}} \sum_{i<j} |v_i - v_j|^\alpha \left[F(\boldsymbol{v}_N, t) - [F]^{(i,j)}(\boldsymbol{v}_N, t)\right] \end{aligned} \tag{11}$$

where

$$[F]^{(i,j)}(\boldsymbol{v}_N, t) = \int_{S^2} b\left(\sigma \cdot \frac{v_i - v_j}{|v_i - v_j|}\right) F(R_{i,j,\sigma}\boldsymbol{v}_N, t)\, d\sigma \tag{12}$$

and

$$(R_{i,j,\sigma}\boldsymbol{v})_k = \begin{cases} v_i^*(\sigma) & k = i, \\ v_j^*(\sigma) & k = j, \\ v_k & k \neq i, j. \end{cases}$$

The case of hard spheres is $\alpha = 1$ but it is of advantage to keep the power $0 \le \alpha \le 2$. Since the collisions preserve energy and momentum, it suffices to consider distributions on the surface given by the equation

$$\frac{1}{N}\sum_{j=1}^{N}|v_j|^2 = E\,, \quad \frac{1}{-N}\sum_{j=1}^{N}v_j = p\,.$$

where E resp. p is the kinetic energy resp. momentum per particle. These equations determine a sphere in $\mathbb{R}^{3N}$ of co-dimension 4 and we denote this space by $\mathcal{S}_{N,E,p}$. As before, we denote the uniform probability measure on this space by $d\sigma_{N,E,p}$.

Propagation of chaos is much harder to prove for this model. The difficulty is that the factor $|v_i - v_j|^\alpha$ can grow with N. This problem, was first solved in a qualitative fashion in [27], and later in a quantitatively by Mischler and Mouhot in their seminal paper [26]. Likewise, to find a lower bound on the gap that is uniform in the number of particles is much harder. Ironically, one of the reasons for this difficulty is that the factor $|v_i - v_j|^\alpha$ can be small and suppress collisions. The gap is defined as

$$\Delta_{N,\alpha}(E,p) = \inf\left\{\mathcal{E}(F,F)\ :\ \langle F,1\rangle_{L^2(\mathcal{S}_{N,E,p})} = 0 \quad \text{and} \quad \|F\|^2_{L^2(\mathcal{S}_{N,E,p})} = 1\right\}, \tag{13}$$

where $\mathcal{E}(F,F) = \langle F, L_{N,\alpha}F\rangle_{L^2(\mathcal{S}_{N,E,p})}$, i.e.

$$\mathcal{E}(F,F) = \frac{N}{2}\binom{N}{2}^{-1}\sum_{i<j}\int_{\mathcal{S}_{N,E,p}}\int_{\mathbb{S}^2}|v_i - v_j|^\alpha b\left(\sigma\cdot\frac{v_i - v_j}{|v_i - v_j|}\right)\left[F(\mathbf{v}_N) - F(R_{i,j,\sigma}\mathbf{v}_N)\right]^2 d\sigma\, d\sigma_N\,. \tag{14}$$

By a unitary transformation mapping $L^2(\mathcal{S}_{N,1,0})$ to $L^2(\mathcal{S}_{N,E,p})$ (see [9]) one finds that

$$\Delta_{N,\alpha}(E,p) = (E - |p|^2)^{\alpha/2}\Delta_{N,\alpha}(1,0)\,, \tag{15}$$

and we call $\Delta_{N,\alpha}(1,0)$ the 'spectral gap for the Kac model'. The following theorem is proved in [9].

Theorem 2 (Spectral gap for the Kac Model with $0 \le \alpha \le 2$) *For each continuous non-negative even function b on $[-1,1]$ satisfying* (10)*, and for each $\alpha \in [0,2]$, there is a strictly positive constant K depending only on b and α, and explicitly computable, such that*

$$\Delta_{N,\alpha} \ge K > 0 \tag{16}$$

for all N. In particular, this is true with b given by (10) *and $\alpha = 1$, the* 3*-dimensional hard sphere Kac model.*

3 Entropy

Unfortunately, the gap is not a very good notion to measure the rate of approach to equilibrium. It is generically the case that the L^2 norm of a probability distribution on $L^2(\sqrt{N}\mathbb{S}^{N-1})$ increases like c^N where $c > 1$. This is easy to see for chaotic sequences. The preferred notion is entropy which for the simple Kac model explained at the beginning is given by

$$H(F|1) = \int_{S_N} F \log F \, d\sigma_N \, .$$

$H(F|1)$ is called 'entropy' which deviates from the usual notion in physics by a sign. The notation $H(F|1)$ also indicates that one is considering a relative entropy, in our case with respect to the 1 distribution. We note that in contrast to the L^2 norm, the entropy is extensive, i.e. for a state that is a product state $F = \prod_{j=1}^N f(v_j)$ we see that

$$H(F|1) = \sum_{j=1}^N \int_{\sqrt{N}\mathbb{S}^{N-1}} F \log f(v_i) \, d\sigma_N \approx N \int_{\sqrt{N}\mathbb{S}^{N-1}} f(v) \log f(v) \, d\sigma_N \, .$$

It is not very difficult to see that the entropy decreases to zero along the Kac flow, although little is known about the rate. Returning to the simplest version of the Kac model, it was shown in [28] that

$$H(F(t)|1) \le e^{-\frac{2t}{N-1}} H(F_0|1) \, .$$

The proof consists of the analysis of entropy production. Taking the derivative

$$\frac{d}{dt}\Big|_{t=0} H(F(t)|1) = -N \int_{\sqrt{N}\mathbb{S}^{N-1}} (I - Q) F \log F \, d\sigma_N$$

one obtains the negative rate of dissipation and the entropy production is then defined by

$$\Gamma_N = \inf_F \frac{N \int_{\sqrt{N}\mathbb{S}^{N-1}} (I - Q) F \log F d\sigma_N}{H(F|1)} \, .$$

Villani [28] proved that

$$\Gamma_N \ge \frac{2}{N-1} \, .$$

That this estimate is almost optimal was shown by Einav [15]. Exponential decay of the entropy with a rate that is independent of N is known as Cercignani's conjecture. For details the reader may consult [28]. Although Einav's result does not preclude that the entropy might decrease eventually at an exponential rate, one could consider this as evidence against Cercignani's conjecture for the Kac model. The challenge is to show that the evolution of Einav's trial functions has small entropy decay for large times.

There are modifications of the Kac model with one dimensional collisions worked out by Villani [28] for which Cercignani's conjecture is true. These modifications accelerate the rate of collisions between the particles i and j by replacing the collision rate by $(1 + v_i^2 + v_j^2)$. Unfortunately nothing can be inferred from that about the original model. It is interesting that Villani's result is not known to hold when the collision rate is merely replaced by $v_i^2 + v_j^2$. The problem is that for small velocities the collision rate is sharply depressed. Likewise, to our knowledge, similar results for the three dimensional 'hard sphere' collisions in which $\alpha = 2$ are not known.

There is, however, another interesting link between the Kac-Boltzmann equation and the Kac master equation. In Sect. 1, we mentioned the close connection between the gap and the gap eigenfunction of the Kac model and the linearized Kac-Boltzmann equation. This can be carried further using the notion of entropic chaos. A sequence $\{F_N\}_{N=1}^{\infty}$ is entropically chaotic with limiting marginal f if it is chaotic and in addition satisfies

$$\lim_{N\to\infty} \frac{H(F_N|1)}{N} = H(f|\gamma)\,.$$

In [12] it was shown that for a given probability density f on the real line with

$$\int_{\mathbb{R}} f(v)v^2\,dv = 1\,, \quad H(f|\gamma) < \infty$$

there exists an entropically chaotic sequence with limiting marginal f. Using this connection one is able to transfer entropy decay of the Kac evolution to entropy decay for the Kac-Boltzmann equation. Unfortunately, the current entropy decay estimates are not strong enough to draw any quantitative conclusions.

From a practical point of view, one rarely considers systems that are very far from equilibrium. Usually one is in a situation in which only a part of the system is out of equilibrium. One way to model such a situation, is to couple a system of M particles to a larger one. The simplest possibility, pursued in [6], is to couple the finite system to a thermostat. By definition, a thermostat does not change during the interaction with the system, and this can be modeled by

$$\frac{\partial f}{\partial t} = -\lambda M(I-Q)f - \mu\sum_{j=1}^{M}(I-R_j)f =: -\mathcal{L}_T f\,. \tag{17}$$

where

$$R_j f := \int \frac{1}{2\pi}\int_0^{2\pi} e^{-\pi w_j^{*2}(\theta)} f(\mathbf{v}_M^j(\theta, w))\,d\theta\,dw$$

and

$$\begin{aligned} \boldsymbol{v}_M^j(\theta, w) &= (v_1, ..., v_j \cos(\theta) + w \sin(\theta), \ldots, v_M), \\ w_j^*(\theta) &= -v_j \sin(\theta) + w \cos(\theta). \end{aligned}$$

The idea is that particle j interacts with a particle from the thermostat whose velocity distribution is given by a Gaussian of temperature $2/\pi$. Note, using these units has the advantage that the Gaussian function $e^{-\pi v^2}$ is normalized. It is easy to see that the Gaussian $\gamma_M = e^{-\pi |\boldsymbol{v}_M|^2}$ is the unique equilibrium state for the system. It was shown in [6] that for an initial condition f_0 with finite entropy

$$H(f_t|\gamma) \leq e^{-\mu t/2} H(f_0|\gamma)\,.$$

Since the equilibrium state is a product function, one would hope to obtain a stronger version of propagation of chaos, namely one that is uniform in time. This was shown in [14] with an explicit error term proportional to $N^{-1/3}$.

It is natural to think that the evolution of the system is similar if one couples the system to a finite reservoir with a large number of particles. The system is then described by a probability distribution $F(\boldsymbol{v}, \boldsymbol{w}, t) = (v_1, \ldots, v_M, w_{M+1}, \ldots, w_{M+N})$, where we label the particles from 1 to $M + N$. Thus, one considers the evolution equation

$$\partial_t F = -\left(\lambda_S \mathcal{L}_M F + \lambda_R \mathcal{L}_N + \mu \mathcal{I}_{M,N}\right) F =: -\mathcal{L}_R F \tag{18}$$

where

$$\begin{aligned} \mathcal{L}_M F &:= \tfrac{2}{M-1} \textstyle\sum_{i<j=1}^{M} (I - R_{i,j}) F \\ \mathcal{L}_N F &:= \tfrac{2}{N-1} \textstyle\sum_{i<j=M+1}^{M+N} (I - R_{i,j}) F \\ \mathcal{I}_{M,N} F &:= \tfrac{1}{N} \textstyle\sum_{i=1}^{M} \sum_{j=M+1}^{M+N} (I - R_{i,j}) F \end{aligned}$$

with the initial condition

$$F_0(\boldsymbol{v}, \boldsymbol{w}) = f_0(\boldsymbol{v}) e^{-\pi |\boldsymbol{v}|^2}\,.$$

The kinetic energy is no longer preserved and therefore one has to consider distributions in $\mathbb{R}^{M+N}$. The collisions are here modeled in one dimension as follows

$$R_{i,j} F(\boldsymbol{u}) := \frac{1}{2\pi} \int_0^{2\pi} F(u_1, \ldots, u_i^*(\theta), \ldots, u_j^*(\theta), \ldots, u_{M+N}), \quad \boldsymbol{u} \in \mathbb{R}^{M+N}, \tag{19}$$

where

$$u_i^*(\theta) := u_i \cos\theta + u_j \sin\theta, \quad u_j^*(\theta) := -u_i \sin\theta + u_j \cos\theta\,.$$

The generator $\mathcal{L}_M$ describes the collisions among the M particles in the considered system, $\mathcal{L}_N$ describes the collisions among the N particles in the reservoir and $I_{M,N}$ the collisions between the particles in the system and the reservoir. Note that the collision rate of a particle in the system with any particle in the reservoir is μ. The interesting point about this evolution is that although initially the reservoir is in an equilibrium it does not stay that way in the course of the evolution. This is in contrast to the thermostat. One would, however, expect that the solution $F(\boldsymbol{v}, \boldsymbol{w}, t)$ is well approximated by the solution of the thermostat problem for large N. That this is the case was shown in [4] using the Gabetta-Toscani-Wennberg metric [17]

$$d_2(f, g) := \sup_{\boldsymbol{\xi}\neq 0, \boldsymbol{\eta}\neq 0} \frac{|\widehat{f}(\boldsymbol{\xi}, \boldsymbol{\eta}) - \widehat{g}(\boldsymbol{\xi}, \boldsymbol{\eta})|}{|\boldsymbol{\xi}|^2 + |\boldsymbol{\eta}|^2}. \tag{20}$$

where $\widehat{f}$ is the Fourier transform of f and one has to impose that the first moments of f and g vanish and the second moments of f and g are finite.

Theorem 3 *Let $F(\boldsymbol{v}, \boldsymbol{w})$ be the initial distribution for the system plus reservoir of the form*

$$F_0(\boldsymbol{v}, \boldsymbol{w}) = f_0(\boldsymbol{v})\gamma_N(\boldsymbol{w}).$$

with f_0 symmetric and satisfying the moment conditions mentioned above. Assume moreover that

$$\int v_i^4 f_0(\boldsymbol{v})\, d\boldsymbol{v} = E_4 < \infty, \tag{21}$$

then for every $t > 0$ we have

$$d_2\left(e^{\mathcal{L}_R t} F_0, e^{\mathcal{L}_T t} F_0\right) \leq \frac{KM}{N}\left(1 - e^{-\frac{\mu}{4}t}\right)\sqrt{d_2(f_0, \gamma_N)(F_4 + d_2(f_0, \gamma_N))}\,. \tag{22}$$

with $F_4 = 3\pi^4\left(E_4 + \frac{\pi+2}{\pi^2}\right)$ and $K = 16\sqrt{3}$ and where $\gamma_N(\boldsymbol{w}) = e^{-\pi|\boldsymbol{w}|^2}$.

This theorem shows, that if one starts with an initial condition in which the reservoir is in equilibrium, then the two evolutions, the one with the thermostat and the one with the finite reservoir, stay close uniformly in time at least for the case where $N >> M$. The reservoir does never move far from its equilibrium.

Unfortunately, the above result does not allow to draw any conclusions about the decay of entropy for the finite reservoir problem. A reasonable proposal, which is in some ways closer to Boltzmann's ideas [2], is to find good rates for entropy decay for the marginals of the distributions. Returning to (18) one focuses on the marginal

$$f(\boldsymbol{v}, t) = \int_{\mathbb{R}^N} e^{-\mathcal{L}_R t} F_0(\boldsymbol{v}, \boldsymbol{w})\, d\boldsymbol{w}$$

and considers the relative entropy

$$H(f(\cdot, t)|\gamma_M)\,.$$

Note that γ_M is the equilibrium state for the system coupled to the thermostat. However, it is not the equilibrium state for the system coupled to the reservoir. The following theorem is proved in [5]

Theorem 4 *For any positive integers N, M we have that*

$$H(f(\cdot,t)|\gamma_M) \le \left(\frac{M}{N+M} + e^{-\frac{\mu(N+M)}{2N}t}\frac{N}{N+M}\right) H(f_0(\cdot)|\gamma_M)\,. \tag{23}$$

The entropy relative to the Gaussian cannot converge to zero, because the Gaussian is not the equilibrium state. The theorem, however, states that if the reservoir is large compared to the system, then the entropy relative to the Gaussian decreases at an exponential rate which is uniform in N. In fact, as $N \to \infty$ this converges to the rate of the thermostat.

4 Information

The proof of Theorem 4 in [5] is rather cumbersome and required the sharp form of the Brascamp-Lieb inequalities. In this section, we indicate a relatively simple proof using the notion of information which was published in [3] for a one-dimensional model. In [18], this was generalized to multidimensional models with a different collision mechanism. We explain the ideas for the case of a three dimensional system undergoing momentum preserving collisions. The law governing the collision between particles i and j is given by

$$v_i^*(\sigma) = v_i - (\sigma\cdot(v_i - v_j))\sigma\,, \quad v_j^*(\sigma) = v_j + (\sigma\cdot(v_j - v_i))\sigma\,, \quad \sigma \in \mathbb{S}^2\,, \tag{24}$$

while the velocities of the other particles are fixed. We set

$$R_{i,j}F(\boldsymbol{u}) = \int_{\mathbb{S}^2} F(u_1,\ldots,u_i^*(\sigma),\ldots,u_j^*(\sigma),\ldots,u_{M+N})\,\rho(d\sigma) \tag{25}$$

where $\rho(d\sigma)$ is a probability measure on $\mathbb{S}^2$. Note that this is a much simpler model than the three dimensional hard sphere model. The collision cross section is independent of the momentum transfer.

The model under consideration is the same as in equation (18) except that we replace all the collision operators $R_{i,j}$ of (19) by (25). As before one starts with the initial condition

$$F_0(\boldsymbol{v},\boldsymbol{w}) = f_0(\boldsymbol{v})e^{-\pi|\boldsymbol{w}|^2}$$

and we label the variables as

$$F(\boldsymbol{v},\boldsymbol{w},t) = F(v_1,\ldots,v_M,w_{M+1},\ldots,w_{M+N},t)$$

where $F(\mathbf{v}, \mathbf{w}, t)$ is the probability distribution on $\mathbb{R}^{3M+3N}$ given by $e^{-\mathcal{L}_R t} F_0$. As before we consider the marginal

$$f(\mathbf{v}, t) = \int_{\mathbb{R}^{3N}} F(\mathbf{v}, \mathbf{w}, t)\, d\mathbf{w}$$

and the relative entropy

$$H(f(\cdot, t)|\gamma_M) = \int_{\mathbb{R}^{3M}} f(\mathbf{v}, t) \log \frac{f(\mathbf{v}, t)}{\gamma_M(\mathbf{v})}\, d\mathbf{v}$$

where $\gamma_M(\mathbf{v}) = e^{-\pi|\mathbf{v}|^2}$. The following theorem is proved in [18].

Theorem 5 *Assume that the initial distribution $f_0 \in L^1(\mathbb{R}^M)$ has finite second moment and $\sqrt{f_0} \in H^1\left(\mathbb{R}^M\right)$. Assume further that*

$$\int_{\mathbb{S}^2} \sigma_i \sigma_j\, \rho(d\sigma) = C \cdot \delta_{ij}\,, \tag{26}$$

for some constant $C \in \mathbb{R}$. Then

$$H(f(\cdot, t)|\gamma_M) \le \left(\frac{M}{M+N} + \frac{N}{M+N} e^{-\frac{\mu}{3}\frac{M+N}{N} t} \right) H(f_0|\gamma_M)\,. \tag{27}$$

The time evolution keeps the Gaussian $e^{-\pi[|\mathbf{v}|^2+|\mathbf{w}|^2]}$ fixed and it will be convenient to set

$$F(\mathbf{v}, \mathbf{w}, t) = G(\mathbf{v}, \mathbf{w}, t) e^{-\pi[|\mathbf{v}|^2+|\mathbf{w}|^2]}\,,$$

where

$$G(\mathbf{v}, \mathbf{w}, t) = [e^{-\mathcal{L}_R t} g_0](\mathbf{v}, \mathbf{w})\,.$$

With

$$f_0(\mathbf{v}) = g_0(\mathbf{v}) e^{-\pi|\mathbf{v}|^2}\,, \quad f(\mathbf{v}, t) = g(\mathbf{v}, t) e^{-\pi|\mathbf{v}|^2}$$

we find that

$$g(\mathbf{v}, t) = \int_{\mathbb{R}^{3N}} G(\mathbf{v}, \mathbf{w}, t) e^{-\pi|\mathbf{w}|^2} d\mathbf{w}$$

and

$$H(f(\cdot, t)|\gamma_M) = \int_{\mathbb{R}^{3M}} g(\mathbf{v}, t) \log g(\mathbf{v}, t) e^{-\pi|\mathbf{v}|^2} d\mathbf{v}\,.$$

We also note that the time evolution for G is essentially the same as the one for F. In this new representation, we define the *information*

$$I(g) = \int_{\mathbb{R}^{3M}} \frac{|\nabla g(v)|^2}{g(v)} e^{-\pi|v|^2} dv\,.$$

The following theorem about the decay of information is proven in [18]. A sketch of the proof is given at the end of the section.

Theorem 6 *Assume that the initial distribution $f_0 \in L^1(\mathbb{R}^M)$ has finite second moment and $\sqrt{f_0} \in H^1(\mathbb{R}^M)$. Assume further that*

$$\int_{\mathbb{S}^2} \sigma_i \sigma_j \, \rho(d\sigma) = C \cdot \delta_{ij}\,, \tag{28}$$

for some constant $C \in \mathbb{R}$. Then

$$I(g(\cdot,t)) \le \left(\frac{M}{M+N} + \frac{N}{M+N} e^{-\frac{\mu}{3}\frac{M+N}{N}t} \right) I(g_0)\,. \tag{29}$$

Theorem 5 can be derived from Theorem 6. To see this, consider the Ornstein-Uhlenbeck semigroup on $\mathbb{R}^{3M+3N}$ defined as

$$P_s G(v, w) = \int_{\mathbb{R}^{3M+3N}} G(e^{-s}(v, w) + \sqrt{1-e^{-2s}}(x, y)) e^{-\pi[|x|^2+|y|^2]} \, dx\, dy\,.$$

It is well known and easy to check that

$$H(g) = \int_0^\infty I(P_{M,s} g)\, ds,$$

where we denoted by $P_{M,s}$ the Ornstein-Uhlenbeck semigroup on $\mathbb{R}^{3M}$. To use this connection between entropy and information, we first claim that for all $s, t \ge 0$

$$P_s e^{-\mathcal{L}_R t} = e^{-\mathcal{L}_R t} P_s\,.$$

The collision law is given by a reflection in $\mathbb{R}^{3M+3N}$ which commutes with P_s, i.e. the operators $R_{i,j}$ representing the reflections of a single collision process commute with P_s. Since the time evolution operator is an average over reflections, the claim follows.

Further, we observe that

$$\begin{aligned} &\int_{\mathbb{R}^{3N}} P_s G(v, w) e^{-\pi|w|^2} \, dw \\ &\quad = \int G(e^{-s}v + \sqrt{1-e^{-2s}}x, e^{-s}w + \sqrt{1-e^{-2s}}y) e^{-\pi[|x|^2+|y|^2]} e^{-\pi|w|^2} \, dx\, dy\, dw \end{aligned}$$

which by choosing variables $p = e^{-s}w + \sqrt{1-e^{-2s}}y$ and $q = -\sqrt{1-e^{-2s}}w + e^{-s}y$ equals

$$\int G(e^{-s}v+\sqrt{1-e^{-2s}}x,\, p)e^{-\pi[|x|^2+|p|^2+|q|^2]}\,dx\,dp\,dq$$
$$= \int G(e^{-s}v+\sqrt{1-e^{-2s}}x,\, p)e^{-\pi[|x|^2+|p|^2]}\,dx\,dp$$
$$= P_{M,s}g(v)\,.$$

This enables us to compute

$$H(g(\cdot,t)|\gamma_M) = \int_0^\infty I(P_{M,s}g(v,t))\,ds = \int_0^\infty I\left(\int_{\mathbb{R}^{3N}} P_sG(v,w)\,e^{-\pi|w|^2}\,dw\right)ds$$
$$= \int_0^\infty I\left(\int_{\mathbb{R}^{3N}} \left[P_s e^{-\mathcal{L}_R t} g_0\right](v,w)\,e^{-\pi|w|^2}\,dw\right)ds\,.$$

Using that $P_s e^{-\mathcal{L}_R t} = e^{-\mathcal{L}_R t} P_s$, we get

$$H(g(\cdot,t)|\gamma_M) = \int_0^\infty I\left(\int_{\mathbb{R}^{3N}} \left[e^{-\mathcal{L}_R t} P_s g_0\right](v,w)\,e^{-\pi|w|^2}\,dw\right)ds$$
$$= \int_0^\infty I\left(\int_{\mathbb{R}^{3N}} \left[e^{-\mathcal{L}_R t} P_{M,s} g_0\right](v,w)\,e^{-\pi|w|^2}dw\right)ds.$$

Applying theorem 6 to the function $[P_{M,s}g](v,t)$ yields the desired bound

$$H(g(\cdot,t)|\gamma_M) \le \left(\frac{M}{M+N} + \frac{N}{M+N}e^{-\mu\frac{M+N}{3N}t}\right)\int_0^\infty I(P_{M,s}g_0(v))\,ds$$
$$= \left(\frac{M}{M+N} + \frac{N}{M+N}e^{-\mu\frac{M+N}{3N}t}\right)H(g_0|\gamma_M).$$

Proof (Sketch of a proof of Theorem 6) We write the master equation as

$$\partial_t F = \Lambda[Q-I]F$$

where $\Lambda = \mu M + \lambda_S M + \lambda_R N$ and

$$Q = \frac{2\lambda_S}{(M-1)\Lambda}\sum_{i<j=1}^{M} R_{i,j} + \frac{2\lambda_R}{(N-1)\Lambda}\sum_{i<j=M+1}^{M+N} R_{i,j} + \frac{\mu}{N\Lambda}\sum_{i=1}^{M}\sum_{j=M+1}^{M+N} R_{i,j}$$

which is a convex combination of the $R_{i,j}$. Thus we may abbreviate this as

$$Q = \sum_{\text{pairs}} \lambda^\alpha R_\alpha\,.$$

with $\sum_\alpha \lambda^\alpha = 1$. The time evolution can then be written as a sum over collision histories

$$e^{\mathcal{L}t} = e^{-\Lambda t} \sum_{k=0}^{\infty} \frac{(\Lambda t)^k}{k!} Q^k = e^{-\Lambda t} \sum_{k=0}^{\infty} \frac{(\Lambda t)^k}{k!} \sum_{\underline{\alpha}_k} \lambda^{\underline{\alpha}_k} R_{\underline{\alpha}_k},$$

where we used multi index notation. Each operator $R_{\underline{\alpha}_k} = R_{\alpha_k} \cdots R_{\alpha_1}$ is now a product of pair collision operators $R_\alpha = R_{i,j}$. The action on the initial condition g_0 of k-collisions is given by

$$[R_{\underline{\alpha}_k} g_0](\boldsymbol{v}, \boldsymbol{w}) = \int d\rho(\sigma_1) \cdots d\rho(\sigma_k)\, g_0(M_{\underline{\alpha}_k}(\underline{\sigma}_k)(\boldsymbol{v}, \boldsymbol{w}))$$

where the $3(M+N) \times 3(M+N)$ orthogonal matrix $M_{\underline{\alpha}_k}(\underline{\sigma}_k)$ is a product of collision matrices.[1] We use the notation $\underline{\sigma}_k = (\sigma_k, \ldots, \sigma_1)$. Writing

$$M_{\underline{\alpha}_k}(\underline{\sigma}_k) = \begin{bmatrix} A_{\underline{\alpha}_k}(\underline{\sigma}_k) & B_{\underline{\alpha}_k}(\underline{\sigma}_k) \\ C_{\underline{\alpha}_k}(\underline{\sigma}_k) & D_{\underline{\alpha}_k}(\underline{\sigma}_k) \end{bmatrix},$$

we find, since g_0 is only a function of the variable $\boldsymbol{v}$, that

$$[R_{\underline{\alpha}_k} g_0](\boldsymbol{v}, \boldsymbol{w}) = \int d\rho(\sigma_1) \cdots d\rho(\sigma_k)\, g_0(A_{\underline{\alpha}_k}(\underline{\sigma}_k)\boldsymbol{v} + B_{\underline{\alpha}_k}(\underline{\sigma}_k)\boldsymbol{w})).$$

Note that the last expression does not depend on $C_{\underline{\alpha}_k}(\underline{\sigma}_k)$ and $D_{\underline{\alpha}_k}(\underline{\sigma}_k)$. With the notation

$$g_{0,\underline{\alpha}_k}(\boldsymbol{v}) = \int_{\mathbb{R}^N} [R_{\underline{\alpha}_k} g_0](\boldsymbol{v}, \boldsymbol{w}) \gamma_N(\boldsymbol{w})\, d\boldsymbol{w},$$

we find using the convexity of information that

$$I(g(\cdot, t)) \le e^{-\Lambda t} \sum_{k=0}^{\infty} \frac{(\Lambda t)^k}{k!} \sum_{\underline{\alpha}_k} \lambda^{\underline{\alpha}_k} I(g_{0,\underline{\alpha}_k}(\cdot)).$$

This reduces the problem to showing the decay of information caused by a sequence of collisions. Note that this is now separated from the evolution in time.

A straightforward computation using convexity and the fact that $M_{\underline{\alpha}_k}(\underline{\sigma}_k)$ is an orthogonal matrix, shows that

[1] Note that the collision mechanism (24) is linear, i.e it can be represented by a matrix

$$\begin{pmatrix} v_i^*(\sigma) \\ v_j^*(\sigma) \end{pmatrix} = M_{(i,j)}^\sigma \begin{pmatrix} v_i \\ v_j \end{pmatrix}, \quad M_{(i,j)}^\sigma := \begin{pmatrix} I - \sigma \oplus \sigma & \sigma \oplus \sigma \\ \sigma \oplus \sigma & I - \sigma \oplus \sigma \end{pmatrix}. \tag{30}$$

Extending $M_{(i,j)}^\sigma$ such that it leaves all other particles invariant, we obtain a collision matrix $\underline{M}_{(i,j)}^\sigma$ acting on $\mathbb{R}^{3M+3N}$ that satisfies $F(u_1, \ldots, u_i^*(\sigma), \ldots, u_j^*(\sigma), \ldots, u_{M+N}) = F(\underline{M}_{(i,j)}^\sigma \boldsymbol{u})$.

$$I(g_{\underline{\alpha}_k}(\cdot))$$
$$\leq \int \rho(d\sigma_1)\cdots\rho(d\sigma_k)\int_{\mathbb{R}^{3M}} \frac{(\nabla g_0)^T(\mathbf{v})A_{\underline{\alpha}_k}(\underline{\sigma}_k)^T A_{\underline{\alpha}_k}(\underline{\sigma}_k)(\nabla g_0)(\mathbf{v})}{g_0(\mathbf{v})}\gamma_M(\mathbf{v})\,d\mathbf{v}\,.$$

After a somewhat more difficult calculation using the symmetry assumption on the probability measure ρ, one can see that

$$e^{-\Lambda t}\sum_{k=0}^{\infty}\frac{(\Lambda t)^k}{k!}\sum_{\underline{\alpha}_k}\lambda^{\underline{\alpha}_k}\int \rho(d\sigma_1)\cdots\rho(d\sigma_k)A_{\underline{\alpha}_k}(\underline{\sigma}_k)^T A_{\underline{\alpha}_k}(\underline{\sigma}_k)$$
$$=\left[\frac{M}{M+N}+\frac{N}{M+N}e^{-\frac{\mu}{3}\frac{M+N}{N}t}\right]I_{3M}\,.$$

The last three (in-)equalities prove Theorem 6. For the details we refer the reader to [18]. □

5 A Quantum Kac Model

Pair collisions of molecules, and more generally, pair interactions of microscopic particles, are properly described by quantum mechanics. It is natural to ask how quantum mechanics can be incorporated in Kac type models. Towards this end, let us return to (1) and write it another way, which, while a little more complex, opens an interesting door. It is well known since the work of Koopman [24] and von Neumann [29] that the classical mechanical time evolution can be written in terms of unitary flows on the Hilbert space consisting of the phase space equipped with Liouville measure (which is preserved by the classical flow). This perspective brings the classical and quantum dynamics close together, and is our point of departure.

Introduce the Hilbert space $L^2(\mathbb{R}^n)$, and define for $1 \leq i < j \leq N, \theta \in [0, 2\pi)$, the unitary operator $U_{i,j,\theta}$ by

$$U_{i,j,\theta}\psi(\mathbf{v}_N) = \psi(\mathbf{v}_N(i, j; \theta))$$

where $\mathbf{v}_N(i, j; \theta)$ is the pair rotation of $\mathbf{v}_N$ specified in (2) and (3). To any probability density F on $\mathbb{R}^N$, associate the multiplication operator ρ_F on $L^2(\mathbb{R}^n)$ by

$$\rho_F\psi(\mathbf{v}_N) := F(\mathbf{v}_N)\psi(\mathbf{v}_N).$$

Note that ρ_F is in general an unbounded positive operator. It is then easy to check that

$$U_{i,j,\theta}\rho_F U^*_{i,j,\theta} = \rho_G \quad \text{where} \quad G(\mathbf{v}_N) = F(\mathbf{v}_N(i, j; \theta))\,.$$

In particular, for each i, j and θ, $U_{i,j,\theta}\rho_F U^*_{i,j,\theta}(\mathbf{v}_N)$ has the form ρ_G where G is another probability density on $\mathbb{R}^N$

$$\binom{N}{2}^{-1}\sum_{i<j}\left(\frac{1}{2\pi}\int_0^{2\pi}\rho_F(\mathbf{v}_N(i,j;\theta))d\theta\right)$$
$$=\binom{N}{2}^{-1}\sum_{i<j}\left(\frac{1}{2\pi}\int_0^{2\pi}\left[U_{i,j,\theta}\psi(\mathbf{v}_N)\rho_F U_{i,j,\theta}\psi(\mathbf{v}_N)^*(\mathbf{v}_N)\right]d\theta\right)$$
$$=: \mathcal{Q}(\rho_F)\ . \tag{31}$$

We can then rewrite (1) as

$$\frac{\partial}{\partial t}\rho_{F(t)} = N\left[\mathcal{Q}(\rho_{F(t)}) - \rho_{F(t)}\right]. \tag{32}$$

In kinetic theory, the collisions take place on a much more rapid time scale than in Kac Master Equation, and thus are instantaneous. Each unitary transformation $U_{i,j,\theta}$ in this Koopman-von Neumann picture represents the final phase space transformation associated to one type of a completed collision between particles i and j. The operator $\mathcal{Q}$ represents an average over the output densities for all kinematically possible collisions.

Between collisions, when the particles are not interacting, the energy is simply given by $H := \sum_{j=1}^N v_j^2$, but this does not encode the part of the dynamics that drives the collisions. However, the absolutely crucial point is that with H regarded as a multiplication operator, for each i, j and θ, $U_{i,j,\theta}HU_{i,j,\theta}^* = H$. Thus H commutes with the unitary describing each collision, which is the analytical expression of the fact that the collisions conserve the energy in this Koopman-von Neumann picture.

In [13] this point of view was developed to define a Quantum Markov Semigroup that describes pair collisions of quantum particles. The energy of a single particle is given by a Hamiltonian h on a Hilbert space $\mathcal{H}$ which we take to be finite dimensional for simplicity. A state of the system of N particles is described by a density matrix ρ on $\otimes^N\mathcal{H} = \mathcal{H}_N$, i.e. a self adjoint positive trace class operator with unit trace. The role of the kinetic energy in the classical Kac model is now taken by the N-particle Hamiltonian

$$H_N = \sum_{j=1}^N I\otimes\cdots I\otimes h_i\otimes I\cdots\otimes I\ .$$

One specifies the binary collisions by a family of unitary operators $U(\sigma)$ on the two particle Hilbert space $\mathcal{H}_2 = \mathcal{H}\otimes\mathcal{H}$ that commute with $H_2 = h\otimes I + I\otimes h$. Here σ lives in a measure space $(\mathcal{C},\nu)$. The precise conditions will be given below. Note that if we extend $U(\sigma)$ to the other factors of the tensor product by the identity, then $U(\sigma)$ also commutes with H_N. The collision operator $\mathcal{Q} : \mathcal{B}(\mathcal{H}_2)\to\mathcal{B}(\mathcal{H}_2)$ is given by

$$\mathcal{Q}(A) = \int_{\mathcal{C}} U(\sigma)AU^*(\sigma)\,d\nu(\sigma)$$

where $\mathcal{B}(\mathcal{H}_2)$ denotes the space of bounded operators on $\mathcal{H}_2$. Note the parallel with (31).

The measure ν is a probability measure and it is easily seen that $\mathcal{Q}$ is a trace preserving map that is positivity preserving, in fact completely positive. Since $U(\sigma)$ commutes with H_2 this collision process preserves energy, i.e. if all the eigenstates of A have the same energy so does $\mathcal{Q}(A)$. Naturally, one wants that the collision of particle 1 with particle 2 and the collision of particle 2 with particle 1 leads to the same result. If V denotes the swap operation

$$V(\phi \otimes \psi) = \psi \otimes \phi$$

then one imposes the condition that

$$\{U(\sigma) : \sigma \in \mathcal{C}\} = \{VU(\sigma)V^* : \sigma \in \mathcal{C}\}$$

and the map $\sigma \to \sigma'$ where $VU(\sigma)V^* = U(\sigma')$ is a measurable transformation that leaves ν invariant. It is also desirable that the collision satisfies local reversibility, i.e. that

$$\{U(\sigma) : \sigma \in \mathcal{C}\} = \{U^*(\sigma) : \sigma \in \mathcal{C}\},$$

and the map $\sigma \to \sigma'$ where $U^*(\sigma) = U(\sigma')$ is a measurable transformation that leaves ν invariant. One easily sees that for any two operators A, B on $\mathcal{H}_2$ one has

$$Tr(A^*\mathcal{Q}(B)) = Tr(\mathcal{Q}(A)^*B)$$

i.e. the operation is self adjoint on the Hilbert space $\mathcal{B}(\mathcal{H}_2)$ with inner product $(A, B) = Tr(A^*B)$. We call $(\mathcal{C}, U, \nu)$ satisfying these conditions a collision specification.

The Quantum Kac Master Equation is then

$$\frac{\partial}{\partial t}\rho(t) = N[\mathcal{Q}(\rho)(t) - \rho(t)]\,, \tag{33}$$

which is formally very similar to (32). We shall be interested in the same sort of questions that we asked about solutions of (32). In particular, we shall be concerned with the long time behavior, and in this context, the question of ergodicity is crucial. Denote by $\mathcal{A}_2$ the commutative subalgebra of $\mathcal{B}(\mathcal{H}_2)$ consisting of all operators that are of the form $f(H_2)$ where $f : \sigma(H_2) \to \mathbb{C}$ is a continuous bounded function. Obviously $\mathcal{A}_2$ is a subset of $\{U(\sigma) : \sigma \in \mathcal{C}\}'$, the commutant of $\{U(\sigma) : \sigma \in \mathcal{C}\}$. We shall require that the two particle collisions are ergodic, that is

$$\mathcal{A}_2 = \{U(\sigma) : \sigma \in \mathcal{C}\}'\,.$$

In this case, *at least for* $N = 2$, the steady state solutions of (33) are exactly the density matrices ρ of the form $\rho = f(H_2)$ for some f. A natural question then is when ergodicity at the level of 2 particles implies ergodicity at the level of N particles.

5.1 Example

The following example taken from [13] is useful for understanding these concepts. For the simplest possible example, take $\mathbb{H} = \mathbb{C}^2$, so that $\mathcal{H}_2 = (\mathbb{C}^2)^{\otimes 2}$. Define the single particle Hamiltonian h by $h = \begin{bmatrix} 0 & 0 \\ 0 & 1 \end{bmatrix}$. Identify $\mathbb{C}^2 \otimes \mathbb{C}^2$ with $\mathbb{C}^4$ using the basis

$$\begin{pmatrix} 1 \\ 0 \end{pmatrix} \otimes \begin{pmatrix} 1 \\ 0 \end{pmatrix}, \quad \begin{pmatrix} 0 \\ 1 \end{pmatrix} \otimes \begin{pmatrix} 1 \\ 0 \end{pmatrix}, \quad \begin{pmatrix} 1 \\ 0 \end{pmatrix} \otimes \begin{pmatrix} 0 \\ 1 \end{pmatrix}, \quad \begin{pmatrix} 0 \\ 1 \end{pmatrix} \otimes \begin{pmatrix} 0 \\ 1 \end{pmatrix}.$$

The standard physics notation for this basis is simply

$$|00\rangle\,, \quad |10\rangle\,, \quad |01\rangle\,, \quad |11\rangle\,, \tag{34}$$

which will be useful. With this identification of $\mathbb{C}^2 \otimes \mathbb{C}^2$ with $\mathbb{C}^4$,

$$\begin{bmatrix} a_{1,1} & a_{1,2} \\ a_{2,1} & a_{2,2} \end{bmatrix} \otimes \begin{bmatrix} b_{1,1} & b_{1,2} \\ b_{2,1} & b_{2,2} \end{bmatrix} =: A \otimes B \quad \text{is represented by} \quad \begin{bmatrix} b_{1,1}A & b_{1,2}A \\ b_{2,1}A & b_{2,2}A \end{bmatrix}.$$

(Switching the order of the second and third basis elements swaps the roles of A and B in the block matrix representation of the tensor product $A \otimes B$.)

In this basis,

$$H_2 = \begin{bmatrix} 0 & 0 \\ 0 & 1 \end{bmatrix} \otimes I + I \otimes \begin{bmatrix} 0 & 0 \\ 0 & 1 \end{bmatrix} = \begin{bmatrix} 0 & 0 & 0 & 0 \\ 0 & 1 & 0 & 0 \\ 0 & 0 & 1 & 0 \\ 0 & 0 & 0 & 2 \end{bmatrix}.$$

Therefore, the spectrum of H_2 is $\{0, 1, 2\}$ and

$$P_0 = \begin{bmatrix} 1 & 0 & 0 & 0 \\ 0 & 0 & 0 & 0 \\ 0 & 0 & 0 & 0 \\ 0 & 0 & 0 & 0 \end{bmatrix}, \quad P_1 = \begin{bmatrix} 0 & 0 & 0 & 0 \\ 0 & 1 & 0 & 0 \\ 0 & 0 & 1 & 0 \\ 0 & 0 & 0 & 0 \end{bmatrix} \quad \textit{and} \quad P_2 = \begin{bmatrix} 0 & 0 & 0 & 0 \\ 0 & 0 & 0 & 0 \\ 0 & 0 & 0 & 0 \\ 0 & 0 & 0 & 1 \end{bmatrix}.$$

Now define $\mathcal{C} = \mathbb{S}^1 \times \mathbb{S}^1 \times \mathbb{S}^1 \times \mathbb{S}^1$ identifying each copy of $\mathbb{S}^1$ with the unit circle in $\mathbb{C}$ so that the general point in $\sigma \in \mathcal{C}$ has the form $\sigma = (e^{i\varphi}, e^{i\theta}, e^{i\psi}, e^{i\eta})$. Then define

$$U(\sigma) := \begin{bmatrix} e^{i\theta} & 0 & 0 & 0 \\ 0 & e^{i\psi}\cos\theta & -e^{i\varphi}\sin\theta & 0 \\ 0 & e^{-i\varphi}\sin\theta & e^{-i\psi}\cos\theta & 0 \\ 0 & 0 & 0 & e^{i\eta} \end{bmatrix}$$

Choosing ν to be the uniform probability measure on $\mathcal{C}$ gives us a collision specification $(\mathcal{C}, U, \nu)$.

A simple computation shows that for every operator A on $\mathcal{H}_2 = \mathbb{C}^2 \otimes \mathbb{C}^2$ identified as the 4×4 matrix with entries $a_{i,j}$ using the basis (34),

$$\begin{aligned} \mathcal{Q}(A) = \int_{\mathcal{C}} d\nu(\sigma)\, U(\sigma) A U^*(\sigma) &= \begin{bmatrix} a_{1,1} & 0 & 0 & 0 \\ 0 & \frac{1}{2}(a_{2,2}+a_{3,3}) & 0 & 0 \\ 0 & 0 & \frac{1}{2}(a_{2,2}+a_{3,3}) & 0 \\ 0 & 0 & 0 & a_{4,4} \end{bmatrix} \\ &= a_{1,1}P_0 + \frac{a_{2,2}+a_{3,3}}{2}P_1 + a_{4,4}P_2 \in \mathcal{A}_2 \,. \end{aligned} \tag{35}$$

Therefore,

$$\{U(\sigma) \,:\, \sigma \in \mathcal{C}\}' \subset \mathrm{ran}(\mathcal{Q}) \subset \mathcal{A}_2 \subset \{U(\sigma) \,:\, \sigma \in \mathcal{C}\}',$$

showing that $(\mathcal{C}, U, \nu)$ is ergodic.

Using these preliminaries it is now straightforward to write the corresponding Quantum Master Equation (QME) as

$$\partial_t \rho = -\mathfrak{L}_N(\rho)$$

with

$$\mathfrak{L}_N(A) = N\binom{N}{2}^{-1} \sum_{i<j} [A - \mathcal{Q}_{i,j}(A)] \tag{36}$$

and where the unitaries in the definition of $\mathcal{Q}_{i,j}$ act nontrivially only on the i-th and j-th factors in the tensor product $\otimes^N \mathcal{H}$. This is a trace preserving completely positive map, i.e. a Quantum Evolution.

5.2 Propagation of Chaos

A density matrix is symmetric if it is invariant under the swap operation between any two factors in the tensor product $\otimes^N \mathcal{H}$. A sequence of symmetric density matrices $\{\rho_N\}_{N=1}^{\infty}$ is chaotic with marginal ϱ, or in short ρ-chaotic, if

$$\lim_{N\to\infty} \mathrm{Tr}_{2,\dots,N}\, \rho_N = \varrho \text{ and } \lim_{N\to\infty} \mathrm{Tr}_{k+1,\dots,N}\, \rho_N = \otimes^k \varrho$$

where $\mathrm{Tr}_{k+1,\ldots,N}$ is the trace taken in the factors $k+1,\ldots,N$. A trivial example of a chaotic sequence is $\otimes^N \varrho$, but one can also construct chaotic sequences that have a sharply defined energy for large N.

We have (see [13]).

Theorem 7 *Let $\{U(\sigma)\ \sigma \in C\}$ be a set of collision operators and let ν be a given Borel probability measure on C. Let $\mathfrak{L}_N$ be defined in terms of these as in (36). Then the semigroup $\mathcal{P}_{N,t} = e^{t\mathfrak{L}_N}$ propagates chaos for all t meaning that if $\{\rho_N\}_{N\in\mathbb{N}}$ is a ϱ-chaotic sequence, then for each t, $\{\mathcal{P}_{N,t}\varrho_N\}_{N\in\mathbb{N}}$ is a $\varrho(t)$-chaotic sequence for some $\varrho(t) = \lim_{N\to\infty}(\mathcal{P}_{N,t}\varrho_N)^{(1)}$, where in particular this limit of the one-particle marginal exists and is a density matrix.*

As expected the marginal density matrix $\varrho(t)$ satisfies a Quantum Kac-Boltzmann equation

$$\frac{d}{dt}\varrho(t) = 2(\varrho(t) \star \varrho(t) - \varrho(t))$$

where quite generally for operators in $\mathcal{B}(\mathcal{H})$

$$A \star B = \mathrm{Tr}_2\left[d\nu(\sigma)\,U(\sigma)[A \otimes B]U^*(\sigma)\right] = \mathrm{Tr}_2[\mathcal{Q}(A \otimes B)]$$

is the Quantum Wild Convolution.

5.3 Equilibrium States

An equilibrium density matrix for the evolution (36) is given by all those density matrices ρ_N that satisfy

$$\mathfrak{L}_N(\rho_N) = 0\,.$$

Recall that the N-particle Hamiltonian is $H_N = \sum_{j=1}^N h_j$ where h_j is the single particle Hamiltonian acting on the jth factor. List the eigenvalues of h as $e_1,\ldots,e_K$ counting their multiplicities and denote the corresponding eigenvectors by $\phi_1,\ldots,\phi_K$. Using the multi-index notation $\alpha = (\alpha_1,\ldots,\alpha_N)$ where $\alpha_j \in \{1,\ldots,K\}$, $j = 1,\ldots,N$ the eigenvalues of H_N are given by $E_\alpha = \sum_{j=1}^N e_{\alpha_j}$ and $\Psi_\alpha = \phi_{\alpha_1} \otimes \cdots \otimes \phi_{\alpha_N}$ are the eigenvectors.

It is not very difficult to show that the set $\mathfrak{C}_N$ of equilibrium states forms a commutative von Neumann algebra, and hence it is generated by the minimal projections. The algebra $\mathcal{A}_N$ consisting of all operators of the form $f(H_N)$, where f is a bounded continuous function, is a sub-algebra of $\mathfrak{C}_N$ and it is generated by the spectral projections of the Hamiltonian H_N. Define two multi-indices α, α' to be adjacent if for some pair (i,j), $e_{\alpha_i} + e_{\alpha_j} = e_{\alpha'_i} + e_{\alpha'_j}$ and $\alpha_k = \alpha'_k$, $k \neq i,j$. With this notion of adjacency the multi-indices α form a graph, the adjacency graph $\mathcal{G}_N$. We denote by $\gamma_1,\ldots,\gamma_n$ the connected components of $\mathcal{G}_N$. In [13] the following theorem is proved.

Theorem 8 *The minimal projections of $\mathfrak{C}_N$ are in one to one correspondence with the connected components of the adjacency graph $\mathcal{G}_N$ and are given by*

$$\mathcal{P}_k = \sum_{\alpha \in \gamma_k} |\Psi_\alpha\rangle\langle\Psi_\alpha| \,.$$

Ergodicity in our context is the notion that the only equilibrium states of the Quantum Kac Model are given by the algebra $\mathcal{A}_N$. By the above theorem this is the case if the connected components of the adjacency graph are determined by the energies of the Hamiltonian H_N. The occupation number representation is useful in this context. We write $E_\alpha = \sum_{j=1}^{K} k_j(\alpha)e_j$ where $k_j(\alpha)$ denotes the number of times the index j occurs in α. Thus, if the energies of h, $\{e_1, \dots, e_K\}$ are rationally independent then any eigenvalue of H_N is uniquely determined by the occupation numbers $k_1(\alpha), \dots, k_K(\alpha)$ (see below). Hence, in this case we have that the minimal projections of $\mathfrak{C}_N$ are eigenprojections of H_N and hence $\mathfrak{C}_N = \mathcal{A}_N$.

Here is an example where $\mathfrak{C}_N \neq \mathcal{A}_N$. Assume the single particle Hamiltonian has the eigenvalues $1, 2, 4$ with the corresponding eigenvectors ψ_1, ψ_2, ψ_3. Then pick n_1 to be even integers and set

$$n_2 = N - \frac{3}{2}n_1 \,, \quad n_3 = \frac{1}{2}n_1 \,.$$

Then

$$n_1 + 2n_2 + 4n_3 = 2N \,, n_1 + n_2 + n_3 = N \,.$$

The number $e = 2N$ is an eigenvalue of the Hamiltonian H_N and it is degenerate. The eigenvectors are of the form $\psi_{\alpha_1} \otimes \cdots \otimes \psi_{\alpha_N}$ where $\alpha_j \in \{1, 2, 3\}$. We set $\alpha = (\alpha_1, \dots, \alpha_N)$ and $n_1(\alpha)$ the number of ψ_1 factors, $n_2(\alpha)$ the number of ψ_2 factors and $n_3(\alpha)$ the number of ψ_3 factors. If α and β are adjacent, then the condition $e_{\alpha_k} + e_{\alpha_\ell} = e_{\beta_k} + e_{\beta_\ell}$ implies that either $e_{\alpha_k} = e_{\beta_k}$ and $e_{\alpha_\ell} = e_{\beta_\ell}$ or $e_{\alpha_k} = e_{\beta_\ell}$ and $e_{\alpha_\ell} = e_{\beta_k}$; anything else is not possible. Hence for any of the indices α and β to be adjacent, we must have that $n_1(\alpha) = n_1(\beta)$, $n_2(\alpha) = n_2(\beta)$, $n_3(\alpha) = n_3(\beta)$. Thus, if these triples are different, but with the same N and e, the two states are not adjacent and hence $\mathcal{G}_{e,N}$, the adjacency graph for a fixed energy e is not a connected graph. The number of elements in a connected component of $\mathcal{G}_{e,N}$ is given by

$$\frac{N!}{n_1!n_2!n_3!}$$

where $N = n_1 + n_2 + n_3$.

The Quantum Kac Master Equation (QKME), being a completely positive map can be written in terms of Kraus operators (see [10]). The collision specifications yield that the Kraus operators are self-adjoint and hence the QKME can be brought into a Lindblad form $\partial_t \rho = \sum_k [V_k, [V_k, \rho]]$. An example, closely related to Example 5.1, is the following Lindblad equation $\partial_t \rho = L_N(\rho)$, where

$$L_N(\rho) = \frac{1}{N-1} \sum_{[\alpha,\beta]\in\mathcal{E}_N} [L_{\alpha,\beta}, [L_{\alpha,\beta}, \rho]] .$$

Here, $\mathcal{E}_N$ is the edge set of the graph $\mathcal{G}_N$. With $F_{\alpha,\beta} = |\Psi_\alpha\rangle\langle\Psi_\beta|$, the 'angular' momentum operators $L_{\alpha,\beta}$ are given by

$$L_{\alpha,\beta} = F_{\alpha,\beta} - F_{\beta,\alpha} .$$

Note that in Example 5.1 the operator given by the collision specifications is L_N up to a factor that commutes with the angular momenta $L_{\alpha,\beta}$. The interesting point is that the gap of the generator L_N is given by the gap of the combinatorial or graph Laplacian on $\mathcal{G}_N$. To describe this we shall assume that the eigenvalues of h are rationally independent. The energies of the Hamiltonian H_N are given by

$$E(\alpha) = \sum_{j=1}^{K} k_j(\alpha) e_j$$

where the $k_j(\alpha)$ are integers and $\sum_{j=1}^{K} k_j(\alpha) = N$. Since the e_j's are rationally independent, the eigenvalues of H_N are in one to one correspondence with the 'occupation numbers' $\mathbf{k}(\alpha) = (k_1(\alpha), \ldots, k_K(\alpha))$. Next, note that $\mathbf{k}(\alpha) = \mathbf{k}(\beta)$ if and only if α and β are related by a finite sequence of pair transpositions. Thus, $H\Psi_\alpha = E\Psi_\alpha$ and $H_N\Psi_\beta = E\Psi_\beta$ if and only if α and β are adjacent in $\mathcal{G}_N$. In other words, there is a one to one correspondence between the eigenspaces of H_N and the connected components of $\mathcal{G}_N$. This is precisely the case in Example 5.1. Indeed the energies of $\mathcal{H}_N$ are given by

$$E(\alpha) = k_1(\alpha) \times 0 + k_2(\alpha) \times 1$$

and with $k_1(\alpha) + k_2(\alpha) = N$ the occupation numbers determine $E(\alpha)$ uniquely. The vertices of the graph $\mathcal{G}_N$ are given by multi indices of length N consisting of 1s and 0s and two indices are connected if one can be transformed into the other by a series of transpositions. Thus, in this case multi indices are adjacent if and only if they have the same number of 0s and hence 1s and clearly the occupation numbers determine the connected components of $\mathcal{G}_N$ uniquely. The subgraphs given by the connected components are well known under the name Johnson Graphs or Johnson Association Schemes. In particular the eigenvalues of the graph Laplacian of these graphs are all known as are the eigenvectors [16]. The following theorem is a special case of a result that will appear in [10].

Theorem 9 *Assume that the eigenalues of h are rationally independent and $N > 2$. Then the gap of $\mathcal{L}_N$ is*

$$\frac{2N}{N-1} .$$

Much remains to be understood concerning the Quantum Kac Master Equation. At present, little is known about sharp entropy production inequalities. Beyond this, there are very interesting questions that have no classical analog—these questions concern *entanglement* which Schrödinger singled out as the fundamental feature setting quantum mechanics apart from classical mechanics. The equilibrium states of the Quantum Kac Master Equations studied here are *separable*; that is, free of entanglement. The rate at which entanglement is "broken" along the flow is of physical interest, and has been investigated in related models [20].

We hope this account of various aspects of the Kac model, classical and quantum, inspires further progress in this favorite field of research of Maria Carvalho.

References

1. Bodineau, T., Gallagher, I., Saint-Raymond, L.: The Brownian motion as the limit of a deterministic system of hard-spheres. Invent. Math. **203**(2), 493–553 (2016)
2. Boltzmann, L.: Lectures on gas theory. University of California Press, Berkeley-Los Angeles, California. Translated by Stephen G. Brush (1964)
3. Bonetto, F., Han, R., Loss, M.: Decay of information for the KAC evolution. Ann. Henri Poincaré **22**(9), 2975–2993 (2021)
4. Bonetto, F., Loss, M., Tossounian, H., Vaidyanathan, R.: Uniform approximation of a Maxwellian thermostat by finite reservoirs. Commun. Math. Phys. **351**(1), 311–339 (2017)
5. Bonetto, F., Geisinger, A., Loss, M., Ried, T.: Entropy decay for the KAC evolution. Commun. Math. Phys. **363**(3), 847–875 (2018)
6. Bonetto, F., Loss, M., Vaidyanathan, R.: The KAC model coupled to a thermostat. J. Stat. Phys. **156**(4), 647–667 (2014)
7. Carlen, E., Carvalho, M.C., Loss, M.: Many-body aspects of approach to equilibrium. In Séminaire: Équations aux Dérivées Partielles, 2000–2001, Sémin. Équ. Dériv. Partielles, pages Exp. No. XIX, 12. École Polytech., Palaiseau (2001)
8. Carlen, E.A., Carvalho, M.C., Loss, M.: Determination of the spectral gap for KAC's master equation and related stochastic evolution. Acta Math. **191**(1), 1–54 (2003)
9. Carlen, E., Carvalho, M., Loss, M.: Spectral gaps for reversible Markov processes with chaotic invariant measures: the KAC process with hard sphere collisions in three dimensions. Ann. Probab. **48**(6), 2807–2844 (2020)
10. Carlen, E., Loss, M: In preparation (2023)
11. Carlen, E.A., Carrillo, J.A., Carvalho, M.C.: Strong convergence towards homogeneous cooling states for dissipative Maxwell models. Ann. Inst. H. Poincaré C Anal. Non Linéaire **26**(5), 1675–1700 (2009)
12. Carlen, E.A., Carvalho, M.C., Le Roux, J., Loss, M., Villani, C.: Entropy and chaos in the KAC model. Kinet. Relat. Models **3**(1), 85–122 (2010)
13. Carlen, E.A., Carvalho, M.C., Loss, M.P.: Chaos, ergodicity, and equilibria in a quantum KAC model. Adv. Math. **358**, 106827, 50, (2019)
14. Cortez, R., Tossounian, H.: Uniform propagation of chaos for the thermostated KAC model. J. Stat. Phys. **183**(2), Paper No. 28, 17 (2021)
15. Einav, A.: On Villani's conjecture concerning entropy production for the KAC master equation. Kinet. Relat. Models **4**(2), 479–497 (2011)
16. Filmus, Y.: An orthogonal basis for functions over a slice of the Boolean hypercube. Electron. J. Combin. **23**(1), Paper 1.23, 27 (2016)
17. Gabetta, E., Toscani, G., Wennberg, B.: Metrics for probability distributions and the trend to equilibrium for solutions of the boltzmann equation. J. Stat. Phys. **81**, 901–934 (1995)

18. Hauger, L.: Decay of entropy and information for multidimensional KAC models (2023). arXiv:2301.09592
19. Illner, R., Pulvirenti, M.: Global validity of the Boltzmann equation for two- and three-dimensional rare gas in vacuum. Erratum and improved result: "Global validity of the Boltzmann equation for a two-dimensional rare gas in vacuum" [Comm. Math. Phys. **105**(2), 189–203 (1986); MR0849204 (88d:82061)] and "Global validity of the Boltzmann equation for a three-dimensional rare gas in vacuum" [ibid. **113**(1), 79–85 (1987); MR0918406 (89b:82052)] by Pulvirenti. Commun. Math. Phys. **121**(1), 143–146 (1989)
20. Ippoliti, M., Gullans, M.J., Gopalakrishnan, S., Huse, D.A., Khemani, V.: Entanglement phase transitions in measurement-only dynamics. Phys. Rev. X **11**, 011030 (2021). Feb
21. Janvresse, E.: Spectral gap for KAC's model of Boltzmann equation. Ann. Probab. **29**(1), 288–304 (2001)
22. Kac, M.: Foundations of kinetic theory. In: Proceedings of the Third Berkeley Symposium on Mathematical Statistics and Probability, 1954–1955, vol. III, pp. 171–197. University of California Press, Berkeley-Los Angeles, California (1956)
23. Kac, M.: Probability and related topics in physical sciences, volume 1957 of Lectures in Applied Mathematics (Proceedings of the Summer Seminar, Boulder, Colorado. Interscience Publishers, London-New York. With special lectures by G.E. Uhlenbeck, A.R. Hibbs, and B. van der Pol (1959)
24. Koopman, B.O.: Hamiltonian systems and transformations in Hilbert space. Proc. Nat. Acad. Sci. **17**(5), 315–318 (1931)
25. Lanford, O.E. III.: A derivation of the Boltzmann equation from classical mechanics. In: Probability (Proc. Sympos. Pure Math., vol. XXXI, Univ. Illinois, Urbana, Ill., 1976), pp. 87–89. Amer. Math. Soc., Providence, R.I. (1977)
26. Mischler, S., Mouhot, C.: Kac's program in kinetic theory. Invent. Math. **193**(1), 1–147 (2013)
27. Sznitman, A.-S.: Équations de type de Boltzmann, spatialement homogènes. Z. Wahrsch. Verw. Gebiete **66**(4), 559–592 (1984)
28. Villani, C.: Cercignani's conjecture is sometimes true and always almost true. Commun. Math. Phys. **234**(3), 455–490 (2003)
29. von Neumann, J.: Zur Operatorenmethode in der klassischen Mechanik. Ann. Math. (in German) **33**(3), 587–642 (1932)

Hydrodynamic Limit from the Boltzmann Equation in a Slightly Compressible Regime

Rossana Marra

Abstract We discuss the hydrodynamic limit for the Boltzmann equation under the diffusive scaling with initial and/or boundary non homogeneous conditions for density and temperature with gradients of order 1.

Keywords Boltzmann equation · Hydrodynamic limits

1 Introduction

In this paper we discuss one of the main open problems in the field of hydrodynamic limit for the Boltzmann equation. In kinetic theory a rarefied gas in a container Ω is described by $F(x, v, t) \geq 0$, the probability density on the phase space $\Omega \times R^3$, solution of Boltzmann equation

$$\frac{\partial F}{\partial t} + v \cdot \nabla_x \mathrm{F} = Q(F, F)$$

$$Q(f, g)(v) = \frac{1}{2} \int_{\mathbb{R}^3} dv_* \int_{\{|\omega|=1\}} d\omega B(\omega, v - v_*) \{ f(v') g(v'_*) - f(v) g(v_*) \}$$

$$v' = v - \omega[\omega \cdot (v - v_*)], \quad v'_* = v_* + \omega[\omega \cdot (v - v_*)].$$

with B a cross section. Here, we will consider only the one for hard spheres: $B(\omega, v - v_*) = |(v - v_*) \cdot \omega|$. For simplicity, Ω is a torus.

The diffusive hydrodynamic limit of the Boltzmann equation in the low Mach number regime is usually described by the incompressible Navier-Stokes-Fourier equations and this limit is proved under suitable assumptions on the initial condition.

R. Marra (✉)
Dipartimento di Fisica and Unità INFN, Università di Roma Tor Vergata, 00133 Roma, Italy
e-mail: marra@roma2.infn.it

E. Carlen et al. (eds.), *From Particle Systems to Partial Differential Equations*,
Springer Proceedings in Mathematics & Statistics 465,
https://doi.org/10.1007/978-3-031-65195-3_9

We are interested in constructing a solution $F^\varepsilon(x, v, t) = F(\varepsilon^{-1}x, v, \varepsilon^{-2}t)$ of the rescaled Boltzmann equation under the diffusive scaling

$$\frac{\partial F^\varepsilon}{\partial t} + \frac{1}{\varepsilon} v \cdot \nabla_x F^\varepsilon + = \frac{1}{\varepsilon^2} Q(F^\varepsilon, F^\varepsilon) \tag{1}$$

where ε is the Knudsen number.

Diffusive scaling here is opposite to the hyperbolic scaling in which the time is scaled only as ε^{-1}. In the last case the limiting equations are the Euler equations.

If we consider the low Mach number regime we look for a solution

$$F^\varepsilon = \mu + \varepsilon \sqrt{\mu} f^\varepsilon(x, v; t)$$

under initial conditions

$$F^\varepsilon(x, v; 0) = \mu + \varepsilon \sqrt{\mu} f^\varepsilon(x, v; 0)$$

where μ is a global Maxwellian

$$\mu(x, v) := \frac{1}{(2\pi)^{\frac{3}{2}}} \exp\left(-\frac{|v|^2}{2} \right).$$

Formally, if $f^\varepsilon \to f_0$ in a suitable (weak) sense, then

$$f_0 = \sqrt{\mu} \left[\rho + \mathbf{u} \cdot v + \frac{1}{2} \theta (|v|^2 - 3) \right]$$

with $\rho, \mathbf{u}, \theta$ solving the incompressible Navier-Stokes-Fourier system

$$\begin{aligned} &\nabla_x \cdot \mathbf{u} = 0, \quad \nabla_x(\rho + \theta) = 0, \\ &\frac{\partial \mathbf{u}}{\partial t} + \mathbf{u} \cdot \nabla_x \mathbf{u} + \nabla_x p = \nu \Delta \mathbf{u}, \\ &\frac{\partial \theta}{\partial t} + \mathbf{u} \cdot \nabla_x \theta = \kappa \Delta \theta, \end{aligned} \tag{2}$$

where p is the unknown pressure and $\nu > 0$ and $\kappa > 0$ are the transport coefficients.

There are many papers proving this result under different methodologies and conditions: by entropy methods [18, 19, 24], and by perturbative methods (Hilbert expansion) [9, 14, 20].

For other partial results for particular geometries see [1, 2, 11], (see also [17]).

Consider instead initial conditions for $F^\varepsilon(x, v, t)$ in which the density and the temperature variations are of order 1 and not perturbation of the order of the Knudsen number of constant profiles, while the velocity field is still of order of the Knudsen number (low Mach number regime)

$$F^\varepsilon(x, v; 0) = \mu + \varepsilon \sqrt{\mu} f^\varepsilon(x, v; 0), \tag{3}$$

where μ is a local Maxwellian

$$\mu(x,v) := \frac{\rho(x,0)}{\big(2\pi T(x,0)\big)^{3/2}} \exp\Big(-\frac{|v|^2}{2T(x,0)}\Big). \tag{4}$$

The formal limit is different: Burnett-like terms are present

$$\begin{aligned}
&\partial_t \rho + \nabla\cdot(\rho \mathbf{u}) = 0; \quad \nabla(\rho T) = 0\\
&\frac{3}{2}\partial_t(\rho T) + \frac{5}{2}\rho T \nabla\cdot \mathbf{u} = \nabla\cdot(\lambda \nabla T),\\
&\rho(\partial_t + \mathbf{u}\cdot\nabla)\mathbf{u}_k + \partial_k p = \partial_j\big(\mu(\partial_j \mathbf{u}_k + \partial_k \mathbf{u}_j - \frac{2}{3}\partial_\ell \mathbf{u}_\ell \delta_{kj})\\
&\qquad + \partial_j\Big(\frac{\lambda^2}{P}(K_1\big(\partial_k\partial_j T - \delta_{kj}\frac{1}{3}\Delta T) + \frac{K_2}{T}((\partial_k T)(\partial_j T) - \delta_{kj}\frac{1}{3}|\nabla T|^2)\Big)
\end{aligned} \tag{5}$$

see [5, 9, 27]. The relevance of this new set of equations has been highlighted by Sone (see [27] and reference therein) in a series of analytical and numerical works, where simulations show the presence of new phenomena. These have been called ghost effects, by the idea that a flux velocity of order Kn generates modifications in the temperature profile of order of unity, thus acting as a ghost, whose presence survives in the limit Kn $\to 0$. From a theoretical point of view, the main point is that this new system of equations can not be derived from the phenomenological hydrodynamic equations. Unfortunately, the rigorous proof of the validity of these equations turns out to be very difficult and is currently argument of research. The difficulties are strictly related to the fact that the leading orders in Kn of mass density and temperature are not constant.

We discuss here first the main difficulties for giving a rigorous proof of these limiting equations. Then we consider a simpler model for which the hydrodynamic limit can be proven and finally we discuss some stationary situations.

2 Difficulties

Under the initial conditions (3) we look for a solution to (1) which is a perturbation of a *local* Maxwellian. One technique is to use a Hilbert expansion, e.g.

$$F = \mu + \sum_{n=1}^{6} \varepsilon^n f_n + \varepsilon^3 \sqrt{\mu} R$$

where now μ is a local Maxwellian

$$\mu(\rho, 0, T) := \frac{\rho(x,t)}{\big(2\pi T(x,t)\big)^{3/2}} \exp\Big(-\frac{|v|^2}{2T(x,t)}\Big) \tag{6}$$

and

$$f_1 = -\mathscr{A} \cdot \frac{\nabla_x T}{2T^2} + \mu^{\frac{1}{2}} \left(\frac{\rho_1}{\rho} + \frac{\mathbf{u} \cdot v}{T} + \frac{T_1(|v|^2 - 3T)}{2T^2} \right), \tag{7}$$

$$\bar{\mathscr{A}} := v \cdot \left(|v|^2 - 5T\right) \mu^{\frac{1}{2}} \in \mathbb{R}^3, \quad \mathscr{A} := L^{-1} \bar{\mathscr{A}} \in \mathbb{R}^3, \tag{8}$$

for some $\rho(x, t)$, $\mathbf{u}(x, t)$, $T(x, t)$, where ρ, T, $\mathbf{u}$ are determined as solution of (5) and the other terms in the expansion are solutions of suitable linear equations.

The equation for the remainder R is

$$\frac{\partial R}{\partial t} + \frac{1}{\varepsilon} v \cdot \nabla R + \frac{1}{\varepsilon} \frac{1}{2} \left(v \cdot \nabla + \varepsilon \frac{\partial}{\partial t} \right) \log \sqrt{\mu} R$$

$$= -\frac{1}{\varepsilon^2} L R + \frac{1}{\varepsilon} \Gamma \left(\frac{f_1}{\sqrt{\mu}}, R \right) + \frac{1}{\varepsilon^2} [\Gamma(\phi_\varepsilon, R)] + \varepsilon \Gamma(R, R) + \varepsilon^2 \tilde{A},$$

where L is the linearized Boltzmann operator around the local equilibrium $\mu(\rho, 0, T)$

$$Lf = -\frac{1}{\sqrt{\mu}} Q(\mu, \sqrt{\mu} f); \quad \Gamma(f, f) = \frac{1}{\sqrt{\mu}} \Gamma(\sqrt{\mu} f, \sqrt{\mu} f),$$

$\sqrt{\mu}\phi_\varepsilon = \sum_{s=2}^{6} \varepsilon^s f_s$ and

$$\tilde{A} = -\frac{1}{\sqrt{\mu}} [\frac{\partial}{\partial t} (f_5 + \varepsilon \sqrt{\mu} f_6) - v \cdot \nabla_x f_6 + \sum_{k \geq 6} \varepsilon^{k-6} \sum_{n,m \geq 2, n+m=k} Q(f_n, f_m)].$$

Here the big difference is the term

$$\frac{1}{\varepsilon} \left(\frac{1}{2} \left(v \cdot \nabla + \varepsilon \frac{\partial}{\partial t} \right) \log \sqrt{\mu} \right) R.$$

The starting point is to consider the linear problem:

$$B = \frac{1}{\varepsilon^2} [\Gamma(\phi_\varepsilon, R)] + \varepsilon \Gamma(R, R) + \varepsilon^2 \tilde{A}$$

$$\frac{\partial R}{\partial t} + \frac{1}{\varepsilon} v \cdot \nabla R + \frac{1}{2} \left(\frac{\partial}{\partial t} + \frac{1}{\varepsilon} v \cdot \nabla \right) \log \sqrt{\mu} R + \frac{1}{\varepsilon^2} L R - \frac{1}{\varepsilon} \Gamma \left(\frac{f_1}{\sqrt{\mu}}, R \right) = B.$$

Note that L is self-adjoint in $L^2_\nu(\mathbb{R}^3)$ satisfying the coercivity property

$$\int_{\mathbb{R}^3} f L f \geq c \, \|(\mathbf{I} - \mathbf{P})[f]\|^2_{L^2_\nu}. \tag{9}$$

Here,

$$\|f\|_{L^2_\nu} := \left(\int_{\Omega \times \mathbb{R}^3} \nu(x,v)\, |f(x,v)|^2 \, dv dx \right)^{\frac{1}{2}}, \quad \nu = \sqrt{(1+|v|^2)}. \tag{10}$$

Usually, the starting step of the proof is getting an a-priori estimate in L_2, call it Green's inequality, based on the coercivity property of L.

Taking the scalar product of the equation with R and using the coercivity property for L in $L_2(R^3)$, it is possible to get the differential inequality for the the norm $\| \cdot \|_{2,2}$ in $L_2(\Omega \times \mathbb{R}^3, dxdv)$. We will also use the notation $|| \cdot ||_2$ for the L_2 norm in $L_2(\Omega, dx)$.

Decompose $R = PR + (I-P)R$ where PR is the projector on the null space of L, $PR = \sum_j \psi_j R_j \sqrt{\mu}$, where ψ_j are the components of the vector $\psi = \{\rho, v_j, |v|^2 - T\}$. We have

$$\frac{1}{2}\frac{d}{dt}\|R\|^2_{2,2} + \frac{1}{\varepsilon^2}\|\sqrt{\bar{\nu}}(I-P)R\|^2_{2,2}$$
$$\lesssim C_\gamma \varepsilon^2 \|PR\|^2_{2,2} + \frac{1}{\varepsilon}\int dv dx |R|^2 v \cdot \nabla \log \mu + \int dx |\langle B, R\rangle|.$$

We use the notation $\gtrsim$, $\lesssim$ to indicate an inequality true up to a constant. We have used the estimate

$$\left| \left\langle R, \varepsilon\Gamma\left(\frac{f_1}{\sqrt{\mu}}, PR\right)\right\rangle \right| = \left| \left\langle (1-P)R, \varepsilon\Gamma\left(\frac{f_1}{\sqrt{\mu}}, PR\right)\right\rangle \right|$$

$$\leq \|f_1\|_{2,2}\, \varepsilon \|PR\|_{2,2}\, \|\sqrt{\bar{\nu}}(1-P)R\|_{2,2} \leq C_\gamma \varepsilon \|PR\|_{2,2}\, \|\sqrt{\bar{\nu}}(1-P)R\|_{2,2}$$

$$\leq C_\gamma \varepsilon \left[\varepsilon^2 \|PR\|^2_{2,2} + \frac{1}{\varepsilon^2}\|\sqrt{\bar{\nu}}(1-P)R\|^2_{2,2} \right],$$

where the constant C is of order γ if we assume that the solution of the hydrodynamic equation is of order γ at time zero. In the case in which the middle term on the r.h.s. has no factor $\frac{1}{\varepsilon}$, it can be estimated (for bounded velocities) as before producing a term $C_\gamma \|R\|^2_{2,2}$. At this point the standard trick is to get an estimate of the type,

$$\|PR\|^2_{2,2} \lesssim \frac{1}{\varepsilon^2}\|\sqrt{\bar{\nu}}(I-P)R\|^2_{2,2} + \|B\|^2_{2,2}, \tag{11}$$

and take γ small to conclude

$$\|R\|^2_{2,2} + \int dt \|\sqrt{\bar{\nu}}(I-P)R\|^2_{2,2} \lesssim \|R_0\|^2_{2,2} + \int dt \|B\|^2_{2,2}. \tag{12}$$

Let us remark that, if at initial time $\rho = 1 + \varepsilon\bar{\rho}$, $\theta = 1 + \varepsilon\bar{\theta}$, then we get a constant εC_γ in front of the bad term and we are back in business.

This is essentially the reason why the incompressible Navier-Stokes limit has been proved only for this special initial condition and the problem of extending the result to more general initial conditions for smooth solutions is still open since 1989.

In the previous computation we have supposed that the velocities are bounded, to stress that in this case the real problem is the divergent factor, but there is also a problem in controlling high velocities.

Let us summarize the two different difficulties in this problem:

- In the term $\frac{1}{\varepsilon}\int_{\Omega\times R^3} dvdx|R|^2 v\cdot\nabla\log\mu$ the integrand is unbounded in velocity
- The term has a diverging factor ε^{-1}

For the first difficulty the term to control is

$$\frac{1}{\varepsilon}\int_{\Omega\times R^3} dvdx|(1-P)R|^2 v\cdot\nabla\log\mu$$

because the part with PR is Maxwellian.

There are in literature two approaches to control the high velocity part difficulty: Caflish [8] used a decomposition in low and high velocity part to control this term. The same approach has been used in stationary situations [11]. A different approach was proposed in [21] when studying the acoustic limit.

For the second difficulty, the really dangerous contribution comes from the part with PR because as explained above we can control $\varepsilon^{-1}||(I-P)R||_{2,2}$ by energy inequality.

$$\frac{1}{\varepsilon}\int_{\Omega\times R^3} dvdx\frac{\nabla T}{4T^2}\cdot v(|v|^2-5T)(PR)^2$$

$$=\frac{1}{\varepsilon}\int_{\Omega\times R^3} dvdx\frac{\nabla T}{4T^2}\cdot v(|v|^2-5T)b\cdot v(p+c(|v|^2-5T)$$

$$PR=\sqrt{\mu}[p+b\cdot v+c(|v|^2-5T)].$$

If b is of order ε or zero we are in business. The main point is to control the velocity flux.

In one dimension for the stationary Boltzmann equation one can have an a-priori control of the velocity flux

$$v_z\partial_z R+\frac{1}{\varepsilon}L=g;\quad \partial_z\int dvv_zR=\int dvg$$

and it is possible to prove the hydrodynamic limit. In fact, in [6, 7] it is studied a binary mixture in 1-d and proved the limit. In [3] it is studied the rescaled Boltzmann equation, in a domain between two coaxial rotating cylinders. When the radius of the inner cylinder is suitably sent to infinity, the limiting evolution is expected to converge to a modified Couette flow which keeps memory of the vanishing curvature of the cylinders. This is a particular case of ghost effect. In [10] it is examined a more difficult problem in a slab with the velocity field of order 1.

3 A Model Without Conservation of Momentum

To avoid the problem of controlling the velocity flux, consider a simpler problem: a model without conservation of momentum introduced in [13] as a model for the flow of a gas (or fluid) in a porous medium. It can also be seen as the Grad-Boltzmann limit of a hard sphere system elastically scattered by randomly distributed obstacles.

The system is described by the kinetic equation on a torus Ω for $F(x, v, t)$

$$\partial_t F + v \cdot \nabla F = Q(F),$$

$$Q(F) = Q_B(F, F) + \alpha Q_d(F), \quad \alpha \geq 0$$

where $Q_B(F, H)$ is the symmetrized Boltzmann collision operator for hard sphere and $Q_d(F)$ is a linear operator, modeling elastic collisions with randomly distributed infinite mass scatterers at rest, conserving only mass and energy, not momentum, namely

$$\int_{R_v^3} dv Q_d(F) = 0, \quad \int_{R_v^3} dv |v|^2 Q_d(F) = 0,$$

for any F, but

$$\int_{R_v^3} dv v Q_d(F) \neq 0$$

for some F. As an example

$$Q_d(F) = \int_{\mathbb{S}^2} d\omega [F(v - 2(v \cdot \omega)\omega) - F(v)] |v \cdot \omega|$$

or

$$Q_d(F) = \int_{\mathbb{S}^2} d\omega [F(v - 2(v \cdot \omega)\omega) - F(v)].$$

The properties of the combined operator are:

$$Q_B(F, F) + \alpha Q_d(F) = 0 \quad \text{iff} \quad F = \mu(\rho, 0, T).$$

The entropy dissipation

$$D_d(F) = \int dx \int dv Q_d(F) \log F \leq 0,$$

vanishing if and only if $F = \mu(\rho, 0, T)$, where μ is a generic local Maxwellian with mean zero.

Consider as usual the rescaled equation for $F^\varepsilon(x, v, t) = F(\varepsilon^{-1}x, v, \varepsilon^{-2}t)$

$$\frac{\partial}{\partial t}F^\varepsilon + \varepsilon^{-1}v \cdot \nabla F^\varepsilon = \varepsilon^{-2}Q(F^\varepsilon),$$

and seek for its solution in the form

$$F^\varepsilon = \mu(\rho, 0, T) + \sum_{i=1}^{3} \varepsilon^i F_i + \varepsilon^{5/2}\sqrt{\mu}f.$$

In the limit, the density ρ and the temperature T are solutions of two coupled diffusive equations

$$\frac{\partial}{\partial t}\rho = \nabla \cdot \left[H\frac{\nabla\rho}{\rho}\right] + \nabla \cdot \left[H'\frac{\nabla T}{T}\right] \tag{13}$$

$$\frac{\partial}{\partial t}e = \nabla \cdot \left[\left(TH' + \frac{3}{2}TH\right)\frac{\nabla\rho}{\rho}\right] + \nabla \cdot \left[\left(TH_1' + \frac{3}{2}TH'\right)\frac{\nabla T}{T}\right] \tag{14}$$

where $e = \frac{3}{2}\rho T$ is the internal energy density and the transport coefficients are

$$H_{ij} = \int dv v_j \mathscr{L}^{-1}(\mu v_i); \quad H'_{ij} = \int dv v_j \left(\frac{|v|^2}{2T} - \frac{3}{2}\right)\mathscr{L}^{-1}(\mu v_i);$$

$$H'_{1_{ij}} = \int dv v_j \left(\frac{|v|^2}{2T} - \frac{3}{2}\right)\mathscr{L}^{-1}\left(v_i(\frac{|v|^2}{2T} - \frac{3}{2}\right)\mu).$$

For isotropy reasons $H_{ij} = \delta_{ij}H$; $H'_{1ij} = \delta_{ij}H'_1$; $H'_{ij} = \delta_{ij}H'$. The transport coefficients satisfy the Onsager relations.

H_1' has the usual expression of the thermal conductivity. There are two new coefficients, whose explicit expression can be evaluated for a sligthly different form of the collisions with obstacles. In this case one can show that

$$\frac{H}{\rho} = \frac{1}{2\pi\alpha}T, \qquad \frac{H'}{T} = \frac{1}{2\pi\alpha}\rho$$

and the first equation assumes the form

$$\partial_t\rho = \frac{1}{2\pi\alpha}\Delta P,$$

where $P = \rho T$. When α goes to zero we loose the diffusion on the density. In [12] it is proven the hydrodynamic limit:

Theorem 1 *Let $t > 0$ be fixed and assume that the solution (ρ, T) to the limiting Eqs. (13) and (14) have positive lower bounds and there is $C(\rho, T) \ll 1$ such that, for $0 \le k, \ell \le 4$ with $1 \le k + \ell \le 4$,*

$$\sup_{0 \le s \le t} \sum_{k,\ell} (\|\nabla^k \partial_t^\ell \rho(s)\|_2 + \|\nabla^k \partial_t^\ell T(s)\|_2) < C(\rho, T).$$

Assume also that the initial value of F is positive and matches the expansion

$$F(0) = \mu(0) + \varepsilon F_1(0) + \varepsilon^2 F_1 2(0) + \varepsilon^3 F_3(0) + \varepsilon^{5/2} \sqrt{\mu(0)} f(0).$$

Then, if $\varepsilon \ll 1$, there is a positive solution $F^\varepsilon = \mu + \sum_{i=1}^3 \varepsilon^i F_i + \varepsilon^{5/2} \sqrt{\mu} f$, such that

$$\|\mu^{-1/2}(F - \mu)\|_{2,2} \le C\varepsilon.$$

The strategy of the proof is based on the usual approach of using the energy inequality to get L_2 bound for the linear problem, with a known term g, and then upgrade to L_∞ to deal with the non linear term.

All the new powerful methods developed to study hydrodynamic limits in [14] cannot be fully applied because of the presence of the local Maxwellian.

To deal with the new term unbounded in velocity, following the approach in [21] we introduce a global Maxwellian μ_T with temperature given by the $\min T(x, t)$ assumed strictly positive and bound the high velocity tail of f in terms of the L_∞ norm of $h \sim f\sqrt{\mu}/\sqrt{\mu_T}$. Then, we use a powerful estimate for the L_∞ norm [14]

$$\|f\|_{\sigma,\beta} = \|(1 + |v|^2)^\sigma e^{\beta|v|^2} f(s)\|_\infty$$

$$\sup_{0 \le s \le t} \varepsilon^{3/2} \|f(s)\|_{\sigma,\beta} \le C\Big[\varepsilon^{3/2} \|f_0\|_{\sigma,\beta} + \sup_{0 \le s \le t} \|f\|_{2,2} + \varepsilon^{7/2} \|\frac{1}{1 + |v|^2} g\|_\infty\Big]. \tag{15}$$

where g is the known term in the linear problem. The term to be controlled is

$$\left|\varepsilon^{-1} \int_{\Omega \times R^3} dx dv ((\mathbf{I} - \mathbf{P})f)^2 v \cdot \nabla \log \mu\right| \le \varepsilon^{-1} \int_{v \le \kappa\varepsilon^{-a}} + \varepsilon^{-1} \int_{v \ge \kappa\varepsilon^{-a}}.$$

The first term is bounded as

$$C\varepsilon^{-1} \Big(\frac{\kappa}{\varepsilon^a}\Big)^2 \|(\mathbf{I} - \mathbf{P})f\|_{L^2_\nu}^2; \quad \nu = \sqrt{(1 + |v|^2)}.$$

The high velocity part is bounded as

$$\varepsilon^{-1} C \|(1 + v^2)^{3/2} \sqrt{1 + v^2}^{-\frac{1}{2}} f I_{v \ge \frac{\kappa}{\varepsilon^a}}\|_\infty \|(\mathbf{I} - \mathbf{P})f\|_\nu \le C_\kappa \varepsilon \|f\|_{\sigma,\beta} \|(\mathbf{I} - \mathbf{P})f\|_{L^2_\nu},$$

for σ large enough and $a = \frac{1}{2}$.

$$C_\kappa \varepsilon \|f\|_{\sigma,\beta} \|(\mathbf{I} - \mathbf{P})f\|_{L^2_\nu} \leq C_\kappa \varepsilon^4 \|f\|^2_{\sigma,\beta} + \frac{1}{4} C_\kappa \varepsilon^{-2} \|(\mathbf{I} - \mathbf{P})f\|^2_{L^2_\nu}].$$

The final estimate is: for any finite t

$$\begin{aligned} &\|f(t)\|^2_{2,2} + \varepsilon^{-2} \int_0^t ds \|(\mathbf{I} - \mathbf{P})f(s)\|^2_{L^2_\nu} \\ &\leq \varepsilon^2 \int_0^t ds \|\nu^{-1/2} g(s)\|^2_{L^2_\nu} + C_\kappa C(\rho, T)\varepsilon^4 \int_0^t \|f(s)\|^2_{\sigma,\beta} + 2\|f(0)\|^2_{2,2}. \end{aligned} \tag{16}$$

In this way a term $\|f\|_{\sigma,\beta}$ is present in the energy inequality and to close all the estimates we pay the price of a local in time estimate.

4 Stationary Problem

To get interesting stationary solutions to the stationary Boltzmann equation in a domain Ω

$$v \cdot \nabla_x F = Q(F, F)$$

it is necessary to add boundary conditions. Here we consider the so called *diffuse reflection* modelling the interaction with a reservoir at a given temperature T_w (thermal wall). The boundary conditions have to be prescribed on the *incoming set*, $\gamma_- = \{(x, v) \in \partial\Omega \times R^3 : n(x) \cdot v < 0\}$. Given $T_w(x)$ on a smooth $\partial\Omega$, the boundary condition is

$$F(x, v) = \mathscr{P}_w F(x, v), \quad (x, v) \in \gamma_-$$

where

$$\mathscr{P}_w F(x, v) = M_{T_w(x)}(v) \int_{v' \cdot n(x) > 0} v' \cdot n(x) F(x, v', t) dv',$$

$$M_{T_w} = \frac{1}{2\pi T_w^2} \exp\left[-\frac{|v|^2}{2T_w}\right], \quad \int_{v \cdot n > 0} dv |v \cdot n| M_{T_w}(v) = 1$$

As a consequence, the mass flux on $\partial\Omega$ vanishes (impermeability condition). The rescaled Boltzmann equation for $F^\varepsilon(x, v) = F(\varepsilon^{-1}x, v)$ is

$$v \cdot \nabla_x F^\varepsilon = \frac{1}{\varepsilon} Q(F^\varepsilon, F^\varepsilon). \tag{17}$$

If one imposes $T_w = T_0 + \varepsilon\tau(x)$ and requires $\mathbf{u}$ of order ε then the limiting equations are the Navier—Stokes—Fourier stationary equations [14]. If instead one chooses the wall temperature T_w of order 1, say $T_w = T_0 + \delta\tau(x)$ (δ independent of ε and eventually small) then the formal limiting equation are not the stationary Navier-Stokes-Fourier equations but the so called ghost effect equations

$$\begin{aligned} P &= \rho T, \\ \frac{P}{T}\mathbf{u}\cdot\nabla_x\mathbf{u} + \nabla_x\mathfrak{p} &= \nabla_x\cdot(\tau^{(1)} + \tau^{(2)}), \\ \nabla_x\cdot(\rho\mathbf{u}) &= 0, \\ \nabla_x\cdot\left(\kappa\frac{\nabla_x T}{2T^2}\right) &= 5P(\nabla_x\cdot\mathbf{u}), \end{aligned} \tag{18}$$

with the boundary conditions

$$T\Big|_{\partial\Omega} = T_w, \quad \mathbf{u}\Big|_{\partial\Omega} := \big(u_{\iota_1}, u_{\iota_2}, u_n\big)\Big|_{\partial\Omega} = \Big(\beta_0\partial_{\iota_1}T_w, \beta_0\partial_{\iota_2}T_w, 0\Big). \tag{19}$$

Here $\kappa(T) > 0$, (ι_1, ι_2) are two tangential variables and n is the normal variable, $\beta_0 = \beta_0[T_w]$ is a functional of T_w,

$$\tau_{ij}^{(1)} := \lambda\left(\partial_i u_j + \partial_j u_i - \frac{2}{3}\delta_{ij}\partial_i u_i\right)$$

and

$$\tau_{ij}^{(2)} := \frac{\lambda^2}{P}\left(K_1\left(\partial_i\partial_j T - \delta_{ij}\frac{1}{3}\Delta T\right) + \frac{K_2}{T}(\partial_i T)(\partial_j T) - \delta_{ij}\frac{1}{3}|\nabla T|^2\right)$$

where $\lambda[T] > 0$ and the constants K_1, K_2 are transport coefficients. $\rho, T, \mathbf{u}$ and the boundary conditions are determined through the expansion. In this case div $\mathbf{u}$ is not zero but is related to the gradient of temperature so the fluid is not really incompressible, even if the Mach number is going to zero.

The gradient of temperature along the boundary wall produces a flow called in literature thermal creep. We refer to [22, 23, 25, 26] who pointed out the new thermal effects in stationary situations. These flows were called SNIF, slow non-isothermal flows.

Remark. In a stationary situation we can only scale the space and there is no difference between the Euler and Navier-Stokes time scales. But we think that the stationary regime in presence of diffuse boundary condition can only be achieved as asymptotic time limit of the Boltzmann equation under diffusive rescaling.

We want to construct a solution in the form

$$F = \mu + \varepsilon F_1 + \varepsilon^2 F_2 + \varepsilon f_1^B + \varepsilon\mathscr{R} \tag{20}$$

where μ is defined in (6) with $\rho, T, \mathbf{u}$ to be determined. The new term f_1^B is a boundary layer correction. The need of these boundary layer term comes from the fact that the boundary condition requires the incoming solution to be Maxwellian on the boundary, while this condition can be satisfied surely only from the first term of the expansion. Its explicit expression will determine also the boundary conditions for the hydrodynamic equations.

When $T_w = T_0 + \varepsilon\tau(x)$ the term of order ε, F_1, of the expansion is in the null of the operator and hence can satisfy the boundary conditions. So no boundary layer correction is required at the first order in ε.

In [14] the case of a stationary Boltzmann equation in a 3d bounded domain with diffuse reflection is studied. Under the condition $T_w = T_0 + \varepsilon\tau(x)$ it is possible to construct a solution to the rescaled Boltzmann equation in the form

$$F = \mu + \varepsilon f^\varepsilon$$

and prove that $\lim_{\varepsilon\to 0} f^\varepsilon = F_1$

$$F_1 = \mu^{\frac{1}{2}}\left(\frac{\rho_1}{\rho_0} + \frac{\mathbf{u}_1 \cdot v}{T_0} + \frac{T_1(|v|^2 - 3T_0)}{2T_0^2}\right), \tag{21}$$

with $\rho_1, T_1, \mathbf{u}_1$ solution of the stationary incompressible Navier-Stokes-Fourier equations.

In the case $T_w = T_0 + \delta\tau(x)$ instead F_1 has a part in the orthogonal to the null space and hence the boundary layer correction starts at order ε and this fact is one more difficulty to add to the ones already discussed in the time dependent case. Moreover, the diffuse reflection boundary condition is in itself a new difficulty.

Let's now discuss the case $T_w = T_0 + \delta\tau(x)$ and how one can get rid of the diverging term in the energy inequality. The terms of the expansion (20) can be explicitly constructed [16] and we have that the first term is a Maxwellian $\mu(\rho, 0, T)$ and

$$F_1 = -\mathscr{A} \cdot \frac{\nabla_x T}{2T^2} + \mu^{\frac{1}{2}}\left(\frac{\rho_1}{\rho} + \frac{\mathbf{u} \cdot v}{T} + \frac{T_1(|v|^2 - 3T)}{2T^2}\right), \tag{22}$$

where ρ, u, T are solution of (18) and (19).

Then, the equation for the remainder R is

$$v \cdot \nabla_x \left(\mu^{\frac{1}{2}} R\right) + \varepsilon^{-1} \mu^{\frac{1}{2}} L[R] = \mu^{\frac{1}{2}} S \ \text{ in } \Omega \times \mathbb{R}^3, \tag{23}$$

$$R(x_0, v) = \mathscr{P}[R](x_0, v) \ \text{ for } x_0 \in \partial\Omega \ \text{ and } \ v \cdot n(x_0) < 0. \tag{24}$$

where

$$\mathscr{P}[R](x_0, v) := \mu_w^{-\frac{1}{2}} M_{T_w} \int_{v' \cdot n(x_0) > 0} \mu^{\frac{1}{2}}(x_0, v') R(x_0, v') \left|v' \cdot n(x_0)\right| \mathrm{d}v', \tag{25}$$

and we do not give the explicit expression of S which depends on the higher order terms of the expansion in the bulk, the boundary term and the non linear term (see [15]). In the equivalent form

$$v \cdot \nabla_x R + \left(\mu^{-\frac{1}{2}} \bar{\mathscr{A}} \cdot \frac{\nabla_x T}{4T^2} \right) R + \varepsilon^{-1} L[R] = S \text{ in } \Omega \times \mathbb{R}^3, \tag{26}$$

$$R(x_0, v) = \mathscr{P}[R](x_0, v) \text{ for } x_0 \in \partial\Omega \text{ and } v \cdot n(x_0) < 0, \tag{27}$$

$\mathscr{A}$ and $\bar{\mathscr{A}}$ are defined in (8).

As before we start from the energy inequality. Multiplying $\varepsilon^{-1} R$ on both sides of (26) and integrating over $(x, v) \in \Omega \times \mathbb{R}^3$,we have

$$\varepsilon^{-1} \Big\langle v \cdot \nabla_x R, R \Big\rangle + \varepsilon^{-1} \left\langle \mu^{-\frac{1}{2}} \bar{\mathscr{A}} \cdot \frac{\nabla_x T}{4T^2}, R^2 \right\rangle + \varepsilon^{-2} \Big\langle R, L[R] \Big\rangle = \varepsilon^{-1} \Big\langle R, S \Big\rangle, \tag{28}$$

where

$$\Big\langle h, g \Big\rangle \equiv \int_{\Omega \times \mathbb{R}^3} h(x, v) g(x, v).$$

The strategy of the proof is based on cancellations to get read of the divergent terms.

The first idea is to use a different decomposition of R, instead of decomposing R in a hydrodynamic part PR and non-hydrodynamic one $(I - PR)$, we decompose R as:

$$R = \bar{P} R + (I - \bar{P}) R \tag{29}$$

where

$$\bar{P} R := \mu^{\frac{1}{2}}(v) \Big(p(x) + v \cdot \mathbf{b}(x) + \Big(|v|^2 - 5T \Big) c(x) \Big) + \mathscr{A} \cdot \mathbf{d}, \tag{30}$$

$$\bar{\mathscr{A}} = v \Big(|v|^2 - 5T \Big) \mu^{\frac{1}{2}} \in \mathbb{R}^3, \quad \mathscr{A} = L^{-1}[\bar{\mathscr{A}}] \in \mathbb{R}^3,$$

where $p, \mathbf{b}, c$ are the components of R on the null space of L and $\mathbf{d}$ is the component along the direction $\mathscr{A}$ which is in the orthogonal to the null space. Note that

$$\int_{\mathbb{R}^3} \mathscr{A} L(\mathbf{I} - \bar{\mathbf{P}})[R] = \int_{\mathbb{R}^3} \bar{\mathscr{A}} (\mathbf{I} - \bar{\mathbf{P}})[R] = 0. \tag{31}$$

Then, we further decompose $\mathbf{d}$ as

$$\mathbf{d} = \nabla_x \xi + \mathbf{e},$$

with ξ solving the Poisson equation

$$\nabla_x \cdot (\kappa \nabla_x \xi) = \nabla_x \cdot (\kappa \mathbf{d}) \ \ in \ \ \Omega, \xi = 0 \ \ on \ \ \partial\Omega.$$

We may directly compute

$$\nabla_x \cdot (\kappa \mathbf{e}) = \nabla_x \cdot (\kappa \mathbf{d}) - \nabla_x \cdot (\kappa \nabla_x \xi) = 0. \tag{32}$$

This implies the crucial orthogonality: for any $\eta(x)$ such that $\eta = 0$ on $\partial\Omega$, we have

$$\int_\Omega (\kappa \mathbf{e}) \cdot \nabla_x \eta \mathrm{d}x = -\int_\Omega \nabla_x \cdot (\kappa \mathbf{e}) \eta \mathrm{d}x = 0. \tag{33}$$

Now we replace (29) and (30) in (28).

The new term coming from $\mathbf{d}$, using the orthogonality between $\mathbf{e}$ and $\nabla_x \xi$, becomes

$$\langle \mathbf{d} \cdot \mathscr{A}, L[\mathbf{d} \cdot \mathscr{A}] \rangle = \left\langle \mathbf{d} \cdot \mathscr{A}, \mathbf{d} \cdot \bar{\mathscr{A}} \right\rangle = \int_\Omega \kappa \, |\nabla_x \xi + \mathbf{e}|^2 \, \mathrm{d}x$$

$$= \int_\Omega \kappa \, |\nabla_x \xi|^2 \, \mathrm{d}x + \int_\Omega \kappa \, |\mathbf{e}|^2 \, \mathrm{d}x.$$

Noticing the coercivity property of L, (9), we have

$$\left\langle (\mathbf{I} - \bar{\mathbf{P}})[R], L\Big[(\mathbf{I} - \bar{\mathbf{P}})[R]\Big] \right\rangle \gtrsim \left\| (\mathbf{I} - \bar{\mathbf{P}})[R] \right\|^2_{L^2_\nu}. \tag{34}$$

In conclusion,

$$\begin{aligned} \varepsilon^{-1} \, |(1 - \mathscr{P})[R]|^2_{L^2_{\gamma_+}} &+ \varepsilon^{-2} \left\| (\mathbf{I} - \bar{\mathbf{P}})[R] \right\|^2_{L^2_\nu} + \varepsilon^{-2} \int_\Omega \kappa \, |\nabla_x \xi|^2 \, \mathrm{d}x \\ &+ \varepsilon^{-2} \int_\Omega \kappa \, |\mathbf{e}|^2 \, \mathrm{d}x \\ \lesssim -\varepsilon^{-1} & \int_{\Omega \times \mathbb{R}^3} \left(\mu^{-\frac{1}{2}} \bar{\mathscr{A}} \cdot \frac{\nabla_x T}{4T^2} \right) R^2 + o(1) \, |\mathscr{P}[R]|^2_{L^2_\gamma} + \varepsilon^{-1} \langle R, S \rangle . \end{aligned} \tag{35}$$

where the bulk and boundary norms are

$$\|f\|_{L^r} := \left(\int_{\Omega \times \mathbb{R}^3} |f(x, v)|^r \, \mathrm{d}v \mathrm{d}x \right)^{\frac{1}{r}}, \quad |f|_{L^r_{\gamma_\pm}} := \left(\int_{\gamma_\pm} |f(x, v)|^r \, |v \cdot n| \, \mathrm{d}v \mathrm{d}x \right)^{\frac{1}{r}},$$

and the ν-norm is defined in (10).

The diverging term

$$\int_{\Omega \times \mathbb{R}^3} \mu^{\frac{1}{2}} \left(\bar{\mathscr{A}} \cdot \frac{\nabla_x T}{4T^2} \right) \left(p + v \cdot \mathbf{b} + \left(|v|^2 - 5T \right) c \right)^2 = 5P \int_\Omega \nabla_x T \cdot (\mathbf{b} c)$$

$$
\begin{aligned}
&\int_{\Omega\times\mathbb{R}^3}\left(\mu^{-\frac{1}{2}}\bar{\mathscr{A}}\cdot\frac{\nabla_x T}{4T^2}\right)R^2 = 5P\int_{\Omega}\nabla_x T\cdot(\mathbf{b}c) + \int_{\Omega}\left(\mathbf{d}\cdot\frac{\nabla_x T}{2T^2}\right)(\kappa p+\sigma c)\\
&+\int_{\Omega\times\mathbb{R}^3}\left(\mu^{-\frac{1}{2}}\bar{\mathscr{A}}\cdot\frac{\nabla_x T}{4T^2}\right)\quad\left(2\mathbf{P}[R]+2(\mathbf{d}\cdot\mathscr{A})+(\mathbf{I}-\bar{\mathbf{P}})[R]\right)(\mathbf{I}-\bar{\mathbf{P}})[R] \qquad (36)\\
&\equiv 5P\int_{\Omega}\nabla_x T\cdot(\mathbf{b}c)+\tilde{Z}.
\end{aligned}
$$

As explained before, the really dangerous term is

$$
\varepsilon^{-1}5P\int_{\Omega}\nabla_x T\cdot(\mathbf{b}c).
$$

To cancel it, we use the so-called *conservation laws*. The name comes from the fact that they are related to the quantities conserved by the dynamics. In fact, to get them we multiply the collision invariants $\mu^{\frac{1}{2}}$, $v\mu^{\frac{1}{2}}$, $|v|^2\mu^{\frac{1}{2}}$ on both sides of the equation for R and integrate over $v\in\mathbb{R}^3$.

$$
\begin{aligned}
P\left(\nabla_x\cdot\mathbf{b}\right) &= \left\langle\mu^{\frac{1}{2}},S\right\rangle_v,\\
P\nabla_x p+\nabla_x\cdot\varpi &= \left\langle v\mu^{\frac{1}{2}},S\right\rangle_v,\\
5P\left(\nabla_x\cdot(\mathbf{b}T)\right)+\nabla_x\cdot(\kappa\mathbf{d}) &= \left\langle|v|^2\mu^{\frac{1}{2}},S\right\rangle_v,
\end{aligned}
$$

where $\varpi := \int_{\mathbb{R}^3}(v\otimes v)(\mathbf{I}-\bar{\mathbf{P}})[R]$.

Due to our new decomposition of R, in the previous relation appears the new "component" $\mathbf{d}$. Hence, we need one more relation involving $\mathbf{d}$ (a new "local conservation law").

Multiply the equation for R (23) by $\mathscr{A}\cdot\nabla\xi$ and integrate on velocity and space. By integration by part in the first term

$$
\begin{aligned}
&\int_{\Omega\times\mathbb{R}^3}\left[-v\cdot\nabla_x\left(\nabla_x\xi\cdot\mathscr{A}\right)R+\varepsilon^{-1}\nabla_x\xi\cdot\mathscr{A}LR+\left(\mu^{-\frac{1}{2}}\bar{\mathscr{A}}\cdot\frac{\nabla_x T}{4T^2}\right)R\left(\nabla_x\xi\cdot\mathscr{A}\right)\right]\\
&\qquad = -\int_{\partial\Omega\times\mathbb{R}^3}\left(\nabla_x\xi\cdot\mathscr{A}\right)R(v\cdot n)+\int_{\Omega\times\mathbb{R}^3}\nabla_x\xi\cdot\mathscr{A}S.
\end{aligned}
$$

Using oddness and the orthogonality of $\mathscr{A}$ with $\mathbf{P}[R]$ and $(\mathbf{I}-\bar{\mathbf{P}})[R]$, we get

$$
\begin{aligned}
&-\int_{\Omega}\nabla_x\cdot\left(\kappa\nabla_x\xi\right)c+\varepsilon^{-1}\int_{\Omega}\nabla_x\xi\cdot\kappa\mathbf{d}\\
&=-\int_{\Omega}\nabla_x\xi\frac{\nabla_x T}{2T^2}(\kappa p+\sigma c)-\int_{\partial\Omega\times\mathbb{R}^3}\left(\nabla_x\xi\cdot\mathscr{A}\right)[I-\mathscr{P}]R(v\cdot n) \qquad (37)\\
&+\int_{\Omega\times\mathbb{R}^3}\left[v\cdot\nabla_x\left(\nabla_x\xi\cdot\mathscr{A}\right)(\mathbf{I}-\bar{\mathbf{P}})[R]\right.
\end{aligned}
$$

$$- \nabla_x \xi \cdot \mathscr{A} \left(\mu^{-\frac{1}{2}} \bar{\mathscr{A}} \cdot \frac{\nabla_x T}{2T^2} \right) (\mathbf{I} - \bar{\mathbf{P}}) R + \nabla_x \xi \cdot \mathscr{A} S \Big] \equiv Z,$$

where

$$\kappa \mathbf{1} := \int_{\mathbb{R}^3} \left(\mathscr{A} \otimes \bar{\mathscr{A}} \right) \mathrm{d}v, \quad \sigma \mathbf{1} := \int_{\mathbb{R}^3} \left(|v|^2 - 5T \right) \left(\mathscr{A} \otimes \bar{\mathscr{A}} \right) \mathrm{d}v. \tag{38}$$

To deal with the boundary term and cancel the bad term $\int_{\partial\Omega \times \mathbb{R}^3} \left(\nabla_x \xi \cdot \mathscr{A} \right) \mathscr{P} R (v \cdot n)$ we have used

$$R \mathbf{1}_\gamma = \mathscr{P}[R] \mathbf{1}_\gamma + (1 - \mathscr{P})[R] \mathbf{1}_{\gamma_+}, \tag{39}$$

were the notation is: $\gamma := \left\{ (x_0, v) : x_0 \in \partial\Omega, v \in \mathbb{R}^3 \right\}$, the incoming boundary γ_-, the outgoing boundary γ_+, based on the sign of $v \cdot n(x_0)$. Hence, by orthogonality of $\mathscr{P}[R]$ to $\mathscr{A}$,

$$\int_{\mathbb{R}^3} \left(\nabla_x \xi \cdot \mathscr{A} \right) \mu^{\frac{1}{2}} (v \cdot n) = 0. \tag{40}$$

From the first and third conservation law

$$\begin{aligned} 5P \int_\Omega \nabla_x T \cdot (\mathbf{b} c) &= - \int_\Omega c \Big(\nabla_x \cdot (\kappa \mathbf{d}) \Big) + \left\langle c \sqrt{\mu} \left(|v|^2 - 5T \right), S \right\rangle \\ &= \int_\Omega [-c \nabla \kappa (\nabla \xi)] + \left\langle c \sqrt{\mu} \left(|v|^2 - 5T \right), S \right\rangle \end{aligned} \tag{41}$$

where we have used the definition of ξ: $\nabla_x \cdot (\kappa \nabla_x \xi) = \nabla_x \cdot (\kappa \mathbf{d})$.

From the fourth conservation law,

$$- \int_\Omega c \nabla_x \cdot (\kappa \nabla \xi) = \int_\Omega [-\varepsilon^{-1} \kappa (\nabla \xi)^2] + Z \tag{42}$$

Finally,

$$\varepsilon^{-1} 5P \int_\Omega \nabla_x T \cdot (\mathbf{b} c) = \int_\Omega (-\varepsilon^{-2} \kappa (\nabla \xi)^2) + \varepsilon^{-1} Z + \varepsilon^{-1} \left\langle c \sqrt{\mu} \left(|v|^2 - 5T \right), S \right\rangle$$

So, we obtain, among many other terms, $\int_\Omega \varepsilon^{-2} \kappa \nabla \xi^2$ and observe that this term cancels out with the same term on the lhs of the energy estimate.

The final reduced energy estimate is

$$\begin{aligned} \varepsilon^{-1} \left| (1 - \mathscr{P})[R] \right|^2_{L^2_{\gamma_+}} &+ \varepsilon^{-2} \left\| \mathbf{e} \right\|^2_{L^2} + \varepsilon^{-2} \left\| (\mathbf{I} - \bar{\mathbf{P}})[R] \right\|^2_{L^2_v} \\ &\lesssim -\varepsilon^{-1} Z - \varepsilon^{-1} \left\langle c \left(|v|^2 - 5T \right), S \right\rangle - \varepsilon^{-1} \tilde{Z} \\ &+ o(1) \left| \mathscr{P}[R] \right|^2_{L^2_\gamma} + \varepsilon^{-1} \langle R, S \rangle . \end{aligned} \tag{43}$$

Summarizing, we do not have anymore the bad term of order ε^{-1}, but we pay the price of not having an a-priori control of the full $\varepsilon^{-1}\mathbf{d}$ but only of $\varepsilon^{-1}\mathbf{e}$.

All this is only the starting point of the control of the remainder. Much more is needed, see [15], where we use, in a perturbative regime $\delta \ll 1$, all the new techniques developed in [14] and many new ideas. In particular, we use the weak formulation of the Boltzmann equation using suitably chosen test functions so that we can cancel other divergent terms.

For sake of completeness we report here the theorem proven in [15]. For that, we need some notations.

Define the weighted L^∞ norms for $0 \le \varrho < \frac{1}{2}$ and $\vartheta \ge 0$ and $T_M > 0$

$$T_M < \min_{x\in\Omega} T < \max_{x\in\Omega} T < 2T_M \quad\text{and}\quad \max_{x\in\Omega} T - T_M = o_T, \tag{44}$$

$$\|f\|_{L^\infty_{\varrho,\vartheta}} := \operatorname*{ess\,sup}_{(x,v)\in\Omega\times\mathbb{R}^3}\left(\nu^{\vartheta}\mathrm{e}^{\varrho\frac{|v|^2}{2T_M}}\,|f(x,v)|\right),$$

$$|f|_{L^\infty_{\gamma_\pm,\varrho,\vartheta}} := \operatorname*{ess\,sup}_{(x,v)\in\gamma_\pm}\left(\nu^{\vartheta}\mathrm{e}^{\varrho\frac{|v|^2}{2T_M}}\,|f(x,v)|\right).$$

Let $\|\cdot\|_{W^{k,p}}$ denote the usual Sobolev norm for $x\in\Omega$ and $|\cdot|_{W^{k,p}}$ for $x\in\partial\Omega$. Let $\|\cdot\|_{W^{k,p}L^q}$ denote $W^{k,p}$ norm for $x\in\Omega$ and L^q norm for $v\in\mathbb{R}^3$. The similar notation also applies when we replace L^q by $L^\infty_{\varrho,\vartheta}$ or L^q_γ.

Denote the working space X via the norm

$$\begin{aligned}
\|R\|_X :=\ & \varepsilon^{-1}\|p\|_{L^2} + \varepsilon^{-\frac12}\|\mathbf{b}\|_{L^2} + \|c\|_{L^2} + \varepsilon^{-1}\|\xi\|_{L^2} + \varepsilon^{-\frac12}\|\xi\|_{H^2} + \varepsilon^{-1}\|\mathbf{e}\|_{L^2}\\
&+ \varepsilon^{-1}\left\|(\mathbf{I}-\bar{\mathbf{P}})[R]\right\|_{L^2_\nu} + \|p\|_{L^6} + \|\mathbf{b}\|_{L^6} + \|c\|_{L^6} + \varepsilon^{-1}\|\xi\|_{L^6} + \|\xi\|_{W^{2,6}}\\
&+ |\mathscr{P}[R]|_{L^2_\gamma} + \varepsilon^{-\frac12}|(1-\mathscr{P})[R]|_{L^2_{\gamma_+}} + \left|\mu^{\frac14}(1-\mathscr{P})[R]\right|_{L^4_{\gamma_+}}\\
&+ \|\mathbf{e}\|_{L^6} + \left\|(\mathbf{I}-\bar{\mathbf{P}})[R]\right\|_{L^6} + \varepsilon^{-\frac12}|\nabla_x\xi|_{L^2_{\partial\Omega}} + \varepsilon^{\frac12}\|R_M\|_{L^\infty_{\varrho,\vartheta}} + \varepsilon^{\frac12}|R_M|_{L^\infty_{\gamma,\varrho,\vartheta}},
\end{aligned} \tag{45}$$

where $R_M = \mu_M^{-\frac12}\mu^{\frac12}R$ with

$$\mu_M := \frac{P}{(2\pi)^{\frac32}T_M^{\frac52}}\exp\left(-\frac{|v|^2}{2T_M}\right). \tag{46}$$

Theorem 2 *Assume that Ω is a bounded C^3 domain and*

$$|\nabla T_w|_{W^{3,\infty}} = o(1). \tag{47}$$

Then for any given $P > 0$, there exists $\varepsilon_0 > 0$ such that for any $\varepsilon \in (0,\varepsilon_0)$, there exists a nonnegative solution F to the Eq. (17) represented by (20) satisfying

$$\int_{\Omega} P(x)\mathrm{d}x = 0, \tag{48}$$

and

$$\|R\|_X \lesssim o_T, \tag{49}$$

where the X norm is defined in (46) and o_T is a small constant depending on T_w such that $o_T = o(1) \to 0$ as $|\nabla T_w|_{W^{3,\infty}} \to 0$. Such a solution is unique among all solutions satisfying (48) and (49).

This further yields that in the expansion (20), $\mu + \varepsilon\mu\big(u \cdot v\big)$ is the leading-order terms in the sense of

$$\int_{\Omega}\left|\int_{\mathbb{R}^3}\Big(F - \mu\Big)\right| + \int_{\Omega}\left|\int_{\mathbb{R}^3}\Big(F - \mu\Big)\Big(|v|^2 - 3T\Big)\right| \lesssim \varepsilon, \tag{50}$$

and

$$\int_{\Omega}\left|\int_{\mathbb{R}^3}\Big(F - \mu - \varepsilon\mu\big(u \cdot v\big)\Big)v\right| \lesssim \varepsilon^{\frac{3}{2}}, \tag{51}$$

where (ρ, u, T) is determined by the ghost-effect Eqs. (18) and (19).

References

1. Arkeryd, L., Esposito, R., Marra, R., Nouri, A.: Stability of the laminar solution of the Boltzmann equation for the Benard problem. Bull. Acad. Sin. **3**, 51–97 (2008)
2. Arkeryd, L., Esposito, R., Marra, R., Nouri, A.: Stability for Rayleigh-Benard convective solutions of the Boltzmann equation. Arch. Rat. Mech. **198**, 125–187 (2010)
3. Arkeryd, L., Esposito, R., Marra, R., Nouri, A.: Ghost effect by curvature in planar Couette flow. Kinet. Relat. Model. **4**, 109–138 (2011)
4. Arkeryd, L., Nouri, A.: Asymptotic techniques for kinetic problems of Boltzmann type. In: Proceedings of the 3rd edition of the summer school in "Methods and Models of kinetic theory", Riv. Mat. Univ. Parma **7**, 1–74 (2007)
5. Bobylev, A.V.: Quasistationary hydrodynamics for the Boltzmann equation. J. Stat. Phys. **80**, 1063–1083 (1995)
6. Brull, S.: The stationary Boltzmann equation for a two-component gas in the slab. Math. Methods Appl. Sci. **31**, 153–178 (2008)
7. Brull, S.: Problem of evaporation-condensation for a two component gas in the slab. Kinet. Relat. Models **1**, 185–221 (2008)
8. Caflisch, R.E.: The fluid dynamic limit of the nonlinear Boltzmann equation. Commun. Pure Appl. Math. **33**, 651–666 (1980)
9. De Masi, A., Esposito, R., Lebowitz, J.L.: Incompressible Navier-Stokes and Euler limits of the Boltzmann equation. Comm. Pure Appl. Math. **42**, 1189–1214 (1989)
10. Esposito, R., Lebowitz, J.L., Marra, R.: Hydrodynamic limit of the stationary Boltzmann equation in a slab. Comm. Math. Phys. **160**, 49–80 (1994). The Navier-Stokes limit of stationary solutions of the nonlinear Boltzmann equation. J. Stat. Phys. **78**, 389–412 (1995)

11. Esposito, R., Lebowitz, J.L., Marra, R.: Solutions to the Boltzmann equation in the Boussinesq regime. J. Stat. Phys. **90**, 1129-1178 (1998). Fluid Dynamics, vol. 3. In: Serre, D., Friedlander, F. (eds.). Elsevier (2004)
12. Esposito, R., Garrido, P.L., Lebowitz, J.L., Marra, R.: Diffusive limit for a Boltzmann-like equation with non-conserved momentum. Nonlinearity **32**, 4834–4852 (2019)
13. Garrido, P.L., Lebowitz, J.L.: Diffusion equations from kinetic models with non-conserved momentum. Nonlinearity **32**, 5441–5462 (2018)
14. Esposito, R., Guo, Y., Kim, C., Marra, R.: Stationary solutions to the Boltzmann equation in the hydrodynamic limit. Ann. PDE **4**, 1–119 (2018)
15. Esposito, R., Guo, Y., Marra, R., et al.: Ghost effect from Boltzmann theory: expansion with remainder. Vietnam J. Math. **52**, 883–914 (2024)
16. Esposito, R., Guo, Y., Marra, R., Wu, L.: Expansion and remainder for ghost effect from Boltzmann theory. Preprint (2023). http://arxiv.org/abs/2301.09560
17. Esposito, R., Marra, R.: Stationary non equilibrium states in kinetic theory. J. Stat. Phys **180**, 773–809 (2020)
18. Golse, F., Saint-Raymond, L.: The Navier-Stokes limit of the Boltzmann equation for bounded collision kernels. Inven. Math. **155**, 81–161 (2004)
19. Golse, F., Saint-Raymond, L.: The incompressible Navier-Stokes limit of the Boltzmann equation for hard cutoff potentials. J. Math. Pures Appl. **91**, 508–552 (2009)
20. Guo, Y.: Boltzmann diffusive limit beyond the Navier-Stokes approximation. Commun. Pure Appl. Math. **59**, 626–687 (2006)
21. Guo, Y., Jang, N., Jang, J.: Acoustic limit for the Boltzmann equation in optimal scaling. Commun. Pure Appl. Math. **63**, 337–361 (2010)
22. Kogan, M.N.: On the equations of motion of a rarefied gas. Appl. Math. Mech. **22**, 597–607 (1958)
23. Kogan, M.N., Galkin, V.S., Fridlender, O.G.: Stresses produced in gases by temperature and concentration inhomogeneities. New types of free convection. Sov. Phys. Usp. **19**(5), 420–428 (1976)
24. Saint-Raymond, L.: Hydrodynamic Limits of the Boltzmann Equation. Springer-Verlag, Berlin (2009)
25. Sone, Y.: Thermal creep in rarefied gas. J. Phys. Soc. Jpn **21**, 1836–1837 (1966)
26. Sone, Y.: Flow induced by thermal stress in rarefied gas. Phys. Fluids **15**, 1418–1423 (1972)
27. Sone, Y.: Kinetic Theory and Fluid Dynamics. Birkhäuser, Boston (2002); Molecular Gas Dynamics, Theory, Techniques, and Applications. World Scientific, Birkhäuser, Boston (2007)

Intemperate Lévy Processes and White Noises

Rui Vilela Mendes

Abstract The distributional support of Lévy processes is important for the construction of sparse statistical models, integration in infinite dimensions and the existence of generalized solutions of stochastic partial differential equations driven by Lévy white noise. The white noise associated to a class of Lévy processes without support in $\mathscr{S}'$, the space of tempered distributions, is constructed as a generalized random process. This construction also provides measures on the distribution spaces that support the paths of the Lévy processes.

Keywords Lévy processes · Path support · White noises

1 Introduction

It is with pleasure and regret that I contribute to this volume in memory of São Carvalho. It is with pleasure because it brings to my mind the memories of our friendship and it is with regret because she is not here to share those memories. The same mixed feelings I have when thinking about the subject of this paper, because it concerns an exploration in the theory of stochastic processes. The theory of stochastic processes is a subject whose importance for mathematical physics São and myself understood many years ago in Bielefeld, under the guidance of our good friend Ludwig Streit.

Characterization of the support of paths of Lévy processes is an important issue both for the construction of sparse statistical models [1], for theories of integration in infinite dimensions [2] and for the existence of generalized solutions of stochastic partial differential equations driven by Lévy white noise [3].

All càdlàg Lévy processes, being locally Lebesgue integrable, have support in $\mathscr{D}'$, the space of distributions (dual to the space $\mathscr{D}$ of infinitely differential functions with compact support). However it turns out that the paths of most Lévy processes

R. Vilela Mendes (✉)
CMAFcIO, University of Lisbon, Lisbon, Portugal
e-mail: rvilela.mendes@gmail.com; rvmendes@ciencias.ulisboa.pt

E. Carlen et al. (eds.), *From Particle Systems to Partial Differential Equations*,
Springer Proceedings in Mathematics & Statistics 465,
https://doi.org/10.1007/978-3-031-65195-3_10

have support on a smaller space $\mathscr{S}'$, the space of tempered distributions, dual to the space $\mathscr{S}$ of rapid decrease functions, topologized by the family of norms

$$\|\varphi\|_{p,r} = \sup_{x\in\mathbb{R}} \left| x^p \varphi^{(r)}(x) \right|, \quad p, r \in \mathbb{N}_0 \tag{1}$$

Necessary and sufficient conditions have been obtained for the support of a Lévy process to be in $\mathscr{S}'$ [4, 5]. In particular [5], a Lévy process X_t has support in $\mathscr{S}'$ if there is a $\eta > 0$ such that $\mathbb{E}\left|X_1^\eta\right| < \infty$. Conversely, the process has no support in $\mathscr{S}'$ if $\mathbb{E}\left|X_1^\eta\right| = \infty$ for any $\eta > 0$. To show that the second statement in the Dalang-Humeau theorem is not empty amounts to find a process X_t with Lévy measure $\nu(dx)$

$$\int \left(x^2 \wedge 1\right) \nu(dx) < \infty \tag{2}$$

but for which $\mathbb{E}\left|X_1^\eta\right| = \infty$ for any $\eta > 0$. Assuming that such processes can be found, it would also be interesting to characterize their support in some space, intermediate between $\mathscr{S}'$ and $\mathscr{D}'$. Such processes with paths not supported in the space $\mathscr{S}'$ of tempered distributions are here called *intemperate Lévy processes.*

A family K_α of one-dimensional Lévy processes without support in $\mathscr{S}'$ was found in [6]. The one-dimensional K_α processes are characterized by the triplet $(0, 0, \nu_\alpha)$ with the following Lévy measure

$$\nu_\alpha(dx) = \frac{1}{(1+|x|)\log^{1+\alpha}(1+|x|)} dx \quad 0 < \alpha < 2 \tag{3}$$

with $\nu_\alpha(\mathbb{R}) = \infty$.

The case $\alpha = 1$ was considered by Fristedt [7] for the construction of a counter example but, as far as I know, no further studies have been made on these processes.

Proposition 1 *The paths of a K_α process are a. s. not supported in $\mathscr{S}'$.*

Proof Let

$$K_\alpha(t) = L_\alpha^M(t) + L_\alpha^P(t) \tag{4}$$

be the Lévy-Itô decomposition of the K_α process, L_α^M being the compensated square integrable martingale containing the small jumps of K_α and L_α^P the compound Poisson process containing the large jumps. Because $L_\alpha^M(t)$ is a. s. in $\mathscr{S}'$, it remains to analyze the L_α^P component, with Lévy measure

$$\nu'_\alpha(dx) = \mathbf{1}_{|x|>1}\nu_\alpha(dx). \tag{5}$$

The Lévy measure $\nu'(dx)$ has no positive moments. Let $\eta > 0$

$$
\begin{aligned}
\int_{\mathbb{R}} \mathbf{1}_{|x|>1} |x|^{\eta} \nu_{\alpha}(dx) =& 2\int_{1}^{\infty} \frac{x^{\eta}}{(1+x)\log^{1+\alpha}(1+x)} dx \\
\geq& 2\int_{1}^{x^*} \frac{x^{\eta}}{(1+x)\log^{1+\alpha}(1+x)} dx + 2\int_{x^*}^{\infty} \frac{dx}{(1+x)\log(1+x)} \\
=& 2\int_{1}^{x^*} \frac{x^{\eta}}{(1+x)\log^{1+\alpha}(1+x)} dx + 2\log(\log(1+x))|_{x^*}^{\infty} = \infty
\end{aligned}
$$

x^* being the solution of

$$
x^* = \log^{\frac{\alpha}{\eta}}\left(1+x^*\right).
$$

Then, because $\int_{\mathbb{R}} \mathbf{1}_{|x|\geq 1} |x|^{\eta} \nu_{\alpha}(dx) < \infty$ if and only if $\mathbb{E}\left|K_{\alpha}^{\eta}\right| < \infty$ (Theorem 25.3 in [8]), the result of Dalang and Humeau [5] implies that the K_{α} process is a. s. not supported in $\mathscr{S}'$.

The plan of the paper is the following: In Sect. 2 the distribution spaces that provide a. s. support for the paths of these Lévy processes are described. In Sect. 3, the support of the processses is identified and $K_{\alpha}-$ white noises are constructed as generalized random processes [9], which provide new infinite dimensional integration measures in appropriately defined test function spaces.

2 The $\mathscr{K}_{\beta}'$ Distribution Spaces

The $\mathscr{K}'$ space of distributions of exponential type is dual to the $\mathscr{K}$ space of functions topologized by the norms [10] (see also [11], Sect. 3.6)

$$
\|\varphi\|_p = \max_{0\leq q\leq p} \sup_{x\in\mathbb{R}} \left|\left(e^{p|x|}\varphi^{(q)}(x)\right)\right| \tag{6}
$$

Denoting by $\mathscr{K}_p$ the Banach space for the norm $\|\cdot\|_p$, $\mathscr{K}'$ is the dual of the countably normed space $\mathscr{K} = \cap_{p=0}^{\infty}\mathscr{K}_p$ and is a dense linear subspace of $\mathscr{D}'$. They have been called *distributions of exponential type*.

Fourier transforms of distributions in $\mathscr{K}'$ are in $\mathscr{U}'$, the space of tempered ultra-distributions [12]. A distribution $\mu \in \mathscr{D}'$ is in $\mathscr{K}'$ if and only if it can be represented in the form

$$
\mu(x) = D^r\left(e^{a|x|} f(x)\right) \tag{7}
$$

for some numbers $r \in \mathbb{N}_0$, $a \in \mathbb{R}$ and a bounded continuous function f, with D a derivative in the distributional sense.

A generalization [13] of this construction leads to spaces $\mathscr{K}_{\beta}'$ of *generalized distributions of exponential type*. Define a family of norms

$$
\|\varphi\|_{p,\beta} = \max_{0\leq q\leq p} \sup_{x\in\mathbb{R}} \left|\left(e^{p|x|^{\beta}}\varphi^{(q)}(x)\right)\right|, \quad \beta > 1 \tag{8}
$$

and denote by $\mathscr{K}_{p,\beta}$ the Banach space associated to the norm $\|\cdot\|_{p,\beta}$ and by $\mathscr{K}_\beta$ the countably normed space $\mathscr{K}_\beta = \cap_{p=0}^{\infty} \mathscr{K}_{p,\beta}$. The topological dual of $\mathscr{K}_\beta$ is $\mathscr{K}'_\beta$ the space of *generalized distributions of exponential type*.

A distribution $\mu \in \mathscr{D}'$ is in $\mathscr{K}'_\beta$ if and only if there is a bounded continuous function f and some numbers r, $\beta \in \mathbb{N}_0$, $a \in \mathbb{R}$ such that

$$\mu(x) = D^r \left(e^{a|x|^\beta} f(x) \right) \tag{9}$$

Definition of this type of spaces is easily generalized for $x \in \mathbb{R}^n$. However here, only the $n = 1$ case is considered.

3 Support, White Noise and Infinite-Dimensional Measures

The support of the K_α processes is obtained by constructing upper functions.

Proposition 2 *The paths of the K_α process, for $1 < \alpha < 2$, have a. s. support in $\mathscr{K}'$. In general a K_α process will have support in $\mathscr{K}'_\beta$ whenever $\alpha > 1/\beta$.*

Proof Let $L_\alpha^P(t)$ be the compound Poisson process containing the large jumps of the K_α process and $L_\alpha^{P'}(t)$ the modulus process (sum of the modulus of the large jumps). This process is a subordinator, the tail of its Lévy measure ν'_α being

$$\overline{\nu'_\alpha}(x) = \nu'_\alpha(x, \infty) = \frac{1}{\alpha \log^\alpha(1+x)}. \tag{10}$$

Now one looks for an upper function. That is, for a function f that increases faster than x and computes

$$I_\alpha(f) = \int_1^\infty \overline{\nu'_\alpha}(f(x))\, dx.$$

For $f(x) = e^{cx^\beta}$

$$I_\alpha\left(e^{cx^\beta}\right) = \int_1^\infty \frac{1}{\alpha \log^\alpha\left(1 + e^{cx^\beta}\right)} \tag{11}$$

one finds that this integral diverges if $\alpha\beta \leq 1$ and is finite if $\alpha\beta > 1$. Hence, by Theorem III.13 in [14], $\limsup_{t\to\infty} \left(\left| L_\alpha^P(t) \right| / \exp\left(c|x|^\beta\right) \right) = 0$, $\exp\left(c|x|^\beta\right)$ is an upper function for the process and from the definition (5) of the norms in $\mathscr{K}_\beta$ it follows that the process has a. s. support in $\mathscr{K}'_\beta$ whenever $\alpha > 1/\beta$.

Thus, one obtains a characterization of the support of paths of the K_α processes generalizing Proposition 2 in [6]:

For many applications both the properties of Lévy processes and Lévy white noise are needed. For each event ω in the probability space, K_α−white noise might be defined by the derivative in the sense of distributions

$$\left\langle \dot{K}_{\alpha}(\omega), \varphi \right\rangle := -\left\langle K_{\alpha}(\omega), \varphi' \right\rangle := -\int_{\mathbb{R}_{+}} K_{\alpha}(t,\omega)\,\varphi'(t)\,dt \tag{12}$$

φ being a function in $\mathscr{K}$ or $\mathscr{K}_{\beta}$. Because there is no upper bound on the order of the derivatives in (6) or (5) one concludes that K_{α}−white noise has the same support properties as the K_{α} processes. There is however a better way to define the white noise associated to the K_{α} processes as a generalized random process [9].

Let $\Psi_{\alpha}(\xi)$

$$\Psi_{\alpha}(\xi) = \int_{\mathbb{R}\setminus\{0\}} \left(e^{i\xi x} - 1 - i\xi x \mathbf{1}_{|x|\le 1}\right) \nu_{\alpha}(dx) \tag{13}$$

be the Lévy exponent associated to the Lévy measure ν_{α} in (3) and $\varphi \in \mathscr{K}_{\beta}$, $\mathscr{K}_{\beta}$ being the test function space with $\alpha > \frac{1}{\beta}$. Then the functional

$$\mathscr{C}_{\alpha}(\varphi) = \exp\left(\int_{\mathbb{R}} \Psi_{\alpha}(\varphi(x))\,dx\right) \tag{14}$$

is continuous, $\mathscr{C}_{\alpha}(0) = 1$ and is positive definite because a necessary and sufficient condition for $\mathscr{C}_{\alpha}(\varphi)$ to be positive definite is that $\exp(s\Psi_{\alpha}(x))$ be positive definite for all positive values of s (Theorem 2 in Sect. 4.2 of [9]). This last quantity is positive definite because $\Psi_{\alpha}(x)$ is the exponent of a characteristic function. Then $\mathscr{C}_{\alpha}(\varphi)$ is the characteristic functional of a generalized random process, the white noise associated to the K_{α} process. Arbitrary integer order derivatives of this process are obtained by replacing $\varphi(x)$ in (14) by $(-1)^{n}\,\varphi^{(n)}(x)$.

On the other hand the $\mathscr{C}_{\alpha}(\varphi)$ functional also provides a measure on cylinder sets on $\mathscr{K}'_{\beta}$ $(\alpha > \frac{1}{\beta})$. Cylinder sets in $\mathscr{K}'_{\beta}$ are defined in the usual way. Let B be a finite dimensional subspace of $\mathscr{K}_{\beta}$ and $\mathscr{K}'_{\beta}/B_0$ the factor space of cosets, each coset being the set of elements of $\mathscr{K}'_{\beta}$ taking the same value in B. Now taking $A \subset \mathscr{K}'_{\beta}/B_0$, the cylinder set $X_{A,B}$ with base A and generating space B_0, is the collection of all elements in $\mathscr{K}'_{\beta}$ that are carried into A by the mapping $\mathscr{K}'_{\beta} \to \mathscr{K}'_{\beta}/B_0$.

The functional $\mathscr{C}_{\alpha}(\varphi)$ on $\mathscr{K}_{\beta}$ being positive definite, continuous and with $\mathscr{C}_{\alpha}(0) = 1$, it is the Fourier transform of a cylinder set measure in the distributional space $\mathscr{K}'_{\beta}$ [9]. Inserting the Lévy exponent (13) with (3) one obtains

$$\mathscr{C}_{\alpha}(\varphi) = \exp\left\{-\int_{\mathbb{R}} dx \int_{0}^{\infty} 2\varphi(x)\sin(y\varphi(x))\,\frac{dy}{\alpha\log^{\alpha}(1+y)}\right\}$$

$\varphi \in \mathscr{K}_{\beta}$ $(\alpha > \frac{1}{\beta})$.

Berger [15], by using a generic multiplier $e^{\omega(x)}$ in the definition of the family of norms, as in (5), defined distribution spaces S'_{ω}, which he calls *spaces of tempered ultradistributions*.[1] Then he established an integral condition on ω and on the Lévy

[1] Notice that this designation had been used before for the Fourier image of the space $\mathscr{K}'$ [11, 12].

measure ν that insures the existence of a Lévy white noise with support in S'_ω. In this setting the family of processes and white noises studied in this paper provides a concrete example of such white noises in the case $\omega(x) = c|x|^\beta$.

References

1. Fageot, J., Amini, A., Unser, M.: On the continuity of characteristic functionals and sparse stochastic modeling. J. Fourier Anal. Appl. **20**, 1179–1211 (2014)
2. Di Nunno, G., Øksendal, B., Proske, F.: White noise analysis for Lévy processes. J. Funct. Anal. **206**, 109–148 (2004)
3. Fageot, J., Humeau, T.: The domain of definition of the Lévy white noise. Stoch. Process. Appl. **135**, 75–102 (2021)
4. Lee, Y.-J., Shih, H.-H.: Lévy white noise measures on infinite-dimensional spaces: existence and characterization of the measurable support. J. Funct. Anal. **237**, 617–633 (2006)
5. Dalang, R.C., Humeau, T.: Lévy processes and Lévy white noise as tempered distributions. Ann. Probab. **45**, 4389–4418 (2017)
6. Vilela Mendes, R.: On a family of Lévy processes without support in $\mathcal{S}'$. arXiv:2205.03816
7. Fristedt, B.E.: Sample function behavior of increasing processes with stationary independent increments. Pacific J. Math. **21**, 21–33 (1967)
8. Sato, K.-I.: Lévy processes and infinitely divisible distributions. Cambridge University Press, Cambridge (1999)
9. Gel'fand, I.M., Vilenkin, N.Y.: Generalized Functions, vol. 4: Applications of Harmonic Analysis. Academic Press, New York (1964)
10. Sebastião e Silva, J.: Les fonctions analytiques comme ultra-distributions dans le calcul opérationnel. Math. Ann. **136**, 58–96 (1958)
11. Hoskins, R.F., Pinto, J.S.: Theories of Generalized Functions: Distributions, Ultradistributions and other Generalized Functions, 2nd edn. Woodhead Publishing, Philadelphia (2011)
12. Sebastião e Silva, J.: Les séries de multipôles des physiciens et la théorie des ultradistributions. Math. Ann. **174**, 109–142 (1967)
13. Sampson, G., Zielezny, Z.: Hypoelliptic convolution equations in $\mathcal{K}'_p$, $p>1$. Trans. Am. Math. Soc. **223**, 133–154 (1976)
14. Bertoin, J.: Lévy Processes. Cambridge University Press, Cambridge (1996)
15. Berger, D.: Lévy driven linear and semilinear stochastic partial differential equations (2019). arXiv:1907.01926

Connecting the Deep Quench Obstacle Problem with Surface Diffusion via Their Steady States

Eric A. Carlen, Amy Novick-Cohen, and Lydia Peres Hari

In memory of Maria Conceicao Carvalho and her passion for life and science.

Abstract In modeling phase transitions, it is useful to be able to connect diffuse interface descriptions of the dynamics with corresponding limiting sharp interface motions. In the case of the deep quench obstacle problem (DQOP) and surface diffusion (SD), while a formal connection was demonstrated many years ago, rigorous proof of the connection has yet to be established. In the present note, we show how information regarding the steady states for both these motions can provide insight into the dynamic connection, and we outline tools that should enable further progress. For simplicity, we take both motions to be defined on a planar disk.

Keywords Deep quench obstacle problem · Surface diffusion · Higher order degenerate parabolic equations · Geometric motions · Limiting motions

1 Introduction

Many two component mixtures exist stably at one temperature, but become unstable at lower temperatures; the subsequent instability typically initiates phase separation, leading to the appearance of spatial regions characterized by two different

E. A. Carlen (✉)
Department of Mathematics, Hill Center, Rutgers University, 110 Frelinghuysen Road, Piscataway, NJ 08854-8019, USA
e-mail: carlen@math.rutgers.edu

A. Novick-Cohen · L. P. Hari
Department of Mathematics, Technion-IIT, 32000 Haifa, Israel
e-mail: amync@technion.ac.il

L. P. Hari
e-mail: lydia@technion.ac.il

E. Carlen et al. (eds.), *From Particle Systems to Partial Differential Equations*,
Springer Proceedings in Mathematics & Statistics 465,
https://doi.org/10.1007/978-3-031-65195-3_11

compositional phases. During the early stages of phase separation, the distinction between the two phases is not sharp and a diffuse description is appropriate. As phase separation progresses, the phases become more distinct, and a sharp interface description is appropriate. Accordingly, in physically realistic models, it should be possible to pass from one description to the other. Unfortunately, often there is a gap between what is physically reasonable and what is possible to justify with mathematical rigor. For example, though the diffuse interface Cahn-Hilliard model with a logarithmic potential and a degenerate mobility has been shown using formal asymptotics [6] to yield the sharp interface surface diffusion model, the connection has yet to have been made rigorous. In the present note, we focus on the zero temperature limit of the Cahn-Hilliard model with a logarithmic potential and degenerate mobility, namely on the deep quench obstacle problem with degenerate mobility, which we shall subsequently refer to here simply as the deep quench obstacle problem (DQOP), and its connection with motion by (isotropic) surface diffusion (SD).

After recalling below some relevant background with regard to both models, in Sect. 2 we discuss the steady states in some detail for both models, focusing in particular on the minimum energy steady states, and then in Sect. 3 we outline certain tools that we are using to bridge the two evolutions.

Let us consider both motions, (DQOP) and (SD), to be defined in $\Omega \subset R^2$, where Ω is a disk centered at the origin whose radius, R_0, is $O(1)$. In considering time evolution, we set $\Omega_T := \Omega \times (0, T)$ and $\partial\Omega_T := \partial\Omega \times (0, T)$ with $0 < T \leq \infty$. The deep quench obstacle problem [20] with the degenerate mobility $M(u) = 1 - u^2$, can be expressed as

$$\text{(DQOP)} \quad \begin{cases} u_t = \nabla \cdot M(u)\nabla w, \quad w + u + \varepsilon^2 \Delta u \in \partial I_{[-1,1]}(u), \quad (x,t) \in \Omega_T, \\ n \cdot \nabla u = n \cdot M(u)\nabla w = 0, \quad (x,t) \in \partial\Omega_T, \\ u(x,0) = u_0(x), \quad x \in \Omega, \end{cases}$$

where $\partial I_{[-1,1]}(\cdot)$ denotes the sub-differential of the indicator function $I_{[-1,1]}(\cdot)$ and n denotes the unit exterior normal to $\partial\Omega$. The deep quench obstacle problem, (DQOP), constitutes the formal zero temperature ($\Theta \downarrow 0$) limit of the Cahn-Hilliard equation with degenerate mobility and with a logarithmic potential [9]:

$$\begin{cases} u_t = \nabla \cdot M(u)\nabla w, \quad w = \frac{\Theta}{2}\{\ln(1+u) - \ln(1-u)\} - u - \varepsilon^2 \Delta u, \quad (x,t) \in \Omega_T, \\ n \cdot \nabla u = n \cdot M(u)\nabla w = 0, \quad (x,t) \in \partial\Omega_T, \\ u(x,0) = u_0(x), \quad x \in \Omega. \end{cases}$$

Given a smoothly embedded curve $\Gamma = \Gamma(t) \subset \Omega$, the curve $\Gamma(t)$ is said to evolve by motion by surface diffusion [15] if, up to rescaling by constants,

$$\text{(SD)} \quad \begin{cases} V = -\kappa_{ss}, \quad (x(s,t),t) \in \Omega_T, \\ \Gamma(0) = \Gamma_0, \end{cases}$$

where s denotes an arc-length parametrization of $\Gamma(t)$, and V and κ denote, respectively, the normal velocity and the mean curvature of $\Gamma(t)$, defined in accordance with the exterior normal, n, relative to the arc-length parametrization, s.

Often it is convenient to express the deep quench obstacle problem with degenerate mobility somewhat informally as

$$\begin{cases} u_t = -\nabla \cdot M(u)\nabla(u + \varepsilon^2 \Delta u), \quad -1 \leq u \leq 1, \quad (x,t) \in \Omega_T, \\ n \cdot \nabla u = n \cdot M(u)\nabla(u + \varepsilon^2 \Delta u) = 0, \quad (x,t) \in \partial\Omega_T, \\ u(x,0) = u_0(x), \quad x \in \Omega. \end{cases} \tag{1}$$

As shown in [6], by considering (DQOP) on the slow time scale, $\tau = \varepsilon^2 t$, to leading order the $\varepsilon \downarrow 0$ limit of (DQOP) yields surface diffusion motion (SD) for $\Gamma = \Gamma(t)$, where $\Gamma(t)$ denotes the limit of the $O(\varepsilon)$ width interfaces which partition the composition $u(x,t)$ in the context of (DQOP) into two phases, $u = \pm 1$, in Ω. Off hand, $\Gamma(t)$ may contain one or more components.

With regard to existence, the following is implied by [2, Theorem 1.1].

Theorem 1 *Let $\varepsilon > 0$ and $T > 0$, and let $\Omega \subset R^2$ be a bounded disk centered at the origin. Let $(\cdot,\cdot)$ denote the $L^2(\Omega)$ inner product, and let $\langle\cdot,\cdot\rangle$ denote the $H^1(\Omega)$, $(H^1(\Omega))'$ duality pairing. Suppose that $u_0 \in \mathcal{K} := \{\eta \in H^1(\Omega) \,|\, |\eta| \leq 1\}$. Then there exists a pair $\{u, w\}$, such that $u \in L^2(0,T; H^2(\Omega)) \cap H^1(0,T; (H^1(\Omega))') \cap L^\infty(0,T;\mathcal{K})$, $w \in L^2(\Omega_T)$, with $w \in H^1_{loc}(\{M(u) > 0\})$ for a.e. $t \in (0,T)$, and*

$$\begin{cases} \langle \frac{\partial u}{\partial t}, \eta \rangle + \int_{\{M(u)>0\}} \nabla w \cdot M(u)\nabla\eta \, dx = 0, \quad \forall \eta \in H^1(\Omega) \quad a.e. \quad t \in (0,T), \\ \varepsilon^2(\nabla u, \nabla\eta - \nabla u) - (u, \eta - u) \geq (w, \eta - u), \quad \forall \eta \in \mathcal{K} \quad a.e. \quad t \in (0,T), \\ u(x,0) = u_0, \quad x \in \Omega. \end{cases} \tag{2}$$

With regard to (SD), we follow the discussion in [23]. Let $\Gamma : R \to R^2$ be a regular smooth immersed plane curve, which is periodic and closed with period $P \in (0,\infty)$, so that in fact $\Gamma : S^1 \to R^2$. We shall assume throughout that Γ is parameterized by arc-length. With regard to local existence,

Theorem 2 *Suppose $\Gamma_0 : R \to R^2$ is a regular closed periodic curve parametrized by arc-length, of class $\mathscr{C}^2 \cap W^{2,2}$ with $\|\kappa\|_{L^2(\Gamma_0)} < \infty$. Then there exists $T \in (0,\infty]$ and a unique one-parameter family of immersions parametrised by arc-length, $\Gamma : R \times [0,T) \to R^2$, such that (i) $\Gamma(0,\cdot) = \Gamma_0$, (ii) $V = -\kappa_{ss}$, (iii) $\Gamma(\cdot,t)$ is of class $\mathscr{C}^\infty$ and periodic with period $|\Gamma(\cdot,t)|$, $\forall t \in (0,T)$, and (iv) T is maximal.*

The uniqueness mentioned in Theorem 2 is modulo rotations, translations, changes in orientation, in accordance with the natural group of invariances for geometric flows in general, and for (SD) in particular.

Though the theorem above is formulated for one regular immersed circular curve, clearly Theorem 2 readily generalizes to accommodate a finite union $\cup_{i\in I}\Gamma_i$, $I\subset\mathscr{N}$, of such curves, which suffices for the purpose of the discussion that follows.

Note that while the existence for (DQOP) is guaranteed by Theorem 1 for arbitrary $T>0$, existence for (SD) is guaranteed by Theorem 2 only on some maximal interval. Indeed in the context of (SD) the maximal interval may well be finite for various reasons, for example, due to finite time self-intersection or curvature singularity formation. If we wish to consider and compare the steady states for both evolutions, the following theorem which prescribes conditions guaranteeing global existence for (SD) is helpful, and can be readily adapted for $\cup_{i\in I}\Gamma_i$, $I\subset\mathscr{N}$.

Theorem 3 (see [23, Theorem 1.1]) *Let $\Gamma_0 : S^1\to R^2$ be a regular smooth immersed closed curve with finite enclosed signed area, $\mathscr{A}(\Gamma_0)>0$ with $\int_{\Gamma_0}\kappa\,ds=2\pi$. Then there exists a constant $K^*>0$, such that if*

$$K_{\text{osc}}(\Gamma_0)<K^* \text{ and } I(\Gamma_0)<\exp(K^*/(8\pi^2)),$$

where $K_{\text{osc}}(\Gamma)=|\Gamma|\int_\Gamma(\kappa-\bar\kappa)^2\,ds$, with $\bar\kappa=|\Gamma|^{-1}\int_\Gamma\kappa\,ds$, denotes the normalized oscillation of the curvature and $I=|\Gamma|^2/(4\pi\mathscr{A}(\Gamma))$ is the isoperimetric ratio, the (SD) evolution for $\Gamma : S^1\times[0,T)\to R^2$, with Γ_0 as initial data, exists for all time and converges exponentially fast to a round circle with radius $\sqrt{\mathscr{A}(\Gamma_0)/\pi}$.

Remark 1 We remark that more regularity is required for the initial conditions in Theorems 2 and 3 than appears to be required for the initial conditions in Theorem 1. However in connecting the (SD) flow with the (DQOP) flow, the (SD) curve $\Gamma(t)$ corresponds rather naturally to the (DQOP) level set, $\{u(x,t)=0\,|\,(x,t)\in\Omega_T\}$, whose support lies within the set $\{(x,t)\in\Omega_T\,|\,\mathscr{M}(u(x,t))>0\}$ where the required regularity is guaranteed for $T>0$.

With regard to prominent dynamical features for (DQOP) and (SD), notably

$$E(t):=\frac{1}{\varepsilon|\Omega|}\int_\Omega\{(1-u^2)+\varepsilon^2|\nabla u|^2\}\,dx \quad\text{and}\quad \mathscr{L}(\mathrm{t}):=|(\mathrm{t})|, \tag{3}$$

where $E(t)$ is a scaled free energy[1] for (DQOP), are monotonically non-increasing for (DQOP) and (SD), respectively. Furthermore,

$$\bar u(t):=\frac{1}{|\Omega|}\int_\Omega u(x,t)\,dx \quad\text{and}\quad \mathscr{A}(\mathrm{t}):=-\frac{1}{2}\int_0^{\mathscr{L}(\mathrm{t})}(\mathrm{s,\,t})\mathrm{n\,ds}, \tag{4}$$

where $\bar u(t)$, the mean mass of $u(x,t)$, and $\mathscr{A}(t)$, the signed area enclosed by $\Gamma(t)$, are invariant under the respective (DQOP) and (SD) evolutions. The Gamma limit ($\varepsilon\downarrow 0$) of the constrained mean mass ($\bar u=\bar u_0$) minimizers of $E(t)$ is well known [12, 21] to yield a curve Γ with prescribed enclosed signed area, $\mathscr{A}_0=\mathscr{A}(0)$; this

[1] Here $E(t)$ has been scaled so that typically as equilibrium is approached, $E(t)\propto\frac{L(t)}{|\Omega|}$, where $L(t)$ reflects the length of the interface of between the two phases following phase separation; see [2].

can be readily be shown to hold also for u constrained to lie in $\mathscr{K}$ [16, 21], with

$$\mathscr{A}_0 = \frac{1}{2}|\Omega|(1+\bar{u}_0), \tag{5}$$

where $\bar{u}_0 := \bar{u}(0)$ in the context of our geometric and boundary assumptions. Notably, (DQOP) can be formulated as a conserved H^{-1} gradient flow with respect to the energy functional, $E(t)$, and (SD) can be formulated as an H^{-1} gradient flow with respect to $\mathscr{L}(t)$; see [22] and the discussion in Sect. 3. Moreover, the functionals

$$\text{Ent}(t) := \frac{1}{|\Omega|}\int_\Omega \{(1-u)\ln(1-u) + (1+u)\ln(1+u)\}\, dx,$$

$$K_{\text{osc}}(\Gamma(t)) := |\Gamma| \int_\Gamma (k-\bar{k})^2 ds,$$

are non-increasing along the respective (DQOP) and (SD) flows.[2] For (DQOP), for initial conditions $u_0 \in \mathscr{K}$ which correspond to a perturbation of $u_0 \equiv \bar{u}, \bar{u} \in (-1, 1)$, the dynamics can be characterized in terms of an initial regime of linear instability and a long time coarsening regime, [13, 17, 18]. For further discussion, see [2].

2 Steady States

We assume both motions to be defined within $\Omega \subset R^2$, where Ω is a planar disc centered at the origin with radius R_0, where R_0 is $O(1)$. When considering time evolution, we set $\Omega_T = \Omega \times (0, T)$, where $0 < T \leq \infty$.

Note that in (DQOP), no flux and Neumann boundary conditions are implied. For simplicity, we shall henceforth assume, more specifically, that $u \equiv -1$ in a δ-neighborhood of $\partial\Omega$, with $\varepsilon \ll \delta \ll O(1)$. While this assumption implies the boundary conditions given in (DQOP), it somewhat limits the resultant dynamics and steady states. Similarly, in studying (SD), surface diffusion motions, within Ω, we consider evolving curves $\Gamma(t) \subset \Omega$ which, more specifically, satisfy $\Gamma(t) \subset \mathscr{B}_{R_0}(\delta)$, where $\mathscr{B}_{R_0}(\delta)$ refers to the open planar disc centered at the origin with radius R_0 which has a bounding annular neighborhood with width δ where $u \equiv -1$.

2.1 *Steady States and Energy Minimizers for (SD)*

In considering (SD) energy minimizers, by recalling (3) one may view the energy to be given by the length of the curve, $\mathscr{L}(t)$, to within scaling constants. We allow $\Gamma(t)$ to be comprised of a finite number of nonintersecting components, $\Gamma(t) = \cup_{i \in I}\Gamma_i(t)$,

[2] Here Ent(t) reflects the physical entropy of the system, while $E(t)$ is a (scaled) free energy.

$I \subset \mathcal{N}$ with $\Gamma_i \cap \Gamma_j = \emptyset$ for $i \neq j$. Accordingly, the evolution by (SD) of each component Γ_i, $i \in I$, may be prescribed as

$$V^i = -\kappa^i_{s_i s_i}, \tag{6}$$

where V^i and κ^i denote respectively, the normal velocity and the mean curvature of $\Gamma_i(t)$, and s_i denotes an arc-length parametrization of $\Gamma_i(t)$.

Within the context of these assumptions, it readily follows from (6) that steady states correspond of a finite disjoint union of circular curves. Since the area encompassed by curves evolving by surface diffusion is conserved under the evolution [23], it follows that $\sum_{i \in I} A(\Gamma_i(t))$ is a conserved quantity. Taking into account the isoperimetric inequality and (6), we may now conclude the following:

Theorem 4 *Under the assumptions outlined above, the set of minimum energy steady states for (SD) corresponds to the set of circular curves with radius* $(A^*/\pi)^{1/2}$, *with* $A^* = \sum_{i \in I} A(\Gamma_i(0))$, *which are located somewhere within* $\mathscr{B}_{R_0}(\delta)$. *Any finite union of disjoint circular curves such that* $\sum_{i \in I} A(\Gamma_i(t)) = \sum_{i \in I} A(\Gamma_i(0))$ *also corresponds to a steady state for (SD).*

2.2 Steady States and Energy Minimizers for (DQOP)

The energy minimizing steady states for (DQOP) correspond to the energy minimizers of $E(t)$ within the set $u \in \mathscr{K}$ and which satisfy $\bar{u} = \bar{u}_0$. Given the definition of the energy $E(t)$ and the geometry described at the beginning of this section, considerations of energy symmetrization and regularity [3], lead us to conclude.

Theorem 5 *Under the assumptions outlined above, the minimum energy steady states for (DQOP) are monotonically decreasing with respect to distance from the origin, modulo possible translations within* $\mathscr{B}_{R_0}(\delta)$.

Remark 2 Since the energy $E(t)$, as well as the problem formulated in (DQOP), are invariant under the transformation $u \to -u$, if we set u to equal $+1$ rather than -1 in the $\delta-$annular neighborhood of $\partial\Omega$, then the conclusion in Theorem 5 would have yielded that the minimum energy steady states are monotonically decreasing, modulo translation within $\mathscr{B}_{R_0}(\delta)$. Without the constraint that u equals ± 1 in a $\delta-$annular neighborhood of $\partial\Omega$, the energetics of possible additional steady states would need to be considered. Such additional steady states would include certain energy minimizing steady state solutions with "droplet like" ± 1 concentrations along the boundary, with lower energy than the axi-symmetric energy minimizing steady states with the same mean mass and their translates, discussed above. For simplicity, we focus here on a more limited set of steady states, which provide insight into the more general case.

It follows from Theorem 5 and Remark 2 that we should consider the set of axi-symmetric monotonically increasing steady states for (DQOP) and their translates

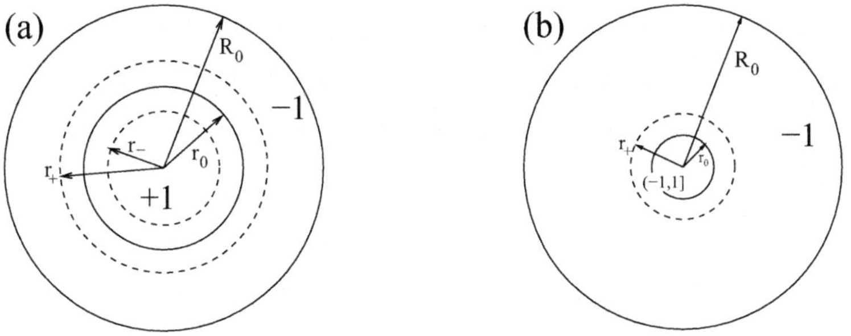

Fig. 1 **a** Schematic portrayal of an annular solution: $u(r) \equiv 1$ for $0 \le r \le r_-$, $-1 < u(r) < 1$ for $r_- < r < r_+$, and $u(r) \equiv -1$ for $r_+ \le r < R_0$. **b** Schematic portrayal of a dimple solution: $a := u(0) = \lim_{r\downarrow 0} u(r) \in (-1, 1]$ with $-1 < u(r) \le a$ for $0 \le r < r_+$, and $u(r) \equiv -1$ for $r_+ \le r < R_0$.

that lie within $\mathscr{B}_{R_0}(\delta)$. Since we are looking for constrained mean mass minimizers, we should explore the set of the monotonically decreasing axi-symmetry steady states with prescribed mean mass, $\bar{u}$. This is undertaken in detail in the two subsections that follow. If a steady state $u \in \mathscr{K}$ equals -1 in a δ−annular neighborhood of $\partial\Omega$ and increases (non-decreases) monotonically, then either (i) $u = 1$ is attained in a circular neighborhood of the origin or (ii) $u \in [-1, 1)$ in Ω except perhaps at the origin. In case (i), the steady states contain unique monotonically decreasing annular transition region, see Fig. 1a and Sect. 2.3 for details. In case (ii), "dimple solutions" are possible, with $-1 < u \le 1$ at the origin and with $-1 \le u < 1$ elsewhere; this possibility is explored in Sect. 2.4. See Fig. 1b.

Before exploring the details of the radial solutions, let us recall that we wish to connect the solutions of (DQOP) with solutions of (SD). In considering the Gamma limit, the set of (DQOP) solutions are compared with (SD) solutions with similar mass. While solutions to (DQOP) depend on the parameter ε, the mean mass constraint is independent of ε. Let us recall (5). If $\bar{u}_0 = -1$, then trivially, $u \equiv -1$. If $\bar{u}_0 = +1$, then $u \equiv +1$ is implied by (5), which does not yield a possible solution due to the requirement that $u = -1$ in a δ−annular neighborhood of $\partial\Omega$. If $\bar{u}_0 \in (-1, 1)$ and $\bar{u}_0$ is not too close to -1, then the monotone axi-symmetric steady state solutions to (DQOP) can be expected to Gamma converge to -1 outside the origin in an annulus with width $R_0 - r_0$, and to converge to $+1$ in the disk centered at the origin with radius r_0, with the circular curve Γ with radius r_0 constituting an axi-symmetric steady state solution to (SD). Recalling (5), we get that if Ω is a disk with radius R_0, then

$$\bar{u}_0 = \frac{1}{|\Omega|}(2\mathscr{A}_0 - |\Omega|) = \frac{2r_0^2 - R_0^2}{R_0^2}. \tag{7}$$

Based on (7), for monotone steady state axisymmetric solutions of (DQOP) with $\bar{u} \in (-1, 1)$, an ε independent **equivalent mean mass condition** is implied, namely

Proposition 1 (Equivalent mean mass) *Let u be a monotonically decreasing axi-symmetric steady state solution to (DQOP) with $\bar{u} = \bar{u}_0 \in (-1, 1)$ which lies within $\Omega = \mathscr{B}_{R_0}(\delta)$, where $\mathscr{B}_{R_0}(\delta)$ denotes the disk with radius R_0 centered at the origin which has a bounding annular neighborhood with width δ where $u \equiv -1$. Then u is radial, $u = u(r)$, and*

$$\int_0^{R_0} u(r) r \, dr = \frac{1}{2} \bar{u} R_0^2 = \frac{1}{2}(2r_0^2 - R_0^2), \tag{8}$$

for some $0 < r_0 < R_0 - \delta$.

Let us consider Fig. 1. Given $\mathscr{B}_{R_0}(\delta)$, it follows from (3) and (7) for $0 < \varepsilon \ll 1$, that if $\bar{u}_0$ is not too close to -1, then the solution can be expected to contain an annular transition region, $r_- < r < r_+$, with $0 < r_- < r_0 < r_+ < R_0 - \delta$ and $r_+ - r_- = O(\varepsilon)$, where $u(r) = -1$ for $r \in [r_+, R_0]$ and $u(r) = +1$ for $r \in [0, r_-]$; we will refer to such solutions as "annular" solutions. If, on the other hand, $\bar{u}_0$ is sufficient close to -1, then any possible (admissible) monotone radial solutions will have $u = -1$ in an annular region, $r_+ \leq r \leq R_0$, with $0 < r_0 < r_+ < R_0$, such that $u \in (-1, 1)$ for $0 < r < r_+$ and $u(0) = \lim_{r \downarrow 0} u(r) \in (-1, 1]$; we will refer to such solutions as "dimple" solutions. In either case, in accordance with (DQOP) and Theorem 1 (see [7]), $u \in C^1(\mathscr{B}_{R_0}(\delta))$ and outside of the annular or circular regions where $u \equiv \pm 1$, $u \in C^2$.

Accordingly, the annular solution should satisfy

$$\begin{aligned} & u + \lambda = -\varepsilon^2 (u_{rr} + \frac{1}{r} u_r), \qquad r_- < r < r_+, \\ & u(r_-) = 1 \,, \quad u(r_+) = -1, \\ & u_r(r_-) = 0 \,, \quad u_r(r_+) = 0, \end{aligned} \tag{9}$$

for some r_-, r_+, with $0 < r_- < r_0 < r_+ < R_0$, and for some $\lambda \in R$. Given the boundary conditions in (9), λ can be considered here as a constant of integration for (6) with vanishing normal velocity. By testing the equation in (9) by $u(r)$, it is readily seen that λ can also be viewed as a mean mass conserving Lagrange multiplier for the free energy $E(t)$; see (11) as well as (14) in the sequel. Note that by (8)

$$\bar{u} = \frac{(2r_0^2 - R_0^2)}{R_0^2} = \frac{2}{R_0^2} \int_0^{R_0} u(r) r \, dr, \tag{10}$$

and from the prescribed structure of annular solutions

$$\int_0^{R_0} u(r) r \, dr = \int_{r_-}^{r_+} u(r) r \, dr + \frac{1}{2}(r_+^2 + r_-^2 - R_0^2).$$

Using (9),

$$\int_{r_-}^{r_+} u(r)r\,dr = -\int_{r_-}^{r_+} [\varepsilon^2(ru_r)_r + \lambda r]\,dr = -\frac{\lambda}{2}(r_+^2 - r_-^2).$$

Hence the equivalent mean mass condition holds for (9) with (10) if and only if

$$\lambda = \frac{r_+^2 + r_-^2 - 2r_0^2}{r_+^2 - r_-^2}. \tag{11}$$

Similarly, the dimple solution should satisfy

$$\begin{aligned} & u + \lambda = -\varepsilon^2(u_{rr} + \frac{1}{r}u_r), \qquad 0 < r < r_+, \\ & -1 < u(0) \le 1\,, \quad u(r_+) = -1, \\ & u_r(0) = 0\,, \quad u_r(r_+) = 0, \end{aligned} \tag{12}$$

for some r_+, with $0 < r_0 < r_+ < R_0$, and for some $\lambda \in R$. Again, by (8) in Proposition 1

$$\bar{u} = \frac{(2r_0^2 - R_0^2)}{R_0^2} = \frac{2}{R_0^2}\int_0^{R_0} u(r)r\,dr, \tag{13}$$

and from the prescribed structure of dimple solutions

$$\int_0^{R_0} u(r)r\,dr = \int_0^{r_+} u(r)r\,dr + \frac{1}{2}(r_+^2 - R_0^2).$$

Using (12),

$$\int_0^{r_+} u(r)r\,dr = -\int_0^{r_+} [\varepsilon^2(ru_r)_r + \lambda r]\,dr = -\frac{\lambda}{2}r_+^2.$$

Hence the equivalent mean mass condition holds for (12) with (13) if and only if

$$\lambda = \frac{r_+^2 - 2r_0^2}{r_+^2}. \tag{14}$$

Before going into the details of the annular and dimple (DQOP) solutions in the next subsections, we pause to point out that it is possible to attain a large class of additional solutions by appropriately pasting together translates of dimple and annular solutions of various sizes, so long as the resulting construction lies in $\mathscr{K}$, for some $\bar{u} \in (-1, 1)$. See, for example, the concentrically ringed solutions identified by Chen [5] as the asymptotic limit of solutions to the Cahn-Hilliard equation.

2.3 Annular Solutions for (DQOP)

We now consider radial "annular" monotonically decreasing solutions in $\Omega = \mathscr{B}_{R_0}(\delta)$, containing a transition between the values ± 1, see Fig. 1a. More specifically we assume that $u(r) \equiv +1$ for $r \in [0, r_-]$, $-1 < u(r) < 1$ for $r \in (r_-, r_+)$, and $u(r) \equiv -1$ for $r \in [r_+, R_0)$, where $r_\pm$ reflect the location of free boundaries with $0 < r_- < r_0 < r_+ < R_0 - \delta$. It is possible to seek non-monotone solutions, but these would have more energy and here we are looking for energy minimizing steady states. Generically we may assume that $r_0 = O(1)$, although, as we shall see, certain values of r_0 with $r_0 = O(\varepsilon)$ are also possible. We shall discuss the generic case first, and treat the general case afterwards.

As we saw in Sect. 2.2, $u(r)$ should satisfy

$$\begin{aligned} &u + \lambda = -\varepsilon^2(u_{rr} + \frac{1}{r}u_r), \qquad r_- < r < r_+, \\ &u(r_-) = 1, \quad u(r_+) = -1, \\ &u_r(r_-) = 0, \quad u_r(r_+) = 0, \end{aligned}$$

where the equivalent mean mass condition holds if and only if

$$\bar{u} = \frac{(2r_0^2 - R_0^2)}{R_0^2}, \qquad \lambda = \frac{r_+^2 + r_-^2 - 2r_0^2}{r_+^2 - r_-^2}, \qquad r_0 = \left[\frac{1+\bar{u}}{2}\right]^{1/2} R_0, \tag{15}$$

for some $0 < r_0 < r_+$, see (9)–(11). Thus $\lambda = \lambda(r_-, r_+, r_0)$ and $r_0 = r_0(\bar{u}, R_0)$. We shall see that annular solutions exist for $-1 + O(\varepsilon^2) < \bar{u} < 1 - 4\delta/R_0 + O(\delta^2, \varepsilon)$ with $O(\varepsilon) < r_0 < R_0 - \delta + O(\varepsilon)$; some further specifics to follow.

The parameters, $r_\pm$, reflecting the location of the free boundaries, are to be determined. Thus, we have four boundary conditions, as well as the monotonicity and range constraints. In solving (9)–(11), we have four degrees of freedom, two from the second order ODE in (9) and two from the parameters $r_\pm$, with $\lambda = \lambda(r_-, r_+, r_0)$. Hence, one would expect the possible annular solutions to be uniquely determined by $\bar{u}$ (or, equivalently, by r_0 and R_0).

Setting

$$q = \frac{r}{\varepsilon}, \quad q_0 = \frac{r_0}{\varepsilon}, \quad q_\pm = \frac{r_\pm}{\varepsilon}, \quad Q_0 = \frac{R_0}{\varepsilon}, \quad \text{and} \quad v(q) := [u(r) + \lambda]|_{r=\varepsilon q}, \tag{16}$$

we get the following problem for $v(q)$,

$$\begin{aligned} &q v_{qq} + v_q + q v = 0, \quad 0 < q_- < q < q_+, \\ &v(q_-) = 1 + \lambda, \quad v(q_+) = -1 + \lambda, \\ &v_q(q_-) = 0, \quad v_q(q_+) = 0, \end{aligned} \tag{17}$$

with

$$\lambda = \frac{q_+^2 + q_-^2 - 2q_0^2}{q_+^2 - q_-^2}. \tag{18}$$

The equation in (17) is a Bessel equation of order zero, [19, 10.2.1], whose general solution ([19, 10.2(iii), 10.6.3]) is

$$v(q) = c_1 J_0(q) + c_2 Y_0(q), \quad q > 0, \qquad c_1, c_2 \in R, \tag{19}$$

with

$$v_q(q) = -c_1 J_1(q) - c_2 Y_1(q), \quad q > 0.$$

The coefficients c_1, c_2, in the solution to (17), should depend on $q_\pm$ and λ, which in turn depend on $\bar{u}$ (or equivalently on r_0), as well as on the underlying parameters R_0 and ε.

Using the polar representation for Bessel functions, [19, 10.18.4, 10.18.6, 10.18.7], we get that

$$v(q) = c_1 J_0(q) + c_2 Y_0(q) = M_0(q)\big(c_1 \cos\theta_0(q) + c_2 \sin\theta_0(q)\big), \quad q > 0, \tag{20}$$

$$-v_q(q) = c_1 J_1(q) + c_2 Y_1(q) = M_1(q)\big(c_1 \cos\theta_1(q) + c_2 \sin\theta_1(q)\big), \quad q > 0, \tag{21}$$

where for $n = 0, 1$, $\theta_n(x) := \arctan(Y_n(x)/J_n(x))$ for $x > 0$, and

$$M_n(x) := \sqrt{J_n^2(x) + Y_n^2(x)} > 0, \quad x > 0,$$

since the Bessel functions J_n, Y_n do not vanish simultaneously.

Setting

$$\cos\varphi = \frac{c_1}{\sqrt{c_1^2 + c_2^2}}, \quad \sin\varphi = \frac{c_2}{\sqrt{c_1^2 + c_2^2}}, \quad \text{and} \quad A := \frac{1}{\sqrt{c_1^2 + c_2^2}}, \tag{22}$$

we can write (20), (21) as

$$v(q) = AM_0(q)\cos\big(\theta_0(q) - \varphi\big), \quad v_q(q) = -AM_1(q)\cos\big(\theta_1(q) - \varphi\big), \quad q > 0. \tag{23}$$

Since M_1 and A are positive, the boundary conditions $v_q(q_-) = v_q(q_+) = 0$ in (17) imply that

$$\cos\big(\theta_1(q_-) - \varphi\big) = \cos\big(\theta_1(q_+) - \varphi\big) = 0. \tag{24}$$

As we are seeking monotone solutions, $v(q)$ should be monotonically decreasing with $v_q < 0$ for $q_- < q < q_+$ by Sturmian theory.[3] The positivity of M_1 and A now implies that $\cos\left(\theta_1(q) - \varphi\right) > 0$ for $q_- < q < q_+$, and thus that

$$\frac{-\pi}{2} < \theta_1(q) - \varphi < \frac{\pi}{2}, \quad 0 < q_- < q < q_+, \tag{25}$$

up to possible translations by $2k\pi$, $k \in Z$, which do not effect the solutions. The function $\theta_1(q)$ is monotonically increasing ([19, 10.18.18]); therefore

$$\theta_1(q_-) - \varphi = -\frac{\pi}{2}, \quad \theta_1(q_+) - \varphi = \frac{\pi}{2}, \quad \text{and} \quad \theta_1(q_+) - \theta_1(q_-) = \pi. \tag{26}$$

Since $\lim_{q\downarrow 0} \theta_1(q) = -\frac{\pi}{2}$ ([19, 10.18.3]), the monotonicity of $\theta_1(q)$ and (26) imply that

$$\theta_1(q_+) > \frac{\pi}{2}, \quad \theta_1(q_-) > -\frac{\pi}{2}, \quad \varphi > 0. \tag{27}$$

2.3.1 Annular Solutions, the Generic Case

We remarked earlier that $-1 + O(\varepsilon^2) < \bar{u} < 1 - 4\delta/R_0 + O(\delta^2, \varepsilon)$ with $O(\varepsilon) < r_0 < R_0 - \delta + O(\varepsilon)$. Thus, generically $r_0 = O(1)$. Considerations of scaling and energy minimization of the energy prescribed in (3) for $0 < \varepsilon \ll 1$ imply that transition widths between the phases for energy minimizing solutions scale as $O(\varepsilon)$, see e.g. [16, 21], and hence, $r_+ - r_- = O(\varepsilon)$ and r_-, r_+ are $O(1)$ when $r_0 = O(1)$. This will also be demonstrated directly in Sect. 2.3.2. Accordingly, the rescalings in (16) imply that

$$q_-, q_0, q_+ = O(\varepsilon^{-1}), \quad q_+ - q_- = O(1), \quad \varepsilon q_0 = O(1) \tag{28}$$

in the generic case. Throughout this subsection, we assume that (28) holds.

For large values of x (see [19, 10.18.18]),

$$\theta_1(x) = x - \frac{3\pi}{4} + \frac{3}{8x} + O\left(\frac{1}{x^3}\right), \quad x \gg 1. \tag{29}$$

Hence, by (26), for large values of q_-, q_+,

$$q_- - \frac{3\pi}{4} + \frac{3}{8q_-} + O\left(\frac{1}{q_0^3}\right) - \varphi = -\frac{\pi}{2}, \quad q_+ - \frac{3\pi}{4} + \frac{3}{8q_+} + O\left(\frac{1}{q_0^3}\right) - \varphi = \frac{\pi}{2}, \tag{30}$$

[3] If v_q vanishes at some point $\bar{q} \in (q_-, q_+)$, then Sturm's Separation Theorem [11] implies that v vanishes at least once in each of the intervals $[q_-, \bar{q}]$, $[\bar{q}, q_+]$, and hence v cannot be monotone.

and thus

$$q_+ - q_- = \pi + \frac{3}{8}\left(\frac{1}{q_-} - \frac{1}{q_+}\right) + O\left(\frac{1}{q_0^3}\right). \tag{31}$$

The boundary conditions $v(q_-) = 1 + \lambda$ and $v(q_+) = -1 + \lambda$ in (17) imply that

$$AM_0(q_-)\cos\left(\theta_0(q_-) - \varphi\right) = 1 + \lambda, \quad AM_0(q_+)\cos\left(\theta_0(q_+) - \varphi\right) = -1 + \lambda. \tag{32}$$

Subtracting the equations in (32),

$$M_0(q_-)\cos\left(\theta_0(q_-) - \varphi\right) - M_0(q_+)\cos\left(\theta_0(q_+) - \varphi\right) = \frac{2}{A}. \tag{33}$$

For large values of x (see [19, 10.18.18]),

$$\theta_0(x) = x - \frac{\pi}{4} - \frac{1}{8x} + O\left(\frac{1}{x^3}\right), \quad x \gg 1.$$

Hence using (30)

$$\theta_0(q_-) - \varphi = -\frac{1}{2q_-} + O\left(\frac{1}{q_0^3}\right), \quad \theta_0(q_+) - \varphi = \pi - \frac{1}{2q_+} + O\left(\frac{1}{q_0^3}\right),$$

and therefore

$$\cos(\theta_0(q_-) - \varphi) = 1 - \frac{1}{8q_-^2} + O\left(\frac{1}{q_0^3}\right), \quad \cos(\theta_0(q_+) - \varphi) = -1 + \frac{1}{8q_+^2} + O\left(\frac{1}{q_0^3}\right). \tag{34}$$

For large values of x (see [19, 10.18.17]),

$$M_0(x) = \sqrt{\frac{2}{\pi x}} + O\left(\frac{1}{x^{5/2}}\right), \quad x \gg 1. \tag{35}$$

Using (34), (35) in (33),

$$\sqrt{\frac{2}{\pi q_-}} \cdot \left(1 - \frac{1}{8q_-^2}\right) - \sqrt{\frac{2}{\pi q_+}} \cdot \left(-1 + \frac{1}{8q_+^2}\right) + O\left(\frac{1}{q_0^{5/2}}\right) = \frac{2}{A},$$

which implies that

$$A = \frac{\sqrt{2\pi}}{\frac{1}{\sqrt{q_-}} + \frac{1}{\sqrt{q_+}}} + O\left(\frac{1}{q_0^{3/2}}\right). \tag{36}$$

Returning to (32) and summing the two equations,

$$AM_0(q_-)\cos\big(\theta_0(q_-)-\varphi\big)+AM_0(q_+)\cos\big(\theta_0(q_+)-\varphi\big)=2\lambda,$$

and then using the approximations in (34), (35),

$$\left(\frac{\sqrt{2\pi}}{\frac{1}{\sqrt{q_-}}+\frac{1}{\sqrt{q_+}}}+O\left(\frac{1}{q_0^{3/2}}\right)\right)\left(\sqrt{\frac{2}{\pi q_-}}-\sqrt{\frac{2}{\pi q_+}}+O\left(\frac{1}{q_0^{5/2}}\right)\right)=2\lambda,$$

which implies that

$$\lambda=\frac{\sqrt{q_+}-\sqrt{q_-}}{\sqrt{q_+}+\sqrt{q_-}}+O\left(\frac{1}{q_0^2}\right). \tag{37}$$

Using now (31),

$$\lambda=\frac{\pi+O\left(\frac{1}{q_0}\right)}{(\sqrt{q_-+\pi}+\sqrt{q_-})^2}=O\left(\frac{1}{q_0}\right),$$

from which we get that

$$q_-=\frac{\pi(1-2\lambda)}{4\lambda}+O\left(\frac{1}{q_0}\right),\quad q_+=\frac{\pi(1+2\lambda)}{4\lambda}+O\left(\frac{1}{q_0}\right), \tag{38}$$

which implies that $q_\pm=O(q_0)$ and $q_+-q_-=O(1)$, in accordance with (28).

Using (38) and (16) in the expression for λ given in (15) and noting that (37) implies that $\lambda>0$, then solving for λ, we obtain that

$$\lambda=\frac{\pi}{4q_0}+O\left(\frac{1}{q_0^2}\right). \tag{39}$$

Substituting the above expression for λ into (38) we get that

$$q_-=q_0-\frac{\pi}{2}+O\left(\frac{1}{q_0}\right)\quad q_+=q_0+\frac{\pi}{2}+O\left(\frac{1}{q_0}\right).$$

In order to obtain an approximate solution $v(q)$ to (17), and subsequently to obtain an approximate solution $u(r)=v(r/\varepsilon)-\lambda$ to (9), based, say on the expression in (19), in the generic case, it remains to identify approximations for the coefficients, c_1, c_2.

The boundary conditions $v(q_+)=-1+\lambda$ and $v_q(q_+)=0$ in (17), imply that

$$c_1=\frac{(-1+\lambda)Y_1(q_+)}{J_0(q_+)Y_1(q_+)+J_1(q_+)Y_0(q_+)},\quad c_2=\frac{(-1+\lambda)Y_0(q_+)}{J_0(q_+)Y_1(q_+)+J_1(q_+)Y_0(q_+)}. \tag{40}$$

We know (see [19, 10.17.3, 10.17.4]) that

$$J_0(x) = \sqrt{\frac{2}{\pi x}}\left(\cos\left(x - \frac{\pi}{4}\right) + O\left(\frac{1}{x}\right)\right), \quad x \gg 1,$$

$$Y_0(x) = \sqrt{\frac{2}{\pi x}}\left(\sin\left(x - \frac{\pi}{4}\right) + O\left(\frac{1}{x}\right)\right), \quad x \gg 1,$$

and that

$$J_1(x) = \sqrt{\frac{2}{\pi x}}\left(\cos\left(x - \frac{3\pi}{4}\right) + O\left(\frac{1}{x}\right)\right), \quad x \gg 1,$$

$$Y_1(x) = \sqrt{\frac{2}{\pi x}}\left(\sin\left(x - \frac{3\pi}{4}\right) + O\left(\frac{1}{x}\right)\right), \quad x \gg 1.$$

Using the approximations above in (40),

$$c_1 = -\sqrt{\frac{q_0\pi}{2}}\frac{[\sin(q_0 - \pi/4)]}{\sin(2q_0)} + O\left(\frac{1}{q_0}\right), \quad c_2 = -\sqrt{\frac{q_0\pi}{2}}\frac{[\cos(q_0 - \pi/4)]}{\sin(2q_0)} + O\left(\frac{1}{q_0}\right), \tag{41}$$

and then returning to (19), we obtain that

$$v(q) = -\sqrt{\frac{q_0}{q}}\frac{\cos(q + q_0)}{\sin(2q_0)} + O\left(\frac{1}{q_0}\right), \quad q_- < q < q_+ \; (q_0 - \frac{\pi}{2} < q < q_0 + \frac{\pi}{2}\;). \tag{42}$$

Recalling (16),

$$u(r) = -\sqrt{\frac{r_0}{r}}\frac{\cos((r + r_0)/\varepsilon)}{\sin(2r_0/\varepsilon)} + O\left(\frac{\varepsilon}{r_0}\right), \quad r_- < r < r_+. \tag{43}$$

2.3.2 Existence and Uniqueness

In this section, we prove.

Theorem 6 *Given ε, δ, and R_0, $0 < \varepsilon \ll \delta \ll 1$, $R_0 = O(1)$. There exists a unique solution to (9), (11) in $\mathscr{B}_{R_0}(\delta)$ for $r_0 \in (r_0^{\inf}, r_0^{\sup}]$, where $r_0^{\inf} = \bar{q}/\varepsilon$ with $\bar{q} := \theta_1^{-1}(\pi/2)$ and $r_0^{\sup} = R_0 - \delta + O(\varepsilon)$.*

Remark 3 In Theorem 6, $r_0^{\inf}$ corresponds to the largest value of r_0 for which dimple solutions exist, to be discussed in detail in Sect. 2.4; for the annular solutions, $r_- \downarrow 0$ as $r_0 \downarrow r_0^{\inf}$. The upper limit, $r_0^{\sup}$, corresponds to the largest possible value of r_0, given the δ−width annulus where $u \equiv -1$ within $\mathscr{B}_{R_0}(\delta)$, with $r_+ \uparrow R_0 - \delta$ as $r_0 \uparrow r_0^{\sup}$.

Proof It is convenient to prove Theorem 6 by using the rescalings defined in (16), and proving unique existence for the equivalent problem prescribed in (17), (18) for $q_0 \in (q_0^{\inf}, q_0^{\sup}]$, where $q_0^{\inf} = \bar{q} = \varepsilon r_0^{\inf}$, $q_0^{\sup} = \varepsilon r_0^{\sup}$. In Proposition 2 below, unique existence is demonstrated for $q_0 \in (\bar{q}, \infty)$, and then the implied ranges indicated in Theorem 6 follows by returning to the original scaling and imposing $r_0^{\sup}$ the upper limit, with $r_0^{\sup}$ corresponding to $r_+^{\sup} = R_0 - \delta$. In the course of the proof of Proposition 2, we shall see that $r_0^{\sup} - r_+^{\sup} = O(\varepsilon)$, which implies that $r_0^{\sup} = R_0 - \delta + O(\varepsilon)$.

Proposition 2 *There exists a unique monotone solution to (17) and (18) for every $q_0 \in (\bar{q}, \infty)$, where $\bar{q} := \theta_1^{-1}(\pi/2)$ corresponds to the first positive root of $J_1(q)$.*

Proof The proof relies on introducing a functional, $D = D(t)$, defined below, which allows us to focus first on existence and then on uniqueness. Let us recall that the general solution to the equation in (17) may be written as in (23), namely as

$$v(q) = AM_0(q)\cos\big(\theta_0(q) - \varphi\big), \quad v_q(q) = -AM_1(q)\cos\big(\theta_1(q) - \varphi\big), \quad q \geq 0,$$

where $A > 0$, $M_0(q)$, $M_1(q) > 0$, for $q \geq 0$. The boundary conditions $v_q(q_-) = v_q(q_+) = 0$ in (17), together with the monotonicity of $\theta_1(x)$ with $\lim_{x\downarrow 0}\theta_1(x) = -\pi/2$, imply that

$$\varphi > 0, \quad \frac{-\pi}{2} < \theta_1(q) - \varphi < \frac{\pi}{2}, \quad 0 < q_- < q < q_+, \quad \theta_1(q_\pm) - \varphi = \pm\frac{\pi}{2}, \tag{44}$$

up to possible translations by $2k\pi$, $k \in Z$, see (25)–(27).

The general solution contains two parameters, A and φ. As we have already accommodated the boundary conditions $v_q(q_-) = v_q(q_+) = 0$, finding a solution to (17), (18) entails identifying q_- and q_+ as functions of A and φ so as to satisfy the two remaining boundary conditions in (17). In view of (44), it is convenient to work with the parameters A and t, rather than with the parameters A and φ, where

$$t = \theta_1(q_0) - \varphi, \quad -\frac{\pi}{2} \leq t \leq \frac{\pi}{2}, \tag{45}$$

up to possible translation by $2k\pi$, $k \in Z$. From (44), (45), we get that

$$\varphi(t) = \theta_1(q_0) - t, \quad \theta_1(q_\pm(t)) = \theta_1(q_0) - t \pm \frac{\pi}{2}, \quad t \in [-\pi/2, \pi/2]. \tag{46}$$

Since $\theta_1(x)$ is monotonically increasing and continuously differentiable for $x > 0$ ([19, 10.18.8]), $\theta_1(x)$ has an inverse function which is also monotonically increasing and continuously differentiable, and

$$q_\pm(t) = \theta_1^{-1}\left(\theta_1(q_0) - t \pm \frac{\pi}{2}\right). \tag{47}$$

Thus φ, q_-, q_+, and well as λ (see (18)), can be viewed as continuously differentiable functions of t, since $\theta_1(q_0) - t - \pi/2 > -\pi/2$ for $t \in [-\pi/2, \pi/2]$, for $q_0 > \theta_1^{-1}(\pi/2) = \bar{q}$.

From (23) and the boundary conditions, $v(q_-) = 1 + \lambda$ and $v(q_+) = -1 + \lambda$,

$$AM_0(q_-)\cos\big(\theta_0(q_-) - \varphi\big) = 1 + \lambda, \quad AM_0(q_+)\cos\big(\theta_0(q_+) - \varphi\big) = -1 + \lambda. \tag{48}$$

As φ, q_-, q_+, and λ are prescribed in terms of t, we get two descriptions for A in terms of t from (48) as long as $\cos\big(\theta_0(q_-) - \varphi\big)$ and $\cos\big(\theta_0(q_+) - \varphi\big)$ do not vanish, since $M_0(x) > 0$ for $x > 0$. Moreover, the two descriptions must be equal for some value of $t \in [-\pi/2, \pi/2]$, if we are to attain a solution to (17), (18).

First we prove

Lemma 1 $\cos\big(\theta_0(q_-) - \varphi\big) > 0$, *and* $\cos\big(\theta_0(q_+) - \varphi\big) < 0$.

Proof First we note that for $x \geq 0$,

$$\begin{aligned} J_0(x) &= M_0(x)\cos\theta_0(x), \quad J_0'(x) = N_0(x)\cos\phi_0(x) = -M_1(x)\cos\theta_1(x), \\ Y_0(x) &= M_0(x)\sin\theta_0(x), \quad Y_0'(x) = N_0(x)\sin\phi_0(x) = -M_1(x)\sin\theta_1(x). \end{aligned} \tag{49}$$

Using (49), the formula [19, 10.18.12], $M_0(x)N_0(x)\sin(\phi_0(x) - \theta_0(x)) = \frac{2}{\pi x}$, $x > 0$, and a little trigonometry,

$$M_0(x)M_1(x)\sin(\theta_0(x) - \theta_1(x)) = \frac{2}{\pi x}, \quad x > 0. \tag{50}$$

Since $M_0(x),\ M_1(x) > 0$ for $x > 0$, (50) implies that $\sin(\theta_0(x) - \theta_1(x)) > 0, \quad x > 0$. Thus

$$0 < \theta_0(x) - \theta_1(x) < \pi, \quad x > 0,$$

up to possible translation by $2k\pi$, $k \in Z$. Therefore,

$$\theta_1(x) < \theta_0(x) < \theta_1(x) + \pi, \quad x > 0. \tag{51}$$

Setting $x = q_-$ in (51), we obtain that $\theta_1(q_-) - \varphi < \theta_0(q_-) - \varphi < \theta_1(q_-) - \varphi + \pi$. Then using (44),

$$-\frac{\pi}{2} < \theta_0(q_-) - \varphi < \frac{\pi}{2}$$

and therefore $\cos\big(\theta_0(q_-) - \varphi\big) > 0$. Similarly, we obtain from (51) and (44) that

$$\frac{\pi}{2} < \theta_0(q_+) - \varphi < \frac{3\pi}{2}$$

and therefore $\cos\big(\theta_0(q_+) - \varphi\big) < 0$. □

Given Lemma 1, we now obtain two expressions for A, namely,

$$A = \frac{1+\lambda}{M_0(q_-)\cos\big(\theta_0(q_-)-\varphi\big)} = \frac{-1+\lambda}{M_0(q_+)\cos\big(\theta_0(q_+)-\varphi\big)}. \tag{52}$$

Using the definition of λ given in (11), and noting that $\theta_1(q_+) - \theta_1(q_-) = \pi$ implies that $q_+ - q_- > 0$, we obtain from (52) that

$$\frac{q_+^2 - q_0^2}{M_0(q_-)\cos\big(\theta_0(q_-)-\varphi\big)} = \frac{q_-^2 - q_0^2}{M_0(q_+)\cos\big(\theta_0(q_+)-\varphi\big)},$$

where all the terms are continuously differentiable functions of t. It remains to verify that the above equation uniquely defines t. Cross-multiplying, we obtain now that roots of $D(t) = 0$, with $t \in [-\pi/2, \pi/2]$, correspond to solutions of (17), (18), where

$$D(t) := \mathscr{D}(q_+(t)) - \mathscr{D}(q_-(t)), \quad \mathscr{D}(q) := (q^2 - q_0^2)M_0(q)\cos\big(\theta_0(q)-\varphi\big), \tag{53}$$

is continuously differentiable for $t \in [-\pi/2, \pi/2]$. Existence now follows from the following lemma.

Lemma 2 $D(-\pi/2) < 0, \quad D(\pi/2) > 0.$

Proof Let us first consider $D(t)$ with $t = -\pi/2$. From (46), $\theta_1(q_-(-\pi/2)) = \theta_1(q_0) > \pi/2$ for $q_0 > \bar{q}$. Recalling the monotonicity of $\theta_1(x)$ for $x > 0$, and using (47),

$$q_-(-\pi/2) = q_0, \quad q_+(-\pi/2) = \theta_1^{-1}\left(\theta_1(q_0) + \pi\right) > q_0. \tag{54}$$

Similarly for $D(t)$ with $t = \pi/2$, we obtain that $\theta_1(q_+(\pi/2)) = \theta_1(q_0) > \pi/2$, and hence

$$q_-(\pi/2) = \theta_1^{-1}\left(\theta_1(q_0) - \pi\right) < q_0, \quad q_+(\pi/2) = q_0. \tag{55}$$

Let us recall Lemma 1 and the positivity of $M_0(x)$ for $x > 0$. Then using (54) and the definition of $D(t)$ in (53),

$$D(-\pi/2) = (q_+^2(-\pi/2) - q_0^2)M_0(q_+(-\pi/2))\cos\big(\theta_0(q_+(-\pi/2)) - \varphi(-\pi/2)\big) < 0.$$

Similarly, using (55) in (53), we obtain that $D(\pi/2) > 0$. □

Uniqueness now follows by proving the following

Lemma 3 $D'(t) > 0$, $t \in (-\pi/2, \pi/2)$.

Proof Our proof is based on calculating $D'(t)$ where $D(t)$, given in (53). We begin by calculating $\varphi'(t)$, $q'_\pm(t)$, $\theta'_0(q_\pm(t))q'_\pm(t)$, and $M'_0(q_\pm(t))q'_\pm(t)$.

We recall the formulas from [19, 10.18.8],

$$M_0^2(x)\theta_0'(x) = \frac{2}{\pi x}, \quad M_1^2(x)\theta_1'(x) = \frac{2}{\pi x}, \quad x > 0. \tag{56}$$

Since $M_0(x)$ and $M_1(x) > 0$ are positive, (56) implies that

$$\theta_0'(x) = \frac{2}{\pi x}\frac{1}{\big(M_0(x)\big)^2}, \quad \theta_1'(x) = \frac{2}{\pi x}\frac{1}{\big(M_1(x)\big)^2} \quad x > 0. \tag{57}$$

It follows from (45) that

$$\varphi'(t) = -1. \tag{58}$$

From (46) and the invertibility of $\theta_1(x)$ for $x > 0$,

$$q_\pm = q_\pm(t) = \theta_1^{-1}\left(\theta_1(q_0) - t \pm \frac{\pi}{2}\right), \tag{59}$$

and therefore $\theta_1'(q_\pm(t))q_\pm'(t) = -1$. Recalling (57), we obtain that

$$q_\pm'(t) = -\frac{\pi q_\pm}{2}\left(M_1^2(q_\pm)\right), \quad \theta_0'(q_\pm(t))q_\pm'(t) = -\left(\frac{M_1^2(q_\pm)}{M_0^2(q_\pm)}\right). \tag{60}$$

□

Let us consider now the formula [19, 10.18.11], $M_0\theta_0'/M_0' = \tan(\phi_0 - \theta_0)$. Recalling (49), we see that $\tan\phi_0 = \tan\theta_1$, and

$$\frac{\sin\phi_0}{\cos\phi_0} = \frac{N_0\sin\phi_0}{N_0\cos\phi_0} = \frac{-M_1\sin\theta_1}{-M_1\cos\theta_1}, \tag{61}$$

and therefore using (60),

$$M_0'(q_\pm) = M_0(q_\pm)\theta_0'(q_\pm)\cot(\theta_1(q_\pm) - \theta_0(q_\pm)),$$

$$M_0'(q_\pm)q_\pm' = \frac{M_1^2(q_\pm)}{M_0(q_\pm)}\cot(\theta_0(q_\pm) - \theta_1(q_\pm)).$$

Since $\theta_0(q_+) - \theta_1(q_+) = \big(\theta_0(q_+) - \varphi\big) - \big(\theta_1(q_+) - \varphi\big) = \big(\theta_0(q_+) - \varphi\big) - \frac{\pi}{2}$, we get that

$$\cot\big(\theta_0(q_+) - \theta_1(q_+)\big) = -\tan\big(\theta_0(q_+) - \varphi\big), \tag{62}$$

and similarly

$$\cot\big(\theta_0(q_-) - \theta_1(q_-)\big) = -\tan\big(\theta_0(q_-) - \varphi\big). \tag{63}$$

Thus

$$M_0'(q_\pm(t))q_\pm'(t) = -\frac{M_1^2(q_\pm)}{M_0(q_\pm)}\tan\big(\theta_0(q_\pm)-\varphi\big). \tag{64}$$

Recalling (53), let us differentiate $\mathscr{D}(q_\pm(t))$. Using the results above,

$$\begin{aligned}\frac{d}{dt}\mathscr{D}(q_\pm(t) &= (q_\pm^2 - q_0^2)M_0(q_\pm)\cos(\theta_0(q_\pm)-\varphi)\\ &= -\pi q_\pm^2 M_0(q_\pm)M_1^2(q_\pm)\cos(\theta_0(q_\pm)-\varphi) - (q^2-q_0^2)M_0(q)\sin(\theta_0(q)-\varphi).\end{aligned} \tag{65}$$

We note that $\sin(\theta_0(q)-\varphi) = \tan(\theta_0(q)-\varphi)\cos(\theta_0(q)-\varphi)$, and using (61)-(63)) we get that $\tan(\theta_0(q)-\varphi) = +\cot(\phi_0(q)-\theta_0(q))$. By formula [19, 10.18.11], $\tan(\phi_0-\theta_0) = 2/(\pi x M_0 M_0')$ for $x > 0$, and hence

$$\tan(\theta_0(q)-\varphi) = \cot(\phi_0(q)-\theta_0(q)) = \tfrac{1}{2}\pi q M_0 M_0' = \tfrac{1}{4}\pi q\left(M_0^2\right)'. \tag{66}$$

Therefore

$$\frac{d}{dt}\mathscr{D}(q_\pm(t)) = -\left(q_\pm M_1^2(q_\pm) + \tfrac{1}{4}(q_\pm^2-q_0^2)\left(M_0^2(q_\pm)\right)'\right)\pi q_\pm M_0(q_\pm)\cos(\theta_0(q_\pm)-\varphi), \tag{67}$$

where $q_\pm = q_\pm(t)$, $\varphi = \varphi(t)$. Let us, then, consider the expression

$$P(x) := x\big(M_1(x)\big)^2 + \tfrac{1}{4}(x^2-q_0^2)\left(M_0(x)^2\right)', \quad x > 0. \tag{68}$$

Since $N_0^2(x) = M_1^2(x)$ for $x \ge 0$ by (49), the formula [19, 10.18.10], $x^2 M_0(x)M_0'(x) + x^2 N_0(x)N_0'(x) + x\big(N_0(x)\big)^2 = 0$, $x \ge 0$, implies that

$$x\big(M_1(x)\big)^2 = -\tfrac{1}{2}x^2(M_0^2(x))' - \tfrac{1}{2}x^2(M_1^2(x))', \quad x \ge 0.$$

Therefore,

$$P(x) = -\tfrac{1}{4}x^2\left(M_0^2(x)\right)' - \tfrac{1}{4}q_0^2\left(M_0^2(x)\right)' - \tfrac{1}{2}x^2\left(M_1^2(x)\right)', \quad x > 0.$$

The claim below implies that $P(x) > 0$ for $x > 0$.

Claim $\left(M_0^2(x)\right)' < 0$ and $\left(M_1^2(x)\right)' < 0$, for $x > 0$.

Proof To prove the claim, we use the Nicholson's Integral Representation (see [19, 10.9.30]), which implies that for $x > 0$,

$$M_0^2(x) = \frac{8}{\pi^2}\int_0^\infty K_0(2x\sinh t)\,dt, \quad M_1^2(x) = \frac{8}{\pi^2}\int_0^\infty \cosh(2t)\,K_0(2x\sinh t)\,dt, \tag{69}$$

and using the formula (see [19, 10.29.3] that

$$K_0'(x) = -K_1(x), \quad x > 0, \tag{70}$$

where in (69), (70), $K_0(x)$ and $K_1(x)$ denote the second standard solutions to the Modified Bessel equation, with $\nu = 0,\ 1$, respectively.

By considering the asymptotic behavior of $K_0(x)$ and $K_1(x)$ for $0 < x \ll 1$ [19, 10.30.2,10.30.3] and $x \gg 1$ [19, 10.25.3], it readily follows that the formal differentiation of the (convergent) representations for $M_0(x)$ and $M_1(x)$ given in (69) is justified,

$$\left(M_0^2(x)\right)' = -\frac{8}{\pi^2}\int_0^\infty K_1(2x\sinh t)\cdot 2\sinh t\,dt, \quad x > 0,$$

$$\left(M_1^2(x)\right)' = -\frac{8}{\pi^2}\int_0^\infty \cosh(2t)\,K_1(2x\sinh t)\cdot 2\sinh t\,dt, \quad x > 0,$$

as the integrals above are convergent uniformly in x, for $x > 0$. As the functions $\sinh(x)$, $\cosh(x)$ and $K_1(x)$ are strictly positive for $x > 0$, the claim follows. □

To complete the proof of Lemma 3, let us recall that $M_0(x) > 0$ for $x > 0$, and $\cos\left(\theta_0(q_+) - \varphi\right) < 0$ and $\cos\left(\theta_0(q_-) - \varphi\right) > 0$ by Lemma 1. Hence the above claim implies that $\pm\frac{d}{dt}\mathscr{D}(q_\pm(t)) > 0$, and therefore $D'(t) > 0$ in accordance with the definition of $D(t)$ in (53). □

This completes the proof of Proposition 2. □

To return now and complete the proof of Theorem 6, let $q_+^{\text{sup}} = (R_0 - \delta)/\varepsilon$ and let us consider the corresponding value of q_0^{sup}. Since $q_+^{\text{sup}} = \theta_1^{-1}(\theta_1(q_0^{\text{sup}} - t + \pi/2))$ for some $t \in [-\pi/1, \pi/2]$ by (47), we obtain that $q_+^{\text{sup}} = q_0^{\text{sup}} + O(1)$ and hence $r_0^{\text{sup}} = r_+^{\text{sup}} + O(\varepsilon)$ as claimed earlier. □

2.4 *Dimple Solutions for (DQOP)*

We now consider radial "dimple" solution in $\Omega = \mathscr{B}_{R_0}(\delta)$, with $-1 < u(r) < 1$ for $r \in (0, r_+)$ and $u(r) \equiv -1$ for $r \in [r_+, R_0)$, where r_+ reflects a free boundary with $0 < r_+ < R_0 - \delta$. As we saw in Sect. 2.2, $u(r)$ should satisfy

$$\begin{aligned}
&u + \lambda = -\varepsilon^2(u_{rr} + \frac{1}{r}u_r), \quad 0 < r < r_+,\\
&u(0) \in (-1, 1]\,, \quad u(r_+) = -1,\\
&u_r(0) = 0\,, \quad u_r(r_+) = 0,
\end{aligned}$$

and the equivalent mean mass condition implies that

$$\bar{u} = \frac{(2r_0^2 - R_0^2)}{R_0^2}, \quad \lambda = \frac{(r_+^2 - 2r_0^2)}{r_+^2}, \quad r_0 = \left[\frac{1+\bar{u}}{2}\right]^{1/2} R_0, \tag{71}$$

for some $0 < r_0 < r_+$, see (12)–(14). Thus $\lambda = \lambda(r_+, r_0)$ and $r_0 = r_0(\bar{u}, R_0)$. We shall see that for $0 < \varepsilon \ll 1$, nontrivial radial dimple solutions exist for all $-1 < \bar{u} = -1 + O(\varepsilon)$ and $0 < r_0 = O(\varepsilon)$ sufficiently small; in particular, as $\varepsilon \downarrow 0$, $\bar{u} \downarrow -1$, $r_0 \downarrow 0$, and $u(r;\varepsilon) \downarrow -1$, $0 < r < R_0$.

The parameter r_+, reflecting the location of the free boundary, is to be determined. The value $u(0) \in (-1, 1]$ is as an additional free parameter to be determined. Thus, we have three boundary conditions and the equivalent mean mass constraint, $\bar{u} = (2r_0^2 - R_0^2)/R_0^2$, as well as the range constraint on $u(0)$. In solving (12)–(14), we have four degrees of freedom, counting the parameters $u(0)$ and r_+, with $\lambda = \lambda(r_+, r_0)$. Hence, from the "count" of the parameters, we would expect the possible dimple solutions to be uniquely determined by $\bar{u}$ (or, equivalently, by r_0).

Setting

$$q = \frac{r}{\varepsilon}, \quad q_0 = \frac{r_0}{\varepsilon}, \quad q_+ = \frac{r_+}{\varepsilon}, \quad Q_0 = \frac{R_0}{\varepsilon}, \quad \text{and} \quad v(q) := [u(r) + \lambda]|_{r=\varepsilon q}, \tag{72}$$

we get the following problem for $v(q)$,

$$\begin{aligned} &q v_{qq} + v_q + q v = 0, \qquad 0 < q < q_+, \\ &v(0) = u(0) + \lambda \in (\lambda - 1, \lambda + 1], \quad v(q_+) = -1 + \lambda, \\ &v_q(0) = 0, \quad v_q(q_+) = 0, \end{aligned} \tag{73}$$

where $\lambda = (q_+^2 - 2q_0^2)/q_+^2$. The equation in (73) is a Bessel's equation of order zero, whose general solution is $v(q) = c_1 J_0(q) + c_2 Y_0(q)$, $c_1, c_2 \in R$, and hence $v_q(q) = -c_1 J_1(q) - c_2 Y_1(q))$. Since $J_0(0) = 1$, $J_1(0) = 0$, and $Y_1(0) \neq 0$, the boundary conditions at $q = 0$ imply that $c_1 = u(0) + \lambda$, $c_2 = 0$. Thus

$$v(q) = (u(0) + \lambda) J_0(q), \quad v_q(q) = -(u(0) + \lambda) J_1(q). \tag{74}$$

The values of J_0 at its sequential minima are increasing [19, 10.3,10.21,10.18], and $v_q(q_+) = 0$ with $v(q_+) = -1 + \lambda$, which corresponds to $u_r(r_+) = 0$, $u(r_+) = -1$. Hence the range constraint, $-1 \le u(r) \le 1$, implies that $q_+ = \bar{q}$, where $\bar{q}$ corresponds to the (unique) first positive zero of the function $J_1(q)$ [19, 10.21]. Thus [1, Tables 9.1, 9.5],

$$q_+ = \bar{q} \approx 3.8 \quad \text{with} \quad J_0(q_+) = J_0(\bar{q}) \approx -0.4. \tag{75}$$

From (73)–(75), we obtain that

$$v(q_+) = v(\bar{q}) = -1 + \lambda = (u(0) + \lambda) J_0(\bar{q}), \tag{76}$$

and using (71), (72), (76),

$$\lambda = \frac{u(0)J_0(\bar{q}) + 1}{1 - J_0(\bar{q})} = \frac{r_+^2 - 2r_0^2}{r_+^2} = \frac{r_+^2 - (1+\bar{u})R_0^2}{r_+^2}. \tag{77}$$

From (75), (77),

$$r_0 = \varepsilon\bar{q}\left[\frac{-(1+u(0))J_0(\bar{q})}{2(1-J_0(\bar{q}))}\right]^{1/2}, \quad \bar{u} = -1 - \varepsilon^2\frac{\bar{q}^2(1+u(0))J_0(\bar{q})}{R_0^2(1-J_0(\bar{q}))}, \tag{78}$$

and recalling (72), (74), (75), (77), (78),

$$u(r) = \frac{(u(0)+1)J_0(r/\varepsilon) - u(0)\,J_0(\bar{q}) - 1}{1 - J_0(\bar{q})}, \quad 0 \le r \le r_+ = \varepsilon\bar{q}, \tag{79}$$

with

$$u(0) + 1 = \frac{(1+\bar{u})R_0^2(J_0(\bar{q}) - 1)}{\varepsilon^2\bar{q}^2 J_0(\bar{q})}. \tag{80}$$

From (77)–(79), we see that there is a one-parameter family of dimple solutions which is uniquely determined by $\bar{u}$, or equivalently by r_0 or by $u(0)$, with $u(0) \in [-1, 1]$.

As $u(0) \downarrow -1$,

$$\lambda \uparrow 1, \quad r_0 \downarrow 0, \quad \bar{u} \downarrow -1, \quad \text{and} \quad u(r) \downarrow -1, \quad \text{fo} \quad r \in [0, R_0].$$

As $u(0) \uparrow +1$,

$$\lambda \downarrow \frac{1+J_0(\bar{q})}{1-J_0(\bar{q})}, \quad r_0 \uparrow \varepsilon\bar{q}\left[\frac{-2J_0(\bar{q})}{2(1-J_0(\bar{q}))}\right]^{1/2}, \quad \bar{u} \uparrow -1 - \varepsilon^2\frac{2\bar{q}^2 J_0(\bar{q})}{R_0^2(1-J_0(\bar{q}))},$$

and $u(r) \uparrow \left[\frac{2J_0(r/\varepsilon) - J_0(\bar{q}) - 1}{1 - J_0(\bar{q})}\right]$ for $r \in [0, r_0]$. Note in particular that

$$\lim_{\varepsilon\downarrow 0}\lim_{u(0)\downarrow -1} u(r;\varepsilon,u(0)) = \begin{cases} -1 & \forall r \in (0, R_0], \\ 1 & r = 0, \end{cases} \qquad \lim_{\varepsilon\downarrow 0}\lim_{u(0)\downarrow -1} r_0(\varepsilon, u(0)) = 0.$$

Summarizing the results above,

Theorem 7 *Given* $0 < \varepsilon \ll 1$ *and* $\mathscr{B}_{R_0}(\delta)$, *with* $R_0 = O(1)$ *and* $\varepsilon \ll \delta \ll 1$. *There exists a unique radial dimple solution, for any* $\bar{u} \in (-1, -1 - 2\varepsilon\bar{q}/R_0)^2 J_0(\bar{q})/(1 - J_0(\bar{q}))$, *where* $\bar{q}$ *denotes the first positive zero of* $J_1(q)$.

Finally, using (12) and recalling (3), it is straightforward to verify that

$$E(t) := \frac{1}{\varepsilon|\Omega|}\int_\Omega \{(1-u^2) + \varepsilon^2|\nabla u|^2\}\,dx = \frac{\varepsilon\bar{q}^2}{R_0^2}(1-\lambda^2), \tag{81}$$

where $\bar{q}$ and λ are prescribed in (75) and (77), respectively.

3 Connecting the Dynamic Problems

At least on a somewhat superficial level, the attractor dynamics for both (SD) and (DQOP) can be seen to be similar in $2D$, under the assumptions outlined in Sect. 2, given $\mathscr{B}_{R_0}(\delta)$ with $0 < \varepsilon \ll \delta \ll 1$ and $R_0 = O(1)$. The steady states for (SD) are given by circular curves centered at the origin, as well as the possible translates of these circular curves that lie within $\mathscr{B}_{R_0}(\delta)$ and the possible nonintersecting union of such curves that lie within $\mathscr{B}_{R_0}(\delta)$. Similarly, the steady states for (DQOP) are given radial solutions which are either annular or dimple solutions, as well as their possible translates within $\mathscr{B}_{R_0}(\delta)$ and solutions obtain via composites of the above given the limitations of the domain $\mathscr{B}_{R_0}(\delta)$. The equivalent mean mass condition for (DQOP) prescribes the mean mass, $\bar{u}$, in terms of an effective radius, r_0, with $0 \leq r_0 < R_0 - \delta$. Accordingly, it is possible to identify a $1-1$ correspondence between the set of radial steady states of (SD), namely circular curves with radius r_0, with $0 < r_0 < R_0 - \delta$, and the set of radial steady states for (DQOP). For simplicity, we may limit our focus to the set of axi-symmetric steady states in both cases, leaving aside for the moment the technical difficulties entailed in taking into account the somewhat larger class of steady states produced by translation. Also, we are neglecting possible radial ringed solutions, composed of concentric circular curves for (DQ) and radially symmetric composite multiple transition (multi-annular solutions, possibly with a dimple solution at the origin).

Thus, if we can identify similar stability properties for both evolutions, we are well on our path to connecting the evolutions. The difficulty arises in considering stability for both evolutions in similar functional analytic settings and in a manner which permits both evolutions to be simultaneously tracked globally in time. We first outline briefly the perhaps easiest and most direct approach, which arises naturally in view of the extant results in the literature for both evolutions, explaining some of the pitfalls in connecting the evolutions. Afterwards, we outline some of the details pertaining to a more robust approach. The more robust approach is based on considering similar minimizing motion evolutionary descriptions for both evolutions, and making a step-by-step connection between the two motions via an appropriate lifting and projecting algorithm.

3.1 Stability

With regard to the stability in the context of (SD) for circular curves, the results of Wheeler [23] are useful. The theory there is based on the following local existence theorem, which is paraphrased below.

Theorem 8 *Suppose that $\Gamma_0 : R \to R^2$ is a periodic regular curve parametrised by arc-length of class $\mathscr{C}^2 \cap W^{2,2}$ with $||\kappa||_2 < \infty$. Then there exists a time $T \in (0, \infty]$ and a unique one-parameter family of immersions $\Gamma : R \times [0, T) \to R^2$ parametrized by arc-length satisfying (SD) such that (i) $\Gamma(0) = \Gamma_0$, (ii) $\Gamma(\cdot, t)$ is* *of class $\mathscr{C}^\infty$ and periodic of period $\mathscr{L}(\Gamma(\cdot,t))$ for every $t \in (0,T)$, and (iii) T is maximal.*

The regularity requirements on the initial data Γ_0 can be somewhat weakened, [23].

The stability results [23, Theorem 1.1], paraphrased below, are formulated in terms of the normalized oscillation of curvature,

$$\kappa_{osc}(\Gamma(\cdot,t)) = \mathscr{L}(\Gamma(\cdot,t)) \int_\Gamma (\kappa - \bar{\kappa})^2\, ds \quad \text{with} \quad \bar{\kappa}(\Gamma(\cdot,t)) = \mathscr{L}^{-1}(\Gamma(\cdot,t)) \int_\Gamma \kappa\, ds,$$

and the isoperimetric ratio, $\mathscr{I}(\Gamma(\cdot,t)) = \mathscr{L}^2(\Gamma(\cdot,t))/[4\pi \mathscr{A}(\Gamma(\cdot,t))]$ where $\mathscr{A}(\Gamma(\cdot,t))$ denotes the area enclosed by $\Gamma(\cdot,t)$.

Theorem 9 *Suppose that $\Gamma_0 : S^1 \to R^2$ is a regular smooth immersed closed curve with $\mathscr{A}(\Gamma_0) > 0$ and $\int_{\Gamma_0} \kappa\, ds = 2\pi$. There exists a constant $\kappa^* > 0$ such that if*

$$\kappa_{osc}(\Gamma_0) < \kappa^* \quad \textit{and} \quad \mathscr{I}(\Gamma_0) < exp\left(\frac{\kappa^*}{8\pi^2}\right),$$

then under (SD) evolution, $\Gamma : S^1 \times [0,T) \to R^2$ with Γ_0 as initial data exists for all time and converges exponentially fast to a round circle with radius $\sqrt{\mathscr{A}(\Gamma_0)/\pi}$.

The results in [23] are quite pleasing. However, one cannot conclude directly from either of these theorems that if $\Gamma_0 \subset \mathscr{B}_{R_0}(\delta)$, then $\Gamma(\cdot,t) \subset \mathscr{B}_{R_0}(\delta)$ for $t \in (0,T)$.

With regard to stability in the (DQOP) context, suppose we wish to demonstrate stability for an annular solution located far from the $\mathscr{B}_{R_0}(\delta)$ boundary. For simplicity, let us consider the stability of an annular solution centered at the origin with $r_0 = O(1)$. We know for (SD) that the encompassed area is maintained, but we do not know off hand that the center of mass does not move. For (DQOP) we similarly know that mass is conserved, but we do not know that the center of mass is time invariant. So, if we are not overly concerned with maintaining the $\mathscr{B}_{R_0}(\delta)$ structure, a reasonable approach is to consider zero mass perturbations, making use of the H^{-1} gradient structure. Within this context establishing a spectral gap should be straightforward, for zero mass perturbations modulo translations of the center of mass. This would enable us to prove stability of the annular solutions, modulo translation, in analogy with the (SD) results above.

In Sect. 2.4, there exist $O(\varepsilon)$ energy dimple solutions which equal -1 except on a circular region with $O(\varepsilon^2)$ area. Clearly a "stray" translate of a dimple (or rather a translate of that part of the dimple solution which differs from -1) could be incorporated into the $O(1)$ region where a generic annular solutions equal -1, with a small alteration in the radii of the annular solution to accommodate the additional mass. Such a small "droplet-like" perturbation would constitute a small energy perturbation, though not covered by the discussion above, as they are not zero mass perturbations.

As such perturbations are natural to consider, we remark here that it is possible to construct a sequence of energy lowering and mass preserving perturbations, which allow the dimple to lose height and to transfer away volume (mass). Details to be published elsewhere, together with the (DQOP) stability results described above.

3.2 Minimizing Motions

Minimizing motion evolution formulation for (SD). In Fonseca et al. [10], De Giorgi's minimizing movement approach is implemented within a framework with H^{-1} gradient flow structure, to prove short time existence, uniqueness, and regularity for the motion of an elastic thin film which evolves by anisotropic surface diffusion. Their approach yields, as a subcase, a proof of short time existence, uniqueness, and regularity for a spatially period $1D$ curve prescribable by the graph of a function, $\Gamma = \{(x, h(x)) \,|\, 0 < x < b\}$, with $b \in (0, \infty)$.

More specifically, starting with b periodic initial data, $h_0 \in H^2_{loc}(R)$, a sequence of approximants $h_{i,N}$ are defined inductively, for $T > 0$, $N \in \mathcal{N}$, $i = 1, \ldots, N$, as a minimizer of $E(h) + \frac{1}{2\tau} d^2(h, h_{i-1,N})$, where $E(h)$ denotes the system energy, $\tau = T/N$, and d measures the H^{-1} distance between h and $h_{i-1,N}$. Their choice for d is based on the following H^{-1} norm for curves Γ,

$$||f||_{H^{-1}(\Gamma)} := \sup_{||\psi||_{H^1(\Gamma)}=1} \int_\Gamma f \psi d\mathcal{H}^1(z),$$

which can be expressed as

$$||f||^2_{H^{-1}(\Gamma)} = \int_\Gamma \left(\Phi(z) - \frac{1}{|\Gamma|} \int_\Gamma \Phi d\mathcal{H}^1 \right)^2 d\mathcal{H}^1(z) + \left(\int_\Gamma f\, d\mathcal{H}^1 \right)^2,$$

where $\Phi(z) := \int_{\Gamma(z_0,z)} f(w)\, d\mathcal{H}^1(w)$, $z_0 = (0, h_{i-1,N}(0))$ and $\Gamma(z_0, z)$ denotes the arc of Γ connecting z_0 with z. Accordingly they base their scheme on the $(H^{-1})^2$ penalization

$$\int_\Gamma \left(\int_{\Gamma(z_0,z)} f(w)\, d\mathcal{H}^1(w) \right)^2 d\mathcal{H}^1(z), \tag{82}$$

for $f = h_i - h_{i-1}$, with two constraints reflecting zero mean and periodicity:

$$\int_\Gamma f\, d\mathcal{H}^1 = 0, \quad \int_\Gamma \int_{\Gamma(z_0,z)} f(w)\, d\mathcal{H}^1(w)\, d\mathcal{H}^1(z) = 0. \tag{83}$$

To implement their approach in our context, an extension of their approach is needed for closed imbedded planar curves, Γ. The resultant penalization with constraints is similar to (82), (83), but incorporates an orientational weight.

Minimizing motion evolution formulation for (DQOP). In [14] a minimizing movement scheme is developed for proving the existence of weak solutions for a class of degenerate parabolic equations of fourth order, which includes (DQOP) given the assumptions outlined in Sect. 1. This class of evolution equations are shown to correspond to a gradient flow with respect to a Wasserstein-like transport metric, W_m, and the weak solutions are obtained via a scheme based on curves of maximal slope.

The metric, W_m, which was shown in [8] to constitute a genuine metric, is neither the L^2-Wasserstein distance not a flat Hilbertian metric; rather it corresponds to a metric tensor based on the following construction: given a tangential vector $v := \partial_s \rho(0)$ to a smooth curve $\rho : (-\varepsilon, \varepsilon) \to L^1(\Omega)$ of strictly positive densities $\rho(s)$ at $\rho_0 = \rho(0)$, it assigns the length

$$||v||^2 = \int_\Omega |D\psi(x)|^2 \, m(\rho_0(x))\, dx, \quad \text{with} \quad -\nabla \cdot (m(\rho_0(x))\nabla\psi(x)) = v, \quad x \in \Omega, \tag{84}$$

with variational boundary conditions on $\partial\Omega$. Within the context of our framework, $\Omega = \mathscr{B}_{R_0}(\delta)$, as discussed in Sect. 1, and the ambient space for the scheme is the metric space $(X(\Omega), W_m)$ where

$$X(\Omega) := \Big\{ u \in L^1(\Omega) \,|\, -1 \le u \le 1 \;\; a.e. \;\; x \in \Omega, \quad \int_\Omega u\, dx = \bar{u} \Big\}.$$

The minimizing movement scheme starts with initial conditions $u_0 \in X(\Omega)$ such that $E[u_0] < \infty$, and approximants are defined by setting

$$u_\tau^0 := u_0, \quad u_\tau^{n+1} := -\arg\min \Phi_\tau^n \in X(\Omega), \quad \Phi_\tau^n(v) := \frac{1}{2\tau} W_m(u_\tau^n, v)^2 + E[v],$$

with $E[u] := \infty$ for $u \notin H^1(\Omega)$. Piecewise constant interpolation is then used to define an approximation $\tilde{u}_\tau(t) : [0, \infty) \to X(\Omega)$.

Lifting and projecting: a Hilbert expansion approach. As we have seen, there exist minimizing motion schemes for both (SD) and (DQOP). To establish a rigorous connection between the two evolutions, (SD) and (DQOP), it makes sense to consider similar time steps, $\tau = T/N$, for both minimizing motions. It is also reasonable to consider similar initial conditions, perhaps similarly perturbed steady state solutions to (SD) and (DQOP); for simplicity, we might consider a circle and an annular solution, respectively, with equivalent mean mass and both centered at the origin. We want to compare the results of the minimizing motion schemes and to demonstrate that they yield the same motion in the limit as $\varepsilon \to 0$ and $\tau \to 0$. Here the difficulty is that at each time step the motion for (SD) is describable in terms of curves Γ_t belonging to $\mathscr{M}$, the set of smooth simple closed curves, and the motion for (DQOP) is described in terms of functions $u(\cdot, t)$ defined for all $x \in \mathscr{B}_{R_0}(\delta)$. Modulo issues of regularity and structure, the zero level set of $u(\cdot, t)$ yields a "projection" of $u(\cdot, t)$ onto the set of curves $\mathscr{M}$. The matter of "lifting" Γ_t to obtain

$u(\cdot, t)$ is more delicate. In the context of connecting the Cahn-Hilliard equation with the Mullins-Sekerka problem, Carlen et al. [4] used a Hilbert expansion approach to construct a globally defined function $u(\cdot, t)$ from a curve $\Gamma_t \in \mathcal{M}$. They constructed an approximation for $u(\cdot, t)$ based on three types of terms, (i) terms depending on Γ_t, (ii) terms reflecting local corrections near Γ_t, and (iii) terms reflecting long range corrections.

Acknowledgements The authors would like to acknowledge support from the Israel Science Foundation #1200/16. EC was partially supported by US NSF Grant DMS 2055282. The authors also thank the referees for their remarks.

References

1. Abramowitz, M., Stegun, I.A.: Handbook of Mathematical Functions. Dover, New York (1972)
2. Bañas, L., Novick-Cohen, A., Nürnberg, R.: The degenerate and non-degenerate deep quench obstacle problem: a numerical comparison. Netw. Heterog. Media **8**, 37–64 (2013)
3. Brothers, J.E., Ziemer, W.P.: Minimal rearrangements of Sobolev functions. J. Reine. Angew. Math. **384**, 153–179 (1988)
4. Carlen, E.A., Carvalho, M.C., Orlandi, E.: Approximate solutions of the Cahn-Hilliard equation via corrections to the Mullins-Sekerka motion. Arch. Ratio. Mech. Anal. **178**, 1–55 (2005)
5. Chen, X.: Global asymptotic limit of solutions of the Cahn-Hilliard equation. J. Differ. Geom. **44**, 262–311 (1996)
6. Cahn, J.W., Elliott, C.M., Novick-Cohen, A.: The Cahn-Hilliard equation with a concentration dependent mobility: motion by minus the Laplacian of the mean curvature. Eur. J. Appl. Math. **7**, 287–301 (1996)
7. de Figueiredo, D.G., dos Santos, E.M., Miyagaki, O.H.: Sobolev spaces of symmetric functions and applications. J. Funct. Anal. **261**, 3735–3770 (2011)
8. Dolbeault, J., Nazaret, B., Savaré, G.: A new class of transport distances between measures. Calc. Var. Part. Differ. **34**, 193–231 (2009)
9. Elliott, C.M., Luckhaus, S.: A generalized diffusion equation for phase separation of a multi-component mixture with interfacial free energy. University of Minnesota, IMA Preprint 887 (1991)
10. Fonseca, I., Fusco, N., Leoni, G., Morini, M.: Motion of elastic thin films by anisotropic surface diffusion with curvature regularization. Arch. Ratio. Mech. Anal. **205**, 425–466 (2012)
11. Hartman, P.: Ordinary Differential Equations, 2nd edn. SIAM, Philadelphia (2002)
12. Jerrard, R.L., Sternberg, P.: Critical points via Γ-convergence: general theory and applications. J. Eur. Math. Soc. (JEMS) **11**, 705–753 (2009)
13. Kohn, R.V., Otto, F.: Upper bounds on coarsening rates. Comm. Math. Phys. **229**, 375–395 (2002)
14. Lisini, S., Matthes, D., Savaré, G.: Cahn-Hilliard and thin film equation with nonlinear mobility as gradient flows in weighted-Wasserstein metrics. J. Differ. Equ. **253**, 814–850 (2012)
15. Mullins, W.W.: Theory of thermal grooving. J. Appl. Phys. **28**, 333–339 (1957)
16. Novick-Cohen, A.: Wetting and prewetting of anti-phase boundaries. Meccanica **40**, 471–483 (2005)
17. Novick-Cohen, A.: Upper bounds for coarsening for the deep quench obstacle problem. J. Stat. Phys. **141**, 142–157 (2010)
18. Novick-Cohen, A., Shishkov, A.: Upper bounds for coarsening for the degenerate Cahn-Hilliard equation. Discret. Contin. Dyn. Syst. **25**, 251–272 (2009)

19. Olver, F.W.J., Olde Daalhuis, A.B., Lozier, D.W., Schneider, B.I., Boisvert, R.F., Clark, C.W., Miller, B.R., Saunders, B.V., Cohl, H.S., McClain, M.A. (eds.): NIST Digital Library of Mathematical Functions. https://dlmf.nist.gov/, Release 1.1.10 of 2023-06-15. (see also: [1])
20. Oono, Y., Puri, S.: Study of phase-separation dynamics by use of cell dynamical systems. I. Modeling. Phys. Rev. A **38**, 434–453 (1988)
21. Sternberg, P.: The effect of a singular perturbation on nonconvex variational problems. Arch. Ratio. Mech. Anal. **101**, 209–260 (1988)
22. Taylor, J.E., Cahn, J.W.: Linking anisotropic sharp and diffuse surface motion laws via gradient flows. J. Stat. Phys. **77**, 183–197 (1994)
23. Wheeler, G.: On the curve diffusion flow of closed plane curves. Ann. Mat. Pura Appl. **4**(192), 931–950 (2013)

Mean Field Limit for the Kac Model and Grand Canonical Formalism

Thierry Paul, Mario Pulvirenti, and Sergio Simonella

Abstract We consider the classical Kac's model for the approximation of the Boltzmann equation, and study the correlation error measuring the defect of propagation of chaos in the mean field limit. This contribution is inspired by a recent paper of the same authors [23] where a large class of models, including quantum systems, are considered. Here we outline the main ideas in the context of grand canonical measures, for which both the evolution equations and the proof simplify.

Keywords Kac model · Boltzmann equation · Mean field scaling · Correlation error

1 Introduction

The kinetic description of particle systems is strongly related to propagation of chaos. This property allows one to substitute the complex dynamics of a huge number of

---

This note is dedicated to the memory of our colleague and friend Maria Conceicao Carvalho. Among her many important contributions to the theory of Kac's model, Sao's work [7, 8] explained the long time behaviour of the homogeneous Boltzmann equation without using the H-theorem nor global properties, but exploiting a Dyson expansion over trajectories. Our contribution is, in spirit, close to this approach.

T. Paul
CNRS Laboratoire Ypatia des Sciences Mathématiques (LYSM), Roma, Italy
e-mail: thierry.paul@sorbonne-universite.fr

M. Pulvirenti
Sapienza Università di Roma, Dipartimento di Matematica G. Castelnuovo, Piazzale A. Moro 5, 00185 Roma, Italy

M. Pulvirenti (✉) · S. Simonella
International Research Center on the Mathematics and Mechanics of Complex Systems MeMoCS, University of L'Aquila, L'Aquila, Italy
e-mail: pulviren@mat.uniroma1.it

S. Simonella
e-mail: sergio.simonella@uniroma1.it

E. Carlen et al. (eds.), *From Particle Systems to Partial Differential Equations*, Springer Proceedings in Mathematics & Statistics 465,
https://doi.org/10.1007/978-3-031-65195-3_12

particles by a single nonlinear partial differential equation for the probability density of a given particle. More precisely, one adopts a statistical description. At time zero, the $N-$particle system is assumed to be “chaotic” in the sense that each particle is distributed identically and independently from the others, at least up to an error which is vanishing when N diverges. The dynamics creates correlations and the independence property is lost at any positive time. However, under suitable scaling limits, the statistical independence of any finite group of particles can be recovered, after averaging, in the limit $N \to \infty$. As a consequence, a given particle evolves according to an effective kinetic equation. The nature of this dynamics is determined by the microscopic details of the system and by the regime of physical parameters. Such a mechanism works in the formal (in a few cases, rigorous) derivation of the most common kinetic equations.

If the system is described by a symmetric probability measure $W^N(Z_N, t)$ at time t, where $Z_N = (z_1, \ldots, z_N)$ is a configuration of the system, being $z_i = (x_i, v_i)$ position and velocity of the particle i, the probability density of finding the first j particles in the microscopic state Z_j is given by the marginal

$$f_j^N(Z_j, t) := \int dz_{j+1} \cdots dz_N W^N(Z_N, t).$$

In case of a two-body interaction, we have that f_1^N depends on f_2^N, which depends on f_3^N and so on, and these marginals are not tensorized. Indeed, the dynamics creates correlations even though at time zero it is assumed that

$$W^N = (f_0)^{\otimes N}.$$

In many cases, the physical regime of interest can be expressed in terms of suitable scaling limits ($N \to \infty$, together with additional prescriptions), which one can hope to use to prove that

$$f_j^N(t) \to f(t)^{\otimes j}, \tag{1}$$

where $f(t)$ is a solution of the aforementioned kinetic equation.

In three dimensions, fundamental scaling limits for Hamiltonian systems are the following.

- Mean-field limit for classical systems: $N \to \infty$ and the coupling constant of the two-body interaction is scaled by $\frac{1}{N}$. Namely, a system of point particles with the mean-field Hamiltonian

$$H_N = \frac{1}{2}\sum_{i=1}^{N} v_i^2 + \frac{1}{2N}\sum_{i \neq j} \varphi(x_i - x_j),$$

  where φ is a two-body potential.
  The corresponding expected kinetic equation is, in this case, the Vlasov equation.

- Low-density or Boltzmann-Grad limit: $N \to \infty$ and the diameter of hard-sphere like particles is $\sim N^{-\frac{1}{2}}$. The corresponding kinetic equation is the Boltzmann equation.

- Weak-coupling limit: $N \to \infty$ and $\varepsilon \sim N^{-\frac{1}{3}}$ is a scale parameter coding the interaction length. The radial interaction potential—say, smooth and decaying—is scaled as $\varphi(\frac{r}{\varepsilon})$, so that the Hamiltonian of the system is

$$H_N = \frac{1}{2}\sum_{i=1}^{N} v_i^2 + \frac{\sqrt{\varepsilon}}{2}\sum_{i\neq j}\varphi\left(\frac{|x_i - x_j|}{\varepsilon}\right).$$

  The corresponding kinetic equation is the Landau equation.

- High-density limit: $N \to \infty$ and $\varepsilon \sim N^{-\frac{1}{4}}$, where the Hamiltonian is

$$H_N = \frac{1}{2}\sum_{i=1}^{N} v_i^2 + \frac{\varepsilon}{2}\sum_{i\neq j}\varphi\left(\frac{|x_i - x_j|}{\varepsilon}\right).$$

  The corresponding kinetic equation is the Lenard-Balescu equation.

For a recent analysis of these scalings we refer to [20, 21], where allowed classes of interaction potentials are also discussed. Scaling limits and propagation of chaos have a long history starting from the pioneering work of Bogoliubov [2], see e.g. [24, 26] for surveys.

The mechanism for which we expect that propagation of chaos actually holds is deeply different in the above cases. For the mean-field limit, the force exerted by particle j on particle i is $O(\frac{1}{N})$ so that the effect of one particle on the other is triviallynegligible. Hence z_i and z_j can be considered as independent random variables in the limit $N \to \infty$, if they were so at time zero. At variance, in the case of the Boltzmann-Grad limit, the forces are strong; but due to the low density regime, the probability of two *given* particles colliding in $(0, t)$ is small, although any given particle is subject to a collision with strictly positive rate. Finally in the weak-coupling limit, the forces are of order $\frac{1}{\sqrt{\varepsilon}} \sim N^{\frac{2}{3}}$, but the potential range is small and the interaction takes place in a time interval of order ε. Overall the variation of momentum of particle i due to the interaction with particle j is of order $\sqrt{\varepsilon}$, thus negligible. A similar argument holds in the high-density case.

Going further in this analysis, one may ask about quantitative estimates on the defect of chaos. In recent times, the authors have approached this problem in [23] by the method of "correlation errors" (also called "v-functions"), which goes back to previous work in the context of kinetic limits [3, 17, 25]. Correlation errors are a tool to measure the tendency of f_j^N to factorize and converge in the mean field limit. We will give a precise definition in the next section (Definition 1 below; see also Definition 2.1 in [23]). In [23], they have been used to prove a result on the Kac's model, which we recall next.

We consider N particles following a stochastic dynamics: to each particle, say particle i, we associate a velocity $v_i \in \mathbb{R}^3$ and the vector

$$V_N = \{v_1, \ldots, v_N\}$$

changes by means of two-body collisions at random times, with random scattering angle. The probability density $W^N(V_N, t)$ evolves according to the master equation (forward Kolmogorov equation)

$$\partial_t W^N = \frac{1}{N} T_N W^N \tag{2}$$

where

$$T_N W^N = \sum_{i<j} T_{i,j} W^N,$$

$$T_{i,j} W^N(V_N) = \int d\omega B(\omega; v_i - v_j)\{W^N(V_N^{i,j}) - W^N(V_N)\}, \tag{3}$$

and where $V_N^{i,j} = \{v_1, \ldots, v_{i-1}, v_i', v_{i+1}, \ldots, v_{j-1}, v_j', v_{j+1}, \ldots, v_N\}$ and the pair v_i', v_j' gives the outgoing velocities after a collision with scattering (unit) vector ø and incoming velocities v_i, v_j. $\frac{B(\omega; v_i - v_j)}{|v_i - v_j|}$ is the differential cross-section of the two-body process and we shall assume here, for simplicity, that B is bounded. The resulting kinetic equation is the homogeneous Boltzmann equation

$$\begin{aligned} \partial_t f(v) &= \int dv_1 \int d\omega B(\omega; v - v_1)\{f(v')f(v_1') - f(v)f(v_1)\} \\ &=: Q(f, f)(v) \end{aligned} \tag{4}$$

Such a model has been introduced and studied by Kac in [14, 15] and has given rise to a wide literature later on (see e.g. [1, 11–13, 18, 19, 27]). For instance in [23], we prove the following.

Theorem 1 ([23]) *For all $t > 0$ and all $j = 1, \ldots, N$, the marginals satisfy*

$$\| f_j^N(t) - f(t)^{\otimes j} \|_{L_1} \leq C_2 e^{C_1 t} \frac{j^2}{N} \tag{5}$$

where $C_1, C_2 > 0$ are explicitly computable constants.

In the present paper, we are going to recover a similar result, albeit in a different setting which we refer to as *grand canonical ensemble*. This is a standard formalism in statistical mechanics, adapted here to the kinetic problem. With respect to the model introduced above, the novelty of the grand canonical framework is twofold: (i) we allow the total number of particles N to be a (Poisson) random variable with intensity $\mu \to \infty$, and consequently (ii) we do not label the particles from 1 to N

(as done in the definition of marginal), but compute only averages of symmetrized quantities. This leads to the notion of *correlation functions*, denoted below $\left\{f_j^\mu\right\}_{j=1}^\infty$, which play the role of the marginals $\left\{f_j^N\right\}_{j=1}^\infty$ in this context.

The advantange of the grand-canonical formalism is that the correlations due to conservation of energy and mass are zero by assumption, and one can focus more easily on the correlations having a purely dynamical origin. In this spirit we will reformulate the method of [23] leading to a result analogous to (5), in terms of correlation functions (Theorem 2 below). We shall see that this leads to simplifications of the corresponding proof in the canonical ensemble. Indeed, even if f_j^N and f_j^μ are asymptotically equivalent, the latter functions encode fewer correlations. This is reflected eventually in a smaller number of terms in the basic evolution equations for the error (see Remark 1 below).

The paper is organized as follows. In the next section, we introduce the Kac's model in the grand canonical ensemble, and define the correlation errors. In Sect. 3 we discuss the evolution equations, whose derivation is postponed to the appendix. In Sect. 3 we establish the main result, Theorem 2, and its proof is presented in the last section.

2 Kac Model and Grand Canonical Ensemble

We start by recalling the main ensembles which can be adopted to describe a collisional dynamics, in the spirit of the abovementioned Kac's model.

- *Microcanonical.* The binary collisions preserve energy, and the stochastic dynamics lives in the energy surface

$$E = \frac{1}{N}\sum_i v_i^2.$$

- *Canonical.* E is random and only N is fixed, as in the model discussed in the Introduction.
- *Grand canonical.* N is random. We introduce a sequence of initial density distributions $\{W_0^n\}_{n\geq 0}$ on the phase space $\bigcup_{n\geq 0}\mathbb{R}^{3n}$. By definition, $\frac{1}{n!}W_0^n(V_n)$ is the symmetric probability density of the configuration V_n, being n the total number of particles. The average number of particles is then

$$\mu := \sum_{n\geq 0} n \int \frac{1}{n!} W_0^n(V_n)dV_n.$$

Here $\mu > 0$ is a free parameter (eventually $\mu \to \infty$) and the distributions do depend on μ, although this dependence is dropped in our notation ($W_0^n = W_0^{n,\mu}$).

At time $t > 0$, the state of the system is described by the time-evolved measure $\{W^n(t)\}_{n\geq 0}$ solving (cf. (2))

$$\partial_t W^n = \frac{1}{\mu} T_n W^n. \tag{6}$$

Focusing now on the grand canonical case, let us introduce a natural description of correlations and propagation of chaos. In order to examine correlations due exclusively to the dynamics, we choose a perfectly tensorized and still symmetric initial state, namely

$$W_0^n = f_0^{\otimes n} e^{-\mu} \mu^n \tag{7}$$

(however most of our discussion could be extended to a larger class of initial measures).

Instead of computing probability densities of particles with labels $\{1, \ldots, j\}$, we define rescaled correlation functions by

$$f_j^\mu(V_j, t) = \mu^{-j} \sum_{k\geq 0} \frac{1}{k!} \int dv_{j+1} \cdots dv_{j+k} W^{j+k}(V_{j+k}, t).$$

Loosely speaking, this corresponds to computing the amount of $j-$tuples of particles in the configuration V_j at time t. For a purely factorized state the average total number of $j-$tuples is μ^j, which explains the rescaling factor. In particular with the choice (7), one checks that $\| f_j^\mu \|_{L^1} = 1$ (initially and for all times).

Suppose that

$$\lim_{N\to\infty} f_j^\mu = f(t)^{\otimes j}, \tag{8}$$

say in L^1-norm. We may ask about convergence rates. We observe that, in general (e.g. for $j \sim \mu$) $f_j^\mu(t)$ is not expected to be asymptotically equivalent to $(f_1^\mu(t))^{\otimes j}$, even if it is assumed to be so at time zero. One can then pose the question: how large can $j = j(\mu)$ be in such a way that (8) holds true?

Having at our disposal the estimate

for some $\alpha \in (0, 1)$ and $C > 0$, we cannot make better predictions than $j(\mu) \sim \log \mu$. On the other hand if for a moderately large j, say $j \approx \mu^\gamma$, $\gamma \in (0, 1)$, $f_j^\mu \approx (f_1^\mu(t))^{\otimes j}$, then it is natural to consider

$$(f_1^\mu(t) - f(t))^{\otimes j} =: E_j; \tag{9}$$

namely the product of the differences rather than the difference of the products. Expanding (9) one finds that

$$E_j(t) := \sum_{K\subset J} (-1)^{|K|} f_{J\setminus K}^\mu(t) f(t)^{\otimes K} \tag{10}$$

where $J = \{1, 2, \ldots, j\}$ is the set of the first j indices, K is any subset of J, $J \backslash K$ is the relative complement of K in J and $|K|$ is the cardinality of K. $f_A^\mu(t)$ stands for the $|A|-$marginal $f_{|A|}^\mu(t)$ computed in the configuration $(z_i)_{i\in A}$. Similarly $f(t)^{\otimes K} = f(t)^{\otimes|K|}$ evaluated in $(z_i)_{i\in K}$. The formula (10) has an inverse formula, (see [23]) that is

$$f_j^\mu(t) = \sum_{K\subset J} E_{J\backslash K}(t) f(t)^{\otimes K} \tag{11}$$

where the notation $E_{J\backslash K}$ is the same one as for $F_{J\backslash K}^\mu$, and where we set $E_0 = E_\emptyset = 1$.

This motivates the following

Definition 1 (*Correlation error (grand-canonical)*) For any $j \geq 1$, setting $J = \{1, 2, \ldots, j\}$, we define the "correlation error" of order j by

$$E_j(t) := \sum_{K\subset J} (-1)^{|K|} f_{J\backslash K}^\mu(t) f(t)^{\otimes K}, \tag{12}$$

where the terms $K = \emptyset$ and $K = J$ have to be interpreted as $f_J^\mu = f_j^\mu$ and $(-1)^j f^{\otimes J} = (-1)^j f^{\otimes j}$, respectively.

For instance

$$\begin{aligned} E_1(v_1) &= f_1^\mu(v_1) - f(v_1) \\ E_2(v_1, v_2) &= f_2^\mu(v_1, v_2) - f_1^\mu(v_1) f_1(v_2) - f(v_1) f_1^\mu(v_1) + f(v_1) f(v_2) \\ E_3(v_1, v_2, v_3) &= f_3^\mu(v_1, v_2, v_3) - f_2^\mu(v_1, v_2) f(v_3) \cdots + f_1^\mu(v_1) f(v_2) f(v_3) \cdots \\ &\quad - f(v_1) f(v_2) f(v_3). \end{aligned}$$

3 Dynamical Equations

It is well known, and can be easily checked from (6), that the sequence $\{f_j^\mu\}$ satisfies the following hierarchy of equations (called BBKGY in analogy with the Hamiltonian case):

$$\partial_t f_j^\mu = \frac{T_j}{\mu} f_j^\mu + C_{j+1} f_{j+1}^\mu \tag{13}$$

for $j = 1, \ldots, \infty$, where

$$C_{j+1} f_{j+1}^\mu(V_j) = \sum_{i=1}^{j} C_{i,j+1} f_{j+1}^\mu(V_j) \tag{14}$$

and

$$C_{i,j+1}f^{\mu}_{j+1}(V_j) = \int dv_{j+1}\int d\omega B(\omega; v_i - v_j)\{f^{\mu}_{j+1}(V^{i,j+1}_{j+1}) - f^{\mu}_{j+1}(V_{j+1})\} \tag{15}$$

with the same notations as in (3).

This hierarchy must be compared with the limiting hierarchy (in the formal limit $\mu \to \infty$)

$$\partial_t f_j = C_{j+1} f_{j+1} \tag{16}$$

which is satisfied by products $f_j(t) = f(t)^{\otimes j}$, being $f(t)$ solution of the homogeneous Boltzmann equation (4).

Introducing correlation errors given by (12), we deduce from (13) the corresponding evolution equations. We find (see the Appendix)

$$\partial_t E_j = \frac{T_j}{\mu}E_j + D_j E_j + D^1_j\left(E_{j+1}\right) + D^{-1}_j\left(E_{j-1}\right) + D^{-2}_j\left(E_{j-2}\right) \tag{17}$$

where $D^1_j = C_{j+1}$ and the operators D_j, D^{-1}_j, D^{-2}_j are defined by:

$$\begin{aligned} D_j E_j &= \sum_{i\in J} C_{i,j+1}[f^{\otimes\{i\}}E_{J^i\cup\{j+1\}} + f^{\otimes\{j+1\}}E_J] \\ D^{-1}_j E_{j-1} &= \frac{1}{\mu}\sum_{i,s\in J} T_{i,s} f^{\otimes\{i\}} E_{J^i} \\ D^{-2}_j E_{j-2} &= \frac{1}{2\mu}\sum_{i,s\in J} T_{i,s} f^{\otimes\{i,s\}} E_{J^{i,s}}, \end{aligned} \tag{18}$$

where J is the set of the first j indices, $J^i = J\backslash\{i\}$, $J^{i,s} = J\backslash\{i,s\}$ and

$$f^{\otimes\{i\}} = f(v_i), \quad f^{\otimes\{i,s\}} = f(v_i)f(v_s).$$

Here we use the convention the $D^{-2}_1 = 0$. Moreover these operators may depend on time, although we will drop out this dependence from the notation.

Note that the first line contains operators which do not change the particle number, C_{j+1} is an operator decreasing by one the number of particles, while D^{-1}_j and D^{-2}_j are operators increasing the number of particles by one and two respectively.

Remark 1 The marginals in the canonical setting satisfy the same hierarchy of equations as in (13) (with μ replaced by N), up to a combinatorial factor $(N-j)/N$ in front of the collision operator C_{j+1}. This slight difference produces several spurious terms in the definitions of the operators D_j, D^{-1}_j, D^{-2}_j (see Eqs. (42)–(44) in [23]), which are absent in the definition (18). Therefore the grand-canonical equations single out the contributions having a purely dynamical interpretation, in terms of variation of clusters (groups of particles) with j mutually correlated particles. Since the interactions are binary, such a cluster can be created in five ways only, corresponding respectively to the five terms in (17), and depending on how many independent particles enter the game through the described collision.

We conclude this section by establishing our main result.

Theorem 2 *Let us assume $f_j^\mu(0) = f_0^{\otimes j}$ being f_0 the initial datum for the Boltzmann equation. Then for all $t > 0$, there exists $G > 1$ such that, for $\frac{j}{\sqrt{\mu}}$ sufficiently small,*

$$\| f_j^\mu(t) - f^{\otimes j}(t) \|_{L_1} \leq \frac{Gj^2}{\mu}. \tag{19}$$

Here we are assuming a full factorization at time zero. Looking at the proof it will be clear that this condition can be relaxed. Here we prefer to work in the simplest context to outline the main ideas.

Let us remark that, as it can be shown from of the proof of Theorem 2 the "$\frac{j}{\sqrt{\mu}}$ sufficiently small" can be quantified as "$\frac{j}{\sqrt{\mu}} \leq \alpha e^{-\beta t}$ for some contants $\alpha, \beta > 0$".

4 Towards the Boltzmann Equation

Starting from (13), by the Dyson (Duhamel) expansion we have:

$$f_j^\mu(t) = \sum_{n=0}^{\infty} \int_0^t dt_1 \int_0^{t_1} dt_2 \ldots \int_0^{t_{n-1}} dt_n \tag{20}$$
$$U_j^\mu(t - t_1) C_{j+1} \ldots U_{j+n-1}^\mu(t_{n-1} - t_n) C_{j+n} U_{j+n}^\mu(t_n) f_{0,n+j},$$

where $f_{0,j} = f_0^{\otimes j}$ and

$$U_j^\mu(t) = e^{\frac{T_j}{\mu} t}.$$

On the other hand we also have, for $f_j(t) = f(t)^{\otimes j}$, being $f(t)$ the solution to the Boltzmann equation

$$f_j(t) = \sum_{n=0}^{\infty} \int_0^t dt_1 \int_0^{t_1} dt_2 \ldots \int_0^{t_{n-1}} dt_n C_{j+1} \ldots C_{j+n} f_{0,n+j} \tag{21}$$
$$= \sum_{n=0}^{\infty} \frac{t^n}{n!} C_{j+1} \ldots C_{j+n} f_{0,n+j}.$$

Assuming that B is bounded then we have

$$\| T_j \| \leq C_1 \frac{j(j-1)}{2}, \qquad \| C_{j+1} \| \leq C_1 j. \tag{22}$$

From now on we denote by $\|\cdot\|$ the L^1-norm and by $C_i, i = 1, 2\cdots$ different numerical (positive) constants.

Then since $\|U_j^\mu(t)\| = 1$ both the right hand side of (20) and (21) are bounded by

$$\sum_{n\geq 0} C_2^n \frac{t^n}{n!} j(j+1)\cdots(j+n-1) \leq 2^j \sum_{n\geq 0} (2C_2 t)^n,$$

which is converging for t small. Therefore the rest of the two series is negligible, uniformly in μ, and this allows us to perform the limit $\mu \to \infty$. Indeed realizing that, for all j,

$$\lim_{\mu\to\infty} U_j^\mu(t) = I$$

strongly, we have also the term by term convergence of (20) and (21) in L^1. Once having the convergence for $t < \frac{1}{2C_2}$, we can iterate the procedure since the new initial conditions, $f_j^\mu(t)$ satisfy $\|f_j^\mu(t)\| = 1$. Then we have convergence for any arbitrary time.

What is the *size of chaos* in this case? By size of chaos we mean the maximal α for which we have convergence for all j such that $\frac{j}{\mu^\alpha} \to 0$ as $\mu \to \infty$. Our main Theorem 2 shows that the size of chaos is $\frac{1}{2}$. On the other hand, as we shall see in a moment, we do not expect a size of chaos strictly greater than $\frac{1}{2}$ so we believe that our result can be considered as optimal as regards this feature.

We observe that

$$\lim_{\mu, j\to\infty} U_j^\mu(t) = I$$

only if $\frac{j^2}{\mu} \to 0$. Indeed expressing $U^\mu(t)$ in terms of the series expansion

$$U^\mu(t) f_{0,j} = \sum_n \frac{t^n}{n!}\frac{1}{\mu^n} \sum_{i_1<k_1} \cdots \sum_{i_n<k_n} T_{i_1,k_1}\cdots T_{i_n,k_n} f_{0,j}$$

we realize that the L^1 norm of the right hand side can be bounded by

$$\sum_n \frac{(Ct)^n}{n!}\left(\frac{j(j-1)}{2\mu}\right)^n.$$

Therefore we expect that an estimate of the size of chaos cannot be better than $\frac{1}{2}$ and this can be considered as optimal.

Coming back to the dynamical description given by (17), we estimate the size of the L^1 norms of the operators appearing in (17) by

$$\|D_j\| = O(j),\ \|D_j^1\| = O(j),\ \|D_j^{-1}\| = O\left(\frac{j^2}{\mu}\right),\ \|D_j^{-2}\| = O\left(\frac{j^2}{\mu}\right).$$

Note that (17) is inhomogeneous so that it has nontrivial solutions even for initial data $E_j(0) = 0,\ j > 0$ (namely when the initial state is tensorized) which is the case we consider here.

The control of E passes trough a Dyson expansion around the two parameters semigroup $\bar{V}_j$ generated by $\frac{T_j}{\mu} E_j + D_j E_j$, namely the solution of

$$\partial_t \bar{V}_j(t,s) = \left(\frac{T_j}{\mu} + D_j\right) \bar{V}_j(t,s)$$
$$\bar{V}_j(s,s) = I.$$

We easily derive the following bound

$$\|\bar{V}_j(t,s)\| \le e^{C\left(\frac{j^2}{\mu}+j\right)(t-s)}. \tag{23}$$

The action of the other operators C_{j+1}, D_j^{-1}, D_j^{-2} represents a positive, a negative and a negative double jump respectively in the space of indices.

We recall that initially the distribution factorizes, namely

$$E_j(0) = \delta_{j,0},$$

then

$$E_j(t) = \sum_{n=0}^{\infty} \sum_{k_1 \cdots k_n} \int_0^t dt_1 \int_0^{t_1} dt_2 \ \ldots \int_0^{t_{n-1}} dt_n \tag{24}$$
$$\bar{V}_{s_1}(t,t_1) D^{k_1} \ldots \bar{V}_{s_n}(t_{n-1},t_n) D^{k_n} \bar{V}_{s_{n+1}}(t_n) E_{j_0}(0).$$

where $k_i \in \{-1,-2,1\}$. The sequence $\{s_1 \cdots s_n\}$ is defined recursively by

$$s_{r+1} = s_r + k_r, \qquad s_1 = j.$$

and

$$j_0 = j + \sum_{i=1}^{n} k_i$$

the initial index. Here $D^k = D^k_{s_k}$ and we drop the indices for notational simplicity.

We have to consider the case $j_0 = 0$ but in the following it will be convenient to consider general $j_0 \ge 0$.

Note also that $E_0 = 1$ and the only possible term involving E_0 is

$$D_2^{-2} E_0 = \frac{1}{\mu} T_2 f^{\otimes 2}$$

which is perfectly defined in $L^1(\mathbb{R}^{2d})$ being $f = f(t)$ the solution of the homogeneous Boltzmann equation.

For simplicity and without loss of generality we choose the parameters in such a way that

$$\|T_j\| \le \frac{j^2}{2}, \quad \|D_j\| \le \frac{j}{2}, \quad \|D_j^1\| \le j, \quad \|D_j^{-1}\| \le \frac{j^2}{\mu}, \quad \|D_j^{-2}\| \le \frac{j^2}{\mu}$$

and we can prove

Proposition 1 *Suppose that,*

$$\|E_j(0)\| \le \left(\frac{j^2}{\mu}\right)^{j/2} C_0^j, \quad \textit{for all } j \ge 1, \tag{25}$$

for some $C_0 \ge 1$. Then, there exists t_0 sufficiently small and $A > 0$ (both explicitily computable) such that for any $t \le t_0$,

$$\|E_j(t)\| \le \left(\frac{j^2}{\mu}\right)^{j/2} (AC_0)^j, \quad \textit{for all } j \ge 1. \tag{26}$$

We first observe that we can assume $j \le \frac{2}{C_0}\sqrt{\mu}$ because otherwise

$$\|E_j(t)\| \le 2^j \le 2^j \left(\frac{jC_0}{2\sqrt{\mu}}\right)^j = \left(\frac{jC_0}{\sqrt{\mu}}\right)^j.$$

Next from (24) we have

$$E_j(t) = \sum_{n=0}^{M-1} \sum_{k_1 \cdots k_n} \int_0^t dt_1 \int_0^{t_1} dt_2 \ldots \int_0^{t_{n-1}} dt_n$$
$$\bar{V}_{s_1}(t, t_1) D^{k_1} \ldots \bar{V}_{s_n}(t_{n-1}, t_n) D^{k_n} \bar{V}_{s_{n+1}}(t_n) E_{j_0}(0) +$$
$$\sum_{k_1 \cdots k_M} \int_0^t dt_1 \ldots \int_0^{t_{M-1}} dt_M \bar{V}_{s_1}(t, t_1) D^{k_1} \ldots \bar{V}_{s_M}(t_{M-1} - t_M) D^{k_M} E_{j_0}(t_M)$$

and denote by $\mathscr{T}_j^1(t)$ and $\mathscr{T}_j^2(t)$ respectively the two terms in the right hand side of the above expression.

Moreover we fix $M = \frac{2}{C_0}\sqrt{\mu}$. With this choice

$$j + s_m \le j + M \le \frac{4}{C_0}\sqrt{\mu}$$

so that all the operators D_j^{-1} and D_j^{-2} appearing in the definition of $\mathscr{T}_j^1$ and $\mathscr{T}_j^2$ satisfy the bound

$$\|T_j\| \le j, \quad \|D_j^{-1}\| \le \frac{j^2}{\mu} \le j, \quad \|D_j^{-2}\| \le \frac{j^2}{\mu} \le j$$

provided that μ is large enough. As a consequence, by (23)

$$\|\bar{V}_j(t)\| \le e^{jt}. \tag{27}$$

We first estimate $\mathscr{T}_j^2(t)$. Observe that, for $t_0 \le 1$ for which $e^{jt_0} \le e^j$

$$\|\bar{V}_{s_1}(t, t_1)D^{k_1} \dots \bar{V}_{s_M}(t_{M-1} - t_M)D^{k_M} E_{j_0}(t_M)\| \le e^j e^{Mt}(j + M)^M 2^{j+M}$$

as follows by the estimates

$$e^{s_1(t-t_1)}e^{s_2(t_1-t_2)} \dots e^{s_M(t_{M-1}-t_M)} \le e^{(j+M)t} \le e^j e^{Mt}$$

$$(j + k_1)(j + k_1 + k_2) \cdots (j + k_1 + k_2 + \cdot k_M) \le (j + M)^M \le (2M)^M$$

$$\|E_{j_0}(t_M)\| \le 2^{j+M}.$$

Therefore, setting $\tau = te^t$, using the bound $M^M \le C_1^M M!$ and choosing τ so small that $\tau 8C_1 \le \frac{1}{2}$, we have

$$\|T_j^2(t)\| \le 4^M \frac{\tau^M}{M!} M^M e^j 2^{j+M} \le (2e)^j (\tfrac{1}{2})^M.$$

Finally since

$$\sup_M M^j (\tfrac{1}{2})^M \le (C_2 j)^j$$

we conclude that

$$\|\mathscr{T}_j^2(t)\| \le \left(\frac{C_3 j}{\sqrt{\mu}}\right)^j. \tag{28}$$

To estimate $\mathscr{T}_j^1(t)$ we split it into two contributions $\mathscr{T}_j^{1,<}(t)$ and $\mathscr{T}_j^{1,>}(t)$ relative to the events $j_0 \le j$ and $j_0 > j$ respectively. Then, proceeding as above

$$\|\mathscr{T}_j^{1,<}(t)\| \le \sum_{n \le M-1} \sum_{k_1 \cdots k_n} e^j \frac{\tau^n}{n!} (j + n)^n \left(\frac{j^2}{\mu}\right)^{\frac{j-j_0}{2}} \|E_{j_0}\|.$$

The contribution $(\frac{j^2}{\mu})^{\frac{j-j_0}{2}}$ arises from the fact that to reach j_0 from j we need at least $\frac{j-j_0}{2}$ negative jumps, each of them produces a factor bounded by $\frac{j^2}{\mu}$.

Finally using (25) and the usual arguments we arrive at

$$\|\mathcal{T}_j^{1,<}(t)\| \leq \sum_{n\geq 0}(C_4\tau)^n e^j \left(\frac{j^2}{\mu}\right)^{\frac{j-j_0}{2}} \left(\frac{C_0^2 j_0^2}{\mu}\right)^{\frac{j_0}{2}} \leq \left(\frac{C_5 C_0^2 j^2}{\mu}\right)^{\frac{j}{2}}.$$

Finally

$$\|\mathcal{T}_j^{1,>}(t)\| \leq \sum_{n\leq M-1}\sum_{k_1\cdots k_n} e^j \frac{\tau^n}{n!}(j+n)^n \left(\frac{j_0^2}{\mu}\right)^{\frac{j_0}{2}} C_0^{j_0}.$$

Setting now $\ell = \ell(k_1, k_2 \cdots k_n) = j_0 - j$ we have

$$\begin{aligned}(\frac{(j+\ell)^2}{\mu})^{\frac{j+\ell}{2}} C_0^\ell &= \left(\frac{j^2}{\mu}\right)^{j/2} \left(\frac{(j+\ell)^2}{\mu}\right)^{\frac{\ell}{2}} \left(\frac{j+\ell}{j}\right)^j C_0^\ell \\ &\leq \left(\frac{j^2}{\mu}\right)^{j/2} (\frac{2}{C_0})^\ell e^\ell C_0^\ell \leq \left(\frac{j^2}{\mu}\right)^{j/2} (2e)^n. \end{aligned} \tag{29}$$

where, in the last step we used that (remind that both j and M are bounded by $\frac{2\sqrt{\mu}}{C_0}$)

$$\frac{(j+\ell)}{\sqrt{\mu}} \leq \frac{j+M}{\sqrt{\mu}} \leq \frac{4}{C_0}, \quad \left(\frac{j+\ell}{j}\right)^j \leq e^\ell, \quad \ell \leq n.$$

Thus, for τ small enough:

$$\|\mathcal{T}_j^{1,>}(t)\| \leq e^{2j} C_0^j \sum_{n\geq 0} C_5^n \tau^n \left(\frac{j^2}{\mu}\right)^{\frac{j}{2}} \leq \left(\frac{C_6 C_0^2 j^2}{\mu}\right)^{\frac{j}{2}}$$

and the proof is achieved.

As a Corollary we can easily prove

Proposition 2 *Let us suppose that $E_j(0) = 0$ for $j \geq 1$ ($E_\emptyset(0) = 1$) corresponding to a fully factorized initial state. Then for all $t > 0$ and all $j > 1$, one has*

$$\|E_j(t)\| \leq \left(Ae^{Bt}\right)^j \left(\frac{j}{\sqrt{\mu}}\right)^j \tag{30}$$

where $B \geq 0$ and $A \geq 1$.

Note that from the evolution equation

$$\partial_t E_1 = D_1 E_1 + C_{1,2} E_2$$

and Proposition 2 we also obtain the estimate

$$\|E_1(t)\| \leq tA^2 e^{2Bt} \frac{4}{\mu}$$

which will be useful in a moment.

Proof of Theorem 2

We find that, for some time-dependent constants C_7, C_8, G, using Proposition 2 and the estimate for $E_1(t)$ above,

$$\begin{aligned}
\|f_j^\mu(t) - f^{\otimes j}(t)\| &\leq j\|E_1(t)\| + \sum_{k=2}^{j} \frac{j!}{k!(j-k)!}\|E_K(t)\| \\
&\leq \frac{jC_8}{\mu} + \sum_{k=2}^{j} \frac{j!}{k!(j-k)!}\left(\frac{Ae^{Bt}k}{\sqrt{\mu}}\right)^k \\
&\leq \frac{jC_8}{\mu} + \sum_{k=2}^{j}\left(\frac{C_7 j}{\sqrt{\mu}}\right)^k = \frac{jC_8}{\mu} + \left(\frac{C_7 j}{\sqrt{\mu}}\right)^2 \frac{1}{1-\frac{C_7 j}{\sqrt{\mu}}} \\
&= G\left(\frac{j}{\sqrt{\mu}}\right)^2
\end{aligned}$$

for μ large such that $\frac{C_7 j}{\sqrt{\mu}} < \frac{1}{2}$.

Acknowledgements We thank the "Istituto Nazionale di Alta Matematica" (INdAM) for partial support.

Appendix

In this appendix, we deduce the correlation error equations starting from the BBGKY hierarchy, (13), which we rewrite for the reader's convenience:

$$\partial_t f_j^\mu = \frac{T_j}{\mu} f_j^\mu + C_{j+1} f_{j+1}^\mu \qquad j \geq 1\,.$$

We shall see that the computation is drastically simpler with respect to the analogous derivation in the canonical ensemble reported in the Appendix of [23] (where additional terms appear due to the canonical constraint).

We recall the definition of correlation error

$$E_j := \sum_{K\subset J}(-1)^{|K|} f_{J\setminus K}^\mu f^{\otimes K} \tag{31}$$

together with the following notation. For any set of particle indices $K = \{i_1, \ldots, i_k\}$ of cardinality $k = |K|$, we write

$$
\begin{aligned}
f_K^\mu &= f_k^\mu\left(z_{i_1}, \ldots, z_{i_k}\right), \\
E_K &= E_k\left(z_{i_1}, \ldots, z_{i_k}\right), \\
f^{\otimes K} &= f(z_{i_1}) \cdots f(z_{i_k})
\end{aligned}
$$

with the conventions $f_\emptyset^\mu = E_\emptyset = f^{\otimes\emptyset} = 1$. We also write $Q(f, f)_i = Q(f, f)(v_i)$ for the Boltzmann operator (4) evaluated in particle i.

We start by the simple computation for $j = 1$:

$$
\begin{aligned}
\partial_t E_{\{1\}} &= \partial_t \left(f_{\{1\}}^\mu - f^{\otimes\{1\}}\right) \\
&= C_2 f_{\{1,2\}}^\mu - Q(f, f)_1 \\
&= C_2 \left(f^{\otimes\{1,2\}} + f^{\otimes\{1\}} E_{\{2\}} + E_{\{1\}} f^{\otimes\{2\}} + E_{\{1,2\}}\right) - Q(f, f)_1
\end{aligned}
$$

where in the last step we used the inverse formula (cf. (11)) for $j = 2$. Noticing that, by (14) and (15), $C_2(f^{\otimes\{1,2\}}) = Q(f, f)_1$, we are left with

$$
\partial_t E_{\{1\}} = C_2 \left(f^{\otimes\{1\}} E_{\{2\}} + E_{\{1\}} f^{\otimes\{2\}}\right) + C_2 \left(E_{\{1,2\}}\right)
$$

and the two terms on the right hand side correspond to $D_1 E_{\{1\}}$ and $D_1^1(E_{\{1,2\}})$ respectively. Therefore (17) is verified for $j = 1$.

We assume now that $\{E_k\}_{k \leq j-1}$ satisfy (17), and prove the same thing for $k = j$. Using definition (31) and the evolution equations for f_J^μ and f, we find that

$$
\begin{aligned}
\partial_t E_J &= \partial_t \left(f_J^\mu - \sum_{\substack{K \subset J \\ K \neq \emptyset}} f^{\otimes K} \, E_{J \setminus K} \right) \\
&= \frac{T_j}{\mu} f_J^\mu + \sum_{i \in J} C_{i,j+1} f_{J \cup \{j+1\}}^\mu - \sum_{\substack{K \subset J \\ K \neq \emptyset}} f^{\otimes K} \, \partial_t E_{J \setminus K} \\
&\quad - \sum_{\substack{K \subset J \\ K \neq \emptyset}} \sum_{i \in K} f^{\otimes K \setminus \{i\}} Q(f, f)_i E_{J \setminus K}.
\end{aligned}
\tag{32}
$$

We compute the three terms in the second line, one by one. First, using the inverse formula and denoting $J^i = J \setminus \{i\}$, $J^{i,s} = J \setminus \{i, s\}$ etc.,

$$
\begin{aligned}
\frac{T_j}{\mu} f_J^\mu &= \frac{T_j}{\mu} E_J + \sum_{\substack{i,s \in J \\ i \neq s}} \frac{T_{i,s}}{\mu} \left\{ f^{\otimes\{i\}} E_{J^i} + \frac{1}{2} f^{\otimes\{i,s\}} E_{J^{i,s}} \right\} \\
&\quad + \sum_{\substack{K \subset J \\ K \neq \emptyset}} f^{\otimes K} \frac{T_{j-k}}{\mu} E_{J \setminus K}
\end{aligned}
$$

$$+\sum_{\substack{K\subset J\\|K|\neq 0,1}}\sum_{\substack{i\in K\\ s\in J\setminus K}}\frac{T_{i,s}}{\mu}f^{\otimes\{i\}}f^{\otimes K^i}E_{J\setminus K}+\sum_{\substack{K\subset J\\|K|\neq 0,1,2}}\sum_{\substack{i,s\in K\\ i\neq s}}\frac{T_{i,s}}{2\mu}f^{\otimes\{i,s\}}f^{\otimes K^{i,s}}E_{J\setminus K} \tag{33}$$

and, similarly,

$$\begin{aligned}\sum_{i\in J}C_{i,j+1}f^{\mu}_{J\cup\{j+1\}}&=\sum_{i\in J}C_{i,j+1}\left\{E_J f^{\otimes\{j+1\}}+f^{\otimes\{i\}}E_{J^i\cup\{j+1\}}+E_{J\cup\{j+1\}}\right\}\\&+\sum_{\substack{K\subset J\\|K|\neq 0}}\sum_{i\in J\setminus K}C_{i,j+1}f^{\otimes K}f^{\otimes\{j+1\}}E_{J\setminus K}\\&+\sum_{\substack{K\subset J\\K\neq\emptyset}}\sum_{i\in J\setminus K}C_{i,j+1}f^{\otimes K}\left\{f^{\otimes\{j+1\}}E_{J\setminus K}+f^{\otimes\{i\}}E_{(J\cup\{j+1\}\setminus K)^i}+E_{J\cup\{j+1\}\setminus K}\right\}.\end{aligned} \tag{34}$$

Secondly, using the inductive hypothesis and the explicit definition of the operators D_{j-k}, D^1_{j-k}, D^{-1}_{j-k}, D^{-2}_{j-k}, we obtain that

$$\begin{aligned}-\sum_{\substack{K\subset J\\K\neq\emptyset}}f^{\otimes K}\,\partial_t E_{J\setminus K}&=-\sum_{\substack{K\subset J\\K\neq\emptyset}}f^{\otimes K}\frac{T_{j-k}}{\mu}E_{J\setminus K}\\&-\sum_{\substack{K\subset J\\K\neq\emptyset}}\sum_{\substack{i,s\in J\setminus K\\i\neq s}}\frac{T_{i,s}}{\mu}f^{\otimes K}\left\{f^{\otimes\{i\}}E_{(J\setminus K)^i}+\frac{1}{2}f^{\otimes\{i,s\}}E_{(J\setminus K)^{i,s}}\right\}\\&-\sum_{\substack{K\subset J\\K\neq\emptyset}}\sum_{i\in J\setminus K}C_{i,j+1}f^{\otimes K}\left\{f^{\otimes\{j+1\}}E_{J\setminus K}+f^{\otimes\{i\}}E_{(J\cup\{j+1\}\setminus K)^i}+E_{J\cup\{j+1\}\setminus K}\right\}.\end{aligned} \tag{35}$$

We note that the first line in (33) and the first line in (34) reproduce the right hand side of the desired Eq. (17). Therefore we need to check that the remaining terms cancel out. This is straightforward:

- the second line in (33) cancels with the first term on the right hand side in (35);
- last line in (33) cancels with the second line in (35) (change variables $K \to K\cup\{i\}$ and $K \to K\cup\{i,s\}$);
- second line in (34) cancels with last line in (32);
- last line in (34) cancels with last line in (35).

This concludes the derivation of (17).

References

1. Aoki, K., Pulvirenti, M., Simonella, S., Tsuji, T.: Backward clusters, hierarchy and wild sums for a hard sphere system in a low-density regime. Math. Mod. and Meth. in App. Sci. **25**(5), 995–1010 (2015)
2. Bogolyubov, N.N.: Problems of a Dynamical Theory in Statistical Physics. Moscow: State Technical Press (1946) in Russian; English translation in Studies in Statistical Mechanics I, edited by J. de Boer and G. E. Uhlenbeck, part A, Amsterdam: North-Holland (1962)
3. Boldrighini, C., De Masi, A., Pellegrinotti, A.: Nonequilibrium fluctuations in particle systems modelling reaction-diffusion equations. Stoch. Process. Appl. **42**, 1–30 (1992)
4. Caprino, S., Pulvirenti, M.: A cluster expansion approach to a one-dimensional Boltzmann equation: a validity result. Commun. Math. Phys. **166**(3), 603–631 (1995)
5. Caprino, S., De Masi, A., Presutti, E., Pulvirenti, M.: A derivation of the Broadwell equation. Commun. Math. Phys. **135**(3), 443–465 (1991)
6. Caprino, S., Pulvirenti, M., Wagner, W.: A particle systems approximating stationary solutions to the Boltzmann equation SIAM. J. Math. Anal. **4**, 913–934 (1998)
7. Carlen, E., Carvalho, M.-C., Gabetta, E.: Central limit theorem for Maxwellian molecules and truncation of the Wild expansion. CPAM **53**(3), 370–397 (2000)
8. Carlen, E., Carvalho, M.-C., Gabetta, E.: On the relation between rates of relaxation and convergence of wild sums for solutions of the Kac equation. J. Funct. Anal. **220**(2), 362–387 (2005)
9. De Masi, A., Orlandi, E., Presutti, E., Triolo, L.: Glauber evolution with Kac potentials. I. Mesoscopic and macroscopic limits, interface dynamics. Nonlinearity **7**, 633–696 (1994)
10. De Masi, A., Presutti, E.: Mathematical Methods for Hydrodynamical Limits. In: Lecture Notes in Mathematics, vol. 1501. Springer-Verlag (1991)
11. Graham, C., Méléard, S.: Stochastic particle approximations for generalized Boltzmann models and convergence estimates. Ann. Probab. **25**, 115–132 (1997)
12. Grünbaum, F.A.: Propagation of chaos for the Boltzmann equation. Arch. Rat. Mech. Anal. **42**, 323–345 (1971)
13. Hauray, M., Mischler, S.: On Ka's chaos and related problems. J. Funct. Anal. **266**, 6055–6157 (2014)
14. Kac, M.: Foundations of kinetic theory. In: Proceedings of the Third Berkeley Symposium on Mathematical Statistics and Probability. University of California Press, Berkeley and Los Angeles (1956)
15. Kac, M.: Probability and Related Topics in Physical Sciences. Interscience, London-New York (1959)
16. Masi, A., Orlandi, E., Presutti, E., Triolo, L.: Glauber evolution with Kac potentials. II. Fluctuations. Nonlinearity **9**, 27–51 (1996)
17. Masi, A., Presutti, E., Tsagkarogiannis, D., Vares, M.E.: Truncated correlations in the stirring process with births and deaths. Electron. J. Probab. **17**, 1–35 (2012)
18. Mischler, S., Mouhot, C.: Kac's program in kinetic theory. Inven. Math. **193**, 1–147 (2013)
19. Mischler, S., Mouhot, C., Wennberg, B.: A new approach to quantitative propagation of chaos for drift, diffusion and jump processes. Probab. Theory Relat. Fields **161**(1–2), 1–59 (2015)
20. Nota, A., Velázquez, J.J.L., Winter, R.: Interacting particle systems with long-range interactions: scaling limits and kinetic equations. Atti Accad. Naz. Lincei Rend. Lincei Mat. Appl. **32**(2), 335–377 (2021)
21. A. Nota, J.J.L. Velázquez, Winter, R.: Interacting particle systems with long-range interactions: approximation by tagged particles in random fields. Atti Accad. Naz. Lincei Cl. Sci. Fis. Mat. Natur. **33**(2), 439–506 (2022)
22. Paul, T., Pulvirenti, M.: Asymptotic expansion of the mean-field approximation. Discret. Contin. Dyn. Syst. A **39**, 1891–1921 (2019)
23. Paul, T., Pulvirenti, M., Simonella, S.: On the size of chaos in the mean field dynamics. Arch. Rat. Mech. Anal. **231**, 285–317 (2019)

24. Pulvirenti, M., Simonella, S.: A Brief Introduction to the Scaling Limits and Effective Equations in Kinetic Theory. In: Albi, G., Merino-Aceituno, S., Nota, A., Zanella, M. (eds.) Trails in Kinetic Theory. SEMA SIMAI Springer Series, vol. 25, pp. 183–208 (2021)
25. Pulvirenti, M., Simonella, S.: The Boltzmann-Grad limit of a hard sphere system: analysis of the correlation error. Inven. Math. **207**(3), 1135–1237 (2017)
26. Spohn, H.: Large Scale Dynamics of Interacting Particles. In: Texts and Monographs in Physics. Springer-Verlag, Heidelberg (1991)
27. Sznitman, A.-S.: Topics in propagation of chaos. In: École d'été de Probabilités de Saint-Flour XIX 1989. Lecture Notes in Mathematics, vol. 1464, pp. 165–251. Springer, Berlin (1991)

Magnetic Reconnection: Some Basic Ideas and a Review on a Collisionless Model

Valeria Ricci, Bruno Coppi, and Renato Spigler

Abstract Basic ideas and definitions are surveyed concerning the theory of *magnetic reconnection* in plasmas, a phenomenon occurring under rather various and different conditions, both in astrophysical systems and in laboratory experiments. Approaches to the description of this phenomenon are reviewed, discussing in the last part a linearized analysis for collisionless regimes.

Keywords Magnetic reconnection · Collisionless plasmas · Magnetically confined plasmas

1 Introduction

A wide variety of processes that involve a drastic change of magnetic field topology is generically referred to as *Magnetic Reconnection*.

Those processes are relevant in the physics of magnetized plasmas, since they may produce a rapid conversion of magnetic energy into other forms of energy (thermal or kinetic, and hence possibly also particle acceleration): this conversion may be

---

To the memory of Maria Conceição Carvalho

V. Ricci (✉)
Dipartimento di Matematica e Informatica, Università degli Studi di Palermo,
Via Archirafi 34, 90123 Palermo, Italy
e-mail: valeria.ricci@unipa.it

B. Coppi
Department of Physics and American Academy of Arts and Sciences,
Massachusetts Institute of Technology, 77, Mass. Ave., Cambridge, MA 02139, USA
e-mail: coppi@psfc.mit.edu

R. Spigler
Department of Mathematics and Physics, Roma Tre University, 1, Largo S. Leonardo Murialdo, 00146 Rome, Italy
e-mail: spigler@mat.uniroma3.it

E. Carlen et al. (eds.), *From Particle Systems to Partial Differential Equations*,
Springer Proceedings in Mathematics & Statistics 465,
https://doi.org/10.1007/978-3-031-65195-3_13

at the origin of several impulsive events observed in astrophysics as well as within thermonuclear laboratory plasmas.

There is a vaste literature on this subject, written since the time when the basic concept of magnetic reconnection was introduced in 1950 [10] (later mentioned in [11]), to explain the coupling of mass, energy, and momentum from the solar wind into Earth's magnetosphere (the first and still valid theory of the origin of auroral substorms, based on collisionless reconnection, is given in [8]).

In the broad variety of cases where magnetic reconnection has been observed, e.g. in space physics (solar wind), geophysics (auroras), astrophysics and thermonuclear laboratory plasmas, there are several concurrent phenomena to identify, and in numerous cases a convincing theory of magnetic reconnection remains to be formulated.

Although magnetic reconnection develops often in collisionless plasmas (so that in principle the correct level of description should be at least the kinetic one), almost all mathematical models of this phenomenon are formulated through magnetohydrodynamics (MHD) equations. In this way, the mathematical description is remarkably simplified, but leaves some doubts about the validity of the theory.

In this paper we review a mathematical model of magnetic reconnection which mix different scales of description, although the general description is focussed on the MHD picture which is commonly used in the literature.

In Sect. 2, we discuss briefly a formal mathematical definition of reconnection process. In Sect. 3, we give a schematic classification based on the main factors producing magnetic reconnection (connected to the collisional regime in which reconnection occurs) and discuss the two cases of *exogenous* and *endogenous* reconnection.

Finally, in Sect. 4, we discuss in some more detail reconnection process in collisionless plasmas on the basis of the analysis performed in [9].

2 Basic Equations for the Onset of Magnetic Reconnection

Giving a rough picture of the phenomenon, magnetic reconnection occurs when some field lines of a magnetic field (with different orientation), in their time evolution, are cut and reconnected in a different way, thus changing the topology of the field and leading to observable global changes in the plasma.

The local rearrangement originated by the reconnection of the field lines may indeed produce a relaxation in the global magnetic configuration. In particular, a localized process which, nevertheless, can produce global effects, is involved.

There are some difficulties in providing a proper definition of magnetic reconnection which could be applied in any symmetry, in particular being independent of the space dimension of the plasma configuration model.

According to Refs. [1, 12], *reconnection occurs if there are points originally located on the same field line which end up on different field lines in the evolution.*

In this section we shall reconsider a macroscopic mathematical description of a multispecies magnetized plasma and illustrate the criterion given in [1, 12].

2.1 Theoretical Model for a Multispecies Plasma

The evolution of a multispecies magnetized plasma, containing N charged species, can be described through the set of mesoscopic equations (here and in what follows, we use Gaussian units):

$$\partial_t f_\alpha + \mathrm{div}_x(\mathbf{v} f_\alpha) + \mathrm{div}_v \left[\left(q_\alpha \mathbf{E} + \frac{q_\alpha}{c}\mathbf{v}\wedge\mathbf{B}\right) f_\alpha\right] = \sum_{\beta=1}^{N} C_{\alpha\beta}(f_\alpha, f_\beta) \quad (1)$$

coupled to the Maxwell's equations for the electromagnetic field (**E**, **B**),

$$\mathrm{curl}\mathbf{E} = -\frac{1}{c}\partial_t \mathbf{B} \quad (2)$$

$$\mathrm{curl}\mathbf{B} = \frac{1}{c}[\partial_t \mathbf{E} + 4\pi \mathbf{J}] \quad (3)$$

$$\mathrm{div}\,\mathbf{E} = 4\pi\rho \quad (4)$$

$$\mathrm{div}\,\mathbf{B} = 0 \quad (5)$$

Here c is the speed of light, $(\mathbf{x}, \mathbf{v})$ denotes the space and velocity coordinates, $\alpha = 1, \ldots, N$ is the species index, q_α and $f_\alpha = f_\alpha(t, \mathbf{v}, \mathbf{x})$ are charge and number density distribution of the species α, and ρ and $\mathbf{J}$ are respectively the total charge density and current, defined as

$$\rho = \sum_\alpha q_\alpha n_\alpha \quad (6)$$

$$\mathbf{J} = \sum_\alpha q_\alpha n_\alpha \mathbf{v}_\alpha, \quad (7)$$

where $n_\alpha = \int d\mathbf{v} f_\alpha(t, \mathbf{v}, \mathbf{x})$ is the number density of the species α and $n_\alpha \mathbf{v}_\alpha = \int d\mathbf{v}\mathbf{v} f_\alpha(t, \mathbf{v}, \mathbf{x})$, so that $\mathbf{v}_\alpha$ denotes the macroscopic velocity of the species α.

From the multispecies Boltzmann formulation (1), by integrating over $\mathbf{v}$ its product by $(1, m_\alpha \mathbf{v}, \frac{1}{2} m_\alpha \mathbf{v}^2)$, we derive the set of hydrodynamics equations describing the transport in the plasma,

$$\partial_t n_\alpha + \mathrm{div}(n_\alpha \mathbf{v}_\alpha) = 0 \quad (8)$$

$$m_\alpha n_\alpha (\partial_t \mathbf{v}_\alpha + \mathbf{v}_\alpha \cdot \nabla_x \mathbf{v}_\alpha) = -\nabla p_\alpha - \mathrm{div}\,\Pi_\alpha + q_\alpha n_\alpha \left(\frac{\mathbf{v}_\alpha \wedge \mathbf{B}}{c}\right) + \mathbf{R}_\alpha \quad (9)$$

$$\frac{3}{2} n_\alpha [\partial_t (n_\alpha k T_\alpha) + \mathrm{div}_x (n_\alpha k T_\alpha \mathbf{v}_\alpha)] + p_\alpha \,\mathrm{div}_x \mathbf{v}_\alpha = -\,\mathrm{div}\,\mathbf{q}_\alpha - \Pi_\alpha : \nabla_x \mathbf{v}_\alpha + Q_\alpha \quad (10)$$

where m_α is the mass of the species α, $P_\alpha = p_\alpha \,\mathrm{Id} + \Pi_\alpha$ and $\mathbf{q}_\alpha$ its pressure tensor (whit $p_\alpha = n_\alpha k T_\alpha$ defining the temperature of the species α) and heat flux, $\mathbf{R}_\alpha$ its momentum variation due to collisions, Q_α its energy variation due to collisions. Equations (8)–(10), together with (2)–(5), constitute the macroscopic model which we adopt to describe the plasma in most parts of this paper.

The plasma is assumed to satisfy the quasi-neutrality condition, $\rho \approx 0$ (we shall therefore assume $\rho = 0$).

From Eq. (9), we obtain, upon some manipulations, the *Generalized Ohm's Law*:

$$\mathbf{E} + \frac{\mathbf{V} \wedge \mathbf{B}}{c} = -\frac{1}{c}\frac{\sum_\alpha \omega_\alpha^2 (\mathbf{v}_\alpha - \mathbf{V}) \wedge \mathbf{B}}{\sum_\alpha \omega_\alpha^2} + \frac{4\pi c}{\sum_\alpha \omega_\alpha^2} \nabla_x \left(\sum_\alpha \frac{\Omega_\alpha P_\alpha}{|\mathbf{B}|} \right) \tag{11}$$

$$- \frac{4\pi c}{|\mathbf{B}|} \left(\sum_\alpha \frac{\Omega_\alpha \mathbf{R}_\alpha}{\sum_\alpha \omega_\alpha^2} \right) + \frac{4\pi c}{\sum_\alpha \omega_\alpha^2} [\partial_t \mathbf{J} + \mathrm{div}(\mathbf{V} \otimes \mathbf{J} + \mathbf{J} \otimes \mathbf{V} + \sum_\alpha n_\alpha q_\alpha (\mathbf{v}_\alpha - \mathbf{V})^{\otimes 2})]$$

where $\mathbf{V} = \frac{\sum_\alpha m_\alpha n_\alpha \mathbf{v}_\alpha}{\sum_\alpha m_\alpha n_\alpha}$ is the average velocity of the plasma, and Ω_α and ω_α are respectively the cyclotron and the plasma frequencies of the species α,

$$\Omega_\alpha = \frac{q_\alpha |\mathbf{B}|}{m_\alpha c}$$

$$\omega_\alpha^2 = \frac{4\pi n_\alpha q_\alpha^2}{m_\alpha}.$$

We recall that the *Ideal Frozen-in Law* is defined as

$$\mathbf{E} + \frac{\mathbf{V} \wedge \mathbf{B}}{c} = 0, \tag{12}$$

so that the right-hand side of equation (11), denoted by $\mathbf{F}$ in what follows, collects the non-ideal effects (dissipative or not) in the plasma, and we shall call *Generalised Ohm's Law* the equation

$$\mathbf{E} + \frac{\mathbf{V} \wedge \mathbf{B}}{c} = \mathbf{F}. \tag{13}$$

2.2 Magnetic Reconnection: A Local Formal Criterion

As reported in [1, 12], the condition under which the topology of the magnetic field $\mathbf{B}(t, \mathbf{x})$ is preserved through its time evolution is that there exist a vector field, $\mathbf{u}(t, \mathbf{x})$, and a scalar field, $\sigma(t, \mathbf{x})$, such that

$$\partial_t \mathbf{B} - \mathrm{curl}(\mathbf{u} \wedge \mathbf{B}) = \mathbf{B}(\sigma + \mathrm{div}_x\, \mathbf{u}) \tag{14}$$

In particular, this is true for ideal flows such that (12) is satisfied, since we can define $\mathbf{u} = \mathbf{V}$ and $\sigma = -\operatorname{div}\mathbf{V}$. Therefore, ideal flows cannot exhibit magnetic reconnection.

We can use (2) to define through (14) a vector $\mathbf{S} = \mathbf{E} + \frac{1}{c}(\mathbf{u} \wedge \mathbf{B})$, which represents the non-ideal force associated to the flow $\mathbf{u}$, so that (14) becomes

$$- c \operatorname{curl}\mathbf{S} = \mathbf{B}(\sigma + \operatorname{div}\mathbf{u}) \tag{15}$$

We are now ready to report the criterion given in [1], and in a different form in [12]. Assume that $\mathbf{S}_{\parallel}$ is the component of $\mathbf{F}$ along the magnetic field, $\mathbf{B}$,

$$\mathbf{S}_{\parallel} = \frac{\mathbf{F}\cdot\mathbf{B}}{|\mathbf{B}|}\,\hat{\mathbf{B}} = \mathbf{E}\cdot\hat{\mathbf{B}}\;\hat{\mathbf{B}}. \tag{16}$$

($\mathbf{F}_{\perp} = \mathbf{F} - \mathbf{S}_{\parallel}$ can be included in $\mathbf{u}$)

Then, the magnetic reconnection may occur if, in a localised region where the Ideal Ohm's Law is broken (i.e. $\mathbf{F} \neq 0$), called *reconnection region*, we have

$$\mathbf{B} \wedge \operatorname{curl}\mathbf{S}_{\parallel} \neq 0, \tag{17}$$

so that (15) is not valid.

This condition is still local. We need then a further condition: the reconnection is global if and only if

$$\int \mathbf{S}_{\parallel} ds = \int \mathbf{E}\cdot\hat{\mathbf{B}} ds \neq 0 \tag{18}$$

along a measurable set of magnetic field lines in the reconnection region [12].

3 Classification of Magnetic Reconnection Processes

Reconnection processes can be classified on the basis of different factors involved in them.

Here we shall mention only two among the classifications which can be found in the literature.

The first one is formulated on the basis of the relevant degree of the collisionality. In this respect, we may distinguish among [1]:

- *Collisional Reconnection* In this regime, dissipative effects are relevant, we may use resistive MHD theory and the relevant contribution in Ohm's law is given by the resistivity.
- *Weakly collisional Reconnection* This regime concerns weakly collisional plasmas. Some amount of collisions are still required (isotropic pressure). This regime can still be described by multifluid equations.

- *Collisionless reconnection* This regime, where collisions are absent, requires a kinetic description (Vlasov equation). Wave-particle resonance effects (e.g. Landau damping) are important in this case.

Another classification is based on the linear analysis of the plasma equations: indeed, in the presence of instabilities magnetic reconnection may occur in order to rearrange the configuration.

This classification is made on the basis of the location of the factor which originates the instability (*driving factor*).

With this respect we may distinguish between [4, 5].

- *Endogenous Magnetic Reconnection* occurring when the driving factor exists within the same region where the reconnection takes place.
- *Exogenous Magnetic Reconnection* occurring when the driving factor exists outside the region where the reconnection takes place.

We cannot characterise all classes in the previous classifications on the basis of the dominant terms in (11), but we may try to connect them to some of the classes above.

We analyse the left-hand side of (11) and its projection on the magnetic field **B**.

We assume that

$$\mathbf{R}_\alpha = \mathbf{R}_{\alpha T} + \mathbf{R}_{\alpha w}$$

where $\mathbf{R}_{\alpha T}$ represents the contribution in $\mathbf{R}_\alpha$ due to the temperature gradient $\nabla_x T_\alpha$, and $\mathbf{R}_{\alpha w}$ the contribution given by the velocity diffusion [2].

We observe first that

$$\mathbf{S}_\parallel = \frac{\mathbf{B}}{|\mathbf{B}|}\left\{\frac{4\pi c}{\sum_\alpha \omega_\alpha^2}\,\mathrm{div}_x\left(\mathbf{B}\sum_\alpha \frac{\Omega_\alpha(P_\alpha - \mathbf{R}_{\alpha T})}{|\mathbf{B}|}\right) - \frac{4\pi c}{|\mathbf{B}|}\Big(\sum_\alpha \frac{\Omega_\alpha \mathbf{B}\cdot\mathbf{R}_{\alpha w}}{\sum_\alpha \omega_\alpha^2}\Big)\right.$$
$$\left. + \frac{4\pi c}{\sum_\alpha \omega_\alpha^2}\mathbf{B}\cdot[\partial_t \mathbf{J} + \mathrm{div}(\mathbf{V}\otimes\mathbf{J} + \mathbf{J}\otimes\mathbf{V} + \sum_\alpha n_\alpha q_\alpha (\mathbf{v}_\alpha - \mathbf{V})^{\otimes 2})]\right\} \tag{19}$$

In (19) the Hall term, the first one on the right-hand side of (11), gives no contribution to $\mathbf{S}_\parallel$: this term, in general, preserves the topology. We recall though that the Hall term could affect the magnetic reconnection dynamics and this underlines the limits of the macroscopic definition given by (17), (18).

We are left then with three terms.

The first one represents the contribution to **F** given by the non-homogeneity and anisotropy in space of the species' temperatures and pressures.

The second term is the resistive contribution, generated by the friction between different species of particles in the plasma

The third term, finally, is associated to the species' inertia.

We could try to classify reconnection phenomena on the basis of the terms in (19), which play a dominant role in their generation: with this respect the second term should be associated to collisional reconnection, while the remaining two terms should be associated to weakly collisional and collisionless reconnection (although collisionless reconnection may not be completely described by (19)).

We recall once more that not all reconnection phenomena can be modelled by hydrodynamic equations: indeed this is one of the issues about the mathematical modelling of magnetic reconnection, and we shall describe in next section a model mixing different scales of description.

4 A Model for Magnetic Reconnection in Collisionless Plasmas

In this section we briefly describe a model adopted in [9], to illustrate magnetic reconnection in collisionless plasmas. Our main interest is in the procedure adopted to describe the phenomenon, so that we shall not go into details about the results.

In [9] the authors consider a 2-fluid ($\alpha = e, i$; $q_i = e$) quasi-neutral ($n_e \approx n_i = n$; $\mathbf{V} \approx v_i$) collisionless ($C_{\alpha\beta} = 0$) plasma.

Since reconnection may occur in the presence of an instability, as the process rearranging the configuration in a more stable one, the analysis is focussed on perturbations around a so called "plane one dimensional" equilibrium configuration.

Denoting by (x, y, z) the space coordinates, and $(\hat{\mathbf{x}}, \hat{\mathbf{y}}, \hat{\mathbf{z}})$ the corresponding unitary vectors, $x = \pm a$ are the edges of the plasma current layer and the equilibrium configuration, after a suitable choice of the coordinate system, is given (around $x = 0$) by:

$$\mathbf{B}_{eq} \approx B_0 \left(\hat{\mathbf{z}} + \frac{x}{L_s} \hat{\mathbf{y}} \right) \tag{20}$$

$$\mathbf{J}_{eq} \approx J_0 \left(1 - \frac{x}{L_J} \right) \hat{\mathbf{z}}. \tag{21}$$

The perturbations $\tilde{\mathbf{G}}$ of a function $\mathbf{G}$ with respect to its equilibrium value ($\mathbf{G} = \mathbf{G}_{eq} + \tilde{\mathbf{G}}$) are then chosen of the form

$$\tilde{\mathbf{G}}(t, x, y) = \bar{G}(x) e^{-i\omega t + iky} \tag{22}$$

and magnetic reconnection is then analysed by studying modes which, in the ideal MHD approximation, have frequency $\omega \ll 1$ and are singular at $x = 0$ (*marginally stable modes*). Observe that in this coordinate system, $\mathbf{k} = k\hat{\mathbf{y}}$ and $\mathbf{k} \cdot \tilde{\mathbf{B}}|_{x=0} = 0$.

The domain is then splitted into two parts, the *outermost zone* $\{\delta < |x| < a\}$ and the *reconnection regions* $\{|x| < \delta\}$.

In the outer zone, it is assumed that the ideal Ohm's (12) holds, while in the reconnection region $x < \delta$ we should have

$$\mathbf{E} + \frac{1}{c}\mathbf{V} \wedge \mathbf{B} \approx \frac{\mathbf{J} \wedge \mathbf{B}}{cen} + \frac{m_e c}{e^2 n}\left(\nabla_x \frac{\Omega_e P_e}{|\mathbf{B}|} + \mathrm{div}_x(\mathbf{V} \otimes \mathbf{J} + \mathbf{J} \otimes \mathbf{V} - \frac{\mathbf{J}^{\otimes 2}}{en}\right) \quad (23)$$

and, for the parallel component:

$$\mathbf{E}_{\|} \approx \frac{m_e c}{e^2 n}\hat{\mathbf{B}} \cdot \left(\nabla_x \frac{\Omega_e P_e}{|\mathbf{B}|} + \mathrm{div}_x(\mathbf{V} \otimes \mathbf{J} + \mathbf{J} \otimes \mathbf{V} - \frac{\mathbf{J}^{\otimes 2}}{en}\right)\hat{\mathbf{B}}. \quad (24)$$

The analysis will be performed considering $\delta \ll 1$ (and therefore, when studying the reconnection region, it can be assumed $a \to \infty$)

We analyse now in more detail what happens in the two zones.

4.1 The Outermost Region

When $\delta < x < a$, by summing over $\alpha = e, i$ equation (9), we obtain

$$\tilde{J}_z = \frac{i}{kB_y}\frac{dJ_z}{dx}\tilde{B}_x. \quad (25)$$

We notice that $\frac{dJ_z}{dx}$ is the factor which originates the instability, i.e. the driving factor. and since it is located outside the reconnection region, if reconnection occurs it will be an exogenous reconnection.

Then, from (2), the following PDE for $\tilde{B}_x$ is obtained:

$$\left(\partial_x^2 - k^2 - \frac{1}{B_y}\frac{d^2}{dx^2}B_y\right)\tilde{B}_x = 0, \quad (26)$$

to be solved under the conditions $\tilde{B}_x(|x| = a) = 0$, $\tilde{B}_x(x = 0) = \tilde{B}_{x0} \neq 0$ (this last in the sense of $\delta \ll a$, and $|x| \approx \delta$).

The relevant solution behaves then like

$$\tilde{B}_x = \tilde{B}_{x0}\left(1 + |x|\frac{\Delta}{2}\right), \quad (27)$$

where Δ is derived from the solution of Eq. (26).

4.2 The Reconnection Region: Inner Asymptotic Regions

When $|x| < \delta$, the problem is studied using the variable $\hat{x} = \frac{x}{\delta}$, with a small parameter $\epsilon = \frac{\delta}{a} \ll 1$.

It is assumed that

$$\tilde{J}_z = \frac{i}{kB_y}\frac{dJ_z}{dx}|_{x=0}\tilde{B}_{x0} + \Delta\tilde{J}_z \quad (28)$$

$$\tilde{B}_x = \tilde{B}_{x0}\big(1 + \epsilon\psi(\hat{x})\big), \quad (29)$$

with ψ a bounded function. $\Delta\tilde{J}_z$ is related to Δ by the equation curl $\mathbf{B} \approx \frac{4\pi}{c}\mathbf{J}$ (we notice that in this heuristic model, since $\tilde{B}_y|_x = 0$ and $\partial_x\tilde{B}_x \neq 0$, we have $\text{div}_x\,\tilde{\mathbf{B}} \neq 0$ in the reconnection region).

The problem is better analysed by introducing the electromagnetic potential $(\tilde{\phi}, \tilde{\mathbf{A}})$, so that $\tilde{\mathbf{B}} = \text{curl}\,\tilde{\mathbf{A}}$ and $\tilde{\mathbf{E}}_\parallel = -ik_\parallel\big(\tilde{\phi} - \frac{\omega}{k_\parallel c}\tilde{\mathbf{A}}_z\big)$ (here $k_\parallel = \frac{\mathbf{k}\cdot\mathbf{B}_{eq}}{|\mathbf{B}_{eq}|}$).

Indeed, equations for $(\tilde{\phi}, \tilde{\mathbf{A}})$ can be obtained by solving the Vlasov equations $(\alpha = e, i)$

$$\partial_t f_\alpha + \text{div}_x(\mathbf{v}f_\alpha) + \text{div}_v\left[\Big(q_\alpha\mathbf{E} + \frac{q_\alpha}{c}\mathbf{v}\wedge\mathbf{B}\Big)f_\alpha\right] = 0 \quad (30)$$

in $\frac{|x|}{\delta} < 1$, $\delta \to 0$ with the following procedure.

The equilibrium distribution functions $f_{eq,\alpha}$ is expressed as a function of the integrals of motion of the equilibrium orbits of electrons and ions (in a suitable coordinate system, the shear coordinate system, $(x, s_\perp, s_\parallel)$ perpendicular and parallel to $\mathbf{B}_{eq}$ [3]), and are of the form

$$f_{eq,\alpha} = \mathcal{M}[n_\alpha, T_\alpha, u_\alpha; m_\alpha](1 + \hat{f}(\Omega_\alpha, m_\alpha, n, T_\alpha, \nabla_x n, \nabla_x\,\text{T}) \quad (31)$$

where

$$\mathcal{M}[n_\alpha, T_\alpha, u_\alpha; m_\alpha] = \frac{n_\alpha}{(\frac{2\pi T_\alpha}{m_\alpha})^{\frac{3}{2}}}exp\left\{-\frac{m_\alpha}{2T_\alpha}[v_x^2 + v_\perp^2 + (v_\parallel - u_\alpha)^2]\right\}$$

and u_α are small drift velocities (with $u_i = 0$). The whole solution is therefore $f_\alpha = f_{eq,\alpha} + \tilde{f}_\alpha$.

The perturbed distribution functions can then be expressed as functions of the perturbed electromagnetic potential, and from the quasineutrality condition $\tilde{n}_e = \tilde{n}_i$ and the expression of the component of the perturbed current along $\mathbf{B}_{eq}$, $\tilde{J}_z = \tilde{J}_{z,e} + \tilde{J}_{z,i}$ we recover the equations governing $\tilde{\phi}$ and $\tilde{\mathbf{A}}$ in the reconnection region.

Denoting by

$$\omega_{*\alpha} = \frac{kcT_\alpha}{q_\alpha B_0}\frac{d\ln n}{dx} \tag{32}$$

$$v_\alpha = \sqrt{\frac{2T_\alpha}{m_\alpha}} \tag{33}$$

$$\eta_\alpha = \frac{d\ln T_\alpha}{d\ln n} \tag{34}$$

respectively the diamagnetic frequencies, the thermal velocities and the temperature gradient parameters of the species and by $\lambda_D = \sqrt{\frac{T_e}{4\pi ne^2}}$ and ρ_i the ion gyroradius, the equations governing $\tilde{\phi}$ and $\tilde{\mathbf{A}}$ in the reconnection region (and therefore describing $E_\parallel$) are [9]:

$$\left\{\left[1-\frac{\omega_{*\alpha}}{\omega}(1+\eta_\alpha)\right]\left[1+W(\frac{\omega}{|k_\parallel|v_\alpha})\right]+\frac{1}{2}\frac{\omega_{*\alpha}}{\omega}\eta_\alpha\frac{\omega}{k_\parallel v_\alpha}W'(\frac{\omega}{|k_\parallel|v_\alpha})\right\}\frac{1}{2}\rho_i^2\frac{d^2\tilde{\phi}}{dx^2}$$

$$= \left[\frac{T_i}{T_e}P_e + P_i\right]\frac{E_\parallel(\tilde{\phi},\tilde{A}_\parallel)}{ik_\parallel} \tag{35}$$

$$\Delta\tilde{J}_z = -\frac{c}{4\pi\lambda_D^2}\frac{\omega}{ck_\parallel}\frac{T_i}{T_e}\left[\frac{T_i}{T_e}P_e + P_i\right]\frac{E_\parallel(\tilde{\phi},\tilde{A}_\parallel)}{ik_\parallel} \tag{36}$$

where

$$W(\lambda) = \frac{1}{\sqrt{\pi}}\int_{-\infty}^{\infty}\frac{ve^{-v^2}}{\lambda - v}dv$$

and

$$P_\alpha = \left[1-\frac{\omega_{*\alpha}}{\omega}(1+\frac{1}{2}\eta_\alpha)\right]W(\frac{\omega}{|k_\parallel|v_\alpha})+\frac{1}{2}\frac{\omega_{*\alpha}}{\omega}\eta_\alpha\left[1+2\left(1-\frac{\omega^2}{k_\parallel^2 v_\alpha^2}\right)W(\frac{\omega}{|k_\parallel|v_\alpha})\right].$$

The analysis of these equations are then performed through asymptotic analytical considerations and numerical simulations, first by neglecting the temperature gradients contribution, then adding it (with similar considerations).

This analysis allows to distinguish four asymptotic regions, where the two species of particles contribute in different ways:

- The *Outer Region* defined by

$$\frac{\omega}{|k_\parallel|} < v_i < v_e \tag{37}$$

  here the disturbance is subsonic and $E_\parallel$ is vanishing.

- The *Intermediate region* defined by

$$v_i < \frac{\omega}{|k_{\parallel}|} < v_e \tag{38}$$

  where $E_{\parallel}$ is not negligible.
- The *Electron Region* defined by

$$v_e \lesssim \frac{\omega_{*e}}{|k_{\parallel}|} \tag{39}$$

- The *Ion Region* defined by

$$\frac{\omega}{|k_{\parallel}|} \sim v_i \tag{40}$$

  where $E_{\parallel}$ is not negligible.

The analysis in presence of temperature gradients is similar to the one in absence of temperature gradients: we refer to [9] for details of the analysis.

5 Conclusions

In this paper we tried to review some basic notions and definitions pertaining to the theory of magnetic reconnection, a phenomenon for which there is a limited number of clearly formulated theories that explain important experimental observations [7, 8]. In fact, a macroscopic level of description is adopted in a large part of the literature about the subject. Therefore we reviewed, in the last part of the paper, a linear model, formulated by one of the authors (B.C.) together with J.W-k.Mark, L.Sugiyama and G.Bertin, describing a collisionless plasma, where different asymptotic scales are dealth with.

Moreover, reconnection phenomena are mostly treated by now by linear theories: a complete theory should definitely include nonlinearity.

An acceptable mathematical formulation of magnetic reconnection has been produced in several important cases, where it has provided the correct interpretation of relevant experiments, but it is still missing, at the present time, for numerous phenomena where reconnection is involved.

Acknowledgements V. Ricci acknowledges the support of GNFM-INdAM (Gruppo Nazionale per la Fisica Matematica) and of FFR2021 Valeria Ricci, FFR2023, FFR2018-2020.
R. Spigler work was carried out within the framework of GNFM-INdAM.
This work has been performed in the framework of the IGNITOR program (MUR, Ministero dell'Università e della Ricerca).

References

1. Biskamp, D.: Magnetic Reconnection in Plasmas. Cambridge University Press (2000)
2. Braginskii, S.I.: Transport processes in a plasma. Rev. Plasma Phys. **1**, 205–311 (1965)
3. Coppi, B.: In: Donnelly, R.J., Herman, I., Prigogine, I. (eds.) Non Equilibrium Thermodynamics Variational Techniques and Stability, p. 259. University of Chicago Press, Chicago (1966)
4. Coppi, B., Basu, B.: Endogenous magnetic reconnection and associated high energy plasma processes. Phys. Lett. A **328**, 400–404 (2018)
5. Coppi, B., Basu, B.: Asymmetric and endogenous magnetic reconnection processes in quasi-collisionless plasmas. Phys. Plasmas **26**, 042115 (2019)
6. Coppi, B., Basu, B.: Magneto-thermal reconnection, Electron temperature gradient, Particle density gradient, Thermal energy and particle diffusion coefficients. Phys. Lett. A **397**, 127265 (2021)
7. Coppi, B., Galvao, R., Pellat, R., Rosenbluth, M., Rutherford, P.: Resistive internal kink mode. Fiz. Plasmi **2**, 961–966 (1976). Sov. J. Plasma Phys. **2**, 533–535 (1976)
8. Coppi, B., Laval, G., Pellat, R.: Dynamics of the geomagnetic tail. Phys. Rev. Lett. **16**(6), 1207–1210 (1966)
9. Coppi, B., Mark, J.W.-K., Sugiyama, L., Bertin, G.: Magnetic reconnection in collisionless plasmas. Ann. Phys. **119**, 370–404 (1979)
10. Dungey, J.W.: Some researches in cosmic magnetism. Ph.D. thesis, Cambridge University Press (1950)
11. Dungey, J.W.: Interplanetary magnetic field and the auroral zones. Phys. Rev. Lett. **6**, 47–48 (1961)
12. Priest, E., Forbes, T.: Magnetic Reconnection. Cambridge University Press (2000)

Uniqueness Criteria for the Vlasov–Poisson System and Applications to Semiclassical Analysis

Laurent Lafleche and Chiara Saffirio

Abstract We review some uniqueness criteria for the Vlasov–Poisson system, emerging as corollaries of stability estimates in strong or weak topologies, and show how they serve as a guideline to solve problems arising in semiclassical analysis. More specifically, we look at the Schatten norms and at the quantum analogue of the Wasserstein distance and prove the semiclassical limit from the Hartree equation with Coulomb interaction to the Vlasov–Poisson system in these topologies by mimicking weak-strong uniqueness principles for the Vlasov–Poisson equation in the respective classical topologies. Different topologies allow to treat different classes of quantum states.

Keywords Semiclassical limit · Hartree equation · Vlasov equation · Coulomb potential · Gravitational potential

1 Introduction

We consider the Vlasov–Poisson equation

$$\partial_t f + \xi \cdot \nabla_x f + E_f \cdot \nabla_\xi f = 0, \tag{1}$$

where $f = f(t, x, \xi)$ is a real-valued non-negative time-dependent function defined on the phase space $\mathbb{R}^6$, $E_f(t, x) = (\nabla K * \rho_f)(t, x)$ is a self-induced force field generated by the spatial density $\rho_f(t, x) = \int f(t, x, \xi)\, d\xi$. To simplify the presentation, in the rest of the paper we will focus on $d = 3$ and

L. Lafleche
UMPA, ENS de Lyon, allée d'Italie, Lyon 69364, France
e-mail: laurent.lafleche@ens-lyon.fr

C. Saffirio (✉)
Department of Mathematics and Computer Science, University of Basel, Spiegelgasse 1, 4051 Basel, Switzerland
e-mail: chiara.saffirio@unibas.ch

E. Carlen et al. (eds.), *From Particle Systems to Partial Differential Equations*, Springer Proceedings in Mathematics & Statistics 465,
https://doi.org/10.1007/978-3-031-65195-3_14

$$K(x) = \frac{\pm 1}{|x|}. \tag{2}$$

In this setting, the well-posedness theory for the Cauchy problem associated with Eq. (1) has been established by Pfaffelmoser [20] and Lions and Perthame [17], respectively proving global existence and uniqueness of classical solutions with compactly supported initial data, and existence of global weak solutions assuming that the initial data f^{in} are such that $f^{\mathrm{in}} \in L^1(\mathbb{R}^6) \cap L^\infty(\mathbb{R}^6)$ and $\int |\xi|^m f^{\mathrm{in}}\, \mathrm{d}x\, \mathrm{d}\xi$ is finite for some $m > 3$. A sufficient condition for uniqueness of measure-valued solutions with spatial density satisfying $\rho_f \in L^\infty([0,T], L^\infty(\mathbb{R}^3))$ was proven by Loeper in [18], and more recently generalized by Miot [19] including spatial densities such that $\sup_{p\geq 1} \frac{1}{p} \left\|\rho_f\right\|_{L^p}$ is finite, uniformly in time, by Holding and Miot [6] considering spatial densities in some Orlicz space, and further extended in [3].

The above mentioned uniqueness criteria deal with assumptions at positive time on the spatial density ρ_f. In this paper we collect three stability estimates for the Vlasov–Poisson equation leading to uniqueness criteria (non optimal in terms of conditions on ρ_f) in *weak-strong* sense. By weak-strong stability we indicate that some regularity conditions are assumed only on one of the two solutions involved in the stability estimate, thus allowing the other solution to satisfy only the minimal existence requirements. We will show how such a weak-strong form in our estimates (cf. Propositions 1 and 3) can be used to deduce results in the context of the semiclassical limit from the Hartree equation (Eq. (6) below) towards the Vlasov–Poisson system (1). We will also look at the weak topology induced by the 2-Wasserstein distance (cf. Sect. 4) and show how one can get a semiclassical estimate as an application of a stability result à la Loeper. The weak topology allows to treat a class of quantum states that were not included in the analysis with the weak-strong method and that are relevant at zero temperature (so-called pure states), however it does not provide a weak-strong estimate. More precisely, we will review the results in [2, 10, 14] and [10, 11], giving a unified picture of the methods.

1.1 Notations

For functions on the phase space of the form $f = f(x,\xi)$, we use the shorthand notation $L^p_x L^{q,r}_\xi = L^p(\mathbb{R}^3, L^{q,r}(\mathbb{R}^3))$, where $L^p(\mathbb{R}^3)$ and $L^{q,r}(\mathbb{R}^3)$ denote respectively the Lebesgue and Lorentz spaces. We also use the notation $\mathscr{P} = \mathscr{P}(\mathbb{R}^6)$ for the set of probability measures on $\mathbb{R}^6$. Let $\alpha \in \mathbb{N}^6$ be a multi-index with $|\alpha| = \sum_{i=1}^6 \alpha_i$, and $\partial^\alpha = \partial^{\alpha_2}_{x_1}\partial^{\alpha_2}_{\xi_1} \ldots \partial^{\alpha_5}_{x_3}\partial^{\alpha_6}_{\xi_3}$, then the weighted Sobolev norm in the Sobolev space $W^{k,p}_n$ can be defined by the formula

$$\|f\|_{W^{k,p}_n} = \Big(\sum_{|\alpha|\leq k} \left\| \langle\xi\rangle^n \, \partial^\alpha f \right\|^2_{L^p}\Big)^{\frac{1}{2}}$$

with $\langle z\rangle = \sqrt{1+|z|^2}$, with the usual associated notations $H^k_n = W^{k,2}_n$ and $L^p_n = W^{0,p}_n$.

Using the correspondence of classical and quantum mechanics and using the symbol $\{\cdot,\cdot\}$ and $[\cdot,\cdot]$ to denote respectively the Poisson brackets and the commutator, we recall that the quantum analogue of

$$\nabla_x f = \{-\xi, f\} \qquad \text{and} \qquad \nabla_\xi f = \{x, f\}$$

are respectively

$$\nabla_x \boldsymbol{\rho} := \big[\nabla, \boldsymbol{\rho}\big] \qquad \text{and} \qquad \nabla_\xi \boldsymbol{\rho} := \Big[\frac{x}{i\hbar}, \boldsymbol{\rho}\Big],$$

where $\hbar = h/(2\pi)$ is the reduced Plank constant.

The quantum analogue of Lebesgue norms and weighted Sobolev norms of $\boldsymbol{\rho}$, a bounded operator acting on $L^2(\mathbb{R}^3)$, are the quantum Lebesgue and Sobolev norms

$$\begin{aligned}\|\boldsymbol{\rho}\|_{\mathcal{L}^p} &= h^{\frac{3}{p}}\,\mathrm{Tr}\big(|\boldsymbol{\rho}|^p\big)^{\frac{1}{p}}\\ \|\boldsymbol{\rho}\|_{\mathcal{W}^{k,p}_n} &= \|\mathsf{m}\boldsymbol{\rho}\|_{\mathcal{L}^p} + \|\mathsf{m}\nabla_x\boldsymbol{\rho}\|_{\mathcal{L}^p} + \big\|\mathsf{m}\nabla_\xi\boldsymbol{\rho}\big\|_{\mathcal{L}^p}\end{aligned}$$

with $|\boldsymbol{\rho}| = \sqrt{\boldsymbol{\rho}^*\boldsymbol{\rho}}$ and $\mathsf{m} := 1 + |\boldsymbol{p}|^n$, with $\boldsymbol{p} = -i\hbar\nabla$. When $p = \infty$, $\|\boldsymbol{\rho}\|_{\mathcal{L}^\infty}$ denotes the operator norm.

To connect functions on the phase space and operators acting on $L^2(\mathbb{R}^3)$, we define the Weyl quantization as follows: to every $f \in L^1(\mathbb{R}^6)$, the Weyl quantization of f is the operator $\boldsymbol{\rho}^W_\hbar(f)$ with kernel

$$\boldsymbol{\rho}^W_\hbar(f)\,(x,y) = \int_{\mathbb{R}^3} e^{-2i\pi(y-x)\cdot\xi}\, f\Big(\big(\frac{x+y}{2}, h\xi\Big)\,\mathrm{d}\xi.$$

Observe that, if f is a probability density on the phase space (hence nonnegative), its Weyl quantization is not necessarily a non-negative operator. For this reason it is convenient to introduce an alternative quantization of the function f, called the Töplitz quantization, or the Wick quantization, or the Husimi quantization, given by

$$\widetilde{\boldsymbol{\rho}}^W_\hbar(f) = \frac{1}{h^3}\int_{\mathbb{R}^6} f(z)\,|\psi_z\rangle\,\langle\psi_z|\,\mathrm{d}z,$$

where

$$\psi_z(y) := (\pi\hbar)^{-3/4}\, e^{-|y-x|^2/(2\hbar)}\, e^{iy\cdot\xi/\hbar}.$$

It follows from this formula that $\widetilde{\boldsymbol{\rho}}^W_\hbar(f)$ is a positive operator whenever $f \geq 0$.

One can also define the inverse transformation of the Weyl quantization. Namely, let $\boldsymbol{\rho}$ be an operator with a regular kernel $\boldsymbol{\rho}(x,y)$, then

$$f_{\rho}(x,\xi) = \int_{\mathbb{R}^3} e^{-2\pi i\,\xi\cdot y/h}\,\boldsymbol{\rho}\left(x+\frac{y}{2}, x-\frac{y}{2}\right)\mathrm{d}y$$

is the Wigner transform of the operator $\boldsymbol{\rho}$.

If not otherwise stated, we will denote by C a positive constant independent of $\hbar$ and will use that, for $a, b \geq 0$ $a \lesssim b$ if and only if there exists $C > 0$ such that $a \leq C\,b$.

2 L^1 Weak-Strong Uniqueness Criterion

In this section we present a L^1 weak-strong stability estimate for the Vlasov–Poisson equation, where regularity is required only on one of the two solutions of (1) and show an application of this criterion to the semiclassical limit from the Hartree equation to the Vlasov–Poisson system.

2.1 L^1 Weak-Strong Stability Criterion for the Vlasov–Poisson Equation

Proposition 1 *Let f_1 and f_2 be two positive solutions of the Vlasov–Poisson equation* (1) *verifying $f_1,\ f_2 \in L^\infty([0,T], L^1(\mathbb{R}^6))$, for some $T > 0$. Then, if*

$$\lambda_{f_2}(t) := \left\|\nabla_\xi f_2\right\|_{L^{3,1}_x L^1_\xi} \tag{3}$$

is finite for every $t \in [0,T]$, the following stability estimate holds

$$\|f_1 - f_2\|_{L^1(\mathbb{R}^6)} \leq \left\|f_1^{\mathrm{in}} - f_2^{\mathrm{in}}\right\|_{L^1(\mathbb{R}^6)} e^{\Lambda_{f_2}(t)},$$

where $\Lambda_{f_2}(t) = C\int_0^t \lambda_{f_2}(s)\,ds$ with $C = \|\nabla K\|_{L^{\frac{3}{2},\infty}}$.

Remark 1 Notice that the regularity condition (3) on f_2 reads

$$\int_{\mathbb{R}^3} \left|\nabla_\xi f_2\right| \mathrm{d}\xi \in L^{3,1}_x.$$

In particular, for each $t \in [0,T]$ and for some $\varepsilon \in (0,2]$, it holds by real interpolation

$$\left\|\int_{\mathbb{R}^3} \left|\nabla_\xi f_2\right| \mathrm{d}\xi\right\|_{L^{3,1}_x} \leq \left\|\int_{\mathbb{R}^3} \left|\nabla_\xi f_2\right| \mathrm{d}\xi\right\|_{L^{3+\varepsilon}_x} + \left\|\int_{\mathbb{R}^3} \left|\nabla_\xi f_2\right| \mathrm{d}\xi\right\|_{L^{3-\varepsilon}_x}. \tag{4}$$

Indeed it is known that solutions of Vlasov–Poisson are regular (see e.g. [7, 17, 20, 21]). This is one of the advantages of the weak-strong method, that requires regularity only on one of the solutions.

Proof (*Proof of Proposition* 1) Let $f := f_1 - f_2$ and define for $k = 1, 2$ $\rho_k = \int_{\mathbb{R}^3} f_k \,\mathrm{d}\xi$ $\rho := \rho_1 - \rho_2$ and $E_k = -\nabla K * \rho_k$. Then it holds

$$\partial_t f + \xi \cdot \nabla_x f + E_1 \cdot \nabla_\xi f = (E_2 - E_1) \cdot \nabla_\xi f_2,$$

whence

$$\begin{aligned}\partial_t \|f\|_{L^1} &= \iint_{\mathbb{R}^6} \left(\nabla K * \rho \cdot \nabla_\xi f_2\right) \operatorname{sign}(f) \,\mathrm{d}x\,\mathrm{d}\xi \\ &= -\int_{\mathbb{R}^3} \rho \,\nabla K \,\dot{*} \left(\int_{\mathbb{R}^3} \operatorname{sign}(f)\, \nabla_\xi f_2 \,\mathrm{d}\xi\right) \mathrm{d}x\end{aligned}$$

where the notation $\dot{*}$ indicates that we perform the scalar product of vectors inside the convolution. This implies

$$\partial_t \|f\|_{L^1} \leq \|f\|_{L^1} \left\| \nabla K * \int_{\mathbb{R}^3} \left|\nabla_\xi f_2\right| \mathrm{d}\xi \right\|_{L^\infty}. \tag{5}$$

By Hölder's inequality for Lorentz spaces (cf. [8, Formula (2.7)]), since $\nabla K \in L^{\frac{3}{2},\infty}$, we have

$$\left\| \nabla K * \left\|\nabla_\xi f_2\right\|_{L^1_\xi} \right\|_{L^\infty} \leq \sup_{z\in\mathbb{R}^3} \int_{\mathbb{R}^3} \left|\nabla K(z-x) \left\|\nabla_\xi f_2\right\|_{L^1_\xi}\right| \mathrm{d}x \leq C \left\|\nabla_\xi f_2\right\|_{L^{3,1}_x L^1_\xi}$$

and Grönwall's Lemma concludes the proof.

2.2 Applications to the Limit from Hartree to Vlasov–Poisson: Convergence in Trace Norm

In this section we consider the time-dependent Hartree equation

$$i\,\hbar\,\partial_t \boldsymbol{\rho} = \left[H_{\boldsymbol{\rho}}, \boldsymbol{\rho}\right] \tag{6}$$

describing the time evolution of $\boldsymbol{\rho} = \boldsymbol{\rho}(t)$, the nonnegative and bounded density operator on $L^2(\mathbb{R}^3)$ such that $\|\boldsymbol{\rho}\|_{\mathcal{L}^1} = 1$. The Hamiltonian is given by the sum of the Laplace operator and the multiplication operator $V_{\boldsymbol{\rho}} = K * \operatorname{diag}(\boldsymbol{\rho})$

$$H_{\boldsymbol{\rho}} = -\frac{\hbar^2}{2}\Delta + V_{\boldsymbol{\rho}}, \tag{7}$$

where $\hbar = h/(2\pi)$ is the reduced Planck constant and

$$\operatorname{diag}(\boldsymbol{\rho})\,(x) := h^3 \boldsymbol{\rho}(x,x).$$

We recall the following result from [14].

Theorem 1 ([14, Theorem 1.1]) *Let f be a nonnegative solution to the Vlasov–Poisson equation* (1) *and $\boldsymbol{\rho}$ a solution to the Hartree equation* (6) *with initial data*

$$f^{\text{in}} \in W^{\sigma+1,\infty}_m(\mathbb{R}^6) \cap H^{\sigma+1}_\sigma(\mathbb{R}^6)$$
$$\boldsymbol{\rho}^{\text{in}} \in \mathcal{L}^1$$

with $m > 3$ and $\sigma > m + 6$. Then there exist two nonnegative real-valued continuous functions $\Lambda_f(t)$ and $C_f(t)$ independent of $\hbar$ such that

$$\left\|\boldsymbol{\rho} - \boldsymbol{\rho}^W_\hbar(f)\right\|_{\mathcal{L}^1} \le \left(\left\|\boldsymbol{\rho}^{\text{in}} - \boldsymbol{\rho}^W_\hbar(f^{\text{in}})\right\|_{\mathcal{L}^1} + C_f(t)\,\hbar\right) e^{\Lambda_f(t)}.$$

Remark 2 Recalling that for $(y, w) \in \mathbb{R}^6$

$$\mathcal{F}[f_{\boldsymbol{\rho}}](y,w) = h^3 \mathrm{Tr}\!\left(e^{-2i\pi\,(y\cdot x + w\cdot \boldsymbol{p})}\boldsymbol{\rho}\right)$$

where x and $\boldsymbol{p}$ are the position and momentum operator respectively and $\mathcal{F}[g]$ denotes the Fourier transform of the function g, we obtain

$$\left\|\mathcal{F}[f_{\boldsymbol{\rho}}] - \mathcal{F}[f]\right\|_{L^\infty} \le \left\|\boldsymbol{\rho} - \boldsymbol{\rho}^W_\hbar(f)\right\|_{\mathcal{L}^1}.$$

As a corollary of Theorem 1, this gives convergence in L^∞ for the Fourier transform of the difference between the Wigner transforms of the solution to the Hartree equation and the solution to the Vlasov–Poisson equation in the semiclassical limit, if the initial data are close for $\hbar \to 0$, providing an explicit rate of convergence, that is supposed to be optimal in $\hbar$. This is a generalization in a stronger topology with explicit rate of the result obtained by Lions and Paul in [16].

The rest of this section is devoted to highlighting the analogy between the proof of Proposition 1 and the proof of Theorem 1, aiming at conveying the idea that, given a weak-strong uniqueness stability estimate, there is a quite general method leading to the proof of the semiclassical limit in the correspondent topology for operators.

2.2.1 Proof of Theorem 1 Part I: Reduction to Stability Estimates

Let $\boldsymbol{\rho}^W_\hbar(f)$ be the Weyl quantization of f and $U_{t,s}\, t > s$, be the two parameter semigroup generated by the Hartree Hamiltonian (6), i.e.

$$i\,\hbar\,\partial_t U_{t,s} = H_\rho(t)\,U_{t,s}, \qquad U_{s,s} = 1.$$

By conjugating the difference of $\boldsymbol{\rho}$ and $\boldsymbol{\rho}_\hbar^W(f)$ by $U_{t,s}$ and performing the time derivative we obtain

$$i\,\partial_t U_{t,s}^*(\boldsymbol{\rho} - \boldsymbol{\rho}_\hbar^W(f))\,U_{t,s} = U_{t,s}^*\left[K * (\rho - \rho_f), \boldsymbol{\rho}_\hbar^W(f)\right]\,U_{t,s} + U_{t,s}^*\,B\,U_{t,s}$$

with B_t the time-dependent operator with integral kernel

$$B(x,y) = \left((K * \rho_f)(x) - (K * \rho_f)(y) - (\nabla K * \rho_f)\left(\frac{x+y}{2}\right)\cdot(x-y)\right)\boldsymbol{\rho}_\hbar^W(x,y) \tag{8}$$

and $\rho_f(x) = \mathrm{diag}(\boldsymbol{\rho}_\hbar^W(f))(x)$. By Duhamel's formula and using the unitarity of $U_{t,s}$ we get

$$\begin{aligned}\left\|\boldsymbol{\rho} - \boldsymbol{\rho}_\hbar^W(f)\right\|_{\mathcal{L}^1} &\leq \left\|\boldsymbol{\rho}^{\mathrm{in}} - \boldsymbol{\rho}_\hbar^W(f^{\mathrm{in}})\right\|_{\mathcal{L}^1} + \frac{1}{\hbar}\int_0^t \|B_s\|_{\mathcal{L}^1}\,\mathrm{d}s \\ &\quad + \frac{1}{\hbar}\int_0^t\int_{\mathbb{R}^3}\left|\rho(x) - \rho_f(x)\right|\left\|\left[K(\cdot - x), \boldsymbol{\rho}_\hbar^W(f)\right]\right\|_{\mathcal{L}^1}\mathrm{d}x\,\mathrm{d}s.\end{aligned} \tag{9}$$

The structure of Eq. (9) is clear in light of the proof of Proposition 1: the second line in Eq. (9) is the quantum analogue of Inequality (5) and has to be treated in the same spirit of a weak-strong stability estimate, whereas the second term in the right-hand side of the first line represents the error due to the fact that we are considering two different dynamics, namely the Hartree and the Vlasov ones. The error term is easily bounded in [14, Proposition 4.4] by

$$\|B\|_{\mathcal{L}^1} \leq C\,\hbar^2\left\|\rho_f\right\|_{L^1\cap H^{\frac{3}{2}}}\left\|\nabla_\xi^2 f\right\|_{H^{2n}_{2n}}, \qquad n > \frac{3}{2}.$$

2.2.2 Proof of Theorem 1 Part II: The Commutator Inequality

To bound the second line in Eq. (9), we proceed as in the proof of Proposition 1, taking the L^∞ norm in the variable z in the commutator term. This yields

$$\begin{aligned}\int_0^t\int_{\mathbb{R}^3}&\left|\rho(x) - \rho_f(x)\right|\left\|\left[K(\cdot - x), \boldsymbol{\rho}_\hbar^W(f)\right]\right\|_{\mathcal{L}^1}\mathrm{d}x\,\mathrm{d}s \\ &\leq \sup_{x\in\mathbb{R}^3}\left\|\left[K(\cdot - x), \boldsymbol{\rho}_\hbar^W(f)\right]\right\|_{\mathcal{L}^1}\left\|\boldsymbol{\rho} - \boldsymbol{\rho}_\hbar^W(f)\right\|_{\mathcal{L}^1}.\end{aligned} \tag{10}$$

As is the proof of Proposition 1, we need to bound the commutator $\left[K(\cdot - x), \boldsymbol{\rho}_\hbar^W(f)\right]$ in $\mathcal{L}^1$, uniformly in x and in $\hbar$. This is achieved in [14, Theorem 4.1], that we report here for completeness.

Proposition 2 ([14, Theorem 4.1]) *Let $\boldsymbol{\rho}$ be a self-adjoint bounded operator and $K(x) = \frac{1}{|x|}$, then for any $\varepsilon \in (0, 2]$, there exists $C > 0$ such that*

$$\left\|\left[K(\cdot - x), \boldsymbol{\rho}\right]\right\|_{\mathcal{L}^1} \leq C\,\hbar\,\left(\left\|\mathrm{diag}(|\nabla_\xi \boldsymbol{\rho}|)\right\|_{L^{3-\varepsilon}} + \left\|\mathrm{diag}(|\nabla_\xi \boldsymbol{\rho}|)\right\|_{L^{3+\varepsilon}}\right).$$

Applying Proposition 2 to bound the right-hand side of Inequality (10), we are left with bounding $\mathrm{diag}(|\nabla_\xi \boldsymbol{\rho}_\hbar^W(f)\,|)$ in L^p, for $p \in \{3-\varepsilon, 3+\varepsilon\}$. This is the quantum analogue of Inequality (4). Mimicking the classical interpolation inequality

$$\left\|\nabla_\xi f\right\|_{L^p_x(L^1_\xi)} \leq \left\|\nabla_\xi f\right\|_{L^\infty_{x,\xi}}^{\frac{1}{p'}} \left(\iint_{\mathbb{R}^6} |\nabla_\xi f|\,|\xi|^{3(p-1)}\,\mathrm{d}x\,\mathrm{d}\xi\right)^{\frac{1}{p}},$$

the boundedness of the operator $\mathrm{diag}(|\nabla_\xi \boldsymbol{\rho}_\hbar^W(f)\,|)$ and the propagation of quantum moments $h^3\mathrm{Tr}(|\nabla_\xi \boldsymbol{\rho}_\hbar^W(f)\,|\,|\boldsymbol{p}^n|)$, for some integer n, yield

$$\left\|\mathrm{diag}\left(\left|\nabla_\xi \boldsymbol{\rho}_\hbar^W(f)\right|\right)\right\|_{L^p} \leq C\,\left\|\nabla_\xi \boldsymbol{\rho}_\hbar^W(f)\right\|_{\mathcal{L}^\infty}^{\frac{1}{p'}} \left(h^3\mathrm{Tr}\left(\left|\nabla_\xi \boldsymbol{\rho}_\hbar^W(f)\right|\,|\boldsymbol{p}|^{3(p-1)}\right)\right)^{\frac{1}{p}}. \quad (11)$$

By the Boulkhemair generalization of the Calderón–Vaillancourt theorem for the Weyl operators [1], it holds

$$\left\|\nabla_\xi \boldsymbol{\rho}_\hbar^W(f)\right\|_{\mathcal{L}^\infty} \leq C\,\left\|\nabla_\xi f\right\|_{W^{4,\infty}}$$

uniformly in $\hbar$, for some $C > 0$. Furthermore, by [14, Proposition 3.3], the quantum velocity moments are bounded by

$$h^3\mathrm{Tr}\left(\left|\nabla_\xi \boldsymbol{\rho}_\hbar^W(f)\right|\,|\boldsymbol{p}|^{3(p-1)}\right) \leq C\,\left\|\nabla_\xi f\right\|_{H^\sigma_\sigma}.$$

Grönwall's Lemma concludes the proof.

3 L^2 Weak-Strong Uniqueness Criterion

In this section we review the proof of [2] to obtain an optimal bound in L^2 for the semiclassical limit from the Hartree to the Vlasov–Poisson by mimicking the proof of a weak-strong stability criterion for solutions to the Vlasov–Poisson system in L^2.

3.1 L^2 Weak-Strong Stability Criterion for the Vlasov–Poisson Equation

Proposition 3 *Let f_1 and f_2 be two positive solutions of the Vlasov–Poisson equation verifying $\|f_1\|_{L^\infty}, \|f_2\|_{L^\infty} \leq C_\infty$. If*

$$\lambda_{f_2}(t) = \left\|\rho_{f_2}\right\|_{L^\infty_x}^{\frac{1}{2}} \left\|\nabla_\xi \sqrt{f_2}\right\|_{L^3_x L^2_\xi} + C_\infty^{\frac{1}{2}} \left\|\nabla_\xi \sqrt{f_2}\right\|_{L^{3,1}_x L^1_\xi}$$

is finite for each $t \in [0, T]$, the following stability estimate holds

$$\left\| \sqrt{f_1} - \sqrt{f_2} \right\|_{L^2} \leq \left\| \sqrt{f_1^{\text{in}}} - \sqrt{f_2^{\text{in}}} \right\|_{L^2} e^{\Lambda_{f_2}(t)}$$

where $\Lambda_{f_2}(t) = C \int_0^t \lambda_{f_2}(s)\, ds$, for some $C > 0$. In particular,

$$\| f_1 - f_2 \|_{L^2} \leq 2\, C_\infty^{\frac{1}{2}} \left\| f_1^{\text{in}} - f_2^{\text{in}} \right\|_{L^1}^{\frac{1}{2}} e^{\Lambda_{f_2}(t)}.$$

Proof Let $v_1 = \sqrt{f_1}\, v_2 = \sqrt{f_2}$ and $v := v_1 - v_2$. Then v satisfies the equation

$$\left(\partial_t + \xi \cdot \nabla_x + E_{v_1^2} \cdot \nabla_\xi\right) v = \left(E_{v_2^2} - E_{v_1^2}\right) \cdot \nabla_\xi v_2 = \left(\nabla K * \rho_{(v_2+v_1)v}\right) \cdot \nabla_\xi v_2,$$

where we use the notations $\rho_g = \int g\, d\xi$ and $E_g = \nabla K * \rho_g$, for any function g. The fact that the vector field $(\xi, E_{v_1^2})$ is divergence-free yields

$$\begin{aligned}
\frac{1}{2} \frac{\mathrm{d}}{\mathrm{d}t} \|v\|_{L^2}^2 &= \iint_{\mathbb{R}^6} v \left(\nabla K * \rho_{(v_2+v_1)v}\right) \cdot \nabla_\xi v_2 \, \mathrm{d}x\, \mathrm{d}\xi \\
&= \iint_{\mathbb{R}^{12}} \left(\left|v(z')\right|^2 + 2\, v_2(z')\, v(z') \right) \nabla K(x - x') \cdot \left(v(z)\, \nabla_\xi v_2(z)\right) \mathrm{d}z\, \mathrm{d}z' \\
&=: I_1 + 2\, I_2.
\end{aligned}$$

We bound the first term by Hölder's inequality

$$I_1 \leq \left\| \nabla K * \int_{\mathbb{R}^3} v\, \nabla_\xi v_2 \, \mathrm{d}\xi \right\|_{L_x^\infty} \|v\|_{L^2}^2 \leq \left\| \, |\nabla K| * \int_{\mathbb{R}^3} \left|\nabla_\xi v_2\right| \mathrm{d}\xi \right\|_{L_x^\infty} \|v\|_{L^\infty} \|v\|_{L^2}^2,$$

and by applying Hölder's inequality for the Lorentz spaces we get

$$I_1 \leq \|\nabla K\|_{L_x^{\frac{3}{2},\infty}} \left\|\nabla_\xi v_2\right\|_{L_x^{3,1} L_\xi^1} \|v\|_{L^\infty} \|v\|_{L^2}^2 .$$

To bound I_2 we use the Hardy–Littlewood–Sobolev inequality and the Cauchy–Schwarz inequality, to obtain the bound

$$\begin{aligned}
I_2 &\leq C \left\| \int_{\mathbb{R}^3} v\, v_2 \, \mathrm{d}\xi \right\|_{L_x^2} \left\| \int_{\mathbb{R}^3} v\, \nabla_\xi v_2 \, \mathrm{d}\xi \right\|_{L_x^{\frac{6}{5}}} \\
&\leq C \left\| \|v\|_{L_\xi^2} \|v_2\|_{L_\xi^2} \right\|_{L_x^2} \left\| \|v\|_{L_\xi^2} \left\|\nabla_\xi v_2\right\|_{L_\xi^2} \right\|_{L_x^{\frac{6}{5}}} .
\end{aligned}$$

By Hölder's inequality, we finally obtain

$$I_2 \leq C \, \|v_2\|_{L_x^\infty L_\xi^2} \left\|\nabla_\xi v_2\right\|_{L_x^3 L_\xi^2} \|v\|_{L^2}^2 = C \left\|\rho_{f_2}\right\|_{L_x^\infty}^{\frac{1}{2}} \left\|\nabla_\xi v_2\right\|_{L_x^3 L_\xi^2} \|v\|_{L^2}^2 ,$$

that combined with the bound on I_1 leads to

$$\frac{\mathrm{d}}{\mathrm{d}t}\|v\|_{L^2}^2 \leq C\left(\left\|\rho_{f_2}\right\|_{L_x^\infty}^{\frac{1}{2}}\left\|\nabla_\xi v_2\right\|_{L_x^3 L_\xi^2}+\|v\|_{L^\infty}\left\|\nabla_\xi v_2\right\|_{L_x^{3,1} L_\xi^1}\right)\|v\|_{L^2}^2 .$$

Using that $\|v\|_{L^\infty} \leq \|f_1\|_{L^\infty}^{\frac{1}{2}} + \|f_2\|_{L^\infty}^{\frac{1}{2}} = 2\,C_\infty^{\frac{1}{2}}$, Grönwall's lemma concludes the proof.

3.2 *Application to the Limit from Hartree to Vlasov–Poisson: Optimal L^2Bound*

Theorem 2 ([2, Theorem 1.2]) *Let $\boldsymbol{\rho} \geq 0$ be a solution to the Hartree equation* (6) *with initial condition $\boldsymbol{\rho}^{\text{in}} \in \mathcal{L}^1 \cap \mathcal{L}^\infty$ and $f \geq 0$ $f \in L^1(\mathbb{R}^6)$ be a solution to the Vlasov–Poisson equation* (1) *with initial condition f^{in} such that*

$$f^{\text{in}},\ \sqrt{f^{\text{in}}} \in W_4^{4,\infty} \cap H_4^4 \quad \text{and} \quad \iint_{\mathbb{R}^6} f^{\text{in}}\,|\xi|^{n_1}\,dx\,d\xi < \infty \tag{12}$$

for $n_1 > 6$. Then there exist time-dependent functions $\Lambda, C_1, C_2 \in C^0(\mathbb{R}_+, \mathbb{R}_+)$, independent of $\hbar$ and depending on the initial conditions of Eqs. (6) *and* (1), *such that*

$$\left\|\boldsymbol{\rho} - \boldsymbol{\rho}_\hbar^W(f)\right\|_{\mathcal{L}^2} \leq C_\infty^{\frac{1}{2}}\left(\left\|\widetilde{\boldsymbol{\rho}}_\hbar^W\!\left(\sqrt{f^{\text{in}}}\right) - \sqrt{\boldsymbol{\rho}^{\text{in}}}\right\|_{\mathcal{L}^2} + C_1(t)\,\hbar\right) e^{\Lambda(t)} + C_2(t)\,\hbar. \tag{13}$$

3.2.1 Proof of Theorem 2 Part I: Auxiliary Dynamics

We consider the Weyl quantization of (1) and rewrite it as

$$\begin{aligned} i\,\hbar\,\partial_t \boldsymbol{\rho}_\hbar^W(f) &= \left[H_f, \boldsymbol{\rho}_\hbar^W(f)\right] - B_f(\boldsymbol{\rho}_\hbar^W(f)) \\ \boldsymbol{\rho}_\hbar^W(f)\,(0) &= \boldsymbol{\rho}_\hbar^W(f^{\text{in}}), \end{aligned}$$

with $H_f = -\frac{\hbar^2}{2}\Delta + V_f$, being V_f the multiplication operator $V_f(x) = (K * \rho_f)(x)$ and B the operator with kernel (8). To mimic the L^2 weak-strong uniqueness criterion given in Proposition 3, we want to consider square roots of operators, that are well-defined only for nonnegative operators. For this reason we introduce the Wick quantization of $\sqrt{f^{\text{in}}}$ (also called Töplitz operator)

$$\widetilde{\boldsymbol{\rho}}_\hbar^W\!\left(\sqrt{f^{\text{in}}}\right) = \boldsymbol{\rho}_\hbar^W\!\left(g_\hbar * \sqrt{f^{\text{in}}}\right)$$

where $g_\hbar(z) := (\pi\,\hbar)^{-3} e^{-|z|^2/\hbar}$ $z \in \mathbb{R}^6$, and consider its time-evolution $\widetilde{\boldsymbol{\rho}}$ solving the linear Hartree-type equation

$$\begin{cases} i\hbar\partial_t\widetilde{\boldsymbol{\rho}} &= \left[H_f, \widetilde{\boldsymbol{\rho}}\right], \\ \widetilde{\boldsymbol{\rho}}^{\text{in}} &:= \widetilde{\boldsymbol{\rho}}_\hbar^W\left(\sqrt{f^{\text{in}}}\right)^2. \end{cases} \tag{14}$$

We therefore bound the difference of $\boldsymbol{\rho}$ and $\boldsymbol{\rho}_\hbar^W(f)$ using the auxiliary dynamics (14) as follows

$$\left\|\boldsymbol{\rho} - \boldsymbol{\rho}_\hbar^W(f)\right\|_{\mathcal{L}^2} \leq \|\boldsymbol{\rho} - \widetilde{\boldsymbol{\rho}}\|_{\mathcal{L}^2} + \left\|\widetilde{\boldsymbol{\rho}} - \boldsymbol{\rho}_\hbar^W(f)\right\|_{\mathcal{L}^2}. \tag{15}$$

The second term on the right-hand side of Inequality (15) measures the error in considering the evolution of the Töplitz quantization instead of the Weyl quantization. Grönwall's Lemma yields

$$\left\|\widetilde{\boldsymbol{\rho}} - \boldsymbol{\rho}_\hbar^W(f)\right\|_{\mathcal{L}^2} \leq \left\|\widetilde{\boldsymbol{\rho}}^{\text{in}} - \boldsymbol{\rho}_\hbar^W(f^{\text{in}})\right\|_{\mathcal{L}^2} + C\hbar\int_0^t \|\nabla E_f\|_{L_x^\infty} \left\|\nabla_\xi^2 f\right\|_{L^2} \mathrm{d}s. \tag{16}$$

The second term on the right-hand side of (16) is bounded under the assumptions (12) on f^{in}. This follows indeed by the propagation of Sobolev regularity for solutions to the Vlasov–Poisson equation (1) (cf. [14]). As for the first term on the right-hand side of Inequality (16), we use that

$$\begin{aligned} \left\|\widetilde{\boldsymbol{\rho}}^{\text{in}} - \boldsymbol{\rho}_\hbar^W(f^{\text{in}})\right\|_{\mathcal{L}^2} &= \left\|\widetilde{\boldsymbol{\rho}}_\hbar^W\left(\sqrt{f^{\text{in}}}\right)^2 - \boldsymbol{\rho}_\hbar^W(f^{\text{in}})\right\|_{\mathcal{L}^2} \\ &\leq \left\|\widetilde{\boldsymbol{\rho}}_\hbar^W\left(\sqrt{f^{\text{in}}}\right)^2 - \widetilde{\boldsymbol{\rho}}_\hbar^W(f^{\text{in}})\right\|_{\mathcal{L}^2} + \left\|\widetilde{\boldsymbol{\rho}}_\hbar^W(f^{\text{in}}) - \boldsymbol{\rho}_\hbar^W(f^{\text{in}})\right\|_{\mathcal{L}^2}. \end{aligned}$$

By standard properties of the Töplitz quantization (cf. for instance [2, 16])

$$\left\|\widetilde{\boldsymbol{\rho}}_\hbar^W(f^{\text{in}}) - \boldsymbol{\rho}_\hbar^W(f^{\text{in}})\right\|_{\mathcal{L}^2} \lesssim \hbar \left\|\nabla^2 f^{\text{in}}\right\|_{\mathcal{L}^2},$$

whereas $\left\|\widetilde{\boldsymbol{\rho}}_\hbar^W\left(\sqrt{f^{\text{in}}}\right)^2 - \widetilde{\boldsymbol{\rho}}_\hbar^W(f^{\text{in}})\right\|_{\mathcal{L}^2}$ can be bounded by interpolation using that (cf. [2, Lemma 3.1] for a detailed proof)

$$\left\|\widetilde{\boldsymbol{\rho}}_\hbar^W\left(\sqrt{f^{\text{in}}}\right)^2 - \widetilde{\boldsymbol{\rho}}_\hbar^W(f^{\text{in}})\right\|_{\mathcal{L}^1} \lesssim \hbar \left\|\nabla f^{\text{in}}\right\|_{L^2}^2$$

and

$$\left\|\widetilde{\boldsymbol{\rho}}_\hbar^W\left(\sqrt{f^{\text{in}}}\right)^2 - \widetilde{\boldsymbol{\rho}}_\hbar^W(f^{\text{in}})\right\|_{\mathcal{L}^\infty} \lesssim \hbar \left\|\nabla f^{\text{in}}\right\|_{L^\infty}^2.$$

3.2.2 Proof of Theorem 3.1 Part II: L^2 Approach

We now focus on the first term on the right-hand side of Inequality (15), where we aim at reproducing for operators the strategy adopted in Proposition 3. We have

$$\|\rho - \widetilde{\rho}\|_{\mathcal{L}^2} \leq \left\|\sqrt{\rho} - \sqrt{\widetilde{\rho}}\right\|_{\mathcal{L}^2} \left(\left\|\sqrt{\rho}\right\|_{\mathcal{L}^\infty} + \left\|\sqrt{\widetilde{\rho}}\right\|_{\mathcal{L}^\infty}\right). \tag{17}$$

If $f^{\text{in}} \geq 0$, then $\widetilde{\rho}_\hbar^W\left(\sqrt{f^{\text{in}}}\right) \geq 0$ and therefore its time-evolution given by Eq. (14) $\widetilde{\rho} \geq 0$. Hence the right-hand side of Inequality (17) is well-defined. Furthermore, $\sqrt{\rho}$ and $\sqrt{\widetilde{\rho}}$ are bounded operators by assumption. Whence, $\|\rho - \widetilde{\rho}\|_{\mathcal{L}^2} \lesssim \left\|\sqrt{\rho} - \sqrt{\widetilde{\rho}}\right\|_{\mathcal{L}^2}$. We are now in the position of mimicking the L^2 weak-strong uniqueness criterion. Performing the time derivative on $\sqrt{\rho} - \sqrt{\widetilde{\rho}}$ and using that if ρ solves Eq. (6) then for any smooth function ϕ, the operator $\phi(\rho)$ solves

$$i\,\hbar\,\partial_t \phi(\rho) = \left[H_\rho, \phi(\rho)\right],$$

Grönwall's Lemma yields

$$\left\|\sqrt{\rho} - \sqrt{\widetilde{\rho}}\right\|_{\mathcal{L}^2} \lesssim \left\|\sqrt{\rho^{\text{in}}} - \sqrt{\widetilde{\rho}^{\text{in}}}\right\|_{\mathcal{L}^2} e^{\Lambda(t)} + \hbar \int_0^t C(s)\, e^{(\Lambda(t)-\Lambda(s))}\, ds,$$

with $C(s)$ and $\Lambda(s)$ independent of $\hbar$ and finite under the assumptions on f^{in}. This latter follows by propagation of regularity for solutions to the Vlasov–Poisson equation (1) (cf. [14]) and to the linear auxiliary dynamics (cf. [2]).

4 Quantum Optimal Transport Pseudo-metrics

The strategies of Theorems 1 and 2 are quite general and they generalizes to the relativistic case (cf. [15]). Application of these methods to other models are certainly possible. Although quite general, the weak-strong type criterion requires some regularity on the solution of the Vlasov–Poisson equation (1), that is not always compatible with certain classes of quantum states, such as pure states. To treat this important set of initial conditions, one can look at weaker topologies, in which almost no regularity conditions on derivatives of solutions to the Vlasov–Poisson are required.

4.1 Classical Optimal Transport Stability Estimate

Among the classical metrics on the set of probability measures are the Wasserstein–(Monge–Kantorovich) distances. For f_1 and f_2 two probability distributions on the phase space $\mathbb{R}^6$, one defines the set of couplings by $C(f_1, f_2)$ as the set of probability measures γ on $\mathbb{R}^6 \times \mathbb{R}^6$ with first marginal f_1 and second marginal f_2, that is $\int_{\mathbb{R}^6} \gamma(\cdot, \mathrm{d}y) = f_1$ and $\int_{\mathbb{R}^6} \gamma(\mathrm{d}x, \cdot) = f_2$. Then, for any $p \geq 1$, the Wasserstein distance of order p is defined by

$$W_p(f_1, f_2) = \left(\inf_{\gamma \in C(f_1, f_2)} \iint_{\mathbb{R}^6} |z_1 - z_2|^p \, \gamma(\mathrm{d}z_1 \, \mathrm{d}z_2) \right)^{1/p}.$$

It gives a natural distance to compare two solutions of the Vlasov equation. To explain this, let us define the characteristic flow $z(t, z_0) = (x(t, z_0), \xi(t, z_0))$ associated to a solution f of the Vlasov equation by $z(0, z_0) = z_0$ and

$$\begin{cases} \partial_t x(t, z_0) = \xi(t, z_0) \\ \partial_t \xi(t, z_0) = E_f(x(t, z_0)) \end{cases}$$

and it verifies for any bounded continuous function φ

$$\int_{\mathbb{R}^6} \varphi(z) \, f(t, z) \, \mathrm{d}z = \int_{\mathbb{R}^6} \varphi(x(t, z_0), \xi(t, z_0)) \, f^{\mathrm{in}}(z_0) \, \mathrm{d}z_0.$$

Then, if E_{f_1} and E_{f_2} are sufficiently close and regular, one observes that two initially close characteristic trajectories $z_1 = z_1(t, z_0)$ following the flow of f_1 and $z_2 = z_2(t, z_0')$ following the flow of f_2 will remain close since

$$\begin{aligned} \partial_t \, |z_1 - z_2|^2 &= 2\,(x_1 - x_2) \cdot (\xi_1 - \xi_2) + 2\,(\xi_1 - \xi_2) \cdot \left(E_{f_1}(x_1) - E_{f_2}(x_2)\right) \\ &\leq \left(1 + \omega_{f_1}(x_1, x_2)^2\right) |x_1 - x_2|^2 + 3\,|\xi_1 - \xi_2|^2 + \left|E_{f_1}(x_2) - E_{f_2}(x_2)\right|^2 \end{aligned}$$

where

$$\omega_{f_1}(x_1, x_2) = \sup_{x_1, x_2} \frac{\left|E_{f_1}(x_1) - E_{f_1}(x_2)\right|}{|x_1 - x_2|}$$

so that one should expect that $|z_1(t) - z_2(t)|$ does not grow too fast. Hence, to control an initially small quantity of the form

$$\mathcal{E}_p = \iint_{\mathbb{R}^6} |z_1 - z_2|^p \, \gamma(t, \mathrm{d}z_1 \, \mathrm{d}z_2) \tag{18}$$

one can choose γ such that

$$\iint_{\mathbb{R}^6 \times \mathbb{R}^6} \varphi(z_1, z_2) \, \gamma(t, z_1, z_2) \, \mathrm{d}z_1 \, \mathrm{d}z_2 = \iint_{\mathbb{R}^6 \times \mathbb{R}^6} \varphi(z_1(t, z), z_2(t, z')) \, \gamma^{\mathrm{in}}(z, z') \, \mathrm{d}z \, \mathrm{d}z'$$

or equivalently, γ solving

$$\partial_t \gamma + \left(\xi_1 \cdot \nabla_{x_1} + E_{f_1} \cdot \nabla_{\xi_1}\right) \gamma + \left(\xi_2 \cdot \nabla_{x_2} + E_{f_2} \cdot \nabla_{\xi_2}\right) \gamma = 0.$$

with initial condition $\gamma(0, \cdot, \cdot) = \gamma^{\mathrm{in}}(\cdot, \cdot)$. Taking the initial coupling γ^{in} to be the optimal coupling then ensures that (18) is initially small if f_1 and f_2 are initially close in W_p. In the case of the Coulomb potential and when ρ_{f_1} and ρ_{f_2} are uniformly bounded, then E_{f_1} is log-Lipshitz, that is $\omega_{f_1}(x_1, x_2) \leq C \max(1, -\ln(|x_1 - x_2|))$, and it was proved by Loeper [18] that $\left|E_{f_1}(x_2) - E_{f_2}(x_2)\right|$ is controlled by $W_2(f_1, f_2)$. As proved in [18], this leads to an inequality of the form

$$\partial_t \mathcal{E}_2(t) \leq C(t)\, \mathcal{E}_2(t) \max(1, -\ln(\mathcal{E}_2(t)))\,, \tag{19}$$

where $C(t) > 0$ depends on the growth of $\left\|\rho_{f_1}\right\|_{L^\infty} + \left\|\rho_{f_2}\right\|_{L^\infty}$. One then concludes by Gronwall's lemma to a stability estimate of the form

$$W_2(f_1, f_2) \leq W_2(f_1^{\mathrm{in}}, f_2^{\mathrm{in}})^{e^{\theta\lambda(t)}}\, e^{e^{\lambda(t)}}$$

where $\theta = \operatorname{sign}(\ln W_2(f_1, f_2))$ and $\lambda(t) > 0$ depends on $C(t)$, and this implies uniqueness for the Vlasov–Poisson equation. Let us mention that using a different weight for the phase space variables (see [9]), one can also obtain an inequality with a square root of the logarithm instead of a logarithm in Inequality (19).

4.2 Quantum Optimal Transport

To obtain an application to semiclassical estimates, one can use the semiclassical analogue of the optimal transport distance defined by Golse and Paul [4]. Let $\mathcal{P}$ be the set of density operators, that is the set of positive trace class operators verifying $h^3 \mathrm{Tr}(\boldsymbol{\rho}) = 1$. As in the classical case, one first introduces the notion of a coupling between a density operator $\boldsymbol{\rho}$ and a classical kinetic density f as an operator valued function $\boldsymbol{\gamma} \in L^1(\mathbb{R}^6, \mathcal{P})$ such that

$$h^3 \mathrm{Tr}(\boldsymbol{\gamma}(z)) = f(z), \qquad \text{and} \qquad \int_{\mathbb{R}^6} \boldsymbol{\gamma}(z)\, \mathrm{d}z = \boldsymbol{\rho}.$$

Denoting by $C(f, \boldsymbol{\rho})$ the set of semiclassical couplings, the semiclassical optimal transport pseudometric of order 2 is defined by

$$W_{2,\hbar}(f, \boldsymbol{\rho}) := \left(\inf_{\boldsymbol{\gamma} \in C(f, \boldsymbol{\rho})} \int_{\mathbb{R}^6} h^3 \mathrm{Tr}(\mathbf{c}_\hbar(z)\, \boldsymbol{\gamma}(z))\, \mathrm{d}z\right)^{\frac{1}{2}}, \tag{20}$$

where for any $z \in \mathbb{R}^6$ $\boldsymbol{c}_\hbar(z)$ is an operator defined by it action on functions φ of the form $\boldsymbol{c}_\hbar(z)\varphi(y) = (|x-y|^2 + |\xi - \boldsymbol{p}|^2)\varphi(y)$, with $z = (x, \xi)$ and $\boldsymbol{p} = -i\hbar\nabla_y$. This is not a distance, as it verifies for example $W_{2,\hbar}(f, \boldsymbol{\rho})^2 \geq 3\,\hbar$, but it is comparable to the classical Wasserstein distance W_2 between the Husimi transform $\tilde{f}_\rho$ of the quantum density operator and the normal kinetic density f in the sense that it verifies [4]

$$W_2(f, \tilde{f}_\rho)^2 \leq W_{2,\hbar}(f, \boldsymbol{\rho})^2 + 3\,\hbar, \tag{21}$$

and in the special case when $\boldsymbol{\rho}$ is the Wick quantization of some function g,

$$W_{2,\hbar}(f, \widetilde{\boldsymbol{\rho}}_\hbar^W(g))^2 \leq W_2(f, g)^2 + 3\,\hbar. \tag{22}$$

Moreover, it verifies approximate triangle inequalities [5]. For general density matrices, it can be proved that $W_{2,\hbar}(f, \boldsymbol{\rho}) \leq W_2(f, \tilde{f}_\rho) + \sqrt{3\hbar} + \hbar\,\|\nabla\sqrt{\boldsymbol{\rho}}\|_{\mathcal{L}^2}$ (see [13]). In particular, this allows to treat more singular states such as some classes of projection operators verifying $\boldsymbol{\rho}^2 = \boldsymbol{\rho} \geq 0$ for which in general one can obtains bounds of the form $\hbar\,\|\nabla\sqrt{\boldsymbol{\rho}}\|_{\mathcal{L}^2} = \hbar\,\|\nabla\boldsymbol{\rho}\|_{\mathcal{L}^2} \leq C\,\hbar^\theta$ for some $\theta \in [0, 1/2]$ (see [12]).

4.3 Application to the Semiclassical Limit

The same strategy as in the classical case was used in [10, 11] to prove the following theorem.

Theorem 3 ([10, Theorem 5]) *Let f be a solution of the Vlasov–Poisson equation* (1) *and $\boldsymbol{\rho}$ be a solution of the Hartree equation* (6) *equation with respective initial conditions f^{in} and $\boldsymbol{\rho}^{\text{in}}$ verifying*

$$f^{\text{in}} \in \mathscr{P} \cap L^1_{n_0} \cap L^\infty_n \qquad \text{and } \boldsymbol{\rho}^{\text{in}} \in \mathcal{P} \cap \mathcal{L}^1_{16} \cap \mathcal{L}^\infty_4.$$

uniformly in $\hbar$ for some $(n_0, n) \in \mathbb{R}^2_+$ verifying $n_0 > 6$ and $n > 3$. Then there exists $T > 0$ such that

$$h^3 Tr\big(|\boldsymbol{p}|^{16}\,\boldsymbol{\rho}\big) \in L^\infty([0, T]) \qquad \text{and} \qquad \boldsymbol{\rho} \in L^\infty([0, T], \mathcal{L}^\infty_4), \tag{23}$$

uniformly in $\hbar$, and there exists a constant C_T depending only on the initial conditions and independent of $\hbar$ such that

$$W_{2,\hbar}(f, \boldsymbol{\rho}) \leq W_{2,\hbar}(f^{\text{in}}, \boldsymbol{\rho}^{\text{in}})^{e^{\theta\lambda(t)}}\, e^{e^{\lambda(t)}} \tag{24}$$

where $\theta = \operatorname{sign}(\ln W_{2,\hbar}(f, \boldsymbol{\rho}))$ and $\lambda(t) \in C^0([0, T], \mathbb{R}_+)$ is an increasing function of time depending on the growth of $\left\|\rho_f\right\|_{L^\infty(\mathbb{R}^3)} + \|\rho\|_{L^\infty(\mathbb{R}^3)}$ and verifying $\lambda(0) = 0$.

The function $\lambda(t)$ is detailed in [11, Theorem 5]. The above theorem implies in particular that if initially there exists a constant C independent of $\hbar$ such that

$W_{2,\hbar}(f^{\text{in}}, \boldsymbol{\rho}^{\text{in}}) \leq C\sqrt{\hbar} \leq 1$, then

$$W_2(f, \tilde{f}_\rho) \leq \hbar^{e^{-\lambda(t)}} e^{e^{\lambda(t)}}.$$

Proof (*Steps of the proof*) Several steps are needed to get the complete result. As in the classical case, the stability estimate in the optimal transport metric uses the fact that both ρ_f and ρ are bounded (uniformly in $\hbar$). To get this property for the Hartree equation, one proceeds in three steps.

- First, one proves the propagation of moments of the form $h^3\mathrm{Tr}(|\boldsymbol{p}|^n\,\boldsymbol{\rho})$ with $n \in 2\mathbb{N}$ [10, Theorem 3]. The proof uses the fact that moments of order two are bounded by the energy conservation, the fact that $\|\boldsymbol{\rho}\|_{\mathcal{L}^1} + \|\boldsymbol{\rho}\|_{\mathcal{L}^\infty}$ is conserved and Lieb–Thirring type inequalities. This is at this step that only a local in time propagation is proved, which prevents the theorem to hold for all times $T > 0$.
- The second point is to prove the propagation of weighted Schatten norms of the form $\left\|\boldsymbol{\rho}\,|\boldsymbol{p}|^n\right\|_{\mathcal{L}^p}$ uniformly with respect to p and $\hbar$ [10, Proposition 5.3] using the previously proved boundedness of moments.
- The last point consists in taking $p \to \infty$ to get that $\left\|\boldsymbol{\rho}\,(1 + |\boldsymbol{p}|^n)\right\|_{\mathcal{L}^\infty}$ is bounded uniformly in $\hbar$, from which one deduces that $\|\rho\|_{L^\infty}$ is bounded uniformly in $\hbar$.

Once these points are proved, the proof of the theorem follows as in [4] by adapting the classical proof of stability for the Vlasov equation but replacing one of the solutions by a solution of the Hartree equation. More precisely, for any coupling $\boldsymbol{\gamma}^{\text{in}} \in \mathcal{C}(f^{\text{in}}, \boldsymbol{\rho}^{\text{in}})$, one looks at the solution $\boldsymbol{\gamma} = \boldsymbol{\gamma}(t, z)$ to

$$\partial_t \boldsymbol{\gamma} + \left(\xi \cdot \nabla_x + E_f \cdot \nabla_\xi\right) \boldsymbol{\gamma} - \frac{1}{i\hbar}\left[H_\rho, \boldsymbol{\gamma}\right] = 0$$

and obtains a Grönwall's type inequality on (the logarithm of)

$$\mathcal{E}_{2,\hbar}(t) := \int_{\mathbb{R}^6} h^3\mathrm{Tr}(\mathbf{c}_\hbar(z)\,\boldsymbol{\gamma}(t, z))\,\mathrm{d}z.$$

Since $W_{2,\hbar}(f, \boldsymbol{\rho}) \leq \mathcal{E}_{2,\hbar}$, the result follows by minimizing over all couplings $\boldsymbol{\gamma}^{\text{in}}$.

Acknowledgements C.S. acknowledges the NCCR SwissMAP, the support of the SNSF through the Eccellenza project PCEFP2_181153 and of the the Swiss State Secretariat for Research and Innovation through the ERC Starting Grant "AEQUA". L.L. has received funding from the European Research Council (ERC) under the European Union's Horizon 2020 research and innovation program (grant agreement No 865711).

References

1. Boulkhemair, A.: L^2 estimates for Weyl Quantization. J. Funct. Anal. **165**(1), 173–204 (1999)
2. Chong, J.J., Lafleche, L., Saffirio, C.: On the L^2 rate of convergence in the limit from the Hartree to the Vlasov-Poisson equation. J. Ec. polytech. Math. **10**, 703–726 (2023)
3. Crippa, G., Inversi, M., Saffirio, C., Stefani, G.: Existence and stability of weak solutions of the Vlasov–Poisson system in localized Yudovich spaces, pp. 1–22 (2023). arXiv:2306.00523
4. Golse, F., Paul, T.: The Schrödinger equation in the mean-field and semiclassical regime. Arch. Ration. Mech. Anal. **223**(1), 57–94 (2017)
5. Golse, F., Paul, T.: Quantum and Semiquantum Pseudometrics and Applications, pp. 1–33 (2021). arXiv:2102.05184
6. Holding, T., Miot, E.: Uniqueness and stability for the Vlasov-Poisson system with spatial density in Orlicz spaces. In: Mathematical Analysis in Fluid Mechanics: Selected Results. Contemporary Mathematics, vol. 710, pp. 145–162. American Mathematical Society, Providence, RI (2018)
7. Horst, E.: Global strong solutions of Vlasov's equation - necessary and sufficient conditions for their existence. Banach Cent. Publ. **19**(1), 143–153 (1987)
8. Hunt, R.A.: On $L(p, q)$ Spaces. L'Enseignement Mathématique **12**, 249–276 (1966)
9. Iacobelli, M.: A new perspective on Wasserstein distances for kinetic problems. Arch. Ration. Mech. Anal. **244**(1), 27–50 (2022)
10. Lafleche, L.: Propagation of moments and semiclassical limit from Hartree to Vlasov equation. J. Stat. Phys. **177**(1), 20–60 (2019)
11. Lafleche, L.: Global semiclassical limit from Hartree to Vlasov equation for concentrated initial data. Ann. Inst. H. Poincaré C Anal. Non Linéaire **38**(6), 1739–1762 (2021)
12. Lafleche, L.: Optimal Semiclassical Regularity of Projection Operators and Strong Weyl Law, pp. 1–17 (2023). arXiv:2302.04816
13. Lafleche, L.: Quantum Optimal Transport and Weak Topologies, pp. 1–5 (2023). arXiv:2306.12944
14. Lafleche, L., Saffirio, C.: Strong semiclassical limits from Hartree and Hartree-Fock to Vlasov-Poisson equations. Anal. PDE **16**(4), 891–926 (2023)
15. Leopold, N., Saffirio, C.: Propagation of moments for large data and semiclassical limit to the relativistic Vlasov equation. To appear in SIAM J. Math. Anal. 1–29 (2022)
16. Lions, P.-L., Paul, T.: Sur les mesures de Wigner. Rev. Mat. Iberoam. **9**(3), 553–618 (1993)
17. Lions, P.-L., Perthame, B.: Propagation of moments and regularity for the 3-dimensional Vlasov-Poisson system. Invent. Math. **105**(2), 415–430 (1991)
18. Loeper, G.: Uniqueness of the solution to the Vlasov-Poisson system with bounded density. J. Math. Pures Appl. **86**(1), 68–79 (2006)
19. Miot, E.: A Uniqueness Criterion for Unbounded Solutions to the Vlasov-Poisson System. Commun. Math. Phys. **346**(2), 469–482 (2016)
20. Pfaffelmoser, K.: Global classical solutions of the Vlasov-Poisson system in three dimensions for general initial data. J. Differ. Equ. **95**(2), 281–303 (1992)
21. Schaeffer, J.: Global existence of smooth solutions to the Vlasov-Poisson system in three dimensions. Commun. Partial Differ. Equ. **16**(8–9), 1313–1335 (1991)

The Extended Phase Space Method in Kinetic Theory

Francesco Salvarani

To the memory of São Carvalho (1956–2021).

Abstract This article describes the extended phase space technique in the study of kinetic equations and provides a review of some relevant contributions appeared in the literature.

Keywords Kinetic equations · Extended phase space

1 Introduction

A kinetic equation is a macroscopic description of a population composed of individual entities which are subject to microscopic laws.

The choice of the system's independent variables is essential in defining a kinetic equation or system. The set of independent variables has a heterogeneous nature. Indeed, the independent variables can be divided, essentially, into two categories. Some variables are "external", such as position $x \in \mathbb{R}^3$ and time $t \in \mathbb{R}_+$. They are defined with respect to an absolute reference frame and an absolute clock. There are moreover other independent variables, called "internal variables" such as, for example, the velocity of a particle $v \in \mathbb{R}^3$ or its internal energy $I \in \mathbb{R}_+$.

By taking into account these independent variables, it is possible to capture some crucial effects of the system evolution. However, the main macroscopically relevant quantities are obtained through moments of the distribution function with respect to these internal variables.

F. Salvarani (✉)
Léonard de Vinci Pôle Universitaire, Research Center, 92916 Paris La Défense, France
e-mail: francesco.salvarani@unipv.it

Dipartimento di Matematica, Università degli Studi di Pavia, 27100 Pavia, Italy

E. Carlen et al. (eds.), *From Particle Systems to Partial Differential Equations*,
Springer Proceedings in Mathematics & Statistics 465,
https://doi.org/10.1007/978-3-031-65195-3_15

Because of the analogy with the traditional phase space, in kinetic theory the set of macroscopic and internal variables is referred to as "phase space". In this setting, both macroscopic and internal variables have their own physical meaning.

A new approach for studying the homogenization of linear kinetic equation is the extended phase space technique and has been introduced into the literature in 2010 [5], after being announced in 2008 [4]. This strategy consists in defining some additional independent variables, which may or may not have physical significance, and studying the problem in the new setting. In particular, in most of the results obtained so far, there is only one extra variable, which is essentially a time variable and denotes the time since the last jump in the stochastic process driving the linear Boltzmann equation.[1] However, since the method is completely general in the choice of the supplementary variables, it is not excluded that other problems can be successfully studied by means of an extended phase space in which the supplementary variables have other physical meanings (or have no physical meaning at all).

This paper will describe the extended phase space technique and its consequences, both theoretically and numerically.

The structure of the article is the following.

In Sect. 2 we sketch the outline of the extended phase space method. Then, in Sect. 3 we review some theoretical results on the extended phase space technique available in the literature. In particular, we describe the arguments developed in [4, 5], devoted to the study of the Boltzmann-Grad limit for the two dimensional periodic Lorentz gas, and in [1], which studies the homogenization problem of the linear Boltzmann equation in a domain with holes following a periodic distribution. We then review [2], and show that some kinetic equations lose their semigroup properties in the homogenization limit, but that they do not if the homogenization procedure is set in a suitably extended phase space. Finally, we compare the homogenization problem of the linear Boltzmann equation in the extended phase space, studied in [3] with the corresponding problem in the standard phase space, described in [9].

In Sect. 4, we review a result showing that the extended phase space technique can be successfully used for designing an efficient numerical strategy for studying kinetic equations with rapidly oscillating terms [11], and provide an example of this numerical approach.

2 The Extended Phase Space Strategy

The extended phase space technique consists in adding at least one extra variable to the original problem, and studying the limiting behaviour of the new problem in the extended phase space. The typical stategy can be summarized as follows.

Let $t \in \mathbb{R}_+$, $x \in \Omega \subseteq \mathbb{R}^d$ and $v \in \mathscr{V} \subseteq \mathbb{R}^d$. Consider an evolution equation in the phase space, satisfying the semigroup property, whose unknown is a function

[1] The literature reports some examples in which the supplementary variables have other physical meanings: for instance, the two additional variables in [4, 5] are the rescaled distance to the next collision and the corresponding impact parameter.

$f_\varepsilon : \mathbb{R}_+ \times \Omega \times \mathscr{V} \to \mathbb{R}$, and suppose that the problem depends on some oscillating terms of period $1/\varepsilon$.

The study of the homogenized problem, i.e. the limiting behavior of the equation and of its solution as $\varepsilon \to 0^+$, is obtained thanks to the following steps.

- Denote with s an additional variable to the phase space (which can be, in principle, either a scalar or a vector) in which the evolution equation is defined, and then consider a function $F_\varepsilon(t, s, x, v)$ satisfying

$$\int_0^\infty F_\varepsilon(t, s, x, v)\mathrm{d}s = f_\varepsilon(t, x, v).$$

- Prove that F_ε satisfies an evolution equation in the extended phase space, which satisfies the semigroup property.
- Deduce that F_ε converges to a limit F, which solves the homogenized problem, and that the homogenized problem itself satisfies the semigroup property.
- Prove that

$$f_\varepsilon \to f := \int_0^\infty F \mathrm{d}s.$$

- If possible, write the evolution equation satisfied by f. We underline that this is not always the case. A closed form for the equation governing the limit f may not exist at all as, for instance, in the Boltzmann-Grad limit of the periodic Lorentz gas in a domain with a periodic distribution of holes [7].

The method can be summarized in the following diagram:

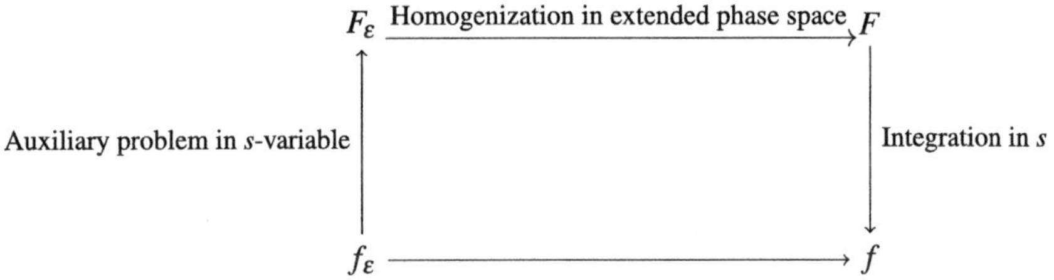

Hence, the limiting behaviour of the solution of the starting problem is recovered as the zeroth-order moment, with respect to the additional variable, of the limit solution of the homogenized problem in the extended phase space.

Therefore, the macroscopic homogenization limit for the linear kinetic equations described in this article involve a more microscopic description of the system than the usual phase space, i.e. the standard phase space together with the additional variable s [1].

3 Theoretical Results

In this Section, we review four possible applications of the extended phase space method: the Boltzmann-Grad limit for the two-dimensional Lorentz gas, the homogenization of the linear Boltzmann equation in a domain with a periodic distribution of holes, the homogenization of the linear radiative transfer equation with rapidly oscillating opacities in frequency and the homogenization problem for the linear Boltzmann equation with a rapidly oscillating cross section.

3.1 The Boltzmann-Grad Limit for the Two-Dimensional Lorentz Gas

In [4] and [5], the authors study the Boltzmann-Grad limit for the Lorentz gas. They restrict their analysis to the case of a two-dimensional Euclidean space, with a periodic system of obstacles. The obstacles are disks of radius r centred at each point of $\mathbb{Z}^2$. The domain available to the motion of the gas particles is the set

$$Z_r = \{x \in \mathbb{R}^2 \mid \text{dist}\,(x, \mathbb{Z}^2) > r\}, \qquad \text{where } 0 < r < \frac{1}{2}.$$

Note that particles move at speeds whose modulus is equal to 1. The trajectory of a particle starting from $x \in Z_r$ with velocity $\omega \in \mathbb{S}^1$ at time $t = 0$ is denoted by $t \mapsto (X_r, \Omega_r)(t; x, \omega) \in \mathbb{R}^2 \times \mathbb{S}^1$. A particle moves in uniform rectilinear motion until it encounters an obstacle. It is then specularly reflected, by following the laws of classical optics:

$$\begin{cases} \dfrac{\mathrm{d}}{\mathrm{d}t} X_r(t) = \Omega_r(t), \\ \dfrac{\mathrm{d}}{\mathrm{d}t} \Omega_r(t) = 0, \end{cases} \qquad \text{whenever } X_r(t) \in Z_r,$$

while

$$\begin{cases} X_r(t+0) = X_r(t-0), \\ \Omega_r(t+0) = \mathscr{R}[X_r(t)]\Omega_r(t-0), \end{cases} \qquad \text{whenever } X_r(t) \in \partial Z_r.$$

The operator $\mathscr{R}[X_r(t)]$ represents the specular reflection on ∂Z_r at a given point $X_r(t) = X_r(t \pm 0)$.

The Boltzmann-Grad limit of the previous equation assumes that $r \ll 1$ and that the initial position x and direction ω of the particle are jointly distributed in $Z_r \times \mathbb{S}^1$ under some slowly varying (with respect to the x-variable) density of the form $f^{in}(rx, \omega)$. By defining

$$f_r(t, x, \omega) := f^{in}(rX_r(-t/r; x, \omega), \Omega_r(-t/r; x, \omega)) \quad \text{whenever } x \in Z_r, \tag{1}$$

the Boltzmann-Grad limit studied by the authors consists in finding the limit of f_r as $r \to 0^+$.

Let $(x, \omega) \in Z_r \times \mathbb{S}^1$. The strategy of the article is the following.

First, the authors define some quantities which will be relevant in the sequel. In particular, they introduce the sequence of collision times on the billiard trajectory in Z_r starting from x with velocity ω: $0 < t_0 < t_1 < t_2 < \ldots$, i.e.

$$\{t_j \mid j \in \mathbb{N}\} = \{t \in \mathbb{R}_+^* \mid X_r(t; x, \omega) \in \partial Z_r\}.$$

Moreover, they introduce the pair

$$(x_j, \omega_j) := (X_r(t_j + 0; x, \omega), \Omega_r(t_j + 0; x, \omega)), \quad j \geq 0$$

and define the following subsets:

$$\Gamma_r^{\pm} = \{(x, \omega) \in \partial Z_r \times \mathbb{S}^1 \mid \pm\, \omega \cdot n_x > 0\},$$
$$\Gamma_r^0 = \{(x, \omega) \in \partial Z_r \times \mathbb{S}^1 \mid \omega \cdot n_x = 0\}.$$

The next step consists in seeking mapping from $\Gamma_r^+ \cup \Gamma_r^0$ to itself whose iterates transform (x_0, ω_0) into the sequence (x_j, ω_j) defined above.

For any $(x, \omega) \in (Z_r \times \mathbb{S}^1) \cup \Gamma_r^+ \cup \Gamma_r^0$, the authors introduce the exit time $\tau_r(x, \omega)$, which is a classical–but crucial–tool for studying transport problems which are not set in the whole space. This quantity depends on x and on the features of the domain.

Definition 1 The exit time of a point $x \in Z_r$, with respect to the velocity $\omega \in \mathbb{S}^1$, is, when it exists, the real number

$$\tau_r(x, \omega) = \inf\{t > 0 \mid x + t\omega \in \partial Z_r\}.$$

When the above infimum does not exist, then $\tau_r(x, \omega) = +\infty$.

Two other definitions will be useful in the sequel.

Definition 2 Let $(x, \omega) \in \Gamma_r^+ \cup \Gamma_r^0$. The impact parameter $h_r(x, \omega)$ is defined by

$$h_r(x, \omega') = \sin(\widehat{\omega', n_x}),$$

where ω' is the outgoing direction corresponding to the incoming direction ω.

Let $(\Gamma_r^+ \cup \Gamma_r^0)/\mathbb{Z}^2$ be the quotient of $\Gamma_r^+ \cup \Gamma_r^0$ under the action of $\mathbb{Z}^2$ by translations on the x variable. Consider now the map

$$(\Gamma_r^+ \cup \Gamma_r^0)/\mathbb{Z}^2 \ni (x, \omega) \mapsto (h_r(x, \omega), \omega) \in [-1, 1] \times \mathbb{S}^1$$

and let Y_r be its inverse.

Definition 3 For $r \in [0, \frac{1}{2}]$, the transfer map $T_r : [-1, 1] \times \mathbb{S}^1 \to \mathbb{R}_+^* \times [-1, 1]$ is defined by

$$T_r(h', \omega) = (2r\tau_r(Y_r(h', \omega)), h_r((X_r, \Omega_r)(\tau_r(Y_r(h', \omega)) \pm 0; Y_r(h', \omega)))) .$$

For a particle leaving the surface of an obstacle in the direction ω with impact parameter h', the transition map $T_r(h', \omega) = (s, h)$ gives the rescaled distance s to the next collision, and the corresponding pre-collisional impact parameter h. Obviously, each trajectory of the particle can be expressed in terms of the transfer map T_r and their iterates.

The Boltzmann-Grad limit of the periodic Lorentz gas is then obtained as the limiting behavior of T_r as $r \to 0^+$. The main result of [4, 5] is given by the following theorem.

Theorem 1 *Let f^{in} be any continuous, compactly supported probability density on $\mathbb{R}^2 \times \mathbb{S}^1$. Denoting by $R[\theta]$ the rotation of an angle θ, let $F \equiv F(t, x, \omega, s, h)$ be the solution of*

$$\begin{cases} (\partial_t + \omega \cdot \nabla_x - \partial_s) F(t, x, \omega, s, h) = \displaystyle\int_{-1}^{1} 2P(2s, h|h') F(t, x, R[\theta(h')]\omega, 0, h') dh' , \\ F(0, x, \omega, s, h) = f^{in}(x, \omega) \displaystyle\int_{2s}^{\infty} \int_{-1}^{1} P(\tau, h|h') dh' d\tau , \end{cases} \tag{2}$$

where (x, ω, s, h) runs through $\mathbb{R}^2 \times \mathbb{S}^1 \times \mathbb{R}_+^* \times [-1, 1]$, and $\theta(h) = \pi - 2 \arcsin(h)$.

Then the family $(f_r)_{0<r<\frac{1}{2}}$ defined in (1) satisfies

$$f_r \to \int_0^{\infty} \int_{-1}^{1} F(\cdot, \cdot, \cdot, s, h) ds dh \text{ in } L^{\infty}(\mathbb{R}_+ \times \mathbb{R}^2 \times \mathbb{S}^1) \text{ weak} - * \tag{3}$$

as $r \to 0^+$.

The transition probability P which appears in the statement of Theorem 1 has the following structure. For each $(h, h') \in [-1, 1]^2$, introduce

$$\zeta = \tfrac{1}{2}|h + h'| , \qquad \eta = \tfrac{1}{2}|h - h'| ,$$

and, for all $(a, b) \in \mathbb{R}^2$,

$$a \wedge b = \inf(a, b) , \qquad a \vee b = \sup(a, b) .$$

We have that

$$P(S,h|h') = \frac{3}{\pi^2 S\eta}\Big((S\eta)\wedge(1-S)_+ + (\eta S - |1-S|)_+ \\ + \big((S-\tfrac{1}{2}S\eta)\wedge(1+\tfrac{1}{2}S\eta) - (\tfrac{1}{2}S+\tfrac{1}{2}S\zeta)\vee 1\big)_+ \\ + \big((S-\tfrac{1}{2}S\eta)\wedge 1 - (\tfrac{1}{2}S+\tfrac{1}{2}S\zeta)\vee(1-\tfrac{1}{2}S\eta)\big)_+\Big).$$

Equivalently, for each $(S, h, h') \in \mathbb{R}_+ \times [-1, 1[\times] - 1, 1]$ such that $|h'| \leq h$,

$$P(S,h|h') = \tfrac{3}{\pi^2}\left(1 \wedge \frac{1}{h-h'}\left(\frac{2}{S} - (1+h')\right)_+\right).$$

We conclude this Subsection with some remarks. Equation (2) and (3) was first proposed by Caglioti and Golse in [4], under some additional unverified assumptions. Shortly afterwards, Marklof and Strömbergsson [10] provided, independently, a rigorous and complete argument for deriving the same equation, without any additional assumption. Their result is not limited to the two-dimensional Euclidean space, but is valid in any finite space dimension. However, their method of proof is not based on the extended phase space technique, and for this reason it will not be detailed here.

We underline that Theorem 1 has introduced two extra variables which enable more detailed tracking of particle property trajectories, giving a kind of Markov process which allows to pass the limit. In this context, the technique of the extended phase space method naturally appears.

The same theorem confirms moreover that the correct limiting equation for the Boltzmann-Grad limit of the periodic Lorentz gas in a periodic domain cannot be the solution of any linear Boltzmann equation of the form

$$(\partial_t + v\cdot\nabla_x + \sigma)f(t,x,v) = \sigma\int_{|w|=1}\tilde{p}(v|w)f(t,x,w)\mathrm{d}w\,,$$

where $\tilde{p}$ is a continuous, symmetric transition probability density on the unit sphere, a negative result which had been already noticed by Golse in [7], by analyzing the time decay of the solution to the equilibrium solution for the homogenized limit of the Linear Boltzmann equation.

It is worth noticing that the previous result uses some arguments which are specific to the Eucidean two-dimensional space case.

3.2 *The Homogenization of the Linear Boltzmann Equation in a Domain with a Periodic Distribution of Holes*

In [1] the authors study, with the extended phase space method, the homogenization of the Linear Boltzmann equation in a periodic domain with holes. They assume that the reciprocal number of holes per unit surface and the length of the circumference of each hole are asymptotically equivalent small quantities.

The authors assume that the holes are absorbing; it means that a particle falling into a hole cannot escape from it. Therefore, a mass loss is expected.

They consider the linear monokinetic Boltzmann equation

$$\partial_t f_\epsilon + v \cdot \nabla_x f_\epsilon + \sigma(f_\epsilon - K f_\epsilon) = 0 \tag{4}$$

in the two-dimensional Euclidean space.

The operator K is such that, for all $\phi \in L^2(\mathbb{S}^1)$,

$$K\phi(v) := \frac{1}{2\pi} \int_{\mathbb{S}^1} k(v, w)\phi(w)\mathrm{d}w,$$

where $\mathrm{d}w$ is the Lebesgue measure on the unit circle $\mathbb{S}^1$. Moreover,

$$k \in L^2(\mathbb{S}^1 \times \mathbb{S}^1)\,, \quad k(v, w) = k(w, v) \geq 0 \ \text{ a.e. in } v, w \in \mathbb{S}^1$$

and

$$\frac{1}{2\pi} \int_{\mathbb{S}^1} k(v, w)\mathrm{d}w = 1 \ a.e.\ in\ v \in \mathbb{S}^1.$$

The linear Boltzmann equation (4) is defined on the spatial domain

$$Z_\epsilon := \left\{x \in \mathbb{R}^2 \mid \text{dist}\,(x, \epsilon\mathbf{Z}^2) > \epsilon^2\right\}.$$

The absorbing boundary conditions on ∂Z_ϵ are defined as follows:

$$f_\epsilon = 0 \text{ for } (t, x, v) \in \mathbb{R}_+^* \times \partial Z_\epsilon \times \mathbb{S}^1, \quad \text{whenever } v \cdot n_x > 0\,,$$

where n_x is the inward unit normal vector to Z_ϵ at the point $x \in \partial Z_\epsilon$.

The free path length in the direction v for a particle starting from x in Z_ϵ has the same form already detailed in Definition 1, i.e.

$$\tau_\epsilon(x, v) := \inf\{t > 0 \,|\, x - tv \in \partial Z_\epsilon\}\,. \tag{5}$$

The authors assume that the initial data of (Ξ_ϵ) satisfy the condition

$$f^{in} \geq 0 \text{ on } \mathbb{R}^2 \times \mathbb{S}^1 \text{ and } \iint_{\mathbb{R}^2 \times \mathbb{S}^1} f^{in}(x, v) \mathrm{d}x \mathrm{d}v + \sup_{(x,v)\in\mathbb{R}^2\times\mathbb{S}^1} f^{in}(x, v) < +\infty. \tag{6}$$

For each $0 < \epsilon \ll 1$, let f_ϵ be the solution of the initial boundary value problem

$$(\Xi_\epsilon) \begin{cases} \partial_t f_\epsilon + v \cdot \nabla_x f_\epsilon + \sigma(f_\epsilon - K f_\epsilon) = 0, & (x, v) \in Z_\epsilon \times \mathbb{S}^1, t > 0, \\ f_\epsilon = 0, \text{ if } v \cdot n_x > 0, & (x, v) \in \partial Z_\epsilon \times \mathbb{S}^1, \\ f_\epsilon(0, x, v) = f^{in}(x, v), & (x, v) \in Z_\epsilon \times \mathbb{S}^1. \end{cases}$$

A well known result guarantees the existence and uniqueness of a mild solution f_ϵ of the problem (Ξ_ϵ) satisfying

$$\begin{aligned} 0 \leq f_\epsilon(t, x, v) \leq \sup_{(x,v)\in\mathbb{R}^2\times\mathbb{S}^1} f^{in}(x, v) \quad \text{a.e. on } \mathbb{R}_+ \times Z_\epsilon \times \mathbb{S}^1, \\ \iint_{Z_\epsilon \times \mathbb{S}^1} f_\epsilon(t, x, v) \mathrm{d}x \mathrm{d}v \leq \iint_{\mathbb{R}^2 \times \mathbb{S}^1} f^{in}(x, v) \mathrm{d}x \mathrm{d}v \end{aligned} \tag{7}$$

(it is indeed a linear Boltzmann equation).

Let (Σ_ϵ) be the initial boundary value problem

$$(\Sigma_\epsilon) \begin{cases} \partial_t F_\epsilon + v \cdot \nabla_x F_\epsilon + \partial_s F_\epsilon + \sigma F_\epsilon = 0, & t, s > 0, (x, v) \in Z_\epsilon \times \mathbb{S}^1, \\ F_\epsilon(t, s, x, v) = 0, \text{ if } v \cdot n_x > 0, & t, s > 0, (x, v) \in (\partial Z_\epsilon \times \mathbb{S}^1), \\ F_\epsilon(t, 0, x, v) = \sigma \int_0^\infty K F_\epsilon(t, s, x, v) ds, & t > 0, (x, v) \in Z_\epsilon \times \mathbb{S}^1, \\ F_\epsilon(0, s, x, v) = \sigma e^{-\sigma s} f^{in}(x, v), & s > 0, (x, v) \in Z_\epsilon \times \mathbb{S}^1, \end{cases}$$

with unknown $F_\epsilon := F_\epsilon(t, s, x, v)$.

The relationship between the initial boundary value problems (Ξ_ϵ) and (Σ_ϵ) is summarized in the following proposition.

Proposition 1 *Assume that f^{in} satisfies assumption (6). Then*
(a) for each $\epsilon > 0$, the problem (Σ_ϵ) has a unique mild solution such that

$$(t, x, v) \mapsto \int_0^{+\infty} |F_\epsilon(t, s, x, v)| ds \textit{ belongs to } L^\infty([0, T] \times Z_\epsilon \times \mathbb{S}^1)$$

for each $T > 0$;
(b) moreover

$$0 \leq F_\epsilon(t, s, x, v) \leq \|f^{in}\|_{L^\infty(\mathbb{R}^2\times\mathbb{S}^1)} \sigma e^{-\sigma s}$$

a.e. in $t, s \geq 0$, $x \in Z_\epsilon$ and $v \in \mathbb{S}^1$, and

$$\int_0^{+\infty} F_\epsilon(t, s, x, v) ds = f_\epsilon(t, x, v),$$

for a.e. $t \geq 0$, $x \in Z_\epsilon$, $v \in \mathbb{S}^1$, where f_ϵ is the solution of (Ξ_ϵ).

On the other hand, let $F := F(t, s, x, v)$ be the solution of the Cauchy problem

$$(\Sigma)\begin{cases} \partial_t F + v \cdot \nabla_x F + \partial_s F = -\sigma F + \frac{\dot{p}}{p}(t \wedge s)F, & t, s > 0, (x, v) \in \mathbb{R}^2 \times \mathbb{S}^1, \\ F(t, 0, x, v) = \sigma \int_0^{+\infty} KF(t, s, x, v)ds, & t > 0, (x, v) \in \mathbb{R}^2 \times \mathbb{S}^1, \\ F(0, s, x, v) = \sigma e^{-\sigma s} f^{in}(x, v), & s > 0, (x, v) \in \mathbb{R}^2 \times \mathbb{S}^1, \end{cases}$$

where $t \wedge s := \min(t, s)$ and $p : \mathbb{R}_+ \to \mathbb{R}_+$ is a positive decreasing function which describes the distribution of free path length. In particular, for each arc $I \subset \mathbb{S}^1$ and each $t \geq 0$,

$$\text{meas}\left(\left\{(x, v) \in \left(Z_\varepsilon \cap [0, 1]^2\right) \times I \mid \varepsilon\tau_\varepsilon(x, v) > t\right\}\right) \to p(t)|I| \tag{8}$$

as $\varepsilon \to 0^+$, where $|I|$ denotes the length of I and the measure considered in Eq. (8) is the uniform measure on $[0, 1]^2 \times \mathbb{S}^1$.

The unknown F of the problem written above is a density defined on the extended phase space:

$$\left\{(s, x, v) | s \geq 0, x \in \mathbb{R}^2, v \in \mathbb{S}^1\right\}$$

involving the extra variable $s \in \mathbb{R}_+$.

For each function $\varphi \equiv \varphi(x)$ defined a.e. on Z_ϵ, let

$$\{\varphi\}(x) = \begin{cases} \varphi(x) & \text{if } x \in Z_\epsilon, \\ 0 & \text{if } x \notin \overline{Z_\epsilon}, \end{cases}$$

In what follows, the notations $\{f_\epsilon\}$ or $\{F_\epsilon\}$ designate the same extension by 0 inside the holes for functions defined on cartesian products involving Z_ϵ as one of their factors, such as $\mathbb{R}_+ \times Z_\epsilon \times \mathbb{S}^1$ in the case of f_ϵ, and $\mathbb{R}_+ \times \mathbb{R}_+ \times Z_\epsilon \times \mathbb{S}^1$ in the case of F_ϵ.

The main result of [1] is the following.

Theorem 2 *Under the assumptions listed in this subsection, the solution F_ϵ of (Σ_ϵ) converges to F, which is the unique (mild) solution of (Σ), weakly in $L^1_{loc}(\mathbb{R}_+ \times \mathbb{R}_+ \times \mathbb{R}^2 \times \mathbb{S}^1)$ as $\epsilon \to 0^+$. Moreover*

$$\{f_\epsilon\} \rightharpoonup \int_0^{+\infty} F ds$$

in $L^\infty(\mathbb{R}_+ \times \mathbb{R}^2 \times \mathbb{S}^1)$ weak-$*$ as $\epsilon \to 0^+$.

Caglioti and Golse obtained the homogenized equation corresponding to (Ξ_ϵ) in the noncollisional case $\sigma = 0$–see the previous subsection–by explicitly computing the solution of the linear Boltzmann equation for each $0 < \epsilon \ll 1$. In the collisional case, i.e. when $\sigma > 0$, there is no explicit formula for the solution of the linear Boltzmann equation. On the other hand, Papanicolau gave in [12] a semi-explicit formula involving the transport process, which is enough for the purposes of the proof of Theorem 2.

It is worth noticing that the dependence of the solution F_ϵ of the problem (Σ) upon the choice of the initial probability density Π disappears after integrating with respect to the s variable.

It is known that the solution f_ϵ of the linear Boltzmann equation is the deterministic version of a stochastic process involving a jump process in the v variable together with a deterministic drift in the x variable (see, for example, [6] pp. 225–226). Hence, the extra variable s in the extended phase space is the time since the last jump in the v variable of the current velocity v in that process.

A deep analysis of the result proved in [1] has been carried out by Golse in [8].

3.3 The Homogenization of the Linear Radiative Transfer Equation with Rapidly Varying Opacities

In [2], the authors have studied the homogenization problem for the radiative transfer equations in the linear approximation. Radiative transfer describes the behavior of a gas of photons which exchange energy with a background materia by using the kinetic approach. These equations are suitable for studying, for example, plasmas in which radiation plays an important role. The interaction between the gas of photons and the medium results in energy exchanges, originated by various phenomena (such as absorption, scatering or emission of photons by the background). By denoting with $c \in \mathbb{R}_+^*$ the speed of light and $h \in \mathbb{R}_+^*$ the Planck constant, the system is described through the radiative intensity $I(t, x, \omega, \nu)$ (which is $ch\nu$ times the number density of photons), which depends on frequency $\nu \in \mathbb{R}_+^*$, time $t \in \mathbb{R}_+$, position $x \in \Omega \subseteq \mathbb{R}^d$ and velocity direction $\omega \in \mathbb{S}^{d-1}$.

If scattering processes are neglected, the radiative intensity satisfies a particularly simple linear equation in which are involved both the specific radiative intensity $B_\nu(T)$ at frequency ν of a black body at temperature T, and the opacity of the background material at temperature T for an incident radiation with frequency ν.

The quantity $B_\nu(T)$ has the explicit expression

$$B_\nu(T) = \frac{4\pi\hbar\nu^3}{c^2}\frac{1}{e^{2\pi\hbar\nu/k_B T} - 1},$$

where k_B is the Boltzmann constant and $\hbar$ is the reduced Planck constant.

On the other hand, the function $\nu \mapsto \sigma(\nu, T)$ may have rapid oscillations and, moreover, the opacity $\sigma(\nu, T)$ is, in general, not explicitly known. The authors have considered the following equation:

$$\begin{cases} \frac{1}{c}\partial_t I_\epsilon + \omega \cdot \nabla_x I_\epsilon = \sigma_\epsilon(\nu, T)B_\nu(T) - \sigma_\epsilon(\nu, T)I_\epsilon\,, \\ I_\epsilon\big|_{t=0} - I^{in}(x, \omega, \nu)\,, \end{cases}$$

for $(t, x, \omega, \nu) \in \mathbb{R}_+^* \times \mathbb{R}^3 \times \mathbb{S}^2 \times \mathbb{R}_+^*$. The oscillations of the opacity depend on the small parameter ϵ, which represents the typical oscillation wavelength in the variable ν. The result proved in [2] is the following.

Theorem 3 *Suppose that $T \in [\theta, \Theta]$ for some constants $0 < \theta < \Theta$, and that the family $(\sigma_\epsilon(\nu, T))_{\epsilon>0}$ satisfies the uniform bound*

$$0 < m \leq \sigma_\epsilon(\nu, T) \leq M\,, \quad \text{for each } \epsilon, \nu > 0 \text{ and } T \in [\theta, \Theta]\,.$$

Assume moreover that, for each $T > 0$, the family $\sigma_\epsilon(\cdot, T)$ converges in the sense of Young measures[2] *to $(\mu_\nu^T)_{\nu>0}$ as $\epsilon \to 0^+$.*

Hence, in the limit as $\epsilon \to 0^+$,

$$I_\epsilon \rightharpoonup I = \int_0^{+\infty} J\, ds \text{ in } L^\infty(\mathbb{R}_+ \times \mathbb{R}^3 \times \mathbb{S}^2 \times \mathbb{R}_+) \text{ weak-}*\,,$$

where $J \equiv J(t, s, x, \omega, \nu)$ is the solution of

$$\begin{cases} \dfrac{1}{c}\partial_t J + \omega \cdot \nabla_x J - \partial_s J = \dfrac{d^2 \tilde{\mu}_\nu^T}{ds^2} B_\nu(T)\,, \\[2ex] J\Big|_{t=0} = -I^{in}(x, \omega, \nu)\dfrac{d\tilde{\mu}_\nu^T}{ds}(s)\,, \end{cases}$$

for $(t, s, x, \omega, \nu) \in \mathbb{R}_+^* \times \mathbb{R}_+^* \times \mathbb{R}^3 \times \mathbb{S}^2 \times \mathbb{R}_+^*$, where $\tilde{\mu}_\nu^T$ is the Laplace transform of μ_ν^T.

Remark 1 The homogenized problem (3) is a transport equation where the "space" variable is the vector (x, s). The obtained equation defines a semigroup acting on the extended phase space $\mathbb{R}_+ \times \mathbb{R}^3 \times \mathbb{S}^2 \times \mathbb{R}_+ = \{(s, x, \omega, \nu)\}$, instead of the standard phase space $\mathbb{R}^3 \times \mathbb{S}^2 \times \mathbb{R}_+ = \{(x, \omega, \nu)\}$. The semigroup property is however lost in the standard phase space.

3.4 *Homogenization of the Linear Boltzmann Equation with a Highly Oscillating Scattering Term in Extended Phase Space*

In [3], the authors study an homogenization problem for the linear Boltzmann equation defined in the d-dimensional torus $\boldsymbol{T}^d$, with velocity in a bounded set $\mathscr{V} \subset \mathbb{R}^d$. Under the assumptions

---

[2] A family of functions $\sigma_\varepsilon(v)$ converges in the sense of Young measures to a measure $\mu_v \in \mathscr{M}(\mathscr{V})$ if and only if

$$\forall g \in C_0(\mathbb{R}_+), \quad g(\sigma_\varepsilon) \overset{*}{\rightharpoonup} \int_0^{+\infty} g(s)\mu_v(ds) \text{ in } L^\infty(\boldsymbol{T}^d \times \mathscr{V}) \text{ weak-}*\,.$$

$$\kappa \in C(\mathcal{V} \times \mathcal{V}), \quad \kappa \geq 0, \quad \kappa(v, w) = \kappa(w, v), \quad \text{and} \quad \mathcal{K}\mathbf{1}_{\mathcal{V}} = \mathbf{1}_{\mathcal{V}},$$

let $\mathcal{K} \in \mathcal{L}(L^1(\mathcal{V}); L^1(\mathcal{V}))$ be the scattering operator, defined as follows:

$$\mathcal{K}\phi(v) := \int_{\mathcal{V}} \kappa(v, w)\phi(w)\mathrm{d}w, \quad \forall \phi \in L^1(\mathcal{V}).$$

In particular, $\mathcal{K}$ describes in which way the particles interact with the medium. The second key quantity in the linear Boltzmann equation is the scattering coefficient $\sigma_\varepsilon \equiv \sigma_\varepsilon(v)$, which describes the probability of interaction with the background. In [3], the scattering coefficient is supposed to be independent of the spatial position. Moreover, it has a highly oscillatory behavior, encoded by the parameter $\varepsilon > 0$.

The scattering coefficient is bounded, i.e. there exist two constants $0 < c \leq C < +\infty$ such that

$$\forall \varepsilon > 0, \ \forall v \in \mathcal{V}, \quad C \geq \sigma_\varepsilon(v) \geq c. \tag{9}$$

The authors then consider the following family of ε-dependent problems for the unknowns f_ε, which denote, at fixed ε, the density at time $t \in \mathbb{R}_+$ of particles with velocity $v \in \mathcal{V}$ and located at $x \in \boldsymbol{T}^d$, moving in the host medium characterized by the scattering operator and by the scattering coefficient:

$$\begin{cases} \partial_t f_\varepsilon + v \cdot \nabla_x f_\varepsilon + \sigma_\varepsilon(v)(f_\varepsilon - \mathcal{K} f_\varepsilon) = 0 \\ f_\varepsilon|_{t=0} = f^{\text{in}} \in L^\infty(\boldsymbol{T}^d \times \mathcal{V}). \end{cases} \tag{10}$$

Existence and uniqueness of a nonnegative mild solution as well as the maximum principle for the Cauchy problem (10) are classical. In particular,

$$\|f_\varepsilon\|_{L^\infty(\boldsymbol{T}^d \times \mathcal{V})}(t) \leq \left\|f^{\text{in}}\right\|_{L^\infty(\boldsymbol{T}^d \times \mathcal{V})} \ \forall t > 0.$$

Thanks to the uniform boundedness of σ_ε in $L^\infty(\mathcal{V})$, the authors deduce that, up to a subsequence, there exists a measure $\mu_v \in \mathcal{M}(\mathcal{V})$ such that $\sigma_\varepsilon \Rightarrow \mu_v$ in the sense of Young measures. Denoting with $\tilde{\mu}_v$ the Laplace transform of μ_v, it is possible to deduce that

$$\forall n \in \boldsymbol{N}, \quad \sigma_\varepsilon^n e^{-\sigma_\varepsilon s} \overset{*}{\rightharpoonup} (-1)^n \frac{\mathrm{d}^n \tilde{\mu}_v}{\mathrm{d}s^n}(s)$$

in $L^\infty([0, T] \times \boldsymbol{T}^d \times \mathcal{V})$ weak-$*$ as $\varepsilon \to 0^+$.

In the extended phase space, let $F \equiv F(t, s, x, v)$ be the mild solution of

$$\begin{cases} (\partial_t + v \cdot \nabla_x - \partial_s)F = \dfrac{\mathrm{d}^2}{\mathrm{d}s^2}\tilde{\mu}_v(s)\mathcal{K} \displaystyle\int_0^{+\infty} F\mathrm{d}s \\ F(0, s, x, v) = -\dfrac{\mathrm{d}\tilde{\mu}_v}{\mathrm{d}s}(s) f^{\text{in}}(x, v). \end{cases} \tag{11}$$

The following theorem holds.

Theorem 4 *Up to a subsequence,*

$$f_\varepsilon \stackrel{*}{\rightharpoonup} \int_0^{+\infty} F \mathrm{d}s$$

in $L^\infty([0,T] \times \boldsymbol{T}^d \times \mathscr{V})$ *weak-*$*$ *as* $\varepsilon \to 0^+$.

The ε-dependent problem in the extended phase space has been obtained by setting

$$F_\varepsilon(t,s,x,v) := \sigma_\varepsilon(v) e^{-s\sigma_\varepsilon(v)} f_\varepsilon(t,x,v), \quad (t,s,x,v) \in \mathbb{R}_+ \times \mathbb{R}_+ \times \boldsymbol{T}^d \times \mathscr{V}.$$

Hence, F_ε solves the following kinetic equation in the extended phase space:

$$\begin{cases} (\partial_t + v \cdot \nabla_x - \partial_s) F_\varepsilon = \sigma_\varepsilon^2 e^{-\sigma_\varepsilon s} \mathscr{K} \displaystyle\int_0^{+\infty} F_\varepsilon \mathrm{d}s \\ F_\varepsilon(0,s,x,v) = \sigma_\varepsilon(v) e^{-s\sigma_\varepsilon(v)} f^{\mathrm{in}}(x,v). \end{cases} \tag{12}$$

Remark 2 The same result holds in the case of an open bounded domain Ω with absorbing boundary conditions, with a similar strategy of proof, *mutatis mutandis*.

Remark 3 Another approach to the study of the homogenization problem for the linear Boltzmann equation with highly oscillating cross sections in the energy variable is based on the two-scale convergence theory. The corresponding results are described in [9].

4 Numerical Study of Kinetic Equations Based on the Extended Phase Space Technique

The previous Section describes the theoretical results available so far. However, the extended phase space technique may also be of interest in the numerical study of such problems. This aspect may seem, at first sight, strange, since the high dimensionality of the equations can introduce a difficulty in preparing efficient numerical methods. However, in the context of homogenization problems, such as those described in the previous Section, the increase in the problem size can be largely offset by the reduction on the discretization mesh and by the need to homogenize only the initial datum. Moreover, in reconstructing the solution of the homogenized problem, the interval in which the effect of the additional variable is important is relatively small, being weighted with a negative exponential.

4.1 Homogenization of the Linear Radiative Transfer Equation with Highly Oscillating Opacities

In [11], the authors proposed a numerical method inspired from the extended phase space strategy for studying radiative transfer problems with highly oscillating opacities. They considered the same problem theoretically studied in [2], by discretizing (3) in the phase space individuated by the variables (s, x, ω, ν).

They introduced, in the interval of definition of the variable ν, a regular Cartesian grid, $(\nu_j)_{0 \leq j \leq J}$ with $J \geq 1$ and fixed $\Delta\nu > 0$, so that $\nu_j = j\Delta\nu$.

The rapid oscillations of the initial data have been filtered by applying the Laplace transform as indicated in the statement of Theorem 3. Then, the numerical solution of (3) has been obtained by the numerical integration of the transport-like equation by means of a standard method for transport problems.

Finally, the solution of the homogenized problem in the standard phase space is obtained by the numerical integration of the numerical solution of (3), with respect to s. Because of the peculiar structure of the integral, the contributions to the integral for large values of s are negligible (because of the presence of a negative exponential), and hence only small values of s are of interest in the numerical approximation of the limiting solution.

Remark 4 An advantage of this method is that the homogenization of (3) requires to manage the rapid oscillations only at the level of the initial data, whereas the next steps of the numerical approximation do not require to deal with fast oscillations anymore.

Remark 5 In this situation, the method allows to find an approximate solution to the problem without the need to handle the memory terms that appear in the limit equation.

4.2 Practical Implementation of the Numerical Strategy

We conclude this Section with an application of the numerical method described above to a model problem. Consider the initial-value ordinary differential equation for all $z \in \Omega = [0, 1]$, where $\epsilon > 0$ is a small parameter:

$$\begin{cases} \dfrac{du_\epsilon}{dt} + \sin^2\left(\dfrac{z}{\epsilon}\right) u_\epsilon = 0, \quad t > 0,\ z \in \Omega, \\ u_\epsilon(0, z) = u^{in}(z) = \dfrac{1}{2}(1 - \cos(2\pi z)). \end{cases} \tag{13}$$

This problem has one and only one solution, which can be explicitly written as

$$u_\epsilon(t, z) = \frac{1}{2}(1 - \cos(2\pi z))e^{-t\sin^2(z/\epsilon)}, \quad t > 0,\ z \in \Omega,$$

for each $\epsilon > 0$.

In the limit $\epsilon \to 0^+$, $u_\epsilon \rightharpoonup u$ in $L^\infty(\mathbb{R}_+ \times \Omega)$weak-*, where u is given by the formula

$$u(t,z) = \frac{1}{2}(1-\cos(2\pi z))\Phi(t), \quad t>0,\ z\in\Omega, \tag{14}$$

with

$$\Phi(t) = \int_0^1 e^{-t\sin^2(y)}dy\,, \quad t \geq 0. \tag{15}$$

Clearly, u is not the solution of the problem

$$\begin{cases} \dfrac{du}{dt}(t,z) + \dfrac{1}{2}u(t,z) = 0, & t>0,\ z\in\Omega, \\ u(0,z) = \dfrac{1}{2}(1-\cos(2\pi z)), \end{cases} \tag{16}$$

obtained by simply computing the weak limit of the function $z \mapsto sin^2(z/\epsilon)$ as $\epsilon \to 0$.

Indeed, thanks to a remark by Tartar [13, 14], we know that the solution u of the homogenized problem satisfies an integro-differential equation with a non-zero memory term.

The numerical strategy described in this Section is based on the theoretical work studied in [2] on the homogenization in an extended phase space. Define

$$U_\epsilon(t,s,z) := \sin^2\left(\frac{z}{\epsilon}\right) u_\epsilon(t,z)e^{-s\sin^2\left(\frac{z}{\epsilon}\right)}, \quad t,s\geq 0,\ z\in\Omega. \tag{17}$$

Thanks to the explicit structure of U_ϵ, we can deduce that U_ϵ satisfies the following free transport equation, with the variable z acting as a parameter:

$$(\partial_t - \partial_s)U_\epsilon(t,s,z) = \sin^2\left(\frac{z}{\epsilon}\right) e^{-s\sin^2\left(\frac{z}{\epsilon}\right)} \left(\frac{du_\epsilon}{dt}(t,z) + \sin^2\left(\frac{z}{\epsilon}\right) u_\epsilon(t,z)\right).$$

Since u_ϵ solves (13), the right-hand side of the previous equation is equal to zero. Consequently U_ϵ satisfies

$$\begin{cases} \partial_t U_\epsilon - \partial_s U_\epsilon = 0\,, & t,s>0\,,\ z\in\Omega, \\ U_\epsilon(0,s,z) = \dfrac{1}{2}(1-\cos(2\pi z))\sin^2\left(\dfrac{z}{\epsilon}\right) e^{-s\sin^2\left(\frac{z}{\epsilon}\right)}. \end{cases} \tag{18}$$

Hence, once averaged the high-frequency oscillations of the initial condition, the transport equation can be numerically treated with any solver for the free transport equation (or exactly solved, and then interpolated on the grid points in the phase space), and subsequently integrated with respect to the supplementary variable to give the solution of the homogenized problem $u = \lim_{\epsilon\to 0} u_\epsilon$.

In the numerical example described here, we have supposed $\epsilon = 0.01$. After introducing an uniform grid on Ω, composed of $N_x = 10^3 + 1$ points and with grid step $\Delta x = 2 \times 10^{-3}$, we have computed the exact solution of (13) on each grid point at $t = 1$. We have moreover computed, for $t = 1$, the exact solution of the homogenized problem, i.e. Formula (14) and (15), and the exact solution of the initial value problem (16) on each grid point.

In order to compute the numerical solution of the homogenized problem in the extended phase space, we have imposed that $s \in [0, 8]$ and we have introduced a uniform grid on the interval $[0, 8]$, with grid step $\Delta s = 2 \times 10^{-3}$. The time step of our simulations is $\Delta t = \Delta s = 2 \times 10^{-3}$. The bounds of the interval for the s variable allow to keep a reasonable accuracy in the reconstruction of the limit density, because of the rapid decay of U_ϵ for $s \to +\infty$ (Formula (17)), except in the spatial grid points in which the argument of the exponential in (17) vanishes. But, in the same grid points, U_ϵ itself is equal to zero, hence preventing the possible degradation of the precision of the integral with respect to the s variable.

In the next figures, we compare the aforementioned exact results to the numerical results obtained through the extended phase space method.

In Fig. 1 we show the solution of the non-homogenized problem (13) and, in Fig. 2, the numerical solution of the homogenized problem obtained via (18), i.e. with the phase space method, when $t = 1$. We see that neither the numerical solution of (18)

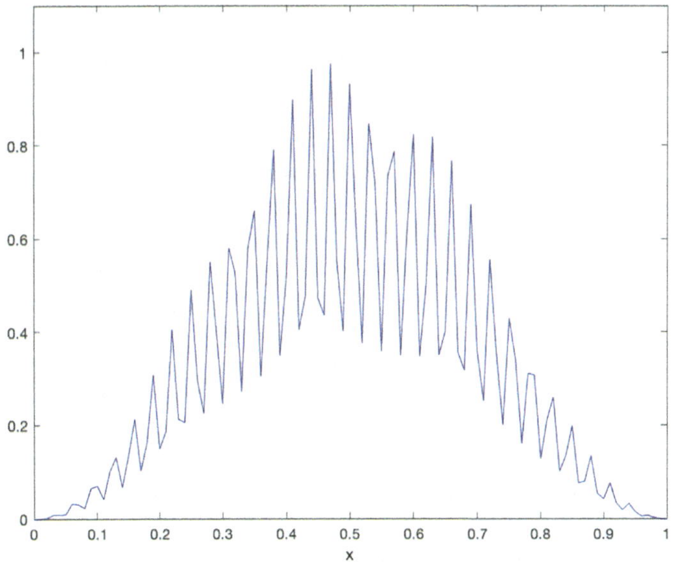

Fig. 1 Solution of (13) when $t = 1$

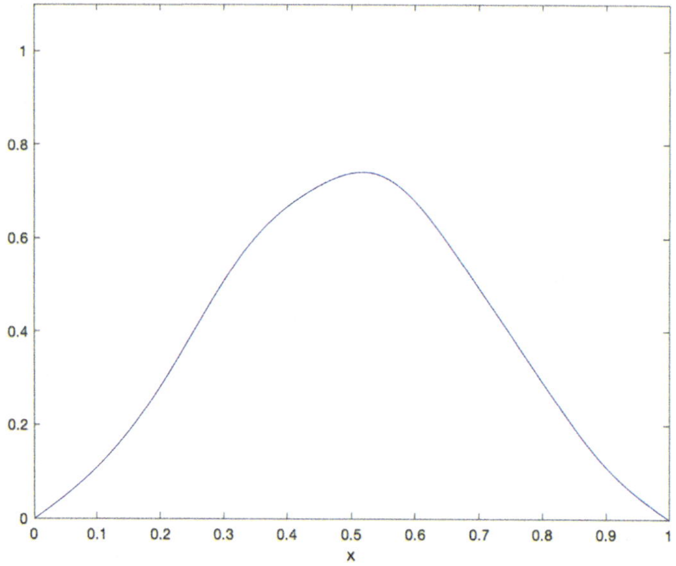

Fig. 2 Solution of the homogenized problem (18), obtained with the phase space method, when $t = 1$

nor the projection of the exact solution of Problem on the spatial grid points are exactly symmetric around $x = 0.5$. This behavior is a consequence of the choices on the discretization step and on the period of the trigonometric functions appearing in the problem.

In Fig. 3 we have plotted the graph of the exact solution of the homogenized problem, given by (14) and (15) (continuous line) and the exact solution of the initial-value problem (16) (dash-dotted line), when $t = 1$. The effect of the memory term on the homogenized solution of (13) is clearly apparent.

Figure 4 visualizes the difference between the exact homogenized solution (14) and (15) and the numerical solution of the homogenized problem obtained with the extended phase space method. We note that the error obtained by using the strategy described in this Section is quite good, despite the low density of grid points.

Finally, in Fig. 5, we show the initial condition of (18), which is the crucial step of the extended phase space method, when $\epsilon = 0.01$. The decay to zero is almost achieved for $s = 8$; this behavior helps in keeping a reasonable size of the numerical discretization, even if the dimensionality of the problem has been increased.

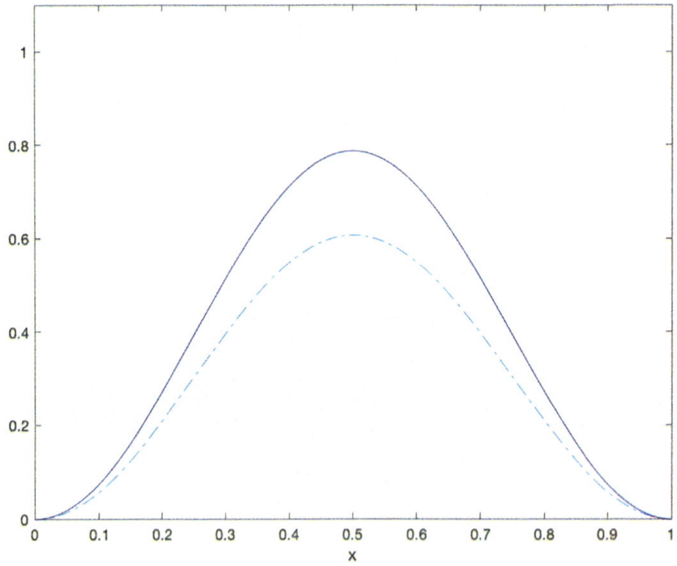

Fig. 3 Comparison between the exact solution of the homogenized problem (14) and (15) (continuous line) and the solution of (16) (dash-dotted line) for $t = 1$

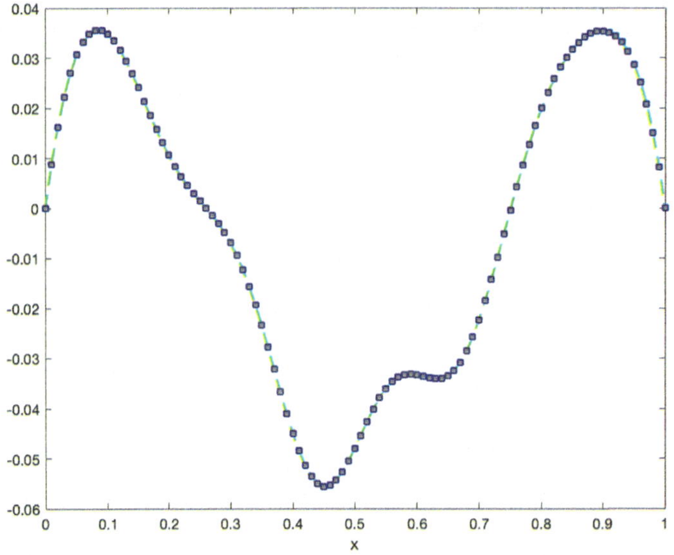

Fig. 4 Difference between the exact homogenized solution and the numerical solution of the homogenized problem obtained with the phase space method, when $t = 1$

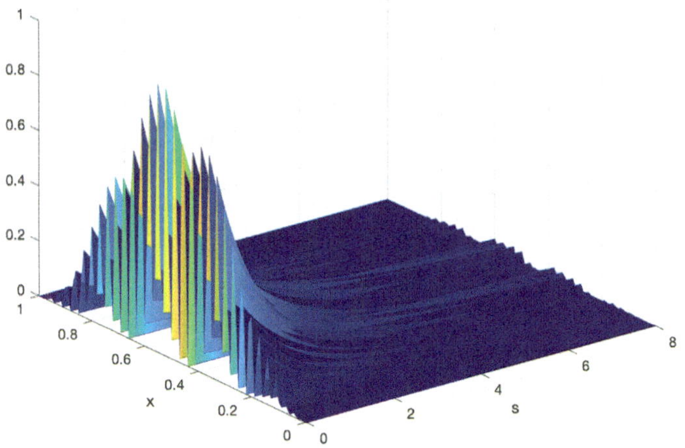

Fig. 5 Initial condition of the transport Eq. (18) in the extended phase space, with $\epsilon = 0.01$

Acknowledgements The author thanks the anonymous referees, whose remarks have helped to improve the article. This article has been written in the framework of the Pessoa project 47871WF "Kinetic theory, particle systems and their hydrodinamic limits" (2022–2023), and under the auspices of the Italian National Institute of Higher Mathematics (INdAM), GNFM group.

References

1. Bernard, E., Caglioti, E., Golse, F.: Homogenization of the linear Boltzmann equation in a domain with a periodic distribution of holes. SIAM J. Math. Anal. **42**(5), 2082–2113 (2010)
2. Bernard, E., Golse, F., Salvarani, F.: Homogenization of transport problems and semigroups. Math. Methods Appl. Sci. **33**(10), 1228–1234 (2010)
3. Bernard, E., Salvarani, F.: Homogenization of the linear Boltzmann equation with a highly oscillating scattering term in extended phase space. Appl. Math. Lett. **143**, 108672 (2023)
4. Caglioti, E., Golse, F.: The Boltzmann-Grad limit of the periodic Lorentz gas in two space dimensions. C. R. Math. Acad. Sci. Paris **346**(7–8), 477–482 (2008)
5. Caglioti, E., Golse, F.: On the Boltzmann-Grad limit for the two dimensional periodic Lorentz gas. J. Stat Phys **141**(2), 264–317 (2010)
6. Dautray, R.: Méthodes probabilistes pour les équations de la physique. In: Collection du Commissariat à l'Énergie Atomique. Série Synthèses. Commissariat à lÉnergie Atomique, Paris (1989)
7. Golse, F.: On the periodic Lorentz gas in the Boltzmann-Grad scaling. Ann. Faculté des Sci. Toulouse (6), **17**(4), 735–749 (2008)
8. Golse, F.: Homogenization and kinetic models in extended phase space. Riv. Math. Univ. Parma (N.S.) **3**(1), 71–89 (2012)
9. Hutridurga, H., Mula, O., Salvarani, F.: Homogenization in the energy variable for a neutron transport model. Asymptot. Anal. **117**(1–2), 1–25 (2020)

10. Marklof, J., Strömbergsson, A.: The Boltzmann-Grad limit of the periodic Lorentz gas. Ann. Math. (2) **174**(1), 225–298 (2011)
11. Mathiaud, J., Salvarani, F.: A numerical strategy for radiative transfer problems with higly oscillating opacities. Appl. Math. Comput. **221**, 249–256 (2013)
12. Papanicolaou, G.C.: Asymptotic analysis of transport processes. Bull. Am. Math. Soc. **81**, 330–392 (1975)
13. Tartar, L.: Nonlocal effects induced by homogenization. In: Partial Differential Equations and the Calculus of Variations, Volume 2 of Progress in Nonlinear Differential Equations and their Applications, pp. 925–938. Birkhäuser Boston, Boston, MA (1989)
14. Tartar, L.: An Introduction to Navier-Stokes Equation and Oceanography. Lecture Notes of the Unione Matematica Italiana, vol. 1. Springer-Verlag, Berlin, Heidelberg (2006)

Mesoscale Mode Coupling Theory for the Weakly Asymmetric Simple Exclusion Process

Gunter M. Schütz

Abstract The asymmetric simple exclusion process and its analysis by mode coupling theory (MCT) is reviewed. To treat the weakly asymmetric case at large space scale $x\varepsilon^{-1}$, large time scale $t\varepsilon^{-\chi}$ and weak hopping bias $b\varepsilon^{\kappa}$ in the limit $\varepsilon \to 0$ we develop a mesoscale MCT that allows for studying the crossover at $\kappa = 1/2$ and $\chi = 2$ from Kardar-Parisi-Zhang (KPZ) to Edwards-Wilkinson (EW) universality. The dynamical structure function is shown to satisfy for all κ an integral equation that is independent of the microscopic model parameters and has a solution that yields a scale-invariant function with the KPZ dynamical exponent $z = 3/2$ at scale $\chi = 3/2 + \kappa$ for $0 \leq \kappa < 1/2$ and for $\chi = 2$ the exact Gaussian EW solution with $z = 2$ for $\kappa > 1/2$. At the crossover point it is a function of both scaling variables which converges at macroscopic scale to the conventional MCT approximation of KPZ universality for $\kappa < 1/2$. This fluctuation pattern confirms long-standing conjectures for $\kappa \leq 1/2$ and is in agreement with mathematically rigorous results for $\kappa > 1/2$ despite the numerous uncontrolled approximations on which MCT is based.

Keywords Weakly asymmetric simple exclusion process · Mode coupling theory · Dynamical structure function · Nonlinear fluctuating hydrodynamics · Dynamical universality classes

1 Introduction

The asymmetric simple exclusion process (ASEP), introduced into the mathematical literature in the seminal paper by Spitzer [91], is arguably the conceptually most important stochastic interacting particle system [32, 59, 66, 88]. Informally this continuous-time Markov process can be described as follows. Each site k of the integer lattice is occupied by at most one particle (exclusion) so that the state of the

G. M. Schütz (✉)
Departamento de Matemática, Instituto Superior Técnico, Av. Rovisco Pais, 1049-001 Lisboa, Portugal
e-mail: gunter.schuetz@tecnico.ulisboa.pt

E. Carlen et al. (eds.), *From Particle Systems to Partial Differential Equations*,
Springer Proceedings in Mathematics & Statistics 465,
https://doi.org/10.1007/978-3-031-65195-3_16

particle system at any time τ is represented by occupation numbers $\eta(k) \in \{0, 1\}$ for $k \in \mathbb{Z}$. All particles have an associated alarm clock. When the clock of a particle on site $k \in \mathbb{Z}$ rings, which happens independently of all other clocks after an exponential random time of mean $\bar{\tau} = 1/(2D)$, the particle attempts to jump to another site. If the target site is occupied the jump attempt fails. In the ASEP only the nearest-neighbour target sites $k \pm 1$ are chosen with probability $[1 \pm B_\varepsilon/(2D)]/2$ resp. where $B_\varepsilon = b\varepsilon^\kappa$ is the bias that defines the strength of the asymmetry. The process considered on large space scales $k = x\varepsilon^{-1}$, large time scales $\tau = t\varepsilon^{-\chi}$ and bias with $\kappa > 0$ is called weakly asymmetric simple exclusion process (WASEP). The process is called symmetric simple exclusion process (SSEP) for $b = 0$ and totally asymmetric simple exclusion process (TASEP) for maximal bias $b = 2D, \kappa = 0$. The parameter D defines a diffusion coefficient.

Over the last five decades the ASEP has become a source of deep insights into generic behaviour of non-equilibrium interacting particle systems in one dimension and is thus sometimes dubbed the Ising model of non-equilibrium physics. The Markov generator is closely related to the quantum Hamiltonian of the spin-1/2 Heisenberg chain [3, 48] which is completely integrable in the Yang-Baxter sense [7, 45]. This link allows for using the algebraic machinery involving non-Abelian symmetries, matrix product approaches, and Bethe ansatz in its various guises to obtain exact results particularly on the microscopic properties of the model [32, 88] and has given prominence to the field of integrable probability [20].

On the other hand, probabilistic methods have led to a detailed and often mathematically rigorous understanding of the emergence of large-scale behaviour [16, 59, 66, 93]. The coarse-grained particle density of the SSEP evolves under scaling of space and time with $\chi = 2$ according to the diffusion equation and also the fluctuations around this deterministic limit are diffusive with the dynamical exponent $z = 2$ of the Edwards-Wilkinson (EW) universality class [38]. The coarse-grained density of the ASEP, on the other hand, evolves under Eulerian scaling ($\chi = 1$) according to the Burgers equation which exhibits shocks. Density fluctuations on the finer scale $\chi = 3/2$ belong to the celebrated superdiffusive Kardar-Parisi-Zhang (KPZ) universality class with dynamical exponent $z = 3/2$ [53, 54]. Over time, a close relationship to directed polymers and growth models has revealed deep links between the ASEP, random matrix theory and the Kardar-Parisi-Zhang equation [26, 51, 61, 70, 97].

The WASEP with $\kappa = 1$ conditioned on carrying an atypically low current exhibits a dynamical phase transition from a shock-like traveling wave behaviour to the flat density profile characteristic for typical currents [10, 19], while for atypically high current in the ASEP one finds a flat profile with algebraically decaying stationary density correlations and dynamical exponent $z = 1$ [50, 72, 94] that have their origin in conformal invariance [56] that describes equilibrium critical phenomena in two-dimensions [23], thus bringing in view yet another seemingly unrelated field of research.

Deep links exist also to quantum mechanical systems. The matrix product ansatz, originally invented for isolated quantum spin chains [1], subsequently further developed for the open ASEP [30], and then generalized to the open Heisenberg spin chain

[79, 80] has revealed intriguing stationary quantum states with density profiles [55, 79] and long-range correlations [21, 79] that are reminiscent of the open SSEP [16, 33, 34, 92]. On the other hand, developing the Bethe ansatz further has led to generalized hydrodynamics [18, 24] which captures the large-scale temporal evolution of the quantum Heisenberg chain and related integrable many-body quantum system on Eulerian scale by taking into account the presence of infinitely many conservation laws [37, 39]. Interestingly, also in the quantum setting fluctuations exhibit scaling behaviour known from the classical KPZ universality class [28, 29, 67, 83, 107].

Remarkably, despite its simplicity the one-dimensional ASEP and its specializations SSEP and TASEP are important models also from an experimental perspective. The SSEP is a model for interface fluctuations in the universality class of the Edwards-Wilkinson equation with numerous experimental realizations [38, 62]. It also serves as a model for reptation dynamics of entangled polymers [36] and has been shown [87] to explain experimental data on the tube-like motion of a single polymer-chain [71]. A single tagged particle in the SSEP has a subdiffusive mean square displacement [6, 100] which has been observed in single-file diffusion in zeolites [63] and colloidal systems [105]. In the presence of pair annihilation the generator of the process can be expressed in terms of free fermions [8, 47, 68, 85] and thus allows for a very detailed rigorous study of diffusion-limited annihilation characterized in one dimension by an anomalous algebraic decay of the particle density which had been discovered earlier in exciton dynamics on polymer chains [78, 90, 103].

The ASEP exhibits shocks both microscopically and on macroscopic Eulerian scale [31, 40, 60, 82] with diffusive fluctuations of the shock position [9, 41] that clarify [86] an apparent slowing down of ribosomes while moving on a mRNA template during protein synthesis, a fundamental biological process for which the ASEP was originally conceived as a qualitative model [69]. Perhaps the most spectacular experimental realization of ASEP behaviour is a series of beautiful experiments on turbulent liquid crystal systems by Takeuchi and Sano where interface fluctuations follow with remarkable accuracy the universal behaviour expected from the KPZ universality class [98, 99]. Predictions from generalized hydrodynamics for integrable quantum systems have been tested in–by now–numerous experiments with cold atoms [57, 64, 104]. A novel feature of the Heisenberg quantum chain that seems to have no analogue in the classical ASEP, but has a mathematical structure in common with shock measures, is the appearance of spin helix states, originally explored theoretically in [75] and recently discovered in experiments on anisotropic Heisenberg chains [52].

This brief survey of exact, rigorous and experimental literature published from 1968 until today can only hint at the reasons why the ASEP continues to fascinate. Indeed, it might come as a surprise that there are still open questions related to the ASEP and its relationship to KPZ universality. From an experimental viewpoint that most pressing one is that despite the mathematically expected robustness of the KPZ universality class the corresponding asymptotic behaviour of experimentally accessible quantities is rarely seen in classical systems. The examples given above (and some more) are notable exceptions. This raises the question of corrections to scaling. Recent numerical evidence suggests that diffusive corrections are important [84], but otherwise this question is essentially completely open.

A second major open problem closely related to the ASEP is the rigorous derivation of dynamical universality classes in processes with more than one conservation law. This issue has been addressed systematically for two conservation laws with the mode coupling approximation [73, 95, 96] which revealed the existence of new universality classes beyond the ubiquitous diffusion and KPZ. Subsequent work on an arbitrary number of conservation laws showed that all these universality classes are the members of a discrete infinite family [74]. This series of dynamical universality classes was dubbed the Fibonacci family because the dynamical exponents are all Kepler ratios of neighbouring Fibonacci numbers F_i. The series begins with Edwards-Wilkinson (EW) ($z = 2 = F_3/F_2$), is then followed by (a) KPZ, (b) a possibly new universality class related to KPZ, and (c) a third universality class named after Lévy, all with $z = 3/2 = F_4/F_3$, then the heat mode exponent $z = 5/3 = F_5/F_4$ [102], and then the remaining series $z_i = F_{i+1}/F_i$ up to the limit value $z_\infty = \varphi := (1 + \sqrt{5})/2$ which is the famous golden mean.

All of these mode coupling studies consider the large-scale limit $x \to \infty$, $t \to \infty$ for strong asymmetry $\kappa = 0$. An open problem is the question is the extent to which these predictions are correct. Numerical Monte-Carlo studies confirm the predicted universal scaling functions [74], but not generally the non-universal scale factors. It therefore remarkable that in one particular case rigorous work on a chain of oscillators [11, 13] fully confirms the mode coupling prediction. In a related series of rigorous work on two conservation laws with weak asymmetry the KPZ and Lévy universality classes with $z = 3/2$ as well as the EW class with $z = 2$ were found [2, 12, 14], but the present state of art of the MCT analysis for strong asymmetry is not applicable and can therefore not be tested against these results.

This leads to a third and more subtle open issue which is the strong universality conjecture and concerns the exact relationship between the KPZ fixed point (characterizing the KPZ universality class) and the WASEP which exhibits a transition from KPZ to EW behaviour at a critical scale of a weak bias [46] made more precise below. It is the purpose of this paper to extend the mode coupling approximation to the WASEP to address this issue from the mode coupling perspective. The main new contributions of the present work are to point out the existence on a mesoscopic scale of generalized mode coupling solutions applicable to the weakly asymmetric case (and then by inference to any system with one conservation law that is governed by nonlinear fluctuating hydrodynamics) and to elucidate under which conditions on the microscopic parameters these solutions are scale invariant.

To put these results into perspective we first continue the review of the ASEP, but with a focus on weak asymmetry (Sect. 2). In Sect. 3 the conventional mode coupling approximation is introduced and the underlying assumptions are made explicit. The mesoscopic MCT is developed in Sect. 4 and the asymptotic solution for the weakly asymmetric case is proved and commented upon. A brief summary and some final remarks are presented in Sect. 5.

2 The (W)ASEP and Its Relation to KPZ and EW Universality

For any bias

$$B_\varepsilon = b\varepsilon^\kappa \tag{1}$$

the unique invariant measure of the ASEP that is also translation invariant is the Bernoulli product measure with density ρ and static compressibility $K(\rho) = \rho(1-\rho)$ [66]. For $B \neq 0$ this invariant measure is not reversible so that the ASEP defines a particle system that is perpetually out of thermal equilibrium with stationary particle current $j_\varepsilon(\rho) = B_\varepsilon \rho(1-\rho)$. The derivatives

$$c_\varepsilon(\rho) := \frac{\mathrm{d}}{\mathrm{d}\rho} j_\varepsilon(\rho) = B_\varepsilon(1-2\rho), \quad H_\varepsilon(\rho) := \frac{\mathrm{d}^2}{\mathrm{d}\rho^2} j_\varepsilon(\rho) = -2B_\varepsilon \tag{2}$$

are the collective velocity $c_\varepsilon(\rho)$ which is the center-of-mass velocity of the density fluctuations [88] and the coupling strength $H_\varepsilon(\rho)$ that is a measure of the strength of the nonlinearity of the current as a function of the particle density ρ. Below we drop the dependence of functions on the microscopic parameters b, κ, ρ and the diffusion coefficient D to avoid excessive notation. However, we mark the dependence on the bias parameter ε by a subscript as at a later stage the limit $\varepsilon \to 0$ is taken.

These collective velocity and the strength of the non-linearity alone say nothing about the fluctuations of the centered occupation numbers $\bar{\eta}(k,t) := \eta(k,t) - \rho$. A quantity of major interest in this context is the large-scale behaviour of the stationary covariance

$$C_\varepsilon^{WASEP}(k,\tau) := \langle \bar{\eta}(k,\tau)\bar{\eta}(0,0) \rangle, \quad k \in \mathbb{Z} \tag{3}$$

with expectation taken w.r.t. the Bernoulli product measure with density ρ. Its Fourier transform

$$S_\varepsilon^{WASEP}(q,\tau) = \frac{1}{\sqrt{2\pi}} \sum_{k\in\mathbb{Z}} \mathrm{e}^{-iqk} C_\varepsilon^{WASEP}(k,\tau) \tag{4}$$

is called the dynamical structure function, abbreviated DSF throughout this paper. Since $\eta(k) \in \{0,1\}$ and $\rho = \langle \eta(k) \rangle \in [0,1]$ the covariance $C_\varepsilon^{WASEP}(k,\tau)$ is trivially uniformly bounded from above and below by ± 1. For $\tau = 0$ the Bernoulli product measure yields $C_\varepsilon^{WASEP}(k,0) = K\delta_{k,0}$ and therefore the DSF satisfies

$$S_\varepsilon^{WASEP}(q,0) = S_\varepsilon^{WASEP}(0,\tau) = \frac{K}{\sqrt{2\pi}} \tag{5}$$

independently of the bias (1). For $\rho = 1/2$ where $c = 0$, reflection symmetry yields

$$S_\varepsilon^{WASEP}(q,\tau) = S_\varepsilon^{WASEP}(-q,\tau) \tag{6}$$

also independently of the bias.

Of particular interest is the large-scale behaviour of the DSF of the WASEP as seen from a shifted coordinate frame moving with the collective velocity c. Since such a shift can be removed by a Galilei transformation we may without loss of generality take $c = 0$ without restricting the consideration to $\rho = 1/2$. For the WASEP the large-scale behaviour is defined by the parameter ε appearing in the bias which, as mentioned above, is taken to define a microscopic length scale. We call the bias strong for $\kappa = 0$ and weak for $\kappa > 0$, with the qualifications "moderately weak" for the range $0 < \kappa < 1/2$, "critically weak" for $\kappa = 1/2$, and "very weak" for $\kappa > 1/2$. After Fourier transformation the large-scale behaviour of the DSF is then characterized by choosing $q = \varepsilon p$ and $\tau = \varepsilon^{-\chi} t$ and taking the limit $\varepsilon \to 0$. The parameter $\chi > 0$, which we call observation exponent, quantifies how much one speeds up the microscopic time relative to the "zooming out" in space by the factor $1/\varepsilon$.

On judiciously chosen large scales one expects scale invariance, i.e., after the Galilei shift and for some critical observation exponent χ^*, the normalized limiting dynamical structure function

$$s^*_{WASEP}(p, t) := \lim_{\varepsilon \to 0} S_\varepsilon^{WASEP}(\varepsilon p, \varepsilon^{-\chi^*} t) / S_\varepsilon^{WASEP}(0, 0) \tag{7}$$

depends non-trivially only on the scaling variable $u = lpt^{1/z}$ with universal dynamical exponent z and non-universal scale parameter l. Universality means that the microscopic details of the model, i.e., the density ρ, the bias parameters b and κ, and the diffusion coefficient D, enter only the scale parameter l, and that this is true not only for just some particular point in the four-dimensional manifold of the microscopic parameters ρ, b, κ, D.

Specifically, for strong asymmetry ($\kappa = 0$) the asymptotic density fluctuations of the ASEP that are quantified by the DSF belong to the KPZ universality class as was first shown using the mode coupling approximation [101]. At observation scale $\chi = 3/2$ the asymptotic DSF obtained in this way is scaling function with dynamical exponent $z = 3/2$. A numerical evaluation of this mode coupling solution [25, 43] shows a clear difference to the Gaussian that one obtains for the SSEP at observation scale $\chi = 2$ and which characterizes the diffusive EW universality class with dynamical exponent $z = 2$. The *exact* asymptotic dynamical structure function of the ASEP, however, remained elusive for a long time. It was eventually derived in a remarkable analysis of the polynuclear growth model [42, 77] which also belongs to the KPZ universality class. The universal scaling form (7) of the DSF is given by the Fourier transform of the Prähofer-Spohn scaling function $f_{PS}(\cdot)$ which is related to Painlevé equations. Remarkbaly, it was been observed with a high degree of accuracy in the experiments on nematic liquid crystals [98, 99] mentioned above.

For very weak asymmetry it has been known for a long time that for $\kappa = 1$ and on diffusive observation scale $\chi = 2$ the limiting density fluctuations of the WASEP converge to a generalized Ornstein-Uhlenbeck process [27, 35]. The corresponding universal dynamical structure function is the Gaussian with scaling variable $u = p\sqrt{Dt}$. This result was later extended to the whole range $\kappa > 1/2$ [46] and shows that generally for very weak asymmetry the density fluctuations belong to the Edwards-Wilkinson universality class [38] with dynamical exponent $z = 2$ like the SSEP where $b = 0$ (or, equivalently, $\kappa \to \infty$).

For the critically weak asymmetry $\kappa = 1/2$ one obtains after a Hopf-Cole transformation [44] an exponential process for which the (Galilei-shifted) limiting fluctuations are given by the stochastic heat equation [17]. Looking directly at the fluctuation field (i.e., without Hopf-Cole transformation) it was proved [46] that the limiting fluctuation field $\phi(x, t)$ satisfies at observation scale $\chi = 2$ (weakly) the stochastic Burgers equation (SBE) which we write loosely in the intuitive form

$$\partial_t \phi = D\partial_x^2 \phi + b(\partial_x \phi)^2 + \partial_x \xi \tag{8}$$

with Gaussian space-time white noise ξ.

The SBE, related to the KPZ equation [54] for the height function $h(x, t)$ via $\phi(x, t) = \partial_x h(x, t)$, is the generic evolution equation for fluctuation fields in systems with one conservation law that are governed by nonlinear fluctuating hydrodynamics [93]. The fundamental underlying assumption is the validity of a Boltzmann-Gibbs principle which asserts that the relaxation of the dynamical structure function is governed by the long-wave length Fourier modes of the conserved field. It should be noted that the fluctuation field $\phi(x, t)$ appearing in the SBE is *not* scale invariant under the transformation $x \to ax, t \to a^{3/2}t$ as one might naively expect from KPZ universality and which is true for the Markov field that defines the KPZ fixed point [70]. However, this scaling is expected to send $\phi(x, t)$ to the true universal fixed point. While universality is by now well-established [5, 81] it is not known how these scaling properties are reflected in the DSF for $\kappa = 1/2$ which was computed only for $\kappa = 0$.

In the moderately weak case $0 < \kappa < 1/2$ one expects the same universal KPZ fluctuations with $z = 3/2$ and Praehofer-Spohn scaling function as for the strongly asymmetric case $\kappa = 0$ [4]. Also, it is not obvious at the which observation scale χ^* this universal behaviour shows up. The exact spectral gap of the (W)ASEP computed from Bethe ansatz [58] suggests that $\chi^* = 3/2 + \kappa$, see also an argument for a system with two conservation laws that has been deduced [22] from the universal KPZ scaling discussed in [81]. A rigorous proof of these conjectures is lacking, however.[1] In the next section we discuss in some detail the mode coupling theory for the ASEP and WASEP to address further below the questions of the proper observation scale and of the associated scaling properties of the dynamical structure function.

[1] One may infer $\chi^* = 3/2 + \kappa$ also from an old result from coupling theory for the one-dimensional ASEP ($\kappa = 0$) [101] but this does not directly predict the range $0 < \kappa < 1/2$ nor the behaviour at $\kappa = 1/2$ unlike the mesoscale mode coupling theory developed below.

3 Mode Coupling Theory for the (W)ASEP

We recall that according to the definition of the ASEP [66] action of the generator $\mathscr{L}^{ASEP}$ on the centered occupation numbers $\bar{\eta}(k) := \eta(k) - \rho$ can be written for any κ as

$$\mathscr{L}^{ASEP}\bar{\eta}(k) = (D\Delta - c_\varepsilon \nabla)\bar{\eta}(k) - H_\varepsilon \bar{\eta}(k)\nabla\bar{\eta}(k) \tag{9}$$

where Δ is the discrete Laplacian defined by $\Delta f(k) := f(k+1) + f(k-1) - 2f(k)$ and ∇ denotes the symmetric discrete derivative defined by $\nabla f(k) := [f(k+1) - f(k-1)]/2$. To discuss the time evolution of the (W)ASEP it is convenient to split the generator $\mathscr{L}^{ASEP}$ in (9) into a linear part $\mathscr{L}_0^{ASEP}$ and a nonlinear part $\mathscr{L}_1^{ASEP}$ that would be absent for non-interacting particles, i.e.,

$$\mathscr{L}^{ASEP} = \mathscr{L}_0^{ASEP} + \mathscr{L}_1^{ASEP}. \tag{10}$$

The action (9) of the generator on the centered occupation numbers then reads

$$\mathscr{L}_0^{ASEP}\bar{\eta}(k) = D\Delta\bar{\eta}(k) - c_\varepsilon \nabla\bar{\eta}(k) \tag{11}$$

and with the asymmetric discrete gradients defined by $\nabla^+ f(k) := f(k+1) - f(k)$, $\nabla^- f(k) := f(k) - f(k-1)$

$$\mathscr{L}_1^{ASEP}\bar{\eta}(k) = -H_\varepsilon \bar{\eta}(k)\nabla\bar{\eta}(k) = -H_\varepsilon \nabla^- \bar{\eta}(k)\bar{\eta}(k+1). \tag{12}$$

One recognizes in (11) and (12) a discrete form of the stochastic Burgers equation (8). The obstacle for discussing the time evolution of the DSF further is the presence of the nonlinear part (12) of the generator. In principle, this problem can be tackled by Bethe ansatz [7, 48] and with the help of random matrix theory [77, 81]. However, this is not only an extremely hard problem, but similar nonlinear terms appear also in the time evolution of models for which no exact methods are known. Hence one appeals to universality and studies the DSF from a another starting point, viz., nonlinear fluctuating hydrodynamics [95, 97] which, as explained in the next Sect. 3.1, is a related stochastic Burgers equation which can then be treated analytically using mode coupling theory (MCT). This leads to an approximation for the evolution equation of the space-time covariances of the associated discrete SBE (Sect. 3.2), or, equivalently, after Fourier and Laplace transform done in Sect. 3.3 for the associated DSF. In the final step this approximate evolution is solved exactly in the large-scale limit.

Hence, mathematically speaking, the strategy is the following:

- Ideally, one would like to prove the large-scale properties (7) of the DSF (4) of the (W)ASEP for the fluctuation field $\bar{\eta}(k, t)$ which evolves by definition as given in (11) and (12), resembling a discrete SBE.
- Instead, one introduces a different fluctuation field $\phi(k, t)$ that evolves according to a similar discrete SBE (given by (15) and (16), see Sect. 3.1).

- To treat this discrete SBE one uses the approximation scheme known as mode coupling theory (Sect. 3.2) which is not rigorous but leads to the closed evolution Eq. (27) for the stationary covariances of the fluctuation field $\phi(k,t)$ (Sect. 3.3).
- In Sect. 4 the *exact* large scale solution of this approximate equation is then derived which is expected to exhibit asymptotic large-scale behaviour at least resembling the asymptotic DSF (7) of the WASEP.

The use of the discretized SBE and of the mode coupling theory is standard and goes back to [101]. The novelty of the present work is the mesoscale approach to treat the mode coupling equations presented in Sect. 4.

3.1 A Discrete Stochastic Burgers Equation

The first step in studying the large-scale behaviour of the space-time covariance (3) is to consider a real-valued random variable $\phi(k)$ with mean 0 instead of the centered binary random number $\bar{\eta}(k)$ and reinterpret the action (11) and (12) on $\bar{\eta}(k)$ as representing the action of a generator $\mathscr{L}^{dSBE} = \mathscr{L}_0^{dSBE} + \mathscr{L}_1^{dSBE}$ of some discrete nonlinear stochastic evolution equation for $\phi(k)$ with an explicit and simple invariant measure. In other words, the idea is to define a generator $\mathscr{L}^{dSBE}$ that acts on $\phi(k)$ in a fashion that mimicks the action of $\mathscr{L}^{ASEP}$ on $\bar{\eta}(k)$. This is achieved with the choice

$$\mathscr{L}_0^{dSBE} := \sum_{j\in\mathbb{Z}} \left[-c_\varepsilon \nabla\phi(j) + D\Delta\left(\phi(j) - \partial_{\phi(j)}\right)\right]\partial_{\phi(j)} \tag{13}$$

$$\mathscr{L}_1^{dSBE} := -H_\varepsilon \sum_{j\in\mathbb{Z}} \left[(\nabla\phi(j))\left(\phi(j) + \frac{1}{3}\Delta\phi(j)\right)\right]\partial_{\phi(j)} \tag{14}$$

which has the desired property

$$\mathscr{L}_0^{dSBE}\phi(k) = -c_\varepsilon\nabla\phi(k) + D\Delta\phi(k) \tag{15}$$

analogous to the linear part of the evolution in the ASEP. The nonlinear part satisfies

$$\mathscr{L}_1^{dSBE}\phi(k) = -H_\varepsilon\left(\phi(j) + \frac{1}{3}\Delta\phi(j)\right)\nabla\phi(k) = -H_\varepsilon\nabla^- F(k) \tag{16}$$

which resembles (12) with the average $[\phi(k-1) + \phi(k) + \phi(k+1)]/3 = (1 + \Delta/3)\phi(k)$ rather than just $\bar{\eta}(k)$ and with

$$F(k) = \frac{\phi^2(k) + \phi(k)\phi(k+1) + \phi^2(k+1)}{3} = \phi(k)\phi(k+1) + \frac{1}{3}\left(\nabla^-\phi(k)\right)^2. \tag{17}$$

This particular discretization ensures that the invariant measure of $\mathscr{L}^{dSBE}$ generator is a product of mean-zero normal distributions. As noted in [95] this is the generator associated with a discretized SBE (8) with Gaussian white noise.

Noting that

$$\mathrm{e}^{\mathscr{L}^{dSBE}\tau} = \mathrm{e}^{\mathscr{L}_0^{dSBE}\tau} + \int_0^\tau \mathrm{d}\sigma \, \mathrm{e}^{\mathscr{L}_0^{dSBE}(\tau-\sigma)} \mathscr{L}_1^{dSBE} \mathrm{e}^{\mathscr{L}^{dSBE}\sigma} \tag{18}$$

the space-time covariance

$$C_\varepsilon^{dSBE}(k,\tau) := \langle \phi(k,\tau)\phi(0,0) \rangle, \quad k \in \mathbb{Z} \tag{19}$$

of the new fluctuation fields $\phi(\cdot,\cdot)$ satisfies the evolution equation

$$\begin{aligned} \frac{\mathrm{d}}{\mathrm{d}\tau} C_\varepsilon^{dSBE}(k,\tau) &= \langle \phi(0)\mathrm{e}^{\mathscr{L}^{dSBE}\tau} \mathscr{L}_0^{dSBE}\phi(k) \rangle + \langle \phi(0)\mathrm{e}^{\mathscr{L}_0^{dSBE}\tau} \mathscr{L}_1^{dSBE}\phi(k) \rangle \\ &+ \int_0^\tau \mathrm{d}\sigma \, \langle \phi(0)\mathrm{e}^{\mathscr{L}_0^{dSBE}(\tau-\sigma)} \mathscr{L}_1^{dSBE} \mathrm{e}^{\mathscr{L}^{dSBE}\sigma} \mathscr{L}_1^{dSBE}\phi(k) \rangle \end{aligned} \tag{20}$$

where the expectation is taken w.r.t. to the Gaussian product measure. Since the action of $\mathscr{L}_1^{dSBE}\phi(k)$ is bilinear in the fields $\phi(\cdot)$ and all odd moments vanish in the expectation (since $\phi(k)$ has mean zero by construction), the linearity of $\mathscr{L}_0^{dSBE}$ implies that the second term on the r.h.s. of the second line vanishes. Next, notice that the adjoint of $\mathscr{L}_1^{dSBE}$ w.r.t. to the flat volume measure has the simple property $(\mathscr{L}_1^{dSBE})^T = -\mathscr{L}_1^{dSBE}$. By using translation invariance and the second equality in (16) the evolution equation reduces to

$$\begin{aligned} \frac{\mathrm{d}}{\mathrm{d}\tau} C_\varepsilon^{dSBE}(k,\tau) &= (-c_\varepsilon \nabla_k + D\Delta_k)\, C_\varepsilon^{dSBE}(k,\tau) \\ &+ H_\varepsilon^2 \sum_{l\in\mathbb{Z}} \int_0^\tau \mathrm{d}\sigma \, \check{C}^{dSBE}(k-l,\tau-\sigma)\Delta_l \Omega(l,0;\sigma) \end{aligned} \tag{21}$$

where the discrete derivatives ∇_m and Δ_m with subscript m act on the coordinate m,

$$\check{C}^{dSBE}(m,\tau-\sigma) := \langle \phi(0)\mathrm{e}^{\mathscr{L}_0^{dSBE}(\tau-\sigma)}\phi(m) \rangle \tag{22}$$

is the linearized covariance given by integrating (15), and

$$\Omega(l,m;\sigma) := \langle F(m)\mathrm{e}^{\mathscr{L}^{dSBE}\sigma} F(l) \rangle = \langle F(l,\sigma)F(m,0) \rangle \tag{23}$$

with $F(\cdot)$ defined in (17). This is an exact evolution equation for the covariance of the discretized SBE defined by $\mathscr{L}^{dSBE}$. Appealing to universality, one expects the large-scale behaviour of the DSF of the discretized SBE to be the same as the DSF (7) of the ASEP [95]. However, the exact evolution Eq. (21) is not straightforwardly tractable due to the appearance of the four-point function (23).

3.2 Mode Coupling Approximations

To proceed we follow [95, 101] and introduce two approximations that lead to the mode coupling equation for the space-time covariance (19).

MCT Approximation 1 (Gaussian):

The idea to handle the four-point function (23) is to treat the random numbers $\phi(\cdot, \cdot)$ as components of a mean-zero Gaussian random vector. The theorem of Isserlis [49, 106] then allows for expressing multi-point correlations between an even number of components in terms of two-point correlations while expectations of an odd number of components are generally zero. This yields the approximation

$$\langle \phi(k_1, \sigma_1) \dots \phi(k_m, \sigma_m) \rangle \approx \sum_{p \in P_n^{(2)}} \prod_{(i,j) \in p} \langle \phi(k_i, \sigma_i) \phi(k_j, \sigma_j) \rangle \tag{24}$$

where $P_n^{(2)}$ is the set of all the distinct pairings (i, j) of the set $\{1, \dots, m\}$. Here and below the symbol $\approx$ for approximate equality means "equal to the given expression on the r.h.s. of the equation plus an unspecified error term". There is no control over the magnitude of the error term and therefore we do not write it out.

It is easy to see that in the subsequent computations the gradient part in (17) generates only subleading contributions to the asymptotic behaviour and can therefore be neglected. Using translation invariance and stationarity and neglecting further subleading gradient terms then yields

$$\Omega(l, 0; \sigma) \approx 2\langle \phi(l, \sigma)\phi(0, 0) \rangle^2. \tag{25}$$

and therefore

$$\begin{aligned} \frac{\mathrm{d}}{\mathrm{d}\tau} C_\varepsilon^{dSBE}(k, \tau) &\approx (-c_\varepsilon \nabla_k + D\Delta_k)\, C_\varepsilon^{dSBE}(k, \tau) \\ &+ 2H_\varepsilon^2 \sum_{l \in \mathbb{Z}} \int_0^\tau \mathrm{d}\sigma\, \check{C}^{dSBE}(k - l, \tau - \sigma) \Delta_l \left[C_\varepsilon^{dSBE}(l, \sigma) \right]^2 \end{aligned} \tag{26}$$

involving both the full covariance and the linear evolution of the covariance.

MCT Approximation 2 (Closure):

The Gaussian approximation made in the previous paragraph still involves two different functions, $C_\varepsilon^{dSBE}(k, \tau)$ and $\check{C}^{dSBE}(k, \tau)$, instead of being a closed equation for the full variance $C_\varepsilon^{dSBE}(k, \tau)$ only. To realize a self-consistent treatment in terms of a closed equation, the linear part $\check{C}_\varepsilon^{dSBE}(k, \tau)$ is replaced *ad hoc* by the full covariance. In other words, we define a new function $C_\varepsilon^{MCT}(k, \tau)$ as being the solution of the mode coupling equation

$$\frac{\mathrm{d}}{\mathrm{d}\tau} C_{\varepsilon}^{MCT}(k,\tau) = (-c_{\varepsilon}\nabla_k + D\Delta_k)\, C_{\varepsilon}^{MCT}(k,\tau) + 2H_{\varepsilon}^2 \sum_{l\in\mathbb{Z}} \int_0^{\tau} \mathrm{d}\sigma\, C_{\varepsilon}^{MCT}(k-l,\tau-\sigma)\Delta_l \left[C_{\varepsilon}^{MCT}(l,\sigma)\right]^2 \quad (27)$$

with initial data $C_{\varepsilon}^{MCT}(k,0) = K\delta_{k,0}$. This nonlinear integro-differential equation is a discrete version of the mode coupling approximation originally obtained in [101] for the microscopic DSF $C_{\varepsilon}^{WASEP}(k,\tau)$ and represents the basic equation of mode coupling theory for nonlinear fluctuating hydrodynamics with one conservation law.

The fundamental hypothesis of this MCT approach–which is supported by numerous Monte-Carlo simulation data for microscopic models (see e.g. [74, 97])–is that the errors generated by the description of the time evolution of the fluctuation fields of the microscopic particle system in terms of a discrete SBE vanish at large space and time scales, while the errors generated by the two mode coupling approximations 1 and 2, are small at large space and time scales so that $s^*_{MCT}(p,t) \approx s^*_{dSBE}(p,t) = s^*_{WASEP}(p,t)$.[2] Hence from now on we shall also refer to the Fourier transform $S_{\varepsilon}^{MCT}(q,\tau)$ of the mode coupling approximation $C_{\varepsilon}^{MCT}(k,\tau)$ as dynamical structure function DSF.

3.3 Fourier and Laplace Transformation

The dynamical structure function of the mode coupling approximation is given by the discrete Fourier transformation

$$S_{\varepsilon}^{MCT}(q,\tau) := \frac{1}{\sqrt{2\pi}} \sum_{k\in\mathbb{Z}} \mathrm{e}^{-iqk} C_{\varepsilon}^{MCT}(k,\tau). \quad (28)$$

with

$$S_{\varepsilon}^{MCT}(q,0) = S_{\varepsilon}^{MCT}(0,\tau) = \frac{K}{\sqrt{2\pi}} \quad \forall \tau \geq 0. \quad (29)$$

These properties are consistent with the corresponding properties (5) of the microscopic DSF of the ASEP. In what follows, we work with the normalized DSF defined by

$$s_{\varepsilon}^{MCT}(q,\tau) := \frac{S_{\varepsilon}^{MCT}(q,\tau)}{S_{\varepsilon}^{MCT}(0,0)} = \frac{\sqrt{2\pi}}{K} S_{\varepsilon}^{MCT}(q,\tau) \quad (30)$$

[2] There is actually one particular model where this hypothesis can be proved rigorously with vanishing error in the mode coupling approximation, see the discussion in [76]. However, there is no general understanding of when MCT is a good approximation or for what type of cases it may break down.

and satisfying the boundary conditions

$$s_\varepsilon^{MCT}(q,0) = s_\varepsilon^{MCT}(0,\tau) = 1 \quad \forall q \in \mathbb{R},\ \tau \in \mathbb{R}_0^+. \tag{31}$$

We recall that on large scales the discrete gradient term in (27) can be removed by a Galilei transformation. Hence it is sufficient to set $c_\varepsilon = 0$ even when $\rho \neq 1/2$. With the quantities

$$G_\varepsilon := KH_\varepsilon = -2bK\varepsilon^\kappa, \quad \vartheta_q := \frac{2[1-\cos(q)]}{q^2} \tag{32}$$

the mode coupling Eq. (27) for $c_\varepsilon = 0$ becomes after Fourier transformation

$$\begin{aligned}\frac{\mathrm{d}}{\mathrm{d}\tau} s_\varepsilon^{MCT}(q,\tau) = &-Dq^2\vartheta_q s_\varepsilon^{MCT}(q,\tau)\\ &-2G_\varepsilon^2 q^2 \vartheta_q \int_0^\tau \mathrm{d}\tau'\, s_\varepsilon^{MCT}(q,\tau-\tau') m_\varepsilon^{MCT}(q,\tau')\end{aligned} \tag{33}$$

with the integral

$$m_\varepsilon^{MCT}(q,\tau') = \int_{-\pi}^{\pi} \frac{\mathrm{d}q'}{2\pi}\, s_\varepsilon^{MCT}(q',\tau') s_\varepsilon^{MCT}(q-q',\tau'). \tag{34}$$

We recall that the dependence of $s_\varepsilon^{MCT}(\cdot,\cdot)$ on ρ, b, κ and D is not written out explicitly.

The Laplace transformation

$$\tilde{s}_\varepsilon^{MCT}(q,\alpha) = \int_0^\infty \mathrm{d}\tau\, \mathrm{e}^{-\alpha\tau} s_\varepsilon^{MCT}(q,\tau) \tag{35}$$

with

$$\tilde{s}_\varepsilon^{MCT}(0,\alpha) = \frac{1}{\alpha} \tag{36}$$

then turns the mode coupling Eq. (33) for $s_\varepsilon^{MCT}(q,\tau)$ into

$$1 = \left[\alpha + Dq^2\vartheta_q + 2G_\varepsilon^2 q^2 \vartheta_q \tilde{m}_\varepsilon^{MCT}(q,\alpha)\right] \tilde{s}_\varepsilon^{MCT}(q,\alpha) \tag{37}$$

with the integral

$$\begin{aligned}\tilde{m}_\varepsilon^{MCT}(q,\alpha) &= \int_0^\infty \mathrm{d}\tau\, \mathrm{e}^{-\alpha\tau} \int_{-\pi}^{\pi} \mathrm{d}q'\, s_\varepsilon^{MCT}(q',\tau) s_\varepsilon^{MCT}(q-q',\tau)\\ &= \frac{|q|}{\alpha} \int_0^\infty \mathrm{d}\nu \mathrm{e}^{-\nu} \int_{-\pi/|q|}^{\pi/|q|} \frac{\mathrm{d}\xi}{2\pi} s_\varepsilon^{MCT}(q\xi,\nu/\alpha) s_\varepsilon^{MCT}(q(1-\xi),\nu/\alpha).\end{aligned} \tag{38}$$

The value 1 on the l.h.s. of (37) arises from the initial condition $s_\varepsilon^{MCT}(q, 0) = 1$ for all $q \in \mathbb{R}$, see (31). The mode coupling equation is trivially satisfied for $q = 0$ and implies the reflection symmetry $s_\varepsilon^{MCT}(q, \tau) = s_\varepsilon^{MCT}(-q, \tau)$ inherited from the microscopic reflection symmetry (6). These properties allow for restricting the discussion to the domain $q, \tau \in \mathbb{R}^+$ of strictly positive momentum and time. The Fourier-Laplace representation (37) of the mode coupling equation is the starting point for developing the mesoscale mode coupling analysis.

4 Mesoscale Mode Coupling Theory

In the original paper [101] treating the ASEP ($\kappa = 0$) as well as in the later work for strongly asymmetric systems with more than one conservation law [73, 74, 95] the discrete mode coupling approximation (27) is further approximated by taking $k \in \mathbb{R}$ and replacing the discrete gradient and discrete Laplacian by the usual partial derivative ∂_k and ∂_k^2 resp. For the FT this means that $q \in \mathbb{R}$ and the integration over q' extends over the whole of $\mathbb{R}$. The second step in these works is then to look for asymptotic scaling solutions for $q \to 0$, $\tau \propto q^{-z} \to \infty$ with dynamical exponent z, corresponding to large space and time scales. Here we take an approach that is different in several respects.

(1) Non-scaling solutions: Rather than looking for asymptotic scaling behaviour for small q and some fixed z we do not make any scaling assumption as an input in the treatment of the mode coupling equation. In our approach scaling emerges as a result.
(2) Weak asymmetry: The asymmetry exponent κ in (1) may take any positive value, including the strongly asymmetric case $\kappa = 0$, thus fully covering the WASEP including the range $0 < \kappa < 1/2$ that is not so well understood.
(3) Observation scale: We introduce a spatial observation scale by choosing $q = \varepsilon p$ and set the temporal observation scale by taking $\tau = \varepsilon^{-\chi} t$ where the exponent $\chi \in \mathbb{R}^+$, which determines the speeding up of time, is taken as a variable rather than taken to be equal to the dynamical exponent z from the outset.
(4) Large-scale behaviour: We consider two distinct large space-time scales: As *mesoscopic* large-scale description we take the limit $\varepsilon \to 0$ keeping $p \in \mathbb{R}^+$, $t \in \mathbb{R}^+$ fixed and study how the behaviour of the DSF as a function of p and t depends on χ. For the *macroscopic* description we fix χ and take a second limit with the choice $p = \check{\varepsilon}\check{p}$, $t = \check{\varepsilon}^{-\check{\chi}}\check{t}$ and $\check{\varepsilon} \to 0$, keeping $\check{p} \in \mathbb{R}^+$, $\check{t} \in \mathbb{R}^+$ fixed and with $\check{\chi} \in \mathbb{R}^+$ taken as a variable. Here $\check{\varepsilon}$ plays the role of a mesoscopic length scale as opposed to the microscopic length scale given by ε. This two-scale approach is in contrast to looking for asymptotic scaling solutions for $q \to 0$ with $q^z\tau$ taken to be some constant.
(5) Change of variables: Instead of the "natural" microscopic variables q, τ we analyse the mode coupling Eq. (33) in terms of judiciously chosen variables r_1, r_2 that depend on q, τ and on the microscopic parameters $\rho, D, b, \varepsilon, \kappa$ of the WASEP.

(6) No continuum approximation: The starting point for investigating solutions of the mode coupling equation is the discrete Eq. (33).

From a physics perspective the most important ingredients of this approach are the link between weak asymmetry defined by (1) and the microscopic length scale ε, the introduction of a variable speeding-up of microscopic time defined by χ, and the distinction between the mesoscopic space-time observation scale defined by the limit $\varepsilon \to 0$ of vanishing microscopic length scale and the macroscopic space-time observation scale defined by the limit $\check{\varepsilon} \to 0$ of vanishing mesoscopic length scale, accompanied by a speeding up of mesoscopic time defined by $\check{\chi}$ for some fixed χ. On the technical level the crucial points are the absence of the scaling assumption as starting point, and the change of variables which anticipates the possibility of scale invariance at large scales without, however, assuming it. Starting from the discrete mode coupling equation rather making a continuum approximation is a technical issue without much physical significance, but helpful for keeping track of the microscopic, mesoscopic and macroscopic observation scales, as discussed below.

We remark that due to the Lieb-Robinson bound [65] the covariance (3) of the (W)ASEP decays quickly at distances far from the center-of-mass position $k^* = [ct]$. Thus the Fourier transform $S_{\varepsilon}^{WASEP}(q, \tau)$ (4) is well-defined and has "nice" properties for all $q \in \mathbb{R}$ and $\tau \in \mathbb{R}_0^+$. By assuming similar properties for the mode coupling DSF (28) the proof of the main result, i.e., the asymptotic solution of the mode coupling equation given in Sects. 4.3.1 and 4.3.3, could be made formally rigorous. However, we refrain from doing so since the necessary technicalities would not add any insight into the original problem of exploring the large-scale properties of the DSF (4) of the (W)ASEP, but would require a significant amount of definitions, technical hypotheses, and notation, and thus obscure the exposition of the essential parts of the exact computations which are presented below and which allow for generalization to systems with more than one conservation law. Therefore we present the main results as informal definitions, propositions and theorems.

4.1 *Observation Scales*

Since in the context of dynamic critical phenomena some subtleties arise, we discuss in some more detail the large-scale behaviour in terms of mesoscopic and macroscopic scales before embarking on developing the mesoscale MCT.

4.1.1 Mesoscopic and Macroscopic Scale

Scale-invariant dynamical structure functions are the same on the mesoscopic and the macroscopic scale, but they are different in the absence of scale invariance.

The mesoscopic description allows for considering this case and for studying crossover between different dynamical universality classes. The distinction between the mesoscopic and the macroscopic scale is made precise as follows.

Definition 1 The normalized mesoscopic DSF is the limit

$$s^*_{MCT}(p,t) := \lim_{\varepsilon\to 0} s^{MCT}_{\varepsilon}(\varepsilon p, \varepsilon^{-\chi} t) \tag{39}$$

of the normalized DSF (30). The normalized macroscopic DSF is the limit

$$s^\bullet_{MCT}(\check{p},\check{t}) := \lim_{\check{\varepsilon}\to 0} s^*_{MCT}(\check{\varepsilon}\check{p}, \check{\varepsilon}^{-\check{\chi}}\check{t}) \tag{40}$$

of the normalized mesoscopic DSF. In the Fourier-Laplace domain the corresponding limits are defined by

$$\tilde{s}^*_{MCT}(p,\omega) := \lim_{\varepsilon\to 0} \varepsilon^{\chi} \tilde{s}^{MCT}_{\varepsilon}(\varepsilon p, \varepsilon^{\chi}\omega) \tag{41}$$

$$\tilde{s}^\bullet_{MCT}(\check{p},\check{\omega}) := \lim_{\check{\varepsilon}\to 0} \check{\varepsilon}^{\check{\chi}} \tilde{s}^*(\check{\varepsilon}\check{p}, \check{\varepsilon}^{\check{\chi}}\check{\omega}). \tag{42}$$

The mesoscale limits and macroscale limits of the DSF for the WASEP and for the discrete SBE are defined analogously, cf. (7) for the mesoscale limit for the WASEP. They should not be confused with the limits above for the DSF arising from mode coupling theory which, as discussed above, are expected to approximate the corresponding WASEP limit. We recall that there is no control over the approximation error which means that the error may be large, small (in some sense) or even zero.

Notice that $s^*_{MCT}(\cdot,\cdot)$ and $\tilde{s}^*_{MCT}(\cdot,\cdot)$ are functions of the mesoscopic observation exponent χ while $s^\bullet_{MCT}(\cdot,\cdot)$ and $\tilde{s}^\bullet_{MCT}(\cdot,\cdot)$ are functions both of χ and the macroscopic observation exponent $\check{\chi}$ by virtue of being a scaling limit of the χ-dependent functions $s^*_{MCT}(\cdot,\cdot)$ and $\tilde{s}^*_{MCT}(\cdot,\cdot)$ respectively. Whether or not the DSF is scale invariant or even trivial is to be determined as a function of the parameters χ and $\check{\chi}$. Moreover, these limits depend on the bias parameters b and κ and also on the density ρ and the diffusion coefficient D. In order to avoid excessive notation we continue to suppress these dependencies. We also drop the qualification "normalized".

4.1.2 A Physical Illustration of the Mesoscopic and Macroscopic Scales

The following *gedankenexperiment* may serve to illustrate the distinction between microscopic, mesoscopic, and macroscopic scales. Imagine a large number of beads such as small polystyrene spheres with a diameter of 1 μm, placed along a long and very narrow channel of about 10 m length, etched a bit more than a micron deep and wide into some material. We assume the beads to perform some kind of Brownian motion, but—being hard physical objects—not being able to penetrate each other. Due to the small diameter of the channel they also cannot pass each other, i.e., they perform a single-file dynamics similar to the colloidal particles investigated in the real experiment reported in [105].

We imagine that the motion of the beads is observed by video imaging with a camera such that one pixel of the image corresponds to a square of 1 μm length of the channel. This experimental setting effectively turns the continuous channel (more precisely, what one observes of the channel) into a lattice of $L = 10^7$ sites and induces a microscopic length scale $\varepsilon = 1\mu$m. Each lattice site represents a channel segment of 1 μm length and is recorded by exactly one pixel. We label the channel segments consecutively by a integer $k \in \{0, \ldots, 10^7 - 1\}$. Since the diameter of the beads is also about 1 μm, each lattice segment can be occupied by at most one particle. We set the contrast of the computer image to be such that the pixel appears black when more than 50% of a bead appears inside segment k and white otherwise. Then one observes at any instant of time a sequence of black or white pixels, representing the discretized positions of the beads along the channel. We also assume that the image has a resolution of 10^7 pixels in horizontal direction so that the whole channel can be observed at all times.

We are now in a position to discuss different observation scales. First consider the case when the computer screen to which the video camera is attached shows not more than, say, 10 consecutive pixels centered around some pixel k. Then, in this very zoomed-in view of the channel one observes a sequence of black or white pixels arising from the random motion of the particles (represented by black or white pixels) due to the Brownian motion and the hard-core repulsion of the beads, very much like the SSEP. Averaging images over many realizations of this process then results in a sequence of 10 gray pixels with a darkness that is proportional to the probability of presence of a particle in the channel segment represented by the respective pixel, ranging from white (no particle ever) to black (always present). Due to the SSEP dynamics the grayness evolves in time approximately according to the discrete diffusion equation. For definiteness let as assume that after a time of 10^{-6} seconds one can observe a perceptible change of the grayness profile. This sets the microscopic time scale of the process. We also assume that after about 10^8 seconds the density has reached equilibrium, i.e., a constant density profile along the whole channel. Hence the equilibration time is 10^{14} microscopic time units.

Next change the zoom of the camera and consider a less zoomed-in picture, with, say, 10^5 consecutive pixels centered around some pixel k appearing on the computer screen. Now the naked eye cannot resolve the screen image in such a way that one would still see individual particles represented by black or white pixels. Instead one sees a grayness profile, even without any averaging over realizations of the process. The grayness, proportional to the coarse-grained particle density, evolves in time approximately according to the usual diffusion equation. Now, in order to watch any evolution on the computer screen, one has to speed up the film, because after 10^{-6} seconds perceptible changes occur on the length scale of 10 μm, which on the computer is nothing but a tiny area making up a fraction of 10^{-4} of the width of the computer screen, i.e., less than a tenth of a millimeter. Diffusive dynamical scaling with dynamical exponent $z = 2$ implies that in order to see the same amount of change in a segment of 10^4 more pixels, the film has to run 10^8 times faster. This defines the mesoscopic time scale of 10^2 seconds. Running the film on microscopic time scales leads to (practically) no evolution on mesocopic space scales. On the other hand, running the film even faster, say 10^{14} times as done in the microscopic observation, means that one sees the constant equilibrium grayness. A smooth variation of the grayness profile can be observed only on the mesoscopic time scale.

Finally, we don't zoom in at all and watch the whole channel as represented by the 10^7 pixels on the computer screen. These are 10^6 more pixels than in the zoomed-in microscopic observation scale and the film has to run 10^{12} times faster in order to see a change that is comparable to the microscopic change. This is the macroscopic time scale of the process. The system equilibrates after 10^2 macroscopic time units. Also on macroscopic scale the grayness evolves in time approximately according to the usual diffusion equation.

With this *gedankenexperiment* one can see that there is indeed a difference between microscopic scale and mesoscopic scale (noisy time evolution with discrete diffusion equation after ensemble averaging vs. deterministic continuous diffusion equation without ensemble averaging), but one may also be led to the conclusion that there is no essential difference between mesoscopic and macroscopic scale as in both cases the grayness evolution is described by the deterministic continuous diffusion equation. However, this lack of difference between the scales is only a consequence of the scale invariance of the diffusion equation. In a different system, where the evolution on the intermediate mesoscopic scale would not be scale invariant, the behaviour on all three scales would be described by different equations.[3]

This is exactly the scenario we will encounter in the WASEP: For $\kappa \neq 1/2$ the evolution on mesoscopic scale is scale invariant and hence there is no essential difference between mesoscopic scale and macroscopic scale. The dynamical exponent is $z = 3/2$ for $\kappa < 1/2$ and $z = 2$ for $\kappa > 1/2$. However, at $\kappa = 1/2$ the evolution on mesoscopic scale is *not* scale invariant, it becomes scale invariant only on the macroscopic scale as defined above. Hence, by not differentiating between mesoscopic scale and macroscopic scale one cannot capture the features of the point $\kappa = 1/2$, which apart from not being scale invariant, has the interesting property of being a cross over point of two different dynamical universality classes.

4.2 Mode Coupling Equation at Observation Scale ε

We first implement items 1 and 5, (Sect. 4.2.1), followed by item 3 as described in the introductory paragraph of this section (Sect. 4.2.2). These steps lead to a reformulation of the mode coupling equation (37) but do not involve a discussion of

[3] This can be exemplified by taking as microscopic scale the atomistic scale of the molecules of which the beads and the channel are composed (with van-der-Waals or similar forces acting between them and Newton's equation governing the dynamics), as mescoscopic scale the description in terms of hard-core beads and the channel as a confining geometry (with Brownian motion and hard-repulsion determining the noisy dynamics, i.e., the microscopic description in the picture above), and as macroscopic scale description with mild zooming in (with the continuous diffusion equation as dynamics, i.e., the mesoscopic description in the picture above).

its solutions. Then, ignoring item 6, we implement in the main Sect. 4.3 item 4 (mesoscale approach) where also item 2 comes in.

4.2.1 Change of Variables

We recall that the microscopic DSF depends not only on the microscopic momentum q and the microscopic time τ but also on the WASEP parameters G_ε and D which appear in the mode coupling Eq. (37). Thus the DSF may alternatively be regarded as a function of another pair of variables (r_1, r_2), which incorporates not only its dependence on the coordinates (q, τ) but also some of the parameter dependence. Since it is not *a priori* clear how such variables should be chosen we do not just present the main result (Proposition 1) of the variable change $(q, \tau) \mapsto (r_1(\varepsilon, q, \tau), r_2(\varepsilon, q, \tau))$ but develop in some detail the ideas that lead to it.

Inspired by dynamical scaling (but not assuming it) we take as alternative variables

$$r_i(\varepsilon, q, \tau) = q(\Lambda_i(\varepsilon)\tau)^{1/z_i}, \quad i \in \{1, 2\} \tag{43}$$

where the quantities $\Lambda_i(\varepsilon)$ are functions of the WASEP parameters and z_i are fixed numbers which are all determined in what follows.[4] This coordinate mapping with the inverse transformation

$$q(\varepsilon, r_1, r_2) = \left(\frac{\Lambda_1(\varepsilon) r_2^{z_2}}{\Lambda_2(\varepsilon) r_1^{z_1}}\right)^{\frac{1}{z_2 - z_1}}, \quad \tau(\varepsilon, r_1, r_2) = \left(\frac{r_1^{z_1}}{\Lambda_1(\varepsilon)}\right)^{\frac{z_2}{z_2 - z_1}} \left(\frac{\Lambda_2(\varepsilon)}{r_2^{z_2}}\right)^{\frac{z_1}{z_2 - z_1}}. \tag{44}$$

is not bijective for $z_2 = z_1$ which is therefore excluded from consideration. Without loss of generality we assume $z_2 > z_1$. We also note that this mapping is singular along the lines $(\varepsilon, q, 0)$, $q \in \mathbb{R}^+$ and $(\varepsilon, 0, \tau)$, $\tau \in \mathbb{R}^+$ which are both mapped onto $0 = r_1(\varepsilon, q, 0) = r_2(\varepsilon, q, 0) = r_1(\varepsilon, 0, \tau) = r_2(\varepsilon, 0, \tau)$. For our purposes this singurality is irrelevant since only $(q, \tau) \in \mathbb{R}^+ \times \mathbb{R}^+$ need to be considered.

With this variable change we express the DSF in terms of r_1 and r_2, i.e., we define the function

$$\sigma_\varepsilon(r_1, r_2) := s_\varepsilon^{MCT}(q(\varepsilon, r_1, r_2), \tau(\varepsilon, r_1, r_2)). \tag{45}$$

which has the limiting behaviour

$$\lim_{a \searrow 0} \sigma_\varepsilon(ax, ay) = \lim_{a \searrow 0} \sigma_\varepsilon(a^{1/z_1} x, a^{1/z_2} y) = 1 \quad \forall \quad (x, y) \in \mathbb{R}^+ \times \mathbb{R}^+ \tag{46}$$

arising from the boundary conditions (31) that a physical solution of the mode coupling Eq. (37) has to satisfy. Conversely,

[4] With a view on studying later the limit $\varepsilon \to 0$, only the ε-dependence of the parameters $\Lambda_i(\varepsilon)$ is indicated.

$$s_\varepsilon^{MCT}(q,\tau) = \sigma_\varepsilon(r_1(\varepsilon,q,\tau), r_2(\varepsilon,q,\tau)) = \sigma_\varepsilon(q(\Lambda_1(\varepsilon)\tau)^{1/z_1}, q(\Lambda_2(\varepsilon)\tau)^{1/z_2}) \tag{47}$$

which implies for $\alpha > 0$

$$\tilde{s}_\varepsilon^{MCT}(q,\alpha) = \frac{1}{\alpha}\int_0^\infty \mathrm{d}\nu\, \mathrm{e}^{-\nu}\sigma_\varepsilon(q(\nu\Lambda_1(\varepsilon)/\alpha)^{1/z_1}, q(\nu\Lambda_2(\varepsilon)/\alpha)^{1/z_2}) \tag{48}$$

$$\begin{aligned}\tilde{m}_\varepsilon^{MCT}(q,\alpha) = {} & \frac{|q|}{\alpha}\int_0^\infty \mathrm{d}\nu\, \mathrm{e}^{-\nu}\int_{-\pi/|q|}^{\pi/|q|}\frac{\mathrm{d}\xi}{2\pi} \\ & \times\sigma_\varepsilon(\xi q(\nu\Lambda_1(\varepsilon)/\alpha)^{1/z_1}, \xi q(\nu\Lambda_2(\varepsilon)/\alpha)^{1/z_2}) \\ & \times\sigma_\varepsilon((1-\xi)q(\nu\Lambda_1(\varepsilon)/\alpha)^{1/z_1}, (1-\xi)q(\nu\Lambda_2(\varepsilon)/\alpha)^{1/z_2})\end{aligned} \tag{49}$$

for the Laplace transforms expressed in terms of $\sigma_\varepsilon(\cdot,\cdot)$.

With a second useful set of new variables, defined for $\alpha > 0$ by

$$\tilde{r}_i(\varepsilon,q,\alpha) := q\left(\frac{\Lambda_i(\varepsilon)}{\alpha}\right)^{1/z_i} \tag{50}$$

with the inverse

$$q(\varepsilon,\tilde{r}_1,\tilde{r}_2) = \left(\frac{\Lambda_1(\varepsilon)\tilde{r}_2^{z_2}}{\tilde{r}_1^{z_1}\Lambda_2(\varepsilon)}\right)^{\frac{1}{z_2-z_1}}, \quad \alpha(\varepsilon,\tilde{r}_1,\tilde{r}_2) = \left(\frac{\Lambda_1(\varepsilon)}{\tilde{r}_1^{z_1}}\right)^{\frac{z_2}{z_2-z_1}}\left(\frac{\tilde{r}_2^{z_2}}{\Lambda_2(\varepsilon)}\right)^{\frac{z_1}{z_2-z_1}}, \tag{51}$$

the variable change allows for rewriting the mode coupling Eq. (37) for $s_\varepsilon^{MCT}(q,\tau)$ in terms of the variables $\tilde{r}_i(\varepsilon,q,\alpha)$. To this end, we also introduce the function

$$f_z(\varepsilon,\tilde{r}_1,\tilde{r}_2) := \frac{(q(\varepsilon,\tilde{r}_1,\tilde{r}_2))^z}{\alpha(\varepsilon,\tilde{r}_1,\tilde{r}_2)} = \left(\frac{\tilde{r}_1^{z_1}}{\Lambda_1(\varepsilon)}\right)^{\frac{z_2-z}{z_2-z_1}}\left(\frac{\tilde{r}_2^{z_2}}{\Lambda_2(\varepsilon)}\right)^{\frac{z-z_1}{z_2-z_1}} \tag{52}$$

and the normalized Laplace transforms

$$\hat{\sigma}_\varepsilon(\tilde{r}_1,\tilde{r}_2) := \alpha_\varepsilon(\tilde{r}_1,\tilde{r}_2)\tilde{s}_\varepsilon^{MCT}(q(\varepsilon,\tilde{r}_1,\tilde{r}_2),\alpha_\varepsilon(\tilde{r}_1,\tilde{r}_2)) \tag{53}$$

$$\hat{\mu}_\varepsilon(\tilde{r}_1,\tilde{r}_2) := \frac{\alpha_\varepsilon(\tilde{r}_1,\tilde{r}_2)}{|q(\varepsilon,\tilde{r}_1,\tilde{r}_2)|}\tilde{m}_\varepsilon^{MCT}(q(\varepsilon,\tilde{r}_1,\tilde{r}_2),\alpha_\varepsilon(\tilde{r}_1,\tilde{r}_2)) \tag{54}$$

which with (51) are given by

$$\hat{\sigma}_\varepsilon(\tilde{r}_1,\tilde{r}_2) = \int_0^\infty \mathrm{d}\nu\, \mathrm{e}^{-\nu}\sigma_\varepsilon(\tilde{r}_1\nu^{1/z_1}, \tilde{r}_2\nu^{1/z_2}) \tag{55}$$

$$\begin{aligned}\hat{\mu}_\varepsilon(\tilde{r}_1,\tilde{r}_2) = {} & \int_0^\infty \mathrm{d}\nu\, \mathrm{e}^{-\nu}\int_{-\pi/|q(\varepsilon,\tilde{r}_1,\tilde{r}_2)|}^{\pi/|q(\varepsilon,\tilde{r}_1,\tilde{r}_2)|}\frac{\mathrm{d}\xi}{2\pi} \\ & \times\sigma_\varepsilon(\xi\tilde{r}_1\nu^{1/z_1}, \xi\tilde{r}_2\nu^{1/z_2}) \\ & \times\sigma_\varepsilon((1-\xi)\tilde{r}_1\nu^{1/z_1}, (1-\xi)\tilde{r}_2\nu^{1/z_2}).\end{aligned} \tag{56}$$

The mode coupling Eq. (37) then becomes

$$1 = \left[1 + \vartheta_{q(\varepsilon,\tilde{r}_1,\tilde{r}_2)}\left(Df_2(\varepsilon,\tilde{r}_1,\tilde{r}_2) + 2G_\varepsilon^2 f_{3/2}^2(\varepsilon,\tilde{r}_1,\tilde{r}_2)\bar{\mu}_\varepsilon(\tilde{r}_1,\tilde{r}_2)\right)\right]\hat{\sigma}_\varepsilon(\tilde{r}_1,\tilde{r}_2) \quad (57)$$

where the quantities $\hat{\sigma}_\varepsilon(\tilde{r}_1,\tilde{r}_2)$ and $\bar{\mu}_\varepsilon(\tilde{r}_1,\tilde{r}_2)$ are the functionals (55) and (56) of the DSF $\sigma_\varepsilon(\tilde{r}_1,\tilde{r}_2)$.

The free parameters $\Lambda_i(\varepsilon)$ that appear in the variables $\tilde{r}_i$ can now be expressed in a convenient way in terms of the WASEP parameters. With the choice

$$z_1 = \frac{3}{2}, \quad z_2 = 2, \quad \Lambda_1(\varepsilon) = \sqrt{2}|G_\varepsilon|, \quad \Lambda_2(\varepsilon) = D \quad (58)$$

one gets

$$Df_2(\varepsilon,\tilde{r}_1,\tilde{r}_2) = \tilde{r}_2^2, \quad 2G_\varepsilon^2 f_{3/2}^2(\varepsilon\tilde{r}_1,\tilde{r}_2) = \tilde{r}_1^3. \quad (59)$$

Thus we are led to the following conclusion.

Proposition 1 *Let for fixed $\varepsilon \in \mathbb{R}_0^+$ the function $\sigma_\varepsilon : \mathbb{R}^+ \times \mathbb{R}^+ \to \mathbb{R}$ be the solution of the integral equation*

$$1 = \left[1 + \vartheta_{q(\varepsilon,x,y)}\left(y^2 + x^3\bar{\mu}_\varepsilon(x,y)\right)\right]\hat{\sigma}_\varepsilon(x,y) \quad (60)$$

with

$$q(\varepsilon,x,y) := \frac{2G_\varepsilon^2}{D^2}\frac{y^4}{x^3} \quad (61)$$

and the functionals

$$\hat{\sigma}_\varepsilon(x,y) := \int_0^\infty \mathrm{d}\nu\, \mathrm{e}^{-\nu}\sigma_\varepsilon(x\nu^{2/3}, y\nu^{1/2}) \quad (62)$$

$$\hat{\mu}_\varepsilon(x,y) := \int_0^\infty \mathrm{d}\nu\, \mathrm{e}^{-\nu}\int_{-\pi/|q(\varepsilon,x,y)|}^{\pi/|q(\varepsilon,x,y)|}\frac{\mathrm{d}\xi}{2\pi} \times\sigma_\varepsilon(\xi x\nu^{2/3}, \xi y\nu^{1/2})\,\sigma_\varepsilon((1-\xi)x\nu^{2/3}, (1-\xi)y\nu^{1/2}) \quad (63)$$

and limits $\lim_{a\searrow 0}\sigma_\varepsilon(ax,ay) = \lim_{a\searrow 0}\sigma_\varepsilon(a^{2/3}x, a^{1/2}y) = 1$ for all $(x,y) \in \mathbb{R}^+ \times \mathbb{R}^+$. Then for $\alpha > 0$ the function

$$\tilde{s}_\varepsilon^{MCT}(q,\alpha) = \frac{1}{\alpha}\hat{\sigma}_\varepsilon(\tilde{r}_1(\varepsilon,q,\alpha), \tilde{r}_2(\varepsilon,q,\alpha)) \quad (64)$$

with

$$\tilde{r}_1(\varepsilon,q,\alpha) = q\left(\frac{\sqrt{2}|G_\varepsilon|}{\alpha}\right)^{2/3}, \quad \tilde{r}_2(\varepsilon,q,\alpha) = q\left(\frac{D}{\alpha}\right)^{1/2} \quad (65)$$

is the Laplace transform of the DSF (30) satisfying the mode coupling Eq. (37).

It should be noted that the functionals $\hat{\sigma}_\varepsilon(x, y)$ and $\hat{\mu}_\varepsilon(x, y)$ are defined on $\mathbb{R}_0^+ \times \mathbb{R}_0^+$ since the limits $\hat{\sigma}_\varepsilon(0, y) := \lim_{x \searrow 0} \hat{\sigma}_\varepsilon(x, y)$, $\hat{\mu}_\varepsilon(0, y) := \lim_{x \searrow 0} \hat{\mu}_\varepsilon(x, y)$, $\hat{\sigma}_\varepsilon(x, 0) := \lim_{y \searrow 0} \hat{\sigma}_\varepsilon(x, y)$, $\hat{\mu}_\varepsilon(x, 0) := \lim_{y \searrow 0} \hat{\mu}_\varepsilon(x, y)$ can be shown to be finite using the Lieb-Robinson bound on the decay of correlations [65]. As will be seen below, the form (60) of the mode coupling equation is more suitable for analyzing the large-scale behaviour of the DSF than the equivalent original form (37) in terms of the variables q, α.

Remark 1 For $(x, y) \in \mathbb{R}^+$ the function $\sigma_0 : \mathbb{R}^+ \times \mathbb{R}^+ \to \mathbb{R}$ is the solution of the integral equation

$$1 = \left[1 + y^2 + x^3 \hat{\mu}_0(x, y)\right] \hat{\sigma}_0(x, y) \tag{66}$$

with the functionals

$$\hat{\sigma}_0(x, y) := \int_0^\infty \mathrm{d}\nu\, \mathrm{e}^{-\nu} \sigma_0(x\nu^{2/3}, y\nu^{1/2}) \tag{67}$$

$$\hat{\mu}_0(x, y) := \int_0^\infty \mathrm{d}\nu\, \mathrm{e}^{-\nu} \int_{-\infty}^{\infty} \frac{\mathrm{d}\xi}{2\pi} \times \sigma_0(\xi x\nu^{2/3}, \xi y\nu^{1/2})\, \sigma_0((1-\xi)x\nu^{2/3}, (1-\xi)y\nu^{1/2}) \tag{68}$$

and limits $\lim_{a \searrow 0} \sigma_0(ax, ay) = \lim_{a \searrow 0} \sigma_0(a^{2/3}x, a^{1/2}y) = 1$ for all $(x, y) \in \mathbb{R}^+ \times \mathbb{R}^+$. However, the solution $\sigma_0(r_1, r_2)$ does not give access to the microscopic limiting DSF $s_0^{MCT}(q, \tau)$ through the the relation (45) since for fixed (r_1, r_2) one has $q(0, r_1, r_2) = 0$ and $\lim_{\varepsilon \to 0} \tau(\varepsilon, r_1, r_2) = \infty$. Obtaining the DSF $s_0^{MCT}(q, \tau)$ from $\sigma_0(r_1, r_2)$ requires taking the limit $q \to 0$ and $\tau \to \infty$ in appropriate fashion as discussed in the following.

4.2.2 Change of Observation Scale

As outlined above, we introduce a (finite) microscopic spatial observation scale by setting $q = p\varepsilon$ with the parameter $\varepsilon \in (0, 1]$ (1) that specifies also the weakness of the asymmetry. The temporal observation scale, i.e., the speeding up of microscopic time, is taken to be $\tau = t\varepsilon^{-\chi}$. With this choice of q and τ and with the renormalized coupling strength

$$g := G_\varepsilon \varepsilon^{-\kappa} = -2\,Kb \tag{69}$$

the mapping $(q, \tau) \mapsto (r_1(\varepsilon, q, \tau), r_2(\varepsilon, q, \tau))$ leads us to define the scaling variables

$$u_i(p, t) := p(\lambda_i t)^{1/z_i} \tag{70}$$

with

$$\lambda_1 = \Lambda_1(\varepsilon)\varepsilon^\kappa = \sqrt{2}|g|, \quad \lambda_2 = \Lambda_2(\varepsilon) = D \tag{71}$$

with z_i and $\Lambda_i(\varepsilon)$ given in (58). The inverse relations read

$$p(u_1, u_2) = \left(\frac{\lambda_1 u_2^{z_2}}{\lambda_2 u_1^{z_1}}\right)^{\frac{1}{z_2-z_1}}, \quad t(u_1, u_2) = \left(\frac{u_1^{z_1}}{\lambda_1}\right)^{\frac{z_2}{z_2-z_1}} \left(\frac{\lambda_2}{u_2^{z_2}}\right)^{\frac{z_1}{z_2-z_1}}. \tag{72}$$

The scaling variables u_i and parameters λ_i are independent of ε and invariant under the scaling transformation $p \to ap, t \to a^{-z_i}t$ which shows that the parameters z_i play the role of dynamical critical exponents while the quantities λ_i are scaling parameters.

To implement the change of observation scale in the DSF it is convenient to introduce the quantities

$$\chi_1 := 3/2 + \kappa, \quad \chi_2 := 2 \tag{73}$$

which yield

$$r_i(\varepsilon, \varepsilon p, \varepsilon^{-\chi} t) = \varepsilon^{(\chi_i - \chi)/z_i} u_i(p, t). \tag{74}$$

Thus (47) leads to

$$s_\varepsilon^{MCT}(\varepsilon p, \varepsilon^{-\chi} t) = \sigma_\varepsilon(\varepsilon^{(\chi_1-\chi)/z_1} u_1(p, t), \varepsilon^{(\chi_2-\chi)/z_2} u_2(p, t)) \tag{75}$$

Similarly, for the Laplace transformation one takes $q = \varepsilon p$ and $\alpha = \varepsilon^\chi \omega$. For the corresponding scaling variables

$$\tilde{u}_i(p, \omega) := p(\lambda_i/\omega)^{1/z_i}, \tag{76}$$

with the inverse relations

$$p(\tilde{u}_1, \tilde{u}_2) = \left(\frac{\lambda_1 \tilde{u}_2^{z_2}}{\lambda_2 \tilde{u}_1^{z_1}}\right)^{\frac{1}{z_2-z_1}}, \quad \omega(\tilde{u}_1, \tilde{u}_2) = \left(\frac{\lambda_1}{\tilde{u}_1^{z_1}}\right)^{\frac{z_2}{z_2-z_1}} \left(\frac{\tilde{u}_2^{z_2}}{\lambda_2}\right)^{\frac{z_1}{z_2-z_1}} \tag{77}$$

one has in the relevant domain $(p, \omega) \in \mathbb{R}^+$

$$\tilde{r}_i(\varepsilon, \varepsilon p, \varepsilon^\chi \omega) = \varepsilon^{(\chi_i - \chi)/z_i} \tilde{u}_i. \tag{78}$$

The Laplace transformation (35) therefore yields

$$\tilde{s}_{\varepsilon}^{MCT}(\varepsilon p, \varepsilon^{\chi}\omega) = \frac{\varepsilon^{-\chi}}{\omega}\int_0^{\infty} \mathrm{d}\nu\, \mathrm{e}^{-\nu}\sigma_{\varepsilon}(\varepsilon^{(\chi_1-\chi)/z_1}\tilde{u}_1\nu^{1/z_1}, \varepsilon^{(\chi_2-\chi)/z_2}\tilde{u}_2\nu^{1/z_2}) \quad (79)$$

$$\begin{aligned}\tilde{m}_{\varepsilon}^{MCT}(\varepsilon p, \varepsilon^{\chi}\omega) &= \frac{\varepsilon^{1-\chi}|p|}{\omega}\int_0^{\infty} \mathrm{d}\nu\, \mathrm{e}^{-\nu}\int_{-\pi/\varepsilon|p|}^{\pi/\varepsilon|p|} \frac{\mathrm{d}\xi}{2\pi} \\ &\times \sigma_{\varepsilon}(\xi\varepsilon^{(\chi_1-\chi)/z_1}\tilde{u}_1\nu^{1/z_1}, \xi\varepsilon^{(\chi_2-\chi)/z_2}\tilde{u}_2\nu^{1/z_2}) \\ &\times \sigma_{\varepsilon}((1-\xi)\varepsilon^{(\chi_1-\chi)/z_1}\tilde{u}_1\nu^{1/z_1}, (1-\xi)\varepsilon^{(\chi_2-\chi)/z_2}\tilde{u}_2\nu^{1/z_2}) \quad (80)\end{aligned}$$

with $\tilde{u}_i$ on the r.h.s. understood as the functions (76) of p and ω.

Conversely, expressing the quantities p and ω as the functions (77) of the variables $\tilde{u}_i$ one obtains with (71)–(74) for any fixed ε the analogue of Proposition 1. To this end, it is convenient to define the critical observation exponent

$$\chi^* := \min(\chi_1, \chi_2), \quad (81)$$

the auxiliary observation exponents

$$\mathring{\theta} := \min(\chi, \chi^*), \quad \psi_i := (\chi_i - \mathring{\theta})/z_i, \quad \mathring{\varphi} := \min(\psi_1, \psi_2), \quad (82)$$

and the normalized Laplace transforms

$$\begin{aligned}\mathring{\sigma}_{\varepsilon}(\tilde{u}_1, \tilde{u}_2) &:= \varepsilon^{\mathring{\theta}}\omega(\tilde{u}_1, \tilde{u}_2)\,\tilde{s}_{\varepsilon}^{MCT}(\varepsilon p(\tilde{u}_1, \tilde{u}_2), \varepsilon^{\chi}\omega(\tilde{u}_1, \tilde{u}_2)) \quad (83)\\ &= \varepsilon^{\mathring{\theta}-\chi}\hat{\sigma}_{\varepsilon}(\varepsilon^{(\chi_1-\chi)/z_1}\tilde{u}_1, \varepsilon^{(\chi_2-\chi)/z_2}\tilde{u}_2) \quad (84)\\ \mathring{\mu}_{\varepsilon}(\tilde{u}_1, \tilde{u}_2) &:= \varepsilon^{\mathring{\theta}+\mathring{\varphi}-1}\frac{\omega(\tilde{u}_1, \tilde{u}_2)}{|p(\tilde{u}_1, \tilde{u}_2)|}\tilde{m}_{\varepsilon}^{MCT}(\varepsilon p(\tilde{u}_1, \tilde{u}_2), \varepsilon^{\chi}\omega(\tilde{u}_1, \tilde{u}_2)) \quad (85)\\ &= \varepsilon^{\mathring{\varphi}+\mathring{\theta}-\chi}\hat{\mu}_{\varepsilon}(\varepsilon^{(\chi_1-\chi)/z_1}\tilde{u}_1, \varepsilon^{(\chi_2-\chi)/z_2}\tilde{u}_2). \quad (86)\end{aligned}$$

With a substitution of the integration variable $\nu \to \varepsilon^{\chi-\mathring{\theta}}\nu$, $\xi \to \varepsilon^{-\mathring{\varphi}}\xi$ to ensure that the integrands have non-zero limits we obtain the following equivalent of Proposition 1.

Proposition 2 *With dynamical exponents z_i given in (58), $p(\cdot,\cdot)$ defined in (71), (77), and the auxiliary observation exponents $\mathring{\theta}$, ψ_i, $\mathring{\varphi}$ defined in (82) let the function $\sigma_{\varepsilon} : \mathbb{R}^+ \times \mathbb{R}^+ \to \mathbb{R}$ be the solution of the integral equation*

$$1 = \left[\varepsilon^{\chi-\mathring{\theta}} + \vartheta_{\varepsilon p(x,y)}\left(\varepsilon^{\chi_2-\mathring{\theta}}y^2 + \varepsilon^{2(\chi_1-\mathring{\theta})-\mathring{\varphi}}x^3\mathring{\mu}_{\varepsilon}(x,y)\right)\right]\mathring{\sigma}_{\varepsilon}(x,y) \quad (87)$$

with the functionals

$$\mathring{\sigma}_{\varepsilon}(x,y) := \int_0^{\infty} \mathrm{d}\nu\, \mathrm{e}^{-\varepsilon^{\chi-\mathring{\theta}}\nu}\sigma_{\varepsilon}(\varepsilon^{\psi_1}x\nu^{1/z_1}, \varepsilon^{\psi_2}y\nu^{1/z_2}) \quad (88)$$

$$\mathring{\mu}_\varepsilon(x, y) := \int_0^\infty \mathrm{d}\nu\, \mathrm{e}^{-\varepsilon^{\chi-\mathring{\theta}}\nu} \int_{-\pi\varepsilon^{\mathring{\varphi}-1}|p(x,y)|^{-1}}^{\pi\varepsilon^{\mathring{\varphi}-1}|p(x,y)|^{-1}} \frac{\mathrm{d}\xi}{2\pi}$$
$$\times\sigma_\varepsilon(\xi\varepsilon^{\psi_1-\mathring{\varphi}}x\nu^{1/z_1}, \xi\varepsilon^{\psi_2-\mathring{\varphi}}y\nu^{1/z_2})$$
$$\times\sigma_\varepsilon((\varepsilon^{\mathring{\varphi}}-\xi)\varepsilon^{\psi_1-\mathring{\varphi}}x\nu^{1/z_1}, (\varepsilon^{\mathring{\varphi}}-\xi)\varepsilon^{\psi_2-\mathring{\varphi}}y\nu^{1/z_2}) \tag{89}$$

and limits $\lim_{a\searrow 0}\sigma_\varepsilon(ax, ay) = \lim_{a\searrow 0}\sigma_\varepsilon(a^{1/z_1}x, a^{1/z_2}y) = 1$ *for all* $(x, y) \in \mathbb{R}^+ \times \mathbb{R}^+$. *Then for any fixed* $\varepsilon \in \mathbb{R}^+$ *the DSF (30) satisfying the mode coupling Eq. (37) with* $q = \varepsilon p$ *and* $\tau = \varepsilon^{-\chi}t$ *is given by (75) with the scaling variables*

$$u_1(p, t) = p\left(\sqrt{2}|g|t\right)^{1/z_1}, \quad u_2(p, t) = p\,(Dt)^{1/z_2} \tag{90}$$

and

$$\tilde{s}_\varepsilon^{MCT}(\varepsilon p, \varepsilon^\chi\omega) = \frac{\varepsilon^{-\mathring{\theta}}}{\omega}\mathring{\sigma}_\varepsilon(\tilde{u}_1(p, \omega), \tilde{u}_2(p, \omega)) \tag{91}$$

with

$$\tilde{u}_1(p, \omega) = p\left(\frac{\sqrt{2}|g|}{\omega}\right)^{1/z_1}, \quad \tilde{u}_2(p, \omega) = p\left(\frac{D}{\omega}\right)^{1/z_2} \tag{92}$$

is the Laplace transform (35) of the DSF (30) with $\alpha = \varepsilon^\chi\omega$.

The limits defining the solution are the limits (31) for $q \to 0$ and $\tau \to 0$ resp. in terms of the variables x and y. Notice that the DSF $\sigma_\varepsilon(x, y)$ has no dependence on g and D, but depends explicitly on ε through the integration range of the integral over ξ in the functional $\hat{\mu}_\varepsilon(\cdot, \cdot)$. On the other hand, the DSF (30) expressed as a function of p, ω does depend on g and D through the dependence of the scaling variables u_i on these parameters which is not written out. The limit $\varepsilon \to 0$ is the mesoscale limit discussed in the next section.

4.3 Large Scale Solution of the Mode Coupling Equation

We recall that the observation exponents and auxiliary observation exponents are functions of the bias exponent κ and the observation exponent χ. Essentially, what Proposition 2 does is to disentangle the joint dependence of the solution of the original mode coupling Eq. (37) on the observation scale set by ε and the bias B_ε to facilitate taking the large-scale limit $\varepsilon \to 0$.

4.3.1 Mesoscale Solution of the Mode Coupling Equation

Lemma 1 *For* $(x, y) \in \mathbb{R}_0^+ \times \mathbb{R}_0^+$ *the mesoscale limit* $\varepsilon \to 0$ *of the mode coupling Eq. (87) is given by*

$$
1 = \begin{cases} \mathring{\sigma}_0(x, y) & \chi < \chi^* \\ \begin{cases} \left[1 + x^3 \mathring{\mu}_0(x, y)\right] \mathring{\sigma}_0(x, y) & \chi_1 < \chi_2 \\ \left[1 + y^2 + x^3 \mathring{\mu}_0(x, y)\right] \mathring{\sigma}_0(x, y) & \chi_1 = \chi_2 \\ \left[1 + y^2\right] \mathring{\sigma}_0(x, y) & \chi_1 > \chi_2 \end{cases} & \chi = \chi^* \\ \begin{cases} x^3 \mathring{\mu}_0(x, y) \mathring{\sigma}_0(x, y) & \chi_1 < \chi_2 \\ \left[y^2 + x^3 \mathring{\mu}_0(x, y)\right] \mathring{\sigma}_0(x, y) & \chi_1 = \chi_2 \\ y^2 \mathring{\sigma}_0(x, y) & \chi_1 > \chi_2 \end{cases} & \chi > \chi^*. \end{cases} \tag{93}
$$

where $\mathring{\sigma}_0(x, y) := \lim_{\varepsilon\to 0} \mathring{\sigma}_\varepsilon(x, y)$ *and* $\mathring{\mu}_0(x, y) := \lim_{\varepsilon\to 0} \mathring{\mu}_\varepsilon(x, y)$.

Proof Since the limits $\mathring{\sigma}_0(x, y)$ and $\mathring{\mu}_0(x, y)$ are finite this follows immediately from $\lim_{\varepsilon\to 0} \vartheta_{\varepsilon|p|} = 1$ for all $p \in \mathbb{R}$ and the definitions (82) of the auxiliary observation exponents $\mathring{\theta}$ and $\mathring{\varphi}$. □

To compute the limiting functionals $\mathring{\sigma}_0(x, y)$ and $\mathring{\mu}_0(x, y)$ we introduce for $x, y \in \mathbb{R}_0^+$ the auxiliary functionals

$$
\hat{\sigma}_>(x, y) := \int_0^\infty \mathrm{d}v\, \sigma_0(x v^{1/z_1}, y v^{1/z_2}) \tag{94}
$$

$$
\begin{aligned} \hat{\mu}_<(x, y) := \int_0^\infty \mathrm{d}v\, \mathrm{e}^{-v} \int_{-\infty}^\infty \frac{\mathrm{d}\xi}{2\pi}\, \sigma_0(\xi x v^{\frac{1}{z_1}}, \xi y v^{\frac{1}{z_2}}) \\ \times \sigma_0(-\xi x v^{\frac{1}{z_1}}, -\xi y v^{\frac{1}{z_2}}) \end{aligned} \tag{95}
$$

$$
\begin{aligned} \hat{\mu}_>(x, y) := \int_0^\infty \mathrm{d}v \int_{-\infty}^\infty \frac{\mathrm{d}\xi}{2\pi}\, \sigma_0(\xi x v^{\frac{1}{z_1}}, \xi y v^{\frac{1}{z_2}}) \\ \times \sigma_0((1-\xi) x v^{\frac{1}{z_1}}, (1-\xi) y v^{\frac{1}{z_2}}). \end{aligned} \tag{96}
$$

where $\sigma_0(\cdot, \cdot)$ is the solution of the MCE (66).

Proposition 3 *For* $\chi > 0$ *the limit*

$$
\mathring{\sigma}^*(\tilde{u}_1, \tilde{u}_2) := \lim_{\varepsilon\to 0} \mathring{\sigma}_\varepsilon(\tilde{u}_1, \tilde{u}_2) \tag{97}
$$

with $\mathring{\sigma}^*_\varepsilon(\cdot, \cdot)$ *defined in (84) is given by*

$$
\mathring{\sigma}^*(\tilde{u}_1, \tilde{u}_2) = \begin{cases} 1 & \chi < \chi^* \\ \begin{cases} \hat{\sigma}_0(\tilde{u}_1, 0) & \chi_1 < \chi_2 \\ \hat{\sigma}_0(\tilde{u}_1, \tilde{u}_2) & \chi_1 = \chi_2 \\ \hat{\sigma}_0(0, \tilde{u}_2) & \chi_1 > \chi_2 \end{cases} & \chi = \chi^* \\ \begin{cases} \hat{\sigma}_>(\tilde{u}_1, 0) & \chi_1 < \chi_2 \\ \hat{\sigma}_>(\tilde{u}_1, \tilde{u}_2) & \chi_1 = \chi_2 \\ \hat{\sigma}_>(0, \tilde{u}_2) & \chi_1 > \chi_2 \end{cases} & \chi > \chi^*. \end{cases} \tag{98}
$$

with the limiting functionals $\hat{\sigma}_0(\cdot,\cdot)$ *and* $\hat{\sigma}_>(\cdot,\cdot)$ *defined in (67), (94).*

Proof This is proved by noting that in (88) the limit $\varepsilon \to 0$ and integration can be interchanged which yields

$$\mathring{\sigma}^*(\tilde{u}_1, \tilde{u}_2) = \int_0^\infty \mathrm{d}\nu \lim_{\varepsilon\to 0} \mathrm{e}^{-\varepsilon^{\chi-\mathring{\theta}}\nu} \sigma_\varepsilon(\varepsilon^{\psi_1}\tilde{u}_1 \nu^{1/z_1}, \varepsilon^{\psi_2}\tilde{u}_2 \tau^{1/z_2}) \tag{99}$$

(i) In the range $0 < \chi < \chi^*$ one has $\mathring{\theta} = \chi$ and therefore $\psi_i > 0$. Thus

$$\mathring{\sigma}^*(\tilde{u}_1, \tilde{u}_2) = \int_0^\infty \mathrm{d}\nu\, \mathrm{e}^{-\nu} \sigma_\varepsilon(0,0) = \hat{\sigma}_\varepsilon(0,0) = 1 \tag{100}$$

(ii) For $\chi = \chi^* = \min(\chi_1, \chi_2)$ one has $\mathring{\theta} = \chi = \chi^*$ so that

$$\begin{aligned} \mathring{\sigma}^*(\tilde{u}_1, \tilde{u}_2) &= \int_0^\infty \mathrm{d}\nu\, \mathrm{e}^{-\nu} \lim_{\varepsilon\to 0} \sigma_\varepsilon(\varepsilon^{\psi_1}\tilde{u}_1 \nu^{1/z_1}, \varepsilon^{\psi_2}\tilde{u}_2 \nu^{1/z_2}) \\ &= \begin{cases} \int_0^\infty \mathrm{d}\nu\, \mathrm{e}^{-\nu} \sigma_0(\tilde{u}_1 \nu^{1/z_1}, 0) & \chi_1 < \chi_2 \\ \int_0^\infty \mathrm{d}\nu\, \mathrm{e}^{-\nu} \sigma_0(\tilde{u}_1 \nu^{1/z_1}, \tilde{u}_2 \nu^{1/z_2}) & \chi_1 = \chi_2 \\ \int_0^\infty \mathrm{d}\nu\, \mathrm{e}^{-\nu} \sigma_0(0, \tilde{u}_2 \nu^{1/z_2}) & \chi_1 > \chi_2 \end{cases} \end{aligned} \tag{101}$$

(iii) The case $\chi > \chi^*$ implies $\mathring{\theta} = \chi^*$ and therefore

$$\mathring{\sigma}^*(\tilde{u}_1, \tilde{u}_2) = \int_0^\infty \mathrm{d}\nu \lim_{\varepsilon\to 0} \mathrm{e}^{-\varepsilon^{\chi-\chi^*}\nu} \sigma_\varepsilon(\varepsilon^{\psi_1}\tilde{u}_1 \nu^{1/z_1}, \varepsilon^{\psi_2}\tilde{u}_2 \nu 1/z_2) \tag{102}$$

$$= \begin{cases} \int_0^\infty \mathrm{d}\nu\, \sigma_0(\tilde{u}_1 \nu^{1/z_1}, 0) & \chi_1 < \chi_2 \\ \int_0^\infty \mathrm{d}\nu\, \sigma_0(\tilde{u}_1 \nu^{1/z_1}, \tilde{u}_2 \nu^{1/z_2}) & \chi_1 = \chi_2 \\ \int_0^\infty \mathrm{d}\nu\, \sigma_0(0, \tilde{u}_2 \nu^{1/z_2}) & \chi_1 > \chi_2 \end{cases} \tag{103}$$

The assertion then follows from the definitions (67) and (94). □

Proposition 4 *For* $\chi > 0$ *the limit*

$$\mathring{\mu}^*(\tilde{u}_1, \tilde{u}_2) := \lim_{\varepsilon\to 0} \mathring{\mu}_\varepsilon(\tilde{u}_1, \tilde{u}_2) \tag{104}$$

with $\mathring{\mu}_\varepsilon(\cdot,\cdot)$ *defined in (89) is given by*

$$\mathring{\mu}^*(\tilde{u}_1, \tilde{u}_2) = \begin{cases} \begin{cases} \hat{\mu}_<(\tilde{u}_1, 0) & \chi_1 < \chi_2 \\ \hat{\mu}_<(\tilde{u}_1, \tilde{u}_2) & \chi_1 = \chi_2 \\ \hat{\mu}_<(0, \tilde{u}_2) & \chi_1 > \chi_2 \end{cases} & \chi < \chi^* \\ \begin{cases} \hat{\mu}(\tilde{u}_1, 0) & \chi_1 < \chi_2 \\ \hat{\mu}(\tilde{u}_1, \tilde{u}_2) & \chi_1 = \chi_2 \\ \hat{\mu}(0, \tilde{u}_2) & \chi_1 > \chi_2 \end{cases} & \chi = \chi^* \\ \begin{cases} \hat{\mu}_>(\tilde{u}_1, 0) & \chi_1 < \chi_2 \\ \hat{\mu}_>(\tilde{u}_1, \tilde{u}_2) & \chi_1 = \chi_2 \\ \hat{\mu}_>(0, \tilde{u}_2) & \chi_1 > \chi_2 \end{cases} & \chi > \chi^*. \end{cases} \tag{105}$$

with the limiting functionals $\hat{\mu}_<(\cdot, \cdot)$, $\hat{\mu}_0(\cdot, \cdot)$, and $\hat{\mu}_>(\cdot, \cdot)$ defined in (95), (68), (96) respectively.

Proof Since by construction $0 \le \mathring{\phi} < 1$ for all $\chi > 0$, the integration range of the integral over ξ extends to $\mathbb{R}$ in the limit $\varepsilon \to 0$. The limit inside the integral depends on χ as follows.
(i) For $\chi < \chi^*$ one has $\mathring{\theta} = \chi$ and $0 < \mathring{\phi} < 1$. This yields (105) for the range $\chi < \chi^*$.
(ii) For $\chi = \chi^*$ one has $\mathring{\theta} = \chi = \chi^*$ and $\mathring{\phi} = 0$ from which (105) for $\chi = \chi^*$ follows.
(iii) For $\chi > \chi^*$ one has $\mathring{\theta} = \chi^*$ and $\mathring{\phi} = 0$ which leads to (105) for the range $\chi > \chi^*$.
Hence with the definitions (68), (95), and (96) the assertion (105) is proved for all $\chi > 0$. □

Corollary 1 *The mesoscale limit $\varepsilon \to 0$ of the mode coupling Eq. (87) is given by*

$$1 = \begin{cases} \hat{\sigma}_0(0, 0) & & \chi < \chi^* \\ \begin{cases} \left[1 + x^3 \hat{\mu}_0(x, 0)\right] \hat{\sigma}_0(x, 0) & \chi_1 < \chi_2 \\ \left[1 + y^2 + x^3 \hat{\mu}_0(x, y)\right] \hat{\sigma}_0(x, y) & \chi_1 = \chi_2 \\ \left(1 + y^2\right) \hat{\sigma}_0(0, y) & \chi_1 > \chi_2 \end{cases} & & \chi = \chi^* \\ \begin{cases} x^3 \hat{\mu}_>(x, 0) \hat{\sigma}_>(x, 0) & \chi_1 < \chi_2 \\ \left[y^2 + x^3 \hat{\mu}_>(x, y)\right] \hat{\sigma}_>(x, y) & \chi_1 = \chi_2 \\ y^2 \hat{\sigma}_>(0, y) & \chi_1 > \chi_2 \end{cases} & & \chi > \chi^*. \end{cases} \tag{106}$$

This follows from the limiting properties established Lemma 1 and in Propositions 3 and 4. Recalling Remark 1 we are now in a position to state and prove the main result.

Theorem 1 *Let the function $\sigma_0(x, y) : R_0^+ \times R_0^+ \to \mathbb{R}$ be the solution of the integral equation*

$$1 = \left[1 + y^2 + x^3 \hat{\mu}_0(x, y)\right] \hat{\sigma}_0(x, y) \tag{107}$$

with the functionals

$$\hat{\sigma}_0(x, y) := \int_0^\infty \mathrm{d}\nu\, \mathrm{e}^{-\nu} \sigma_0(x\nu^{2/3}, y\nu^{1/2}) \tag{108}$$

$$\begin{aligned}\hat{\mu}_0(x, y) := &\int_0^\infty \mathrm{d}\nu\, \mathrm{e}^{-\nu} \int_{-\infty}^\infty \frac{\mathrm{d}\xi}{2\pi} \\ &\times \sigma_0(\xi x\nu^{2/3}, \xi y\nu^{1/2}) \\ &\times \sigma_0((1-\xi)x\nu^{2/3}, (1-\xi)y\nu^{1/2})\end{aligned} \tag{109}$$

having for $(x, y) \in \mathbb{R}^+ \times \mathbb{R}^+$ *finite limits* $\hat{\sigma}_0(x, 0) = \lim_{y\searrow 0} \hat{\sigma}_0(x, y)$, $\hat{\sigma}_0(0, y) = \lim_{x\searrow 0} \hat{\sigma}_0(x, y)$ *and satisfying* $\lim_{a\searrow 0} \hat{\sigma}_0(ax, ay) = 1$, $\lim_{x\searrow 0} x^3 \hat{\mu}_0(x, y) = 0$. *Then with the scaling variables*

$$\tilde{u}_1 \equiv \tilde{u}_1(p, \omega) := \frac{\sqrt{2}|g||p|^{3/2}}{\omega}, \quad \tilde{u}_2 \equiv \tilde{u}_2(p, \omega) := \frac{Dp^2}{\omega} \tag{110}$$

defined in (76) the normalized mesoscopic limit (41) at observation scale set by the exponent χ *is given by*

$$\tilde{s}^*_{MCT}(p, \omega) = \frac{1}{\omega} \begin{cases} 1 & \chi < \chi^* \\ \begin{cases} \hat{\sigma}_0(\tilde{u}_1(p, \omega), 0) & 0 \le \kappa < 1/2 \\ \hat{\sigma}_0(\tilde{u}_1(p, \omega), \tilde{u}_2(p, \omega)) & \kappa = 1/2 \\ \hat{\sigma}_0(0, \tilde{u}_2(p, \omega)) & \kappa > 1/2 \end{cases} & \chi = \chi^* \\ 0 & \chi > \chi^* \end{cases} \tag{111}$$

with the critical observation scale $\chi^* = \min\{(3/2 + \kappa, 2)\}$.

Proof The definition (41) of the mesoscale DSF, the definition (82) of the auxiliary observation scale $\mathring{\theta}$ and Proposition 2 imply that

$$\tilde{s}^*_{MCT}(p, \omega) = \frac{1}{\omega} \lim_{\varepsilon\to 0} \varepsilon^{\chi - \mathring{\theta}} \mathring{\sigma}_\varepsilon(\tilde{u}_1(p, \omega), \tilde{u}_2(p, \omega)) \tag{112}$$

$$= \frac{1}{\omega} \begin{cases} \lim_{\varepsilon\to 0} \mathring{\sigma}_\varepsilon(\tilde{u}_1(p, \omega), \tilde{u}_2(p, \omega)) & \chi \le \chi^* \\ \lim_{\varepsilon\to 0} \varepsilon^{\chi-\chi^*} \mathring{\sigma}_\varepsilon(\tilde{u}_1(p, \omega), \tilde{u}_2(p, \omega)) & \chi > \chi^*. \end{cases} \tag{113}$$

<u>$(A)\chi \le \chi^*$:</u> With the definition (97) the limit in (113) yields by Proposition 3 for any fixed strictly positive p and ω

$$\tilde{s}^*_{MCT}(p, \omega) = \frac{1}{\omega} \begin{cases} \hat{\sigma}_0(0, 0) & \chi < \chi^* \\ \begin{cases} \hat{\sigma}_0(\tilde{u}_1(p, \omega), 0) & \chi_1 < \chi_2 \\ \hat{\sigma}_0(\tilde{u}_1(p, \omega), \tilde{u}_2(p, \omega)) & \chi_1 = \chi_2 \\ \hat{\sigma}_0(0, \tilde{u}_2(p, \omega)) & \chi_1 > \chi_2 \end{cases} & \chi = \chi^*. \end{cases} \tag{114}$$

The definition (73) of the auxiliary observation exponents χ_i then yields (111) in the range $\chi \le \chi^*$ of the observation exponent. On the other hand, according to Corollary 1, the function $\sigma_0(\tilde{u}_1, \tilde{u}_2)$ does indeed satisfy the integral Eq. (107) with the functionals $\hat{\sigma}_0(\tilde{u}_1, \tilde{u}_2)$ and $\hat{\mu}_0(\tilde{u}_1, \tilde{u}_2)$.

$(B)\chi > \chi^*$: According to Corollary 1 the function $\sigma_0(u_1, u_2)$ satisfies the mode coupling equation $1 = \left[\tilde{u}_2^2 + \tilde{u}_1^3 \hat{\mu}_>(\tilde{u}_1, \tilde{u}_2)\right] \hat{\sigma}_>(\tilde{u}_1, \tilde{u}_2)$ which ensures that $\hat{\sigma}_>(\tilde{u}_1, \tilde{u}_2)$ is finite for any fixed $(\tilde{u}_1, \tilde{u}_2) \in \mathbb{R}_0^+ \times \mathbb{R}_0^+ \setminus \{(0, 0)\}$. The limit in (113) yields the assertion for observation exponents in the range $\chi > \chi^*$. □

Corollary 2 *The mesoscale DSF is given by*

$$s^*_{MCT}(p, t) = \begin{cases} 1 & \chi < \chi^* \\ \begin{cases} \sigma_0(u_1(p, t), 0) & 0 \le \kappa < 1/2 \\ \sigma_0(u_1(p, t), u_2(p, t)) & \kappa = 1/2 \\ \sigma_0(0, u_2(p, t)) & \kappa > 1/2 \end{cases} & \chi = \chi^* \\ 0 & \chi > \chi^* \end{cases} \tag{115}$$

with the scaling variables $u_(p, t)$ defined in (70).

4.3.2 Scaling Properties of the Mesoscale DSF

We now elaborate on the physical significance of Theorem 1. On mesoscopic scale the solution is scale invariant for the critical value $\chi = \chi^*$ except at the point $\kappa = 1/2$. One reads off the superdiffusive KPZ dynamical exponent $z = 3/2$ for $0 \le \kappa < 1/2$ and the diffusive EW critical exponent $z = 2$ for $\kappa > 1/2$. The scaling solution is, as expected for dynamical critical behaviour, independent of the microscopic parameters except that they determine the non-universal scale parameters λ_i as given in (71). On the moderately asymmetric KPZ line $0 \le \kappa < 1/2$–which includes the point $\kappa = 0$ of strong asymmetry–the DSF is given by the solution of the integral equation

$$\hat{\sigma}_0(x, 0) = \left[1 + x^3 \hat{\mu}_0(x, 0)\right]^{-1}, \quad x \in \mathbb{R}_0^+ \tag{116}$$

and was studied numerically in [43]. Comparison with the exact Praehofer-Spohn scaling function for the KPZ fixed point shows that the MCT solution $s^*_{MCT}(p, t) = \sigma_0(u_1(p, t), 0)$ is not exact, but numerically not very far from the exact curve $\tilde{s}^*_{WASEP}(u_1(p, t), 0)$ with the same scaling variable $u_1(p, t)$ [77]. On the weakly asymmetric EW line $\kappa > 1/2$ the MCT integral equation reduces to the simple linear equation

$$\hat{\sigma}_0(0, \tilde{u}_2) = \left(1 + \tilde{u}_2^2\right)^{-1}, \quad \tilde{u}_2 \in \mathbb{R}_0^+ \tag{117}$$

which yields the Gaussian

$$s^*_{MCT}(p,t) = \mathrm{e}^{-Dp^2t} \quad u_2 \in \mathbb{R}_0^+ \tag{118}$$

which represents the exact EW critical DSF implied by the rigorous treatment in [46]. The crossover point $\kappa = 1/2$ between the two universality classes coincides with the point where the fluctuations are governed by the SBE [46]. The absence of scale invariance of the SBE is mirrored in the absence of scale invariance of the MCT solution $s^*_{MCT}(p,t) = \sigma_0(u_1(p,t), u_2(p,t))$ (115) obtained above.

The trivial subcritical behaviour simply means that on slow observations scales $\chi < \chi^*$ (short times) the spreading of the initially strictly local correlation is not noticeable. Only the ballistic motion of the density peak with rescaled velocity $v = b(1-2\rho)$ is captured by the DSF and gives rise to $s^*_{MCT}(p,t) = 1$.

On the other hand, on very fast observation scales $\chi > \chi^*$ (long times) the fluctuations are uncorrelated, corresponding to complete relaxation of the initial perturbation of the invariant measure and therefore leading to $s^*_{MCT}(p,t) = 0$.

4.3.3 Macroscale Solution of the DSF

The *macroscopic* limit (42) given by

$$s^\bullet_{MCT}(p,t) = \begin{cases} 1 & \chi < \chi^* \\ \begin{cases} s^*_{MCT}(0, \varepsilon^{1-\hat{\chi}/z_2} u_2) & \kappa > 1/2 \\ s^*_{MCT}(\varepsilon^{1-\hat{\chi}/z_1} u_1, \varepsilon^{1-\hat{\chi}/z_2} u_2) & \kappa = 1/2 \\ s^*_{MCT}(\varepsilon^{1-\hat{\chi}/z_1} u_1, 0) & 0 \le \kappa < 1/2 \end{cases} & \chi = \chi^* \\ 0 & \chi > \chi^* \end{cases} \tag{119}$$

with the scaling variables $u_i = u_i(p,t)$. The macroscopic DSF is scale invariant for all $\hat{\chi} > 0$ with non-trivial scale invariance appearing in the critical regime $\chi = \chi^*$, *including* the point $\kappa = 1/2$. Non-trivial scale invariance emerges for the mesoscopic observation exponent $\chi = \chi^*$.

At the macroscopic observation scale set by $\hat{\chi} = z_1 = 3/2$ this yields

$$s^\bullet_{MCT}(p,t) = \begin{cases} 1 & \kappa > 1/2 \\ s^*_{MCT}(u_1, 0) & \kappa = 1/2 \\ s^*_{MCT}(u_1, 0) & 0 \le \kappa < 1/2 \end{cases} \tag{120}$$

which exhibits at the crossover point $\kappa = 1/2$ the same KPZ-like behaviour as the range $0 \le \kappa < 1/2$ where the WASEP fluctuations are believed to be governed by the KPZ fixed point, see [70, 81] for recent discussions.

On the other hand, for long time scales with $\hat{\chi} = z_2 = 2$ the mesoscale MCT predicts

$$s^\bullet_{MCT}(p,t) = \begin{cases} s^*_{MCT}(0, u_2) & \kappa > 1/2 \\ 0 & \kappa = 1/2 \\ 0 & 0 \le \kappa < 1/2. \end{cases} \tag{121}$$

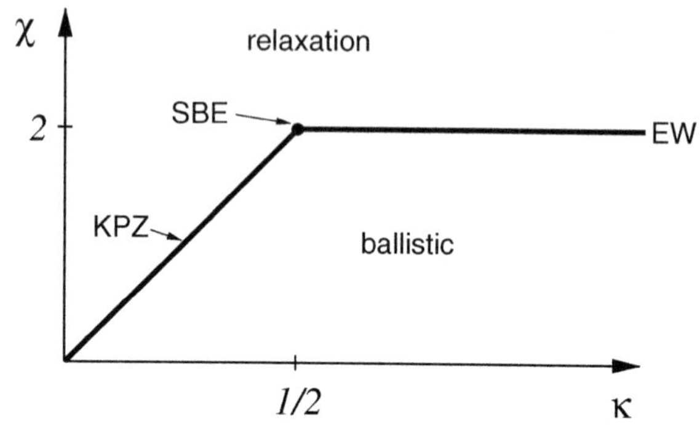

Fig. 1 Mesoscale MCT prediction of the phase diagram of the WASEP as a function of the bias exponent κ and the observation scale set by χ. The critical line (81) corresponds to KPZ universality ($\kappa < 1/2$), the SBE point $\kappa = 1/2$, and EW universality ($\kappa > 1/2$)

Thus at the macroscopic diffusive observation scale set by $\hat{\chi} = 2$ decorrelation has set in at and below the crossover point $\kappa = 1/2$, i.e., for moderate and critical bias (Fig. 1).

5 Conclusions

The main methodical novelties of the mesoscale scale approach to MCT are the absence of an *a priori* scaling assumption and the distinction between a mesoscopic and macroscopic space-time scale (Definition 1). Technically, this is facilitated by working the variables (43) which have simple scaling properties rather than using the traditional microscopic space-time variables momentum q and time τ. As a result we obtain on mesoscopic scale for $\kappa = 1/2$ a solution which is not scale invariant but has features expected from the stochastic Burgers equation which describes the fluctuations at this non-critical crossover point between the (expected) KPZ to the (established) EW universality class. For moderate asymmetry with $\kappa < 1/2$ MCT predicts the critical line (81) of non-trivial scale invariant behavior which is expected to be in the KPZ universality class, while for weak asymmetry $\kappa > 1/2$ one gets EW criticality.

There is no control over the approximation leading to the discrete SBE and the mode coupling approximations (1) and (2). Only the discrete gradient terms neglected in arriving at (27) are easily shown to lead to controlled subleading contributions to the large-scale behaviour. It is therefore encouraging to see that the line (81) is consistent with an exact result, viz., the scaling $L^{-\chi^*}$ of the spectral gap of the generator of the WASEP with system size obtained from Bethe ansatz [58]. Moreover, for $\kappa > 1/2$, MCT yields the EW universality proved mathematically rigorously in [46]. Thus, even though the actual scaling form of the DSF does not match the exact Praehofer-Spohn scaling function in the KPZ regime, it appears that mesoscale MCT can predict correctly various non-trivial features of the DSF. In particular, it seems feasible to study in an approximate sense the crossover between KPZ and EW universality as well as diffusive corrections. The mesoscale approach also lends itself to generalization to weakly nonlinear processes with more than one conserved quantity.

For using the mesoscale MCT to study processes other than the WASEP it should be noted that in Sect. 3 the microscopic mode coupling equations were derived directly from the Markov generator. This shortcut, bypassing nonlinear fluctuating hydrodynamics, is generally possible when the microscopic invariant measure is a product measure since in that case the generator acting on the centered local random variables produces the correct macroscopic current and therefore the exact collective velocity and coupling strength that enter the discrete SBE and consequently the mode coupling approximation. In the presence of correlations, however, one needs to compute the exact current and appeal to nonlinear fluctuating hydrodynamics [95] to postulate the correct discrete stochastic partial differential equation on which MCT is based.

Acknowledgements It is a pleasure to acknowledge stimulating discussions with P. Gonçalves, M. Jara, and H. Spohn. This work is financially supported by FCT/Portugal through CAMGSD, IST-ID, Projects UIDB/04459/2020 and Grant UIDP/04459/2020 and by the FCT Grant 2022.09232.PTDC.

References

1. Affleck, I., Kennedy, T., Lieb, E.H., Tasaki, H.: Rigorous results on valence-bond ground states in antiferromagnets. Phys. Rev. Lett. **59**, 799–802 (1987)
2. Ahmed, R., Bernardin, C., Gonçalves, P., Simon, M.: A microscopic derivation of coupled SPDE's with a KPZ flavor. Ann. Inst. H. Poincaré Probab. Statist. **58**, 890–915 (2022)
3. Alexander, S., Holstein, T.: Lattice diffusion and the Heisenberg ferromagnet. Phys. Rev. B **18**, 301 (1978)
4. Amir, G., Corwin, I., Quastel, J.: Probability distribution of the free energy of the contin-uum directed random polymer in 1+1 dimensions. Commun. Pure Appl. Math. **64**, 466–537 (2011)
5. Arai, Y.: On the KPZ Scaling and the KPZ Fixed Point for TASEP. Math. Phys. Anal. Geom. **27**, 4 (2024)
6. Arratia, R.: The motion of a tagged particle in the simple symmetric exclusion system in Z. Ann. Probab. **11**, 362–373 (1983)
7. Baxter, R.J.: Exactly Solved Models in Statistical Mechanics. Academic, New York (1982)
8. Ben Avraham, D.: Complete exact solution of diffusion-limited coalescence, $A + A - > A$. Phys. Rev. Lett. **81**, 4756–4759 (1998)
9. Belitsky, V., Schütz, G.M.: Diffusion and scattering of shocks in the partially asymmetric simple exclusion process. Electron. J. Probab. **7**, paper 11, 1–21 (2002)
10. Belitsky, V., Schütz, G.M.: Microscopic structure of shocks and antishocks in the ASEP conditioned on low current. J. Stat. Phys. **152**, 93–111 (2013)
11. Bernardin, C., Stoltz, G.: Anomalous diffusion for a class of systems with two conserved quantities. Nonlinearity **25**, 1099–1133 (2012)
12. Bernardin, C., Gonçalves, P.: Anomalous fluctuations for a perturbed Hamiltonian system with exponential interactions. Commun. Math. Phys. **325**, 291–332 (2014)
13. Bernardin, C., Gonçalves, P., Jara, M.: 3/4-fractional superdiffusion in a system of harmonic oscillators perturbed by a conservative noise. Arch. Rat. Mech. Anal. **220**, 505–542 (2016)

14. Bernardin, C., Gonçalves, P., Jara, M.: Weakly harmonic oscillators perturbed by a conservative noise. Ann. Appl. Probab. **28**, 1315–1355 (2018)
15. Bernardin, C., Gonçalves, P., Olla, S.: Space-time fluctuations in a quasi-static limit. arXiv:2305.05319 [math.PR], to appear in Math. Phys. Anal. Geom. (2024)
16. Bertini, L., De Sole, A., Gabrielli, D., Jona Lasinio, G., Landim, C.: Macroscopic fluctuation theory. Rev. Mod. Phys. **87**, 593–636 (2015)
17. Bertini, L., Giacomin, G.: Stochastic Burgers and KPZ equations from particle systems. Commun. Math. Phys. **183**, 571–607 (1997)
18. Bertini, B., Collura, M., De Nardis, J., Fagotti, M.: Transport in out-of-equilibrium XXZ chains: exact profiles of charges and currents. Phys. Rev. Lett. **117**, 207201 (2016)
19. Bodineau, T., Derrida, B.: Distribution of current in non-equilibrium diffusive systems and phase transitions. Phys. Rev. E **72**, 066110 (2005)
20. Borodin, A., Petrov, L.: Lectures on Integrable probability: Stochastic vertex models and symmetric functions. In: Schehr, G., Altland, A., Fyodorov, Y.V., O'Connell, N., Cugliandolo, L.F. (eds.) Stochastic Processes and Random Matrices: Lecture Notes of the Les Houches Summer School: Volume 104, July 2015. Oxford University Press, Oxford (2017)
21. Buča, B., Prosen, T.: Connected correlations, fluctuations and current of magnetization in the steady state of boundary driven XXZ spin chains. J. Stat. Mech. **2016**, 023102 (2016)
22. Cannizzaro, G., Gonçalves, P., Misturini, R., Occelli, A.: From ABC to KPZ (2023). arXiv:2304.02344 [math.PR]
23. Cardy, J.L.: Conformal Field Theory and Statistical Mechanics. In: Jacobsen, J., Ouvry, S., Pasquier, V., Serban, D., Cugliandolo, L. (eds.) Exact Methods in Low-dimensional Statistical Physics and Quantum Computing: Lecture Notes of the Les Houches Summer School: Volume 89, July 2008. Oxford University Press, Oxford (2010)
24. Castro-Alvaredo, O.A., Doyon, B., Yoshimura, T.: Emergent hydrodynamics in integrable quantum systems out of equilibrium. Phys. Rev. X **6**, 041065 (2016)
25. Colaiori, F., Moore, M.A.: Numerical solution of the mode-coupling equations for the Kardar-Parisi-Zhang equation in one dimension. Phys. Rev. E **65**, 017105 (2001)
26. Corwin, I.: The Kardar-Parisi-Zhang equation and universality class. Random Matrices Theory Appl. **1**, 1130001 (2012)
27. De Masi, A., Presutti, E., Scacciatelli, E.: The weakly asymmetric simple exclusion process. Ann. Inst. H. Poincaré Probab. Statist. **25**, 1–38 (1989)
28. De Nardis, J., Medenjak, M., Karrasch, C., Ilievski, E.: Anomalous spin diffusion in one-dimensional antiferromagnets. Phys. Rev. Lett. **123**, 186601 (2019)
29. De Nardis, J., Gopalakrishnan, S., Vasseur, R.: Nonlinear fluctuating hydrodynamics for Kardar-Parisi-Zhang scaling in isotropic spin chains. Phys. Rev. Lett. **131**, 197102 (2023)
30. Derrida, B., Evans, M.R., Hakim, V., Pasquier, V.: Exact solution of a 1D asymmetric exclusion model using a matrix formulation. J. Phys. A Math. Gen. **26**, 1493–1517 (1993)
31. Derrida, B., Janowsky, S.A., Lebowitz, J.L., Speer, E.R.: Exact solution of the totally asymmetric simple exclusion process: shock profiles. J. Stat. Phys. **73**, 813–842 (1993)
32. Derrida, B.: An exactly soluble nonequilibrium system: the asymmetric simple exclusion process. Phys. Rep. **301**, 65–83 (1998)
33. Derrida, B., Lebowitz, J.L., Speer, E.R.: Large deviation of the density profile in the steady state of the open symmetric simple exclusion process. J. Stat Phys. **107**, 599–634 (2002)
34. Derrida, B.: Non-equilibrium steady states: fluctuations, large deviations of the density, of the current. J. Stat. Mech. P07023 (2007)
35. Dittrich, P., Gärtner, J.: A central limit theorem for the weakly asymmetric simple exclusion process. Math. Nachr. **151**, 75–93 (1991)
36. Doi, M., Edwards, S.F.: The Theory of Polymer Dynamics. Oxford University Press, Oxford (1986)
37. Doyon B.: Lecture notes on generalised hydrodynamics. SciPost Phys. Lect. Notes 18 (2020)
38. Edwards, S.F., Wilkinson, D.: The surface statistics of a granular aggregate. Proc. R. Soc. Lond. A **381**, 17–31 (1982)
39. Essler, F.H.L.: A short introduction to generalized hydrodynamics. Phys. A 127572 (2022)

40. Ferrari, P.A., Kipnis, C., Saada, E.: Microscopic structure of travelling waves in the asymmetric simple exclusion process. Ann. Probab. **19**, 226–244 (1991)
41. Ferrari, P.A., Fontes, L.R.G.: Shock fluctuations in the asymmetric simple exclusion process. Probab. Theory Relat. Fields **99**, 305–319 (1994)
42. Ferrari, P.L., Spohn, H.: Scaling limit for the space-time covariance of the stationary totally asymmetric simple exclusion process. Commun. Math. Phys. **265**, 1–44 (2006); Correction in 45–46
43. Frey, E., Täuber, U.C., Hwa, T.: Mode-coupling and renormalization group results for the noisy Burgers equation. Phys. Rev. E **53**, 4424–4438 (1996)
44. Gärtner, J.: Convergence towards Burgers' equation and propagation of chaos for weakly asymmetric exclusion processes. Stoch. Process. Appl. **27**, 233–260 (1988)
45. Gaudin, M.: The Bethe Wave Function. Cambridge University Press, Cambridge (2014)
46. Gonçalves, P., Jara, M.: Crossover to the KPZ equation. Ann. Henri Poincaré **13**, 813–826 (2012)
47. Grynberg, M.D., Stinchcombe, R.B.: Dynamics of adsorption-desorption processes as a soluble problem of many fermions. Phys. Rev. E **52**, 6013–6024 (1995)
48. Gwa, L.H., Spohn, H.: Bethe solution for the dynamical-scaling exponent of the noisy Burgers equation. Phys. Rev. A **46**(2), 844–854 (1992)
49. Isserlis, L.: On a formula for the product-moment coefficient of any order of a normal frequency distribution in any number of variables. Biometrika **12**, 134–139 (1918)
50. Jack, R.L., Thompson, I.R., Sollich, P.: Hyperuniformity and phase separation in biased ensembles of trajectories for diffusive systems. Phys. Rev. Lett. **114**, 060601 (2015)
51. Jara, M., Flores, G.R.M.: Stationary directed polymers and energy solutions of the Burgers equation. Stoch. Process. Appl. **130**, 5973–5998 (2020)
52. Jepsen, P.N., Lee, Y.K., Lin, H., et al.: Long-lived phantom helix states in Heisenberg quantum magnets. Nat. Phys. **18**, 899–904 (2022)
53. Halpin-Healy, T., Takeuchi, K.A.: A KPZ Cocktail-Shaken, not Stirred... J. Stat. Phys. **160**, 794–814 (2015)
54. Kardar, M., Parisi, G., Zhang, Y.-C.: Dynamic scaling of growing interfaces. Phys. Rev. Lett. **56**, 889–892 (1986)
55. Karevski, D., Popkov, V., Schütz, G.M.: Driven isotropic Heisenberg spin chain with arbitrary boundary twisting angle: exact results. Phys. Rev. E **88**, 062118 (2013)
56. Karevski, D., Schütz, G.M.: Conformal invariance in driven diffusive systems at high currents. Phys. Rev. Lett. **118**, 030601 (2017)
57. Kerr, M.L., Kheruntsyan, K.V.: The theory of generalised hydrodynamics for the one-dimensional Bose gas. AAPPS Bull. **33**, 25 (2023)
58. Kim, D.: Bethe ansatz solution for crossover scaling functions of the asymmetric XXZ chain and the KPZ-type growth model. Phys. Rev. E **52**, 3512–3524 (1995)
59. Kipnis, C., Landim, C.: Scaling Limits of Interacting Particle Systems. Springer, Berlin (1999)
60. Krebs, K., Jafarpour, F.H., Schütz, G.M.: Microscopic structure of travelling wave solutions in a class of stochastic interacting particle systems. New J. Phys. **5**, 145.1–145.14 (2003)
61. Kriecherbauer, T., Krug, J.: A pedestrian's view on interacting particle systems, KPZ universality and random matrices. J. Phys. A Math. Theor. **43**, 403001 (2010)
62. Krug, J.: Origins of scale invariance in growth processes. Adv. Phys. **46**, 139–282 (1997)
63. Kukla, V., Kornatowski, J., Demuth, D., Girnus, I., Pfeifer, H., Rees, L.V.C., Schunk, S., Unger, K., Kärger, J.: NMR studies of single-file diffusion in unidimensional channel zeolites. Science **272**, 702–704 (1996)
64. Langen, T., Erne, S., Geiger, R., Rauer, B., Schweigler, T., Kuhnert, M., Rohringer, W., Mazets, I.E., Gasenzer, T., Schmiedmayer, J.: Experimental observation of a generalized Gibbs ensemble. Science **348**, 207–211 (2015)
65. Lieb, E., Robinson, D.: The finite group velocity of quantum spin systems. Commun. Math. Phys. **28**, 251–257 (1972)
66. Liggett, T.M.: Stochastic Interacting Systems: Contact. Voter and Exclusion Processes. Springer, Berlin (1999)

67. Ljubotina, M., Žnidarič, M., Prosen, T., Ljubotina, M., Žnidarič, M., Prosen, T.: Kardar-Parisi-Zhang physics in the quantum Heisenberg magnet. Phys. Rev. Lett. **122**, 210602 (2019)
68. Lushnikov, A.A.: Binary reaction $1 + 1 \to 0$ in one dimension. Phys. Lett. A **120**(3), 135–137 (1987)
69. MacDonald, J.T., Gibbs, J.H., Pipkin, A.C.: Kinetics of biopolymerization on nucleic acid templates. Biopolymers **6**, 1–25 (1968)
70. Matetski, K., Quastel, J., Remenik, D.: The KPZ fixed point. Acta Math. **227**, 115–203 (2021)
71. Perkins, T.T., Smith, D.E., Chu, S.: Direct observation of tube-like motion of a single polymer-chain. Science **264**, 819–822 (1994)
72. Popkov, V., Schütz, G.M.: Transition probabilities and dynamic structure function in the ASEP conditioned on strong flux. J. Stat. Phys. **142**, 627–639 (2011)
73. Popkov, V., Schmidt, J., Schütz, G.M.: Universality classes in two-component driven diffusive systems. J. Stat. Phys. **160**, 835–860 (2015)
74. Popkov, V., Schadschneider, A., Schmidt, J., Schütz, G.M.: Fibonacci family of dynamical universality classes. Proc. Natl. Acad. Sci. (USA) **112**(41), 12645–12650 (2015)
75. Popkov, V., Schütz, G.M.: Solution of the Lindblad equation for spin helix states. Phys. Rev. E **95**, 042128 (2017)
76. Popkov, V., Schütz, G.M.: Quest for the golden ratio universality class. Submitted (2023)
77. Prähofer, M., Spohn, H.: Exact scaling function for one-dimensional stationary KPZ growth. J. Stat. Phys. **115**, 255–279 (2004)
78. Privman, V. (ed.): Nonequilibrium Statistical Mechanics in One Dimension. Cambridge University Press, Cambridge (1997)
79. Prosen, T.: Exact nonequilibrium steady state of a strongly driven open XXZ chain. Phys. Rev. Lett. **107**, 137201 (2011)
80. Prosen, T.: Matrix product solutions of boundary driven quantum chains. J. Phys. A Math. Theor. **48**, 373001 (2015)
81. Quastel, J., Sarkar, S.: Convergence of exclusion processes and the KPZ equation to the KPZ fixed point. J. Am. Math. Soc. **36**, 251–289 (2023)
82. Rezakhanlou, F.: Hydrodynamic limit for attractive particle systems on $\mathbb{Z}^d$. Commun. Math. Phys. **140**, 417–448 (1991)
83. Roy, D., Dhar, A., Khanin, K., Kulkarni, M., Spohn, H.: Universality in coupled stochastic Burgers systems with degenerate flux Jacobian (2024). arXiv:2401.06399 [cond-mat.stat-mech]
84. Schmidt, J., Schütz, G.M., van Beijeren, H.: A lattice gas model for generic one-dimensional Hamiltonian Systems. J. Stat. Phys. **183**, 8 (2021)
85. Schütz, G.M.: Diffusion-annihilation in the presence of a driving field. J. Phys. A Math. Gen. **28**, 3405–3415 (1995)
86. Schütz, G.M.: The Heisenberg chain as a dynamical model for protein synthesis-Some theoretical and experimental results. Int. J. Mod. Phys. B **11**, 197–202 (1997)
87. Schütz, G.M.: Non-equilibrium relaxation law for entangled polymers. Eur. Lett. **48**, 623–628 (1999)
88. Schütz, G.M.: Exactly solvable models for many-body systems far from equilibrium. In: Domb, C., Lebowitz, J. (eds.) Phase Transitions and Critical Phenomena, vol. 19. Academic Press, London (2001)
89. Schütz, G.M.: Fluctuations in stochastic interacting particle systems. In: Giacomin, G., Olla, S., Saada, E., Spohn, H., Stoltz, G. (eds.) Stochastic Dynamics out of Equilibrium. IHPStochDyn 2017. Springer Proceedings in Mathematics & Statistics, vol. 282. Springer, Cham (2019)
90. Shank, C.V., Yen, R., Fork, R.L., Orenstein, J., Baker, G.L.: Picosecond dynamics of photoexcited gap states in polyacetylene. Phys. Rev. Lett. **49**, 1660–1663 (1982)
91. Spitzer, F.: Interaction of Markov processes. Adv. Math. **5**, 246–290 (1970)
92. Spohn, H.: Long-range correlations for stochastic lattice gases in a non-equilibrium steady state. J. Phys. A Math. Gen. **16**, 4275–4291 (1983)
93. Spohn, H.: Large Scale Dynamics of Interacting Particles. Springer, Berlin (1991)
94. Spohn, H.: Bosonization, vicinal surfaces, and hydrodynamic fluctuation theory. Phys. Rev. E **60**, 6411–6420 (1999)

95. Spohn, H.: Nonlinear Fluctuating hydrodynamics for anharmonic chains. J. Stat. Phys. **154**, 1191–1227 (2014)
96. Spohn, H., Stoltz, G.: Nonlinear fluctuating hydrodynamics in one dimension: the case of two conserved fields. J. Stat. Phys. **160**, 861–884 (2015)
97. Spohn, H.: The Kardar-Parisi-Zhang equation: a statistical physics perspective. In: Stochastic Processes and Random Matrices, Lecture Notes of the Les Houches Summer School, Volume 104, July 2015. Oxford University Press, Oxford, UK (2017)
98. Takeuchi, K.A., Sano, M.: Universal fluctuations of growing interfaces: evidence in turbulent liquid crystals. Phys. Rev. Lett. **104**, 230601 (2010)
99. Takeuchi, K.A., Sano, M., Sasamoto, T., Spohn, H.: Growing interfaces uncover universal fluctuations behind scale invariance. Sci. Rep. (Nat.) **1**, 34 (2011)
100. van Beijeren, H., Kehr, K.W., Kutner, R.: Diffusion in concentrated lattice gases. III. Tracer diffusion on a one-dimensional lattice. Phys. Rev. B **28**, 5711–5723 (1983)
101. van Beijeren, H., Kutner, R., Spohn, H.: Excess noise for driven diffusive systems. Phys. Rev. Lett. **54**, 2026–2029 (1985)
102. van Beijeren, H.: Exact results for transport properties of one-dimensional Hamiltonian systems. Phys. Rev. Lett. **108**, 180601 (2012)
103. Vardeny, Z., Strait, V., Moses, D., Chung, T.C., Heeger, A.J.: Picosecond photoinduced dichroism in Trans-(CH)x: direct measurement of soliton diffusion. Phys. Rev. Lett. **49**, 1657–1660 (1982)
104. Vidmar, L., Ronzheimer, J.P., Schreiber, M., Braun, S., Hodgman, S.S., Langer, S., Heidrich-Meisner, F., Bloch, I., Schneider, U.: Dynamical quasicondensation of hard-core bosons at finite momenta. Phys. Rev. Lett. **115**, 175301 (2015)
105. Wei, Q.-H., Bechinger, C., Leiderer, P.: Single-File diffusion of colloids in one-dimensional channels. Science **287**, 625–627 (2000)
106. Withers, C.S.: The moments of the multivariate normal. Bull. Aust. Math. Soc. **32**, 103–107 (1985)
107. Ye, B., Machado, F., Kemp, J., Hutson, R.B., Yao, N.Y.: Universal Kardar-Parisi-Zhang dynamics in integrable quantum systems. Phys. Rev. Lett. **129**, 230602 (2022)

The Two Dimensional Lorentz Gas in the Kinetic Limit: Theoretical and Numerical Results

Bernt Wennberg

Abstract The Lorentz gas is a model for the motion of electrons in a metal, where the motion is dominated by collisions of the electrons with immobile atomic nuclei, the scatterers. The motion depends on the distribution of scatterers, and we focus here on modifications of periodic scatterer distributions in two dimensions, and in the low density, or Boltzmann Grad, limit. Some theoretical results are complemented with numerical illustrations including modified periodic scatterer distributions, and scatterer distributions given by a quasi-crystal and by the zero set of a Gaussian analytic function.

Keywords Lorentz gas · Boltzmann-Grad limit · Boltzmann equation · Quasi-crystal · Gaussian analytic function

1 Introduction

In his paper [18], H. A. Lorentz used ideas from kinetic theory of gases to understand macroscopic properties of metals, for example, heat conduction and electric conductivity. Kinetic theory of gases is also the basis of Drude's paper [11], but Lorentz makes explicit calculations under the hypothesis that the electrons are elastically scattered on hard spheres. He is not very explicit about the scaling of particle size versus density, however, and not on the precise distribution of scatterers, but in writing down a kinetic transport equation with a well defined mean free path, he makes an implicit assumption of the relation between the diameter of the scatterers and their

This paper is written in memory of Maria Conceição Carvalho who has been a friend and colleague since we first met many years ago, and who will always remain in our memories.

B. Wennberg (✉)
Chalmers University of Technology and the University of Gothenburg,
SE41296 Gothenburg, Sweden
e-mail: wennberg@chalmers.se

E. Carlen et al. (eds.), *From Particle Systems to Partial Differential Equations*,
Springer Proceedings in Mathematics & Statistics 465,
https://doi.org/10.1007/978-3-031-65195-3_17

density. This corresponds to the low density limit, or the Boltzmann-Grad limit, if this relation is maintained when the obstacle diameter is allowed to decrease to zero. The kinetic equation formulated by Lorentz is the forward Kolmogorov equation for the Markov jump process describing the path of an electron, and this also corresponds to an implicit assumption that the free path length, i.e. the distance an electron moves between two successive collisions with an obstacle, is an exponentially distributed random variable.

This paper deals only with the kinetic scaling, and the main issue is how the distribution of scatterers influences the distribution of mean free paths. Gallavotti proved in [13, 14] that the Lorentz gas converges to a linear kinetic equation if the obstacle positions follow a Poisson distribution. On the other hand, Golse proved in [15] that in a periodic distribution of scatterers, the Boltzmann-Grad limit does not lead to a linear kinetic equation of the same type. This is due to the fact that free path lengths in a periodic distribution of scatterers does not decay exponentially (see [4]). Marklof and Strömbergsson [24] have shown that the same is true when the scatterers are distributed as in a quasi crystal. However, the phase space may be extended to include information about the previous collision with an obstacle in such a way that the Boltzmann-Grad limit is a kinetic equation. This was shown in [8] under a certain independence assumption, and more generally and without additional assumptions in [22, 23].

On the other hand, if the scatterer distribution is derived from a periodic distribution, by randomly removing or moving obstacles one may still obtain a linear kinetic equation in the small scatterer limit, also if the perturbation is very small. Such results were obtained in [31, 35], and are summarized in Sect. 3.

The Boltzmann-Grad limit of the Lorentz gas thus depends strongly on the distribution of scatterers, and before going into the mathematical details for the case of a perturbed periodic distribution, we illustrate the differences with numerical simulations. The numerical results in Sect. 2 show that both the periodic and quasi periodic distributions give rise to non-exponential decay of the free path lengths, whereas in a Poisson distribution of scatterers, as well as a distribution given by the zero set of a Gaussian analytic function (GAF) do give rise to an exponential distribution of free path lenghts. For the first three cases these are only illustrations of mathematically established results, but as far as I know there are no theoretical results concerning the Lorentz gas in a GAF zero-set.

In the remaining part of this first section we give a more detailed description of the Lorentz model and the particular distribution of scatterers that is studied in detail in Sect. 3, and give some relevant references to the subject. This follows closely the corresponding description in [35].

Let $\mathscr{X}$ be a point set in $\mathbb{R}^2$; it can also be described as a (random) locally finite counting measure, so that

$$\mathscr{X}(A) = \#\{p_j \in A\}, \tag{1}$$

the number of points in the set A, or by the empirical measure

$$\mathscr{X} = \sum \delta_{p_j} . \tag{2}$$

The Lorentz process is an infinite billiard where for each point $p \in \mathscr{X}$ there is a circular obstacle with radius r, the same for all obstacles, and center at p. When hitting an obstacle, the particle path is reflected specularly. Therefore, given an initial position and velocity at $t = 0$, $(x_0, v_0) \in \mathbb{R}^2 \times S^1$, the particle's position at time t is given by

$$(x(t), v(t)) = T^t_{\mathscr{X},r}(x_0, v_0) = \left(x_0 + \sum_{j=1}^{M} (t_j - t_{j-1})v_{j-1} + (t - t_M)v_M, v_M\right), \tag{3}$$

where $t_0 = 0$, and $\{t_1, ..., t_M\}$ is the set of times where the trajectory of the point particle hits an obstacle, and where the v_j are the new velocities that result from the specular reflection. We also set $x_j = x(t_j)$. The notation is clarified in Fig. 1. For a fixed realization of $\mathscr{X}$ the motion is deterministic, but when $\mathscr{X}$ is random, so is the motion of the point particle. If $\mathscr{X}$ is random, for example a Poisson distribution, it may happen that two obstacles overlap. Then Eq. (3) is not valid if the particle path hits the intersection of the two obstacles boundaries. For the point sets considered in Sect. 3 the obstacles cannot overlap and that simplifies the theoretical analysis considerably. For some of the point sets studied numerically in Sect. 2 a fraction of the obstacles do overlap, but this fraction decreases with r, and in the samples of obstacle sets used here we have not observed this.

In the kinetic limit that is the subject matter of this paper the obstacle radius is set to decrease simultaneously as the density of obstacles increases. For a given point set $\mathscr{X}$ we also consider the rescaled set $\mathcal{X}_\epsilon = \sqrt{\epsilon}\mathscr{X}$, so that for any set $A \subset \mathbb{R}^2$

Fig. 1 A trajectory of the Lorentz process [35]

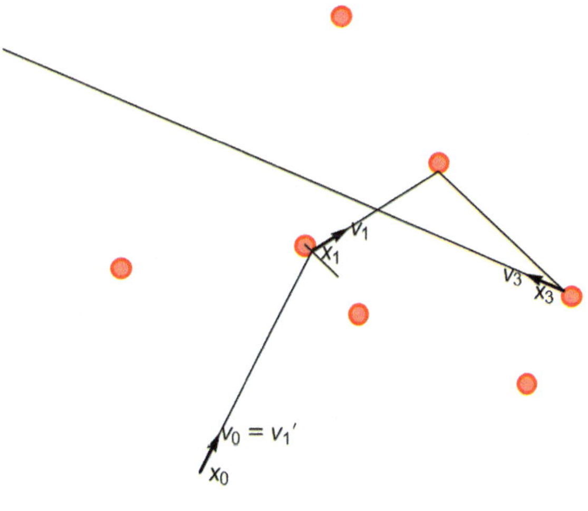

$$\mathcal{X}_\epsilon(A) = \mathscr{X}(\frac{A}{\sqrt{\epsilon}})\,. \tag{4}$$

Then $\mathscr{X}$ defines the domain of the Lorentz gas in the microscopic scale, and $\mathscr{X}_\epsilon$ in the macroscopic scale. The point set $\mathscr{X}_\epsilon$ is assumed to satisfy

$$\lim_{\epsilon\to 0} \epsilon \; \#\{x \in \mathcal{X}_\epsilon \cap A\} = c\, m(A)\,, \tag{5}$$

where $m(A)$ is the Lebesgue measure of an arbitrary open set $A \subset \mathbb{R}^2$ and c is a positive constant. Some additional conditions are needed to ensure that the motion is (almost certainly) well defined, and that the number of encounters with obstacles, M, is finite. For the rest of the paper the obstacle radius is fixed to be equal to ϵ in the macroscopic scaling, and therefore $\sqrt{\epsilon}$ in the microscopic scaling.

In the macroscopic scale, the time t_1 of the first encounter with an obstacle for a typical trajectory $T^t_{\mathcal{X}_\epsilon,\epsilon}(x_0, v_0)$ satisfies $t_1 = \mathcal{O}(1/c)$, i.e. the *mean free path length* of a typical trajectory is of the order $1/c$. Let $f_0 \in L^1_+(\mathbb{R}^2 \times S^1)$ be a probability density of point particles, and consider its evolution under the Lorentz process. At time $t > 0$ the expected distribution is then $f_{\epsilon,t} = f_\epsilon(x, v, t) = \mathbb{E}[f_0(T^{-t}_{\mathcal{X}_\epsilon,\epsilon}(x, v))]$, where the expectation is taken with respect to the probability distribution of the point set $\mathcal{X}_\epsilon$ in case it is random. We are interested in taking the limit $\epsilon \to 0$. This is called the annealed limit, as opposed to the quenched limit, where one takes one fixed realization of the point process $\mathscr{X}$ and considers the limit of $f_\epsilon(x, v, t) = f_0(T^{-t}_{\mathcal{X}_\epsilon,\epsilon}(x, v))$ as $\epsilon \to 0$, and is well described in [19]. To make sense of a quenched limit in the setting considered here one would need to define the perturbation of the periodic distribution of scatterers slightly, for example as in [27], but that lies outside the scope of this article.

The motion of the point particle is reversible for each realization of $\mathscr{X}$, and therefore (the annealed) limit may equivalently be defined weakly so that, for each $g \in C_0(\mathbb{R}^2 \times S^1)$,

$$\int_{\mathbb{R}^2\times S^1} f_\epsilon(x, v, t) g(x, v)\, dxdv = \int_{\mathbb{R}^2\times S^1} f_0(x, v)\mathbb{E}[g(T^t_{\mathcal{X}_\epsilon,\epsilon}(x, v))]\, dxdv. \tag{6}$$

Gallavotti [13, 14] proved that when $\mathscr{X}$ is a Poisson process with unit intensity, and ϵ converges to zero, then f_ϵ converges to a density f which satisfies the linear Boltzmann equation:

$$\partial_t f(x, v, t) + v \cdot \nabla_x f(x, v, t) = -2 f(x, v, t) + \int_{S^1_-} f(x, v', t)|v \cdot \omega|\, d\omega\,. \tag{7}$$

Here $S^1_- = S^1_-(v) = \{\omega \in S^1 \mid v \cdot \omega < 0\}$ and $v' = v - 2(\omega, v)\omega$. Spohn has proven a related, and more general, result in [33]. Boldrighini, Bunimovich and Sinai [3] proved the same result in the quenched setting.

On the other hand, it is also known that when $\mathscr{X} = \mathbb{Z}^2$ (or for that matter many other regular point sets, such as quasi-crystals), then the Lorentz process is not Markovian in the limit and therefore (7) does not hold, see [4, 15]. For a periodic distribution of scatterers in $\mathbb{R}^2$, Caglioti and Golse [8, 9] proved, under a certain independence assumption, that there is a limiting kinetic equation in an enlarged phase space. More general results without additional assumptions were obtained by Marklof and Strömbergsson [20, 21].

Marklof and Strömbergsson have studied this problem in several papers [22–25], and present in [26] a general theorem concerning the Boltzmann-Grad limit of the Lorentz process, where they give a concise set of conditions for a point set $\mathscr{X}$, such that the Lorentz process $T^t_{\mathcal{X}_{\epsilon,\epsilon}}(x, v)$ converges to a free flight process. In that last paper, they work in the quenched setting, i.e. with a fixed realization of the point process.

The problem studied in [35] and reported in an abbreviated form here, is the following: Let $\mathcal{Y}_\epsilon$ be a family of point processes converging to the periodic point set $\mathscr{X}$, where each point p of $\mathscr{X}$ is in the center of a cell of the regular periodic lattice in $\mathbb{R}^2$. Both $\mathcal{Y}_\epsilon$ and $\mathscr{X}$ have exactly one point in each lattice cell, and convergence here means that the points from $\mathcal{Y}_\epsilon$ converge (almost surely) to the center of the cell, where the points from $\mathscr{X}$ are situated. To study this in the Boltzmann-Grad limit, we set

$$\mathcal{Y}_{\epsilon,\epsilon} = \sqrt{\epsilon}\mathcal{Y}_\epsilon \qquad \text{and} \qquad \mathcal{X}_\epsilon = \sqrt{\epsilon}\mathscr{X} . \tag{8}$$

We are then interested in comparing the limits of the corresponding Lorentz processes, $T^t_{\mathcal{Y}_{\epsilon,\epsilon}}$ and $T^t_{\mathcal{X}_\epsilon}$, assuming that the obstacle radius is ϵ, and construct a family of point sets $\mathcal{Y}_\epsilon$ that converges to the periodic distribution, and yet the Lorentz process $T^t_{\mathcal{Y}_{\epsilon,\epsilon}}(x, v)$ converges to the free flight process generated by the linear Boltzmann equation (7). It is in a sense a non-stability result for the periodic Lorentz gas.

The proof follows quite closely the construction in [7, 31], and makes use of a third process, $\widetilde{T}^t_{\mathcal{Y}_{\epsilon,\epsilon}}$ which can be proven to be pathwise close to the free flight process, and also to the Lorentz process. The more general results by Marklof and Strombergsson [26] cover the periodic obstacle distribution and several other cases, and most likely it also covers the perturbed periodic distribution that is studied here. However, proving that the conditions for their general theorem hold for a particular case is not easy, and it may anyway be instructive to give a more direct proof.

The scaling studied here for the Lorentz process is not the only one studied in the literature. A more challenging problem is the long time limit, where the process is studied over a time interval of $[0, t_\epsilon[$, where $t_\epsilon \to \infty$ when $\epsilon \to 0$. Recent results of this kind have been obtained in [19] when the obstacle distribution is Poissonian, and it would be interesting to investigate the same limit in the setting of the present paper. In a different direction it is interesting to study the long time behaviour of the Lorentz gas when the obstacle diameter is kept fixed. Examples are the papers by Bunimovich and Sinai [5, 6] and a recent examples are [28, 30].

2 Numerical Simulation of Free Path Lengths

Here some numerical simulations are presented as illustrations of the Lorentz gas in some particular configurations of scatterers: the periodic Lorentz gas, the perturbed periodic Lorentz gas which is studied in detail in Sect. 3, the Lorentz gas in a Poisson distribution of scatterers, a quasi crystal and in the zero set of a Gaussian analytic function. The simulations were performed by computing a large number of trajectories in one fixed realization of the random distribution of scatterers, and so they correspond to a quenched limit as described in the introduction. The free path length distribution is then approximated by histograms of the first lap of the computed trajectories. And indeed, for the periodic distribution of scatterers, and for the quasi-crystal the simulation results give evidence of a distinct non exponential decay, whereas for all the other examples, the decay is exponential. This is exactly what the theory says. The zero set of a GAF seems not to have been studied in this context before.

2.1 The Zero Set of a Gaussian Analytic Function

There is a rather large literature on Gaussian analytic functions and related functions, and an nice introduction to the topic can be found in the lecture notes by Hough et al. [16]. Here we are interested in one very special GAF, defined as follows. Let $\{a_j\}_{j=0}^{\infty}$ be a sequence of i.i.d normal random variables in $\mathbb{C}$, i.e. random variables with density $\frac{1}{\pi}e^{-|z|^2}$, and let

$$f(z) = \sum_{j=0}^{\infty} \frac{a_j}{\sqrt{j!}} z^j . \tag{9}$$

The series is almost surely convergent and its sum an analytic function. Its value at any point $z \in \mathbb{C}$ is a Gaussian random variable, or more generally for any n and any $(z_1, ..., z_n) \in \mathbb{C}^n$, the complex vector $(f(z_1), ..., f(z_n))$ has mean zero complex Gaussian distribution. The covariance function, which characterizes the distribution of $f(z)$ is $e^{z\bar{w}}$ for $z, w \in \mathbb{C}$. It follows rather easily that the distribution of the zero set of $f(z)$ is translation and rotation invariant (clearly, $f(z)$ itself is not). And with probability one,

$$\lim_{s\to\infty} \frac{1}{4s^2} \# \left\{ z \mid |\Re z| < s, |\Im z| < s \text{ and } f(z) = 0 \right\} = \frac{1}{\pi} . \tag{10}$$

Moreover, the zero sets of $f(z)$ are ergodic under the action of translation and rotation, meaning that if A is an event involving the zeros and A is translation and rotation invariant, then the probability of A is either zero or one. The zero set of this GAF is repulsive, the zeros try to avoid each other. But this is a short range interaction, the proof of ergodicity in [16] shows that the number of zeros in a set $A \subset \mathbb{C}$ and translations $B_w = \{z + w \mid z \in B\}$ are asymptotically independent when $|w| \to \infty$.

In the simulations here we have used an sample of $10^6 + 1$ i.i.d. complex normal r.v. a_j and so approximated $f(z)$ with a polynomial. This gives 10^6 zeros of $f(z)$, of which a very large fraction is situated within a disk of radius of about 1000.

The zeros of the GAF were found using MPSolve, a multi precision polynomial solver developed by Bini and coauthors [1, 2]. It finds all complex zeros of an arbitrary degree polynomial with predetermined precision. The coefficients in (10) were generated with a precision of 100 decimal digits and the requested precision for the zeros was 8 decimal digits (and so the error should be much smaller than the diameter of the obstacle). The computational cost is very high, and besides using the internal controls of MPSolve, it is difficult to verify the precision of the position at each point. Therefore this calculation should be regraded as an illustration rather than a conclusive numerical simulation.

It is because the computed zeros of the truncated GAF-series are essentially distributed in a circular area that the other pointsets too are restricted to a circular area, even though in the other cases it would have been easy to find a point set in an arbitrary region.

2.2 A Quasi-crystal

Since the discovery of crystallographic diffraction patterns with forbidden symmetries the study of quasi-crystals has been an active area of research in physics as well as mathematics. A background to the discovery can be found in the web page of the Nobel Prize organisation [17], and for the mathematical background we have used the book by Senechal [32].

By definition a crystal is a point set with an essentially discrete diffraction pattern, and a quasi-crystal is one that exhibits a forbidden symmetry. There are several ways of constructing such point sets, two of which are by way of a Penrose tiling, and a cut and project method. The latter is the one that Marklof and Strömbergsson refer to in their works cited in the introduction. The construction is as follows:

- Let $\mathcal{L}^p$ be a point lattice in $\mathbb{R}^k$ (usually the standard lattice), and let K be the unit cell in the lattice that contains the origin.
- Let $\mathcal{E}$ be an n-dimensional subspace of $\mathbb{R}^k$, such that $\mathcal{E} \cap \mathcal{L}^p = \{0\}$, and let $\mathcal{E}^\perp$be the orthogonal complement of $\mathcal{E}$.
- Let Π and $\Pi^\perp$ be the orthogonal projections on $\mathcal{E}$ and $\mathcal{E}^\perp$ respectively.
- Let $X = \mathcal{L}^p \cap \left(\Pi^\perp(K) \oplus \mathcal{E}\right)$, i.e. the set of lattice points contained in the strip (or cylinder) obtained by translating the unit cell K along $\mathcal{E}$.

With this construction it is possible to find the points of the quasi-crystal in any part subset of $\mathcal{E}$, and hence a much smaller number of obstacles need to be taken into account when following the path of the Lorentz process.

Yet another method is by using the dual of a so called multigrid which is well described in Senechal's book. That is essentially the method used here to generate a quasi crystal consisting of approximately 10^8 points. The exact algorithm is taken from [12]. A variant of the dual method is also used in the very detailed numerical study performed by Mendoza Sosa and Kraemer in [29]. A one-dimensional version of this was studied numerically in [34]. A "pentagrid" consists of five sets of parallel lines in $\mathbb{R}^2$, each set dividing the plane in strips of unit width:

$$\left\{x \in \mathbb{R}^2 \,|\, x \cdot \omega_j + \gamma_j = k_j,\ j = 1...5,\ k_j \in \mathbb{Z}\right\}, \tag{11}$$

where ω_j are five unit vectors pointing to the vertices of a regular pentagon. The γ_j are translations of each set of lines, determined to guarantee that three or more lines never intersect in one point. The plane is thus divided in polygons with three or more sides, each of which corresponds to a point in the dual graph; these are the points of the quasi-crystal. One can show that with particular choices of the translation parameters γ_j these points correspond to the vertices of a Penrose tiling. This construction by duality is robust, and only a reasonable computational effort was needed to calculate the 10^8 points used in the simulations below.

2.3 Numerical Illustrations

All simulations were performed by brute force, meaning that all positions of the scatterers were kept in memory, and the scattering was calculated exactly (to machine precision).

The difference between the point sets can be seen in Fig. 2, which shows the scatterer centers in a square $|x_j| < 10$ for $x = (x_1, x_2)$.

The simulations were carried out by

- choosing a random initial position x_0 inside a disk with radius one tenth of the radius of the circular domain containing the point set, and a random initial velocity $v_0 \in S^1$.
- finding all scatterer centers inside a strip of width 2ϵ around the half line $x_0 + tv$, and computing the collision point x_1 and v_1 as the velocity after an elastic collision with the obstacle.
- Then iterating this until the particle leaves the circular domain containing the scatterers.

Most illustrations below are the result of 10^5 independent trajectories, and with ϵ chosen to give a mean free path length corresponding to one tenth of the radius of the domain. The length of the first and second lap of the trajectory were recorded, but only the first was used to estimate the free path length.

In the left part of Figs. 3, 4, 5, 6 and 7, the horizontal axis represents the length of the first lap of the a trajectory, rescaled to make the mean free path equal to one.

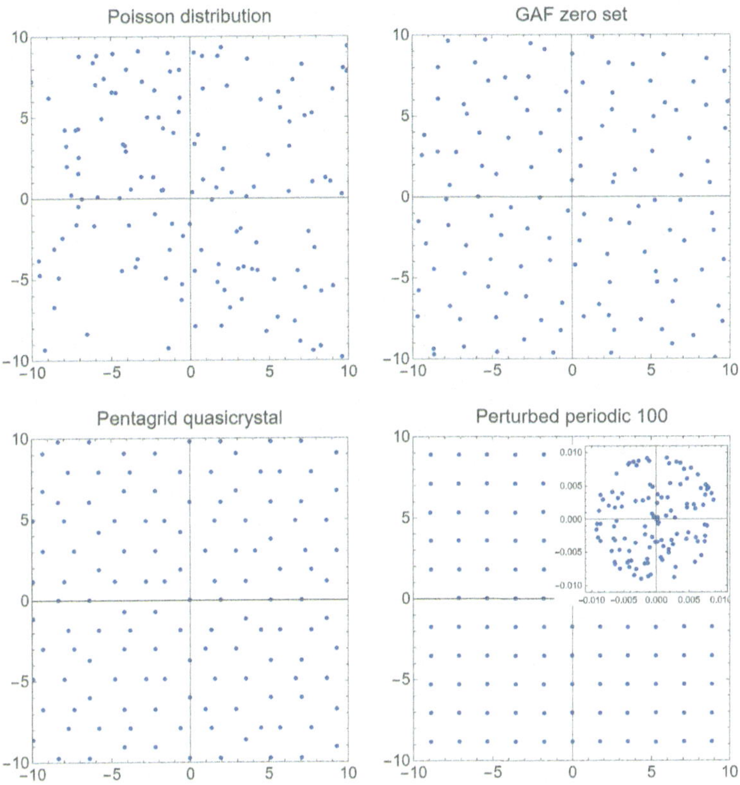

Fig. 2 The four images show a subset of the scatterer positions for the Poisson distribution(upper left), the zero set of a GAF (upper right), a quasi-crystal (lower left), and the perturbed periodic set (lower right). The perturbation is so small that in the figure one cannot distinguish this one from the periodic distribution; the inlay shows same points plotted with respect to their lattice point

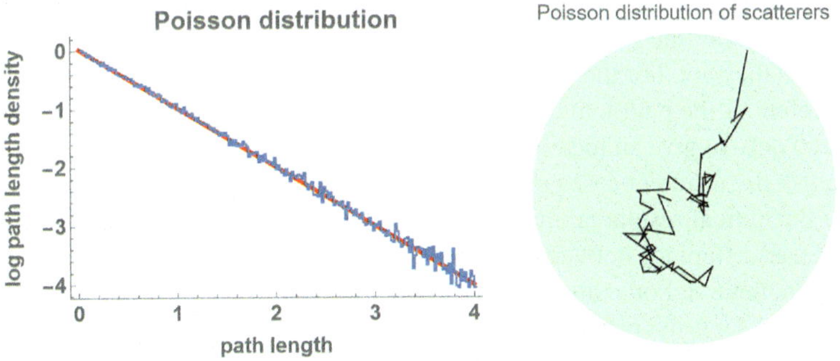

Fig. 3 The logarithm of the free path length density in a Poisson distributed set of obstacles. The image to the right shows a typical trajectory. The (red) straight line is a line fitted to the data

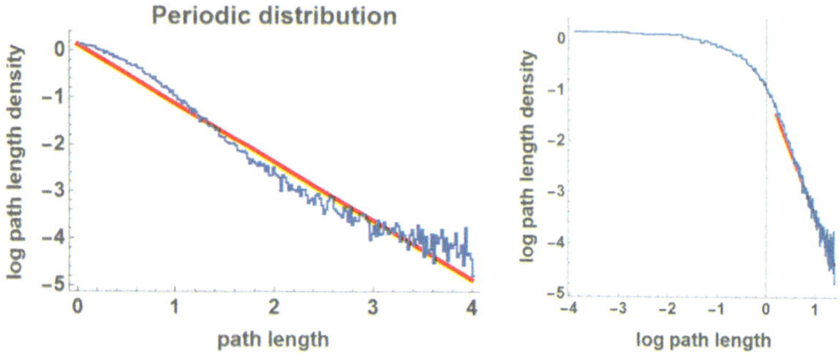

Fig. 4 The free path length density in a periodic distribution of obstacles. The image to the right shows a log-log plot of the same density. The (red) straight lines are fitted linear curves

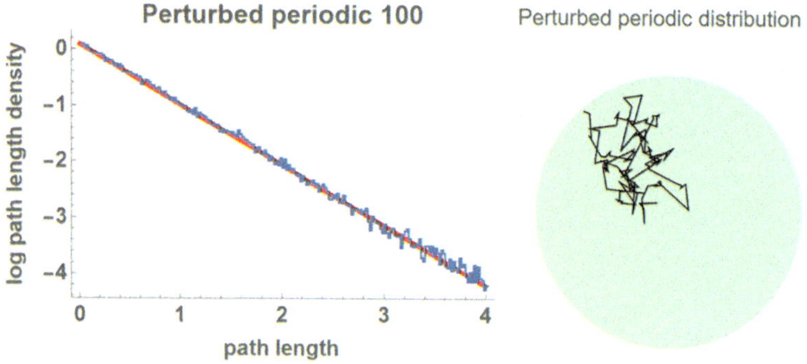

Fig. 5 The free path length density in a perturbed periodic distribution of obstacles. The image to the right shows a typical trajectory

The lengths are recorded in a histogram with bin width 0.02 in these units. The graphs illustrate the well-known facts that in a periodic distribution of scatterers or in a quasi-crystal, the path lengths are not exponentially decaying, whereas in the other cases they are. For the periodic distribution and the quasi-crystal we also show log-log plots of the path length densities. The typical trajectories in some cases are presented only to give an idea of what they may look like. It is not really possible to distinguish the different cases only by looking at a few trajectories.

The theoretical results are asymptotic, and maybe one should be surprised that such a clear difference between the cases can be seen also with an unsophisticated simulation method. For example, it is very difficult to see in the lower right of Fig. 2 that the point set is not periodic, whereas the difference between Figs. 4 and 5 is very clear.

The simulation here indicates that also this scatterer distribution of a Gaussian analytic function results in an exponentially decaying distribution of free path lengths.

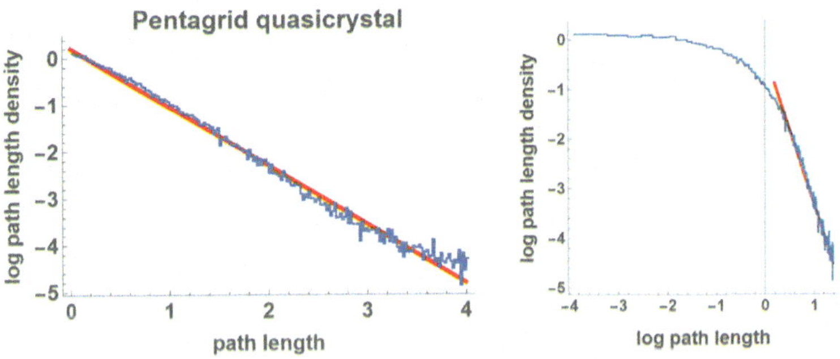

Fig. 6 The free path length density in an obstacle distribution given by a pentagrid quasi crystal, and a log log plot of the same density

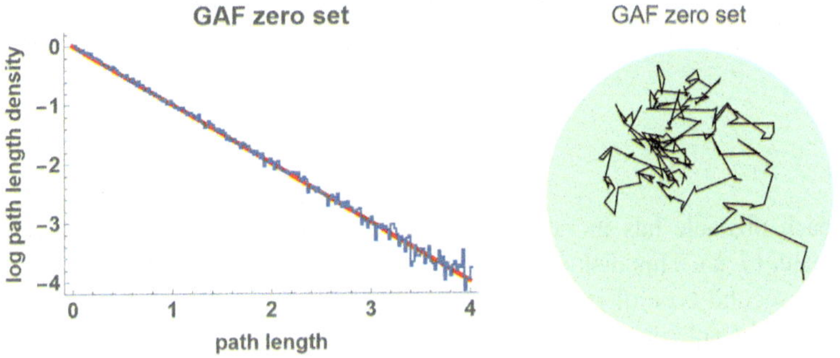

Fig. 7 The free path length density in an obstacle distribution defined by the zero set of a Gaussian analytic function and a typical particle trajectory

3 Random Modifications of a Periodic Distribution of Scatterers

While a perfectly periodic distribution of scatterers gives rise to a distribution of free path lengths that are not compatible with the linear Boltzmann equation as a small scatterer limit, it turns out that a rather small perturbation is enough to get the same limit as for the Poisson distribution of scatters.

One example of such a distribution is one where the obstacle size and lattice size are kept from the purely periodic Boltzmann—Grad setting, but the obstacle centers are moved independently at random from the periodic sites. More precisely let $\mathscr{X} = \mathbb{Z}^2$ and define $\mathscr{Y}_\epsilon$ as a perturbation of $\mathscr{X}$ in the following way: Let ϕ be a rotationally symmetric probability density supported in $|x| < 1$, fix $\nu \in]1/2, 1[$, set

$$\mathscr{Y}_\epsilon = \left\{ (j,k) + \epsilon^{1-\nu} \xi_{j,k} \mid (j,k) \in \mathbb{Z}^2, \xi_{j,k} \text{ i.i.d. with density } \phi \right\} , \qquad (12)$$

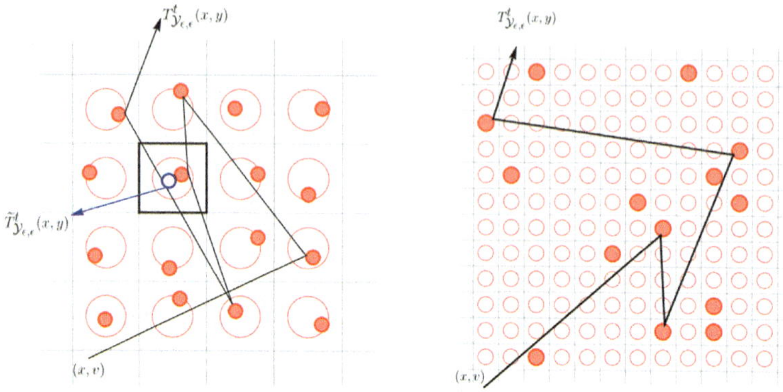

Fig. 8 The perturbed periodic (left, [35]) and diluted periodic (right) distribution, together with trajectories as described in the text

and let

$$\mathcal{Y}_{\epsilon,\epsilon} = \sqrt{\epsilon}\mathcal{Y}_{\epsilon}\,. \tag{13}$$

Thus each obstacle has its center in a disk of radius $\epsilon^{1-\nu}$ centered at an integer coordinate, (j, k). This disk will be called the *obstacle patch* below.

In the second type of perturbation the obstacle size is the same as the obstacle patch. The obstacle distribution is then too dense to give a finite free path length in the limit, but this is achieved by a random thinning of the point set:

$$\mathcal{Y}_{\epsilon} = \left\{2\epsilon^{\nu}(j,k) \mid \xi_{j,k} = 1,\ \xi_{j,k} \text{ i.i.d. Bernoulli with } \mathbb{P}[\xi_{j,k} = 1] = \epsilon^{2\nu-1}\right\}\,. \tag{14}$$

Here it is clear that whenever $\nu > 1/2$ this $\mathcal{Y}_{\epsilon}$ converges in law to the Poisson process with intensity one. The two different cases are shown in Fig. 8, together with a couple of possible trajectories.

The main result in [35] is the following theorem, which concerns the case shown in the left part of Fig. 8:

Theorem 1 *Let $T^t_{\mathcal{Y}_{\epsilon,\epsilon}}(x, v)$ be the Lorentz process obtained by placing a circular obstacle of radius ϵ at each point of $\mathcal{Y}_{\epsilon,\epsilon}$. Let $\bar{t} > 0$, and let $f_0(x, v)$ be a probability density in $\mathbb{R}^2 \times S^1$. Define $f_\epsilon(x, v, t)$ as the function such that for all $t \in [0, \bar{t}]$ and all $g \in C(\mathbb{R}^2 \times S^1)$,*

$$\int_{\mathbb{R}^2\times S^1} f_\epsilon(x, v, t)g(x, v)\,dxdv = \int_{\mathbb{R}^2\times S^1} f_0(x, v)\mathbb{E}[g(T^t_{\mathcal{Y}_{\epsilon,\epsilon}}(x, v))]\,dxdv\,. \tag{15}$$

Then there is a density $f(x, v, t) \in C\left([0, \bar{t}], L^1(\mathbb{R}^2 \times S^1)\right)$ such that for all $t \leq \bar{t}$, $f_\epsilon(x, v, t) \to f(x, v, t)$ in $L^1(\mathbb{R}^2 \times S^1)$ when $\epsilon \to 0$, and such that $f(x, v, t)$ satisfies the linear Boltzmann equation

$$\partial_t f(x,v,t) + v \cdot \nabla_x f(x,v,t) = \kappa \left(\int_{S^1_-} f(x,v',t)|v\cdot\omega|\,d\omega - 2f(x,v,t) \right). \tag{16}$$

The constant κ depends on ϕ and S^1_- and v' are defined as in Eq. (7).

The same result holds if $T^t_{y_{\epsilon,\epsilon}}$ from Eq. (15) is replaced by the $T^t_{y_\epsilon}$ that comes from the diluted periodic distribution, Eq. (14) when $1/2 < \nu < 1$. This was proven in [31]; the proof of Theorem 1 here is very similar. And although the limiting microscopic distributions in the two cases are very different, there is a very strong similarity of the paths in the kinetic limit: the number of obstacle patches that are traversed by a typical trajectory is the same, and the probability of hitting an obstacle in the patch is also necessarily the same if the kinetic limit is to be achieved, and this opens for a more general study involving different scalings of the number of traversed obstacle patches.[1] The case $\nu = 1/2$ is treated in [27], where the authors obtain a power law lower bound for the free path length.

A fully deterministic distribution of scatters that would most likely give exactly the same result as the randomly perturbed periodic distribution is the following: let β_1 and β_2 be fixed irrational numbers, and assume that the lattice cells are enumerated with $(j,k) \in \mathbb{Z}^2$. Then let the obstacle center be put on a circle with radius $\epsilon^{1-\nu}$ around the lattice cell center, and with an angle θ given by $\theta = \pi(j\beta_1 + k\beta_2)$. The proof would have to be modified with respect to the proof here, but it seems feasible to prove at least that the free path lengths are exponentially distributed, as in the case of a random perturbation.

One essential difficulty when proving that the Lorentz gas converges to a Boltzmann equation for any distribution of scatterers is that the Boltzmann equation is the generator of a Markov process, whereas the Lorentz gas has a memory: if a trajectory has hit an obstacle and returns to the same place, it must collide again. And similarly, if a trajectory crosses itself, or comes very close, then there cannot be a jump there that would have been the result of hitting an obstacle that would also have affected the history of the path. To deal with this difficulty we introduce an auxiliary process, a *Markovian Lorentz process*, which first first used in [7]. The difference between the Lorentz process and the Markovian Lorentz process can be seen to the left of Fig. 8, where a trajectory makes a third jump in the lattice cell that is indicated with a solid square, jumps again and then returns to the same obstacle patch. In the Lorentz case, the position of the obstacle is fixed once and for all, and hence the trajectory misses it when it returns to the patch, but in the Markovian Lorentz process, the random position of the obstacle is chosen each time the trajectory enters a lattice cell, and therefore the blue trajectory indicated by $\tilde{T}^t_{y_{\epsilon,\epsilon}}$ is a possible outcome in this case.

So we now have three random processes to study: The Boltzmann process that is generated by the Boltzmann equation, the Lorentz process and the Markovian Lorentz process.

[1] This observation is due to an anonymous referee to [35].

The Boltzmann process is the random flight process $(x(t), v(t)) = T_B^t(x, v)$ generated by Eq. (7). Let $0 = t_0 < t_1 <, ..., t_n < t$ be a sequence of times generated by independent, exponentially distributed increments $t_j - t_{j-1}$ with intensity 2, let $v_0 = v$, and define $v_j = v_{j-1} - 2(\omega_j \cdot v_{j-1})\,\omega_j$, where the $\omega_j \in S^1$ are independent and uniformly distributed. Finally set

$$x(t) = x_n(t) = x + t_1 v + (t_2 - t_1)v_1 + \cdots + (t_n - t_{n-1})v_{n-1} + (t - t_n)v_n$$
$$v(t) = v_n\,. \tag{17}$$

The number n in the sum is then Poisson distributed. The solution of Eq. (1) may be defined weakly by

$$\int_{\mathbb{R}^2\times S^1} f_0(x, v)g(x, v, t)\,dxdv = \int_{\mathbb{R}^2\times S^1} f(x, v, t)g(x, v)\,dxdv\,. \tag{18}$$

The function $g(x, v, t)$ can be expanded as a sum of terms, each representing the paths with a fixed number of jumps:

$$g(x, v, t) = V^t g(x, v) = \sum_{n=0}^{\infty} g_n(x, v, t)\,, \tag{19}$$

with

$$g_n(x, v, t) = (V^t g)_n(x, v) = e^{-2t}\int_0^t\int_{t_1}^t\cdots\int_{t_{n-1}}^t dt_1\cdots dt_n\,\times$$
$$\int_{(S^1_-)^n} d\omega_1\cdots d\omega_n \prod_{k=1}^{n} |\omega_k\cdot v_{k-1}| g(x_n(t), v_n)\,. \tag{20}$$

The Markovian Lorentz process. This process can be defined as in Eq. (18), with the function $g(x, v, t)$ replaced by

$$\tilde{g}_\epsilon(x, v, t) = \widetilde{V}_\epsilon^t g(x, v) = \mathbb{E}\left[g(\widetilde{T}^t_{y_{\epsilon,\epsilon}}(x, v))\right] = \sum_{n=0}^{\infty}\tilde{g}_{\epsilon,n}(x, v, t)\,,$$
$$\left(\widetilde{V}^t_{\epsilon,n}\right) g(x, v) = \tilde{g}_{\epsilon,n}(x, v, t)\,, \tag{21}$$

where $\tilde{g}_{\epsilon,n}(x, v, t)$ is the contribution of trajectories having exactly n velocity jumps in the interval $[0, t]$. All these terms can be computed rather explicitly by counting the number of times a trajectory crosses the obstacle range of one cell.

The Lorentz process. This is the original process described in the introduction. For comparison with the other two processes we just note that in this case $g(x, v, t)$ is directly defined from Eq. (15) by setting

$$g(x, v, t) = \mathbb{E}[g(T^t_{\mathcal{Y}_{\epsilon,\epsilon}}(x, v))]\,. \tag{22}$$

Once the three processes are defined, the main steps of the proof consist of

- proving that a density $\tilde{f}_\epsilon(x, v, t)$ evolving through the Markovian Lorentz process converges to the solution of the linear Boltzmann equation, and
- that the processes $T^t_{\mathcal{Y}_{\epsilon,\epsilon}}(x, v)$ and $\widetilde{T}^t_{\mathcal{Y}_{\epsilon,\epsilon}}(x, v)$ are pathwise close.

The first part will imply that the also the Boltzmann process and the Markovian Lorentz process are close. The meaning of pathwise close here is in the sense of functions in the Skorokhod space $D_{[0,\bar{t}]}(\mathbb{R}^2 \times S^1)$ of *cádlág* functions on $\mathbb{R}^2 \times S^1$. The details of this are left out here, but can be found in [35].

The statement concerning the convergence of the density $\tilde{f}_\epsilon(x, v, t)$ to the solution of the Boltzmann equation is the content of Theorem 2 below, and the second statement follows from an estimate of the measure of initial data (x_0, v_0) that result in a re-collision in a given time interval, uniform over all realizations of $\mathscr{X}$, as stated in Theorem 3.

Theorem 2 *Let $\widetilde{T}^t_{\mathcal{Y}_{\epsilon,\epsilon}}(x, v)$ be the Markovian Lorentz process. Fix $\bar{t} > 0$. For any density $f_0(x, v)$ in $\mathbb{R}^2 \times S^1$, let $\tilde{f}_\epsilon(x, v, t)$ be the unique function such that*

$$\int_{R^2\times S^1} \tilde{f}_\epsilon(x, v, t) g(x, v)\, dxdv = \int_{R^2\times S^1} f_0(x, v) g(\widetilde{T}^t_{\mathcal{Y}_{\epsilon,\epsilon}}(x, v))\, dxdv\,, \tag{23}$$

for all $t \in [0, \bar{t}]$, and any $g(x, v) \in C(\mathbb{R}^2 \times S^1)$. Then there is a function $f(x, v, t)$ such that for all t

$$\tilde{f}_\epsilon(x, v, t) \to f(x, v, t)\,, \tag{24}$$

and $f(x, v, t)$ satisfies the linear Boltzmann equation , Eq. (7).

Proof Let $g \in C_0(\mathbb{R}^2 \times S^1)$, and $t < \bar{t}$. We must prove that

$$\int_{\mathbb{R}^2\times S^1} \left(\tilde{f}_\epsilon(x, v, t) - f(x, v, t)\right) g(x, v)\, dxdv \to 0 \quad \text{when} \quad \epsilon \to 0\,, \tag{25}$$

or, equivalently,

$$\int_{\mathbb{R}^2\times S^1} f_0(x, v) \left(V^t g(x, v) - \tilde{V}^t_\epsilon g(x, v)\right) dxdv \to 0\,. \tag{26}$$

The expression in (26) is bounded by

$$2\|g\|_{L^\infty} \int_{\mathbb{R}^2\times S^1} f_0 \left(\mathbf{1}_{f_0>\lambda} + \mathbf{1}_{|x|\geq M}\right) dxdv + \lambda\|(V^t g - \tilde{V}^t_\epsilon g)\mathbf{1}_{|x|<M}\|_{L^1}\,, \tag{27}$$

and choosing λ and M large enough, depending on f_0, the first term can be made arbitrarily small, smaller than $\varepsilon_0/2$, say, for any $\varepsilon_0 > 0$. It remains to show that the second term can be made smaller than $\varepsilon_0/2$ by choosing ϵ small enough. Both V^t and $\widetilde{V}_\epsilon^t$ are semigroups that are contractions in $L^1 \cap L^\infty$, and therefore, after dividing the interval $[0, t]$ into N equal sub-intervals, $[jt/N, (j+1)t/N]$ one gets

$$V^t g(x, v) - \widetilde{V}_\epsilon^t g(x, v) = \sum_{j=0}^{N-1} \widetilde{V}_\epsilon^{jt/N} \left(V^{t/N} - \widetilde{V}_\epsilon^{t/N}\right) V^{(N-1-j)t/N} g(x, v) . \tag{28}$$

Therefore it is enough to prove that for $N = N_\epsilon$ appropriately chosen

$$N_\epsilon \left\| V^{t/N_\epsilon} g - \widetilde{V}_\epsilon^{t/N_\epsilon} g \right\|_{L^1} \to 0 \tag{29}$$

when $\epsilon \to 0$. With $\tau_\epsilon = t/N_\epsilon$ we find

$$\begin{aligned} &\left|V^{\tau_\epsilon} g(x, v) - \widetilde{V}_\epsilon^{\tau_\epsilon} g(x, v)\right| \le \\ &\qquad \left|(V^{\tau_\epsilon})_0 g(x, v) - (\widetilde{V}_\epsilon^{\tau_\epsilon})_0 g(x, v)\right| + \left|(V^{\tau_\epsilon})_1 g(x, v) - (\widetilde{V}_\epsilon^{\tau_\epsilon})_1 g(x, v)\right| \\ &\qquad + \left|\sum_{k=2}^{\infty} (V^{\tau_\epsilon})_k g(x, v)\right| + \left|\sum_{k=2}^{\infty} (\widetilde{V}_\epsilon^{\tau_\epsilon})_k g(x, v)\right| . \end{aligned} \tag{30}$$

Set $r_\epsilon = \epsilon^{(2\nu-1)/2}\left(1 + \log(t/\sqrt{\epsilon})\right)$. After integrating over $\{|x| \le M\} \times S^1$ and multiplying by N_ϵ we find that the contribution of the last two terms of Eqs. (30)–(29) are bounded by $CM^2 \|g\|_\infty t \tau_\epsilon$, as long as $\tau_\epsilon > r_\epsilon^{1/2}$, and hence converges to zero. The proof is omitted here.

The first term accounts for paths with no jump in the given interval. It follows from Proposition 1, which is presented below, that, after integrating over $\mathbb{R}^2 \times S^1$, it is bounded by

$$\frac{t}{\tau_\epsilon} C M^2 r_\epsilon^{1/2} . \tag{31}$$

This expression converges to zero with ϵ if $r_\epsilon^{1/2}/\tau_\epsilon$ does. The second term accounts for paths with exactly one jump, and according to Proposition 2, also stated below, it is bounded by

$$CM^2 \left(\omega(\tau_\epsilon, g) + r_\epsilon + \|g\|_\infty \sqrt{r_\epsilon} + \|g\|_\infty r_\epsilon/\tau_\epsilon\right) . \tag{32}$$

For this term it is enough that r_ϵ/τ_ϵ converges to zero, but the rate of convergence may depend on the modulus of continuity of the test function g. So therefore all terms that contribute to the left-hand side of Eq. (29) converge to zero if $\tau_\epsilon > r_\epsilon^{1/2}$. And this concludes the proof of Theorem 2. □

From the proof of Theorem 2 we see that it is enough to consider only very small time intervals, and therefore we need only to analyze two cases: a particle path starting at $(x, v) \in \mathbb{R}^2 \times S^1$, i.e. with position $x \in \mathbb{R}^2$ in the direction v, moves without changing velocity during the whole interval, or, hits an obstacle at some point

x' and continues from there in the new direction v', but suffers no more collisions. In the study of these two cases we rely strongly on the periodic structure of the problem.

Consider a path starting in the direction v from a point x in the direction of v. The velocity $v \in S^1$ may also be be defined by an angle $\beta \in [0, 2\pi[$, and it is enough to assume that $\beta \in [0, \pi/4]$ to the vertical axis as in the left of Fig. 9, because all other cases can be treated in the same way just by finite number of rotations and reflections of the physical domain. When a collision takes place, the new velocity v' depends on v and on the impact parameter r as shown to the right in Fig. 9, where $r \in [-1, 1]$ when scaled with respect to the obstacle radius. Define

$$p_0(x, v, t) = \mathbb{P}\Big[\text{ no collision in the interval } [0, t[\\ \text{for a path starting at } (x, v) \in \mathbb{R}^2 \times S^1 \Big], \tag{33}$$

and similarly, let $p_j = p_j(x, v)$ be the probability that a scattering event takes place in the j-passage of a lattice cell. It is a function of the distance from the path to the centre of the cell, ρ_j, and is zero outside $|\rho_j| < \epsilon^{1-\nu} + \epsilon^{1/2}$. Let $A_{x,v,t}$ be any event that depends on a position x and direction v of a path segment of length t. Then the probability that $A = A_{x',v',t'}$ is realized after the first collision can be computed as

$$\mathbb{P}[A] = \sum_{j=1}^{n} p_0(x, v, t_{j-}) p_j(x, v) \mathbb{P}[A_{x',v',t'} \mid j]. \tag{34}$$

Here $\mathbb{P}[A_{x',v',t'} \mid j\,]$ denotes the conditional probability of the event $A_{x',v',t'}$ given that the collision takes place when the particle crosses cell number j along the path, and this also depends on the x and v. The time when the path enters cell number j is denoted t_{j-}, and the terms in the sum contain a factor $p_0(x, v, t_{j-})$ for the probability that no collision has taken place earlier. Similarly t_{j+} denotes the time when the trajectory leaves the cell; that is a random number, but we always have $0 < t_{j+} - t_{j-1} < 2\sqrt{\epsilon}$. In (34), (x', v', t') are random, so $\mathbb{P}[A]$ involves also an integral over these variables, and the notation $\mathbb{P}[A_{x',v',t'} \mid j]$ is intended to include this integration. The expectation $\mathbb{E}[\psi]$ of some function ψ is computed in the same way with $\mathbb{P}[A_{x',v',t'} \mid j]$ replaced by $\mathbb{E}[\psi \mid j]$.

Given a rotationally symmetric density ϕ for the random position of the obstacle define

$$\varphi_0(x_1) = \int_{\mathbb{R}} \phi_0(x_1, x_2)\, dx_2 \,, \tag{35}$$

where (x_1, x_2) is an arbitrary coordinate system in $\mathbb{R}^2$. Note that φ_0 is independent of the orientation of this coordinate system. It is a smooth probability density with support in $[-1, 1]$, and then $\varphi_\epsilon(r) \equiv \epsilon^{\nu-1}\varphi_0(\epsilon^{\nu-1} r)$ is also a smooth probability density with support in $|r| \le \epsilon^{1-\nu}$.

The two expressions in (33) and (34) can be approximated by sums of terms corresponding to one passage of a lattice cell, and such sums can be evaluated using the following rather technical lemma, that depends on the periodic structure of the problem.

Lemma 1 *Consider a lattice cell in the microscopic scale, so that the cell side has length one, and assume that there is an obstacle of radius* $\epsilon^{1/2}$ *with center distributed with a density of the form* $\phi(x) = \epsilon^{2\nu-2}\phi_0(\epsilon^{\nu-1}x)$*, where* $\phi_0 \in C^\infty(\mathbb{R}^2)$ *has support in the unit ball of* $\mathbb{R}^2$*. Let* φ_ϵ *be defined as in Eq. (35) and rescaled with* ϵ*. Let* $\psi_0 \in L^1(\mathbb{R})$ *with* $\psi_0(x) = 0$ *for* $|x| > 1$*, and set* $\psi_\epsilon(x) = \psi_0(x/\sqrt{\epsilon})$*. Consider a particle path traversing the cell n times with an angle* $\beta \in [0, \pi/4]$ *to the vertical cell sides, and set* $y_1, \dots, y_n$ *to be the consecutive points at the lower edge of the cell. We have* $y_1 \in [-1/2, 1/2[$ *and* $y_k = -1/2 + (1/2 + (k-1)\tan(\beta) \mod 1)$*. Then*

$$p[\psi_\epsilon](y, \beta) \equiv \int_{\mathbb{R}} \psi_\epsilon(x_1)\phi_\epsilon(y\cos(\beta) + x_1)\,dx_1 \tag{36}$$

satisfies

$$\sum_{k=1}^{n} p[\psi_\epsilon](y_k, \beta) = \frac{\sqrt{\epsilon}\,n}{\cos(\beta)} \int_0^1 \psi_0(x)\,dx + R_a(y_1, \beta)\,. \tag{37}$$

The remainder term R_a *also depends on n and* ϵ*. It is bounded by a function* $\bar{R}_A(\beta)$*, that depends on* ϕ*, n and* ϵ *but not on* ψ*, and that satisfies*

$$|R_a(n, \epsilon, y_1, \beta)| \le \|\psi_0\|_{L^1} \bar{R}_A(\beta) \qquad \textit{where} \tag{38}$$

$$\int \bar{R}_A(\beta)\,d\beta \le C\epsilon^{\nu-1/2}(1 + \log n)\,. \tag{39}$$

The importance of this lemma is that the sum of the terms coming from traversing the consecutive lattice cells is an integral plus a rest term which, although not pointwise small, is small after integrating over the velocity directions. The number of traversed cells, n, is proportional to $1/\sqrt{\epsilon}$, and therefore the factor multiplying the integral in the right-hand side of (37) is converging to a finite number. The proof is omitted here. It can be found in [35] and a very similar result was used in [31]. And using the lemma, we may prove the two propositions used in the proof of Theorem 2

Proposition 1 *The terms* $(V^t)_0 g(x, v)$ *and* $(\widetilde{V}_\epsilon^t)_0 g(x, v)$ *are given by*

$$(V^t)_0 g(x, v) = e^{-2t} g(x + vt, v) \qquad \textit{and} \tag{40}$$

$$(\widetilde{V}_\epsilon^t)_0 g(x, v) = p_0(x, v, t) g(x + vt, v)\,. \tag{41}$$

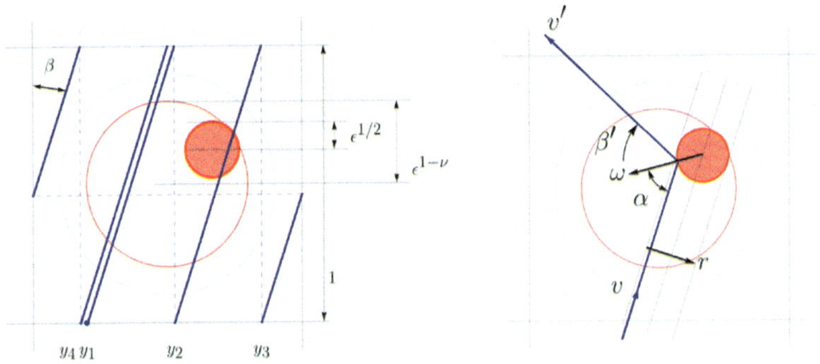

Fig. 9 Left: A particle path traversing the lattice cells. The cell size, diameter of the distribution of the obstacle center and obstacle size are given in the microscopic scale. The macroscopic scale is obtained by multiplication by a factor $\sqrt{\epsilon}$. Right: A collision with an obstacle and the notation used to describe this [35]

where $p_0(x, v, t)$ *is the probability that a trajectory starting at* $x \in \mathbb{R}^2$ *in direction* v *does not hit an obstacle in the interval* $[0, t]$*; this can be computed explicitly. The function* $p_0(x, v, t)$ *satisfies* $p_0(x, v, t) - e^{-2t} = R_b(x, v, t)$ *with*

$$|R_b(x, v, t)| \le \bar{R}_B(\beta) \tag{42}$$

for a function $\bar{R}_B(\beta)$ *that satisfies*

$$\int_{S_1} \bar{R}_B(\beta)\, dv \le C\epsilon^{(2\nu-1)/4}\sqrt{1+\log(t/\sqrt{\epsilon})}\,, \tag{43}$$

and consequently, for any $g \in C\left(\mathbb{R}^2 \times S^1\right)$ *with support in* $|x| \le M$*,*

$$\left\|(V^t)_0 g - (\widetilde{V}_\epsilon^t)_0 g\right\|_{L^1(\mathbb{R}^2\times S^1)} \le CM^2\|g\|_{L^\infty}\epsilon^{(2\nu-1)/4}\sqrt{1+\log(t/\sqrt{\epsilon})}\,. \tag{44}$$

Proof Consider a path starting at $x \in \mathbb{R}^2$ in the direction of v. There is no restriction in assuming that the direction v is clockwise rotated with an angle β as illustrated to the left in Fig. 9, which shows one lattice cell, with an obstacle patch indicated with a red circle, and a blue circle indicating the maximal range for an obstacle, and the red solid disk shows one possible position of the obstacle. In the macroscopic scale, the lattice size is $\epsilon^{1/2}$, the obstacle has radius ϵ, and the obstacle patch has radius $\epsilon^{3/2-\nu}$. These values are indicated within parenthesis, and the microscopic scale, in which the lattice size is 1 is indicated to the left. When $\epsilon \to 0$ the obstacle patch in microscopic scale shrinks to a point, but is is drawn large in the image for clarity. The path under consideration enters a new lattice cell for the first time at y_1, and enters the second lattice cell at y_2, drawn in the same image. A path of length t in macroscopic scale will traverse a number n of lattice cells in the vertical direction. We set $m = \lfloor \frac{t\cos\beta}{2\epsilon^{1/2}} \rfloor$, the integer part of $\frac{t\cos\beta}{2\epsilon^{1/2}}$. Then

$$n = 2m + 1 = \frac{t\cos\beta}{\epsilon^{1/2}} + \zeta\,, \tag{45}$$

where $\zeta \in]-1,1]$, and increasing the interval to $\zeta \in [-2,2]$ one may take into account the uncertainty of the starting and end point inside a cell. Then the probability that a there is no collision in the interval $[0,t[$ is

$$p_0(x,v,t) = \prod_{j=1}^{n}(1 - p(y_j,\beta)) = \exp\left(\sum_{j=1}^{n}\log(1 - p(y_j,\beta))\right) \tag{46}$$

where $p(y,\beta)$ is the probability that a trajectory entering a lattice cell at y with angle β along the lower cell boundary meets an obstacle before leaving the cell on the top. Because $p(y_j,\beta) < c\epsilon^{\nu-1/2} \to 0$ when $\epsilon \to 0$ we may assume that $p(y_j,\beta) < 1/2$, and therefore

$$-(1 + c\epsilon^{\nu-1/2})p(y_j,\beta) < \log(1 - p(y_j,\beta)) < -p(y_j,\beta)\,, \tag{47}$$

which provides an asymptotic expression for $p_0(x,v,t)$ once the sum $\sum_{j=1}^{n} p(y_j,\beta)$ has been evaluated. That $(V^t)_0 g(x,v)$ has the form (40) is evident, and hence it remains to prove the estimate (44). From (47) we get

$$\begin{aligned}\left|e^{-2t} - p_0(x,v,t)\right| &\le \left|e^{-2t} - e^{\sum_{j=1}^{n} p(y_1\beta)}\right| \\ &+ \left|e^{-(1+c\epsilon^{\nu-1/2})\sum_{j=1}^{n} p(y_1,\beta)} - e^{-\sum_{j=1}^{n} p(y_j,\beta)}\right| \\ &\le \left|e^{-2t} - e^{-\sum_{j=1}^{n} p(y_1,\beta)}\right| + c\epsilon^{\nu-1/2}\,.\end{aligned} \tag{48}$$

Referring to the notation in Lemma 1 the probability $p(y_j,\beta)$ can be computed as

$$p(y_j,\beta) = p[\mathbf{1}_\epsilon](y,\beta)\,, \tag{49}$$

that is, ψ_0 is set to one in the interval $[-1,1]$ and zero outside this interval. The same lemma then gives

$$\sum_{j=1}^{n} p(y_1,\beta) = \frac{2\sqrt{\epsilon}n}{\cos(\beta)} + R_a(n,\epsilon,y_1,\beta) = 2t + \hat{R}_a\,, \tag{50}$$

where R_a satisfies the estimate (38),(39), and $|\widehat{R}_a - R_a| \le 2\sqrt{\epsilon}$. Hence $|\hat{R}_a(n,\epsilon,y_1,\beta)| \le C\bar{R}_A(\beta)$ for some constant C, and Markov's inequality says that for any $\lambda > 0$, the inequality $m(\{v \in S^1 \mid \bar{R}_A > \lambda\}) \le \frac{C}{\lambda}\epsilon^{\nu-1/2}(1 + \log(t/\sqrt{\epsilon})$

holds. Therefore

$$\int_{S_1} \left\| e^{-2t} - p_0(\cdot, v, t) \right\|_{L^\infty(\mathbb{R}^2)} dv \le \int_{S^1} \left\| e^{-2t} - e^{-2t+\hat{R}_3} \right\|_{L^\infty(\mathbb{R}^2)} dv + 2\pi c \epsilon^{\nu-1/2}$$
$$\le C \frac{\epsilon^{\nu-1/2}(1+\log(t/\sqrt{\epsilon}))}{\lambda} + C\lambda + C\epsilon^{\nu-1/2} \le C\epsilon^{(2\nu-1)/4}\sqrt{1+\log(t/\sqrt{\epsilon})}\,. \tag{51}$$

where the norms inside the integrals are taken with respect to the variable x. Then (44) follows after integrating the expression over $\mathbb{R}^2$. □

The following proposition concerns the terms $\left(V^t\right)_1 g(x, v)$ and $\left(\widetilde{V}_\epsilon^t\right)_1 g(x, v)$ defined by

$$\left(V^t\right)_1 g(x, v) = e^{-2t} \int_0^t \int_{S^1_-} g(x + \tau v + (t - \tau)v', v')|v \cdot \omega| \, d\omega \, d\tau \tag{52}$$

and

$$\left(\widetilde{V}_\epsilon^t\right)_1 g(x, v) = \mathbb{E}\left[g(x + \tilde{\tau} v + (t - \tilde{\tau})v', v')\mathbf{1}_{J_1(x,v,t)}\right], \tag{53}$$

where $J_1(x, v, t))$ is the event that there is exactly one jump on the trajectory starting at x in the direction of v. The $\tilde{\tau} \in]0, t[$ and $v' \in S^1$ are the random jump time and velocity of the particle after the jump.

Proposition 2 *Let $\omega(\delta, g)$ be the modulus of continuity for g, i.e. a function such that $|g(x, v) - g(x_1, v_1)| \le \omega(|x - x_1| + |v - v_1|, g)$. Then*

$$\left\| \left(V^t\right)_1 g - \left(\widetilde{V}_\epsilon^t\right)_1 g \right\|_{L^1(\mathbb{R}^2 \times S^1)}$$
$$\le CM^2 \left(t\omega(t, g) + r_\epsilon \omega(t, g) + \|g\|_{L^\infty} r_\epsilon + \|g\|_{L^\infty} t r_\epsilon^{1/2}\right) \tag{54}$$

where the support of g is contained in $\{(x, v) \mid |x| < M\}$ and $r_\epsilon = \epsilon^{\nu-1/2}(1 + \log n)$.

The proof is based on calculations similar to the proof of Proposition 1. It is written in full detail in [35], and is not repeated here.

The Lorentz process and the Markovian Lorentz process are equivalent on the set of trajectories that do not return to the same obstacle patch, and it remains to prove that the probability that a path in the Markovian process returns to the same patch vanishes in the limit as $\epsilon \to 0$. This is a geometrical construction which consists in estimating the measure of the set $B_\epsilon \subset \mathbb{R}^2 \times S^1$ of initial values that result in a path returning to the same obstacle patch. This subset depends on the realization of the obstacle positions, but the estimates of the measure $m(B_\epsilon)$ of this set do not, and $m(B_\epsilon) \to 0$ when $\epsilon \to 0$.

Consider then a path passing through an obstacle range, and then returning to the same patch. All such paths can be enumerated by giving the relative (integer) coordinates of the lattice cells where the path changes direction, that is $\xi_j \in \mathbb{Z}^2$

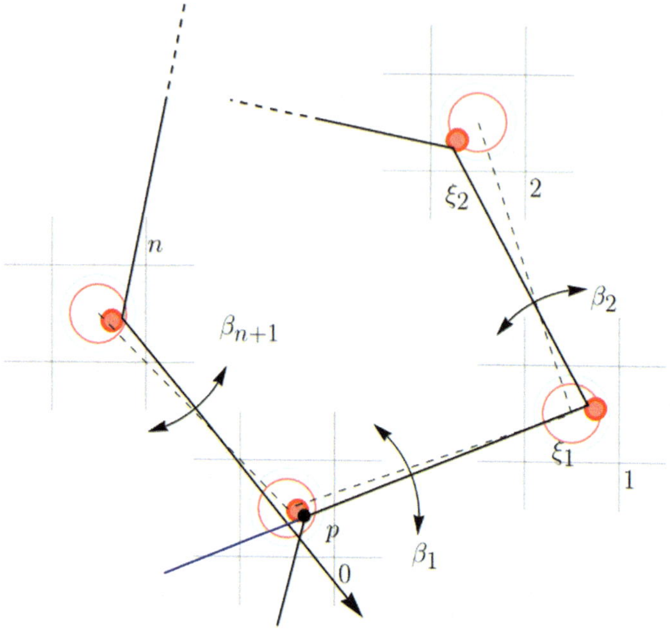

Fig. 10 A path making a loop by returning to the same obstacle range in the lattice cell that here is indexed by $0 \in \mathbb{Z}^2$ [35]

denotes the difference of the integer coordinates of the end point and starting point of a straight line segment of the path. This is illustrated in Fig. 10, where the starting point p of the loop is indicated with a black dot in cell nr. 0, which in this case is a point where the path is reflected, but in most cases it would not be. For a loop with n collisions (not counting the one in cell 0 if there is a collision there), we have

$$\xi_1 + \xi_2 + \ldots + \xi_{n+1} = 0 \in \mathbb{Z}^2\,, \tag{55}$$

where ξ_{n+1} corresponds to the last line segment, going from the nth collision point and back to the starting point of the loop. The list of ξ_i-s satisfy

$$|\xi_i| + |\xi_2| + \cdots + |\xi_{n+1}| \le \bar{t}\epsilon^{-1/2}\,, \tag{56}$$

because the size of a lattice cell is $\sqrt{\epsilon}$ in the scaling we are considering here.

A loop returning to the same obstacle patch allows some freedom in setting the initial position and angle of the path. Following a particle backwards yields the following estimate for the interval of directions that are allowed for a given path:

$$\Delta\beta_1 \le c^n \frac{\epsilon^{n/2}}{|\xi_1||\xi_2| \cdot \ldots \cdot |\xi_n|} \Delta\beta_{n+1} \le c^{n+1} \frac{\epsilon^{n/2}}{|\xi_1||\xi_2| \cdot \ldots \cdot |\xi_n|} \frac{\epsilon^{1-\nu}}{|\xi_{n+1}|}\,. \tag{57}$$

And continuing the trajectory backwards from the starting point of the loop and using the fact that the so called billiard map is measure preserving (see e.g. [10]) shows that the fraction of initial data (x_0, v_0) that result in a loop, i.e. a path coming back to the same obstacle patch is bounded by

$$\bar{t}Ce^{c\bar{t}}\log(\bar{t}/\sqrt{\epsilon})\,\epsilon^{2(\nu-1)}\,. \tag{58}$$

Exactly the same argument holds when there is a collision in the beginning of the loop and when there is not. This calculation may be summarized as a proposition:

Proposition 3 *Let $T^t_{\mathcal{Y}_{\epsilon,\epsilon}}$ be the Lorentz process as in Theorem 1. Denote the event that there is a loop along the path of length $\bar{t}$ starting at (x, v) by $\mathcal{L}(x, v)$. Then*

$$\int_{\mathbb{R}^2\times S^1} f_0(x,v)\mathbb{E}\left[g(T^t_{\mathcal{Y}_{\epsilon,\epsilon}}(x,v))\mathbf{1}_{\mathcal{L}(x,v)}\right]dxdv \to 0 \tag{59}$$

when $\epsilon \to 0$

Proof It is enough to see that the integral is bounded by

$$\|g\|_\infty \int_{B_M\times S^1} f_0(x,v)\mathbf{1}_{\mathcal{L}(x,v)}\,dxdv \tag{60}$$

where B_M is the ball of radius M in $\mathbb{R}^2$, and assumed to contain the support of g. This is bounded by

$$\begin{aligned}
&\|g\|_\infty \int_{B_M\times S^1} f_0(x,v)\mathbf{1}_{\{f_0>\lambda\}}\,dxdv \,+\, \|g\|_\infty\lambda \int_{B_M\times S^1} \mathbf{1}_{\mathcal{L}(x,v)\cap\{f<\lambda\}}\,dxdv\\
&\le \|g\|_\infty \int_{B_M\times S^1} f_0(x,v)\mathbf{1}_{\{f_0>\lambda\}}\,dxdv + \|g\|_\infty\lambda\bar{t}Ce^{c\bar{t}}\log(\bar{t}/\sqrt{\epsilon})\,\epsilon^{2(\nu-1)}\,.
\end{aligned} \tag{61}$$

This expression can be made arbitrarily small by first choosing λ large enough to make the first term as small one wishes, and then the second term can be made equally small by choosing ϵ sufficiently small. All this is uniform in the random positions of the obstacles. □

The last result presented here is a statement of propagation of chaos. Two simultaneous trajectories with initial data $(x_{1,0}, v_{1,0})$ and $(x_{2,0}, v_{2,0})$ do not move quite independently, because there is a positive probability that they meet the same obstacle patch. By calculations similar to the ones used in the proof that a path is not likely to make a loop can be modified to obtain the following result. For the proof we refer to [35].

Theorem 3 *Let $T^t_{\mathcal{Y}_{\epsilon,\epsilon}}$ be the Lorentz process as in Theorem 1. Then for $g_1, g_2 \in C(B_M \times S^1)$*

$$\lim_{\epsilon\to 0}\int_{(\mathbb{R}^2\times S^1)^2} f_0(x,v)f_0(x',v')\Big(\mathbb{E}\left[g_1\otimes g_2(T^t_{\mathcal{Y}_{\epsilon,\epsilon}}(x,v,x',v'))\right]-\\ \mathbb{E}\left[g_1(T^t_{\mathcal{Y}_{\epsilon,\epsilon}}(x,v))\right]\mathbb{E}\left[g_2(T^t_{\mathcal{Y}_{\epsilon,\epsilon}}(x',v'))\right]\Big)\,dxdvdx'dv'=0\,. \tag{62}$$

This says that in the limit as $\epsilon\to 0$ the correlations created by particles hitting the same obstacle vanish, and therefore that the positions in phase space at any time $t>0$ become independent in this limit, given that the initial positions were independent.

Acknowledgements The theoretical results were first presented at the workshop "Frontiers in the interplay between probability and kinetic theory" at ICMS, Edinburgh in April 2022, and in more detail together with the numerical results at "Particle Systems and PDE's X" at the University of Minho in June 2022. I am grateful for having been given the opportunity to participate in these events. I am also grateful for discussions with Mario Pulvirenti, Valeria Ricci, as well as for the very useful comments by anonymous referees.

References

1. Bini, D.A., Fiorentino, G.: Design, analysis, and implementation of a multiprecision polynomial rootfinder. Numer. Algo. **23**(2–3), 127–173 (2000)
2. Bini, D.A., Robol, L.: Solving secular and polynomial equations: a multiprecision algorithm dedicated to prof. Dario A. Bini's mother. J. Comput. Appl. Math. **272**, 276–292 (2014)
3. Boldrighini, C., Bunimovich, L.A., Sinaĭ, Ya.. G.: On the Boltzmann equation for the Lorentz gas. J. Stat. Phys. **32**(3), 477–501 (1983)
4. Bourgain, J., Golse, F., Wennberg, B.: On the distribution of free path lengths for the periodic Lorentz gas. Commun. Math. Phys. **190**(3), 491–508 (1998)
5. Bunimovich, L.A., Sinaĭ, Y.G.: Markov partitions for dispersed billiards. Commun. Math. Phys. **78**(2), 247–280 (1980/81)
6. Bunimovich, L.A., Sinaĭ, Y.G.: Statistical properties of Lorentz gas with periodic configuration of scatterers. Commun. Math. Phys. **78**(4), 479–497 (1980/81)
7. Caglioti, E., Pulvirenti, M., Ricci, V.: Derivation of a linear Boltzmann equation for a lattice gas. Markov Process. Related Fields **6**(3), 265–285 (2000)
8. Caglioti, E., Golse, F.: The Boltzmann-Grad limit of the periodic Lorentz gas in two space dimensions. C. R. Acad. Sci. Paris, Ser. I **346**, 477–482 (2008)
9. Caglioti, E., Golse, F.: On the Boltzmann-Grad limit for the two dimensional periodic Lorentz gas. J. Stat. Phys. **141**(2), 264–317 (2010)
10. Chernov, N., Markarian, R.: Chaotic billiards. Mathematical Surveys and Monographs, vol. 127. American Mathematical Society, Providence, RI (2006)
11. Drude, P.: Zur elektronentheorie der metalle. Annalen der Physik **306**(3), 566–613 (1900)
12. Fang, F., Hammock, D., Irwin, K.: Methods for calculating empires in quasicrystals. Crystals **7**(10), 304, 10 (2017)
13. Gallavotti, G.: Rigorous theory of the Boltzmann equation in the Lorentz gas. Nota Interna, Istituto di Fisica, Università di Roma, (358) (1972)
14. Gallavotti, G.: Statistical Mechanics. Texts and Monographs in Physics. Springer, Berlin (1999). A short treatise
15. Golse, F.: On the periodic Lorentz gas and the Lorentz kinetic equation. Ann. Fac. Sci. Toulouse Math. (6) **17**(4), 735–749 (2008)

16. Ben Hough, J., Krishnapur, M., Peres, Y.l., Virág, B.: Zeros of Gaussian Analytic Functions and Determinantal Point Processes. University Lecture Series, vol. 51. American Mathematical Society, Providence, RI (2009)
17. Ledin, S.: The discovery of quasi crystals (2011). https://www.nobelprize.org/uploads/2018/06/advanced-chemistryprize2011.pdf Cited 10 Feb. 2023
18. Lorentz, H.: Le mouvement des électrons dans les méteaux. Arch. Néerl. **10**, 336–371 (1905)
19. Lutsko, C., Tóth, B.: Invariance principle for the random Lorentz gas–beyond the Boltzmann-Grad limit. Commun. Math. Phys. **379**(2), 589–632 (2020)
20. Marklof, J., Strömbergsson, A.: Kinetic transport in the two-dimensional periodic Lorentz gas. Nonlinearity **21**(7), 1413–1422 (2008)
21. Marklof, J., Strömbergsson, A.: The distribution of free path lengths in the periodic Lorentz gas and related lattice point problems. Ann. Math. (2) **172**(3), 1949–2033 (2010)
22. Marklof, J., Strömbergsson, A.: The Boltzmann-Grad limit of the periodic Lorentz gas. Ann. Math. (2) **174**(1), 225–298 (2011)
23. Marklof, J., Strömbergsson, A.: The periodic Lorentz gas in the Boltzmann-Grad limit: asymptotic estimates. Geom. Funct. Anal. **21**(3), 560–647 (2011)
24. Marklof, J., Strömbergsson, A.: Free path lengths in quasicrystals. Commun. Math. Phys. **330**(2), 723–755 (2014)
25. Marklof, J., Strömbergsson, A.: Generalized linear Boltzmann equations for particle transport in polycrystals. Appl. Math. Res. Express. AMRX **2**, 274–295 (2015)
26. Marklof, J., Strömbergsson, A.: Kinetic theory for the low-density Lorentz gas. Mem. Amer. Math. Soc. **294** (1464) (2024)
27. Markof, J., Vinogradov, I.: Spherical averages in the space of marked lattices. Geom. Dedicata **186**, 75–102 (2017)
28. Melbourne, I., Pène, F., Terhesiu, D.: Local large deviations for periodic infinite horizon Lorentz gases (2021)
29. Mendoza Sosa, A.R., Kraemer, A.S.: Efficient algorithm for simulating particles in true quasiperiodic environments. J. Phys. A: Math. Theor. **55**(24) (2022)
30. Pène, F., Terhesiu, D.: Sharp error term in local limit theorems and mixing for Lorentz gases with infinite horizon. Commun. Math. Phys. **382**(3), 1625–1689, 2 (2021)
31. Ricci, V., Wennberg, B.: On the derivation of a linear Boltzmann equation from a periodic lattice gas. Stochastic Process. Appl. **111**(2), 281–315 (2004)
32. Senechal, M.: Quasicrystals and Geometry. Cambridge University Press, Cambridge (1995)
33. Spohn, H.: The Lorentz process converges to a random flight process. Commun. Math. Phys. **60**(3), 277–290 (1978)
34. Wennberg, B.: Free path lengths in quasi crystals. J. Stat. Phys. **147**(5), 981–990 (2012)
35. Wennberg, B.:. The Lorentz process with a nearly periodic distribution of scatterers. J. Stat. Phys. **190**(7): Paper No 127 (2023)

SPRINGER NATURE

GPSR Compliance

The European Union's (EU) General Product Safety Regulation (GPSR) is a set of rules that requires consumer products to be safe and our obligations to ensure this.

If you have any concerns about our products, you can contact us on ProductSafety@springernature.com

In case Publisher is established outside the EU, the EU authorized representative is:

Springer Nature Customer Service Center GmbH
Europaplatz 3
69115 Heidelberg, Germany

The manufacturer's authorised representative in the EU is Springer
Nature Customer Service Centre GmbH, Europaplatz 3, 69115 Heidelberg,
Germany. If you have any concerns regarding our products, please
contact ProductSafety@springernature.com

Printed and bound by CPI Group (UK) Ltd, Croydon, CR0 4YY
01/10/2026
02225871-0013